Handbook of Meteorology

Handbook of Meteorology

Editor: Aiden Martinez

R CALLISTO REFERENCE

www.callistoreference.com

Callisto Reference,
118-35 Queens Blvd., Suite 400,
Forest Hills, NY 11375, USA

Visit us on the World Wide Web at:
www.callistoreference.com

ISBN: 978-1-64116-265-4 (Hardback)

Cataloging-in-Publication Data

Handbook of Meteorology / edited by Aiden Martinez.
    p. cm.
Includes bibliographical references and index.
ISBN 978-1-64116-265-4
1. Meteorology. 2. Atmospheric science. 3. Climatic changes. I. Martinez, Aiden.
QC861.3 .H36 2020
551.5--dc23

Table of Contents

Preface

This book has been an outcome of determined endeavour from a group of educationists in the field. The primary objective was to involve a broad spectrum of professionals from diverse cultural background involved in the field for developing new researches. The book not only targets students but also scholars pursuing higher research for further enhancement of the theoretical and practical applications of the subject.

Meteorology is a sub-discipline of the atmospheric sciences that is focused on weather forecasting. It encompasses the fields of atmospheric physics and atmospheric chemistry. The science of meteorology seeks to explain observable weather events, and quantifies these in terms of temperature, air pressure, mass flow, water vapour, their interactions and change with respect to time. Weather prediction is done at the microscale, mesoscale, synoptic scale and global scales. Besides weather prediction, meteorological studies are crucial to air traffic management, agricultural management, the study of the distribution of radioactive gases and aerosols in the atmosphere, analysis of industrial pollution, etc. This book contains some path-breaking studies in the field of meteorology. Also included herein is a detailed explanation of the various concepts and applications of this field. Those with an interest in meteorology would find this book helpful.

It was an honour to edit such a profound book and also a challenging task to compile and examine all the relevant data for accuracy and originality. I wish to acknowledge the efforts of the contributors for submitting such brilliant and diverse chapters in the field and for endlessly working for the completion of the book. Last, but not the least; I thank my family for being a constant source of support in all my research endeavours.

Editor

Effects of Landscape Design on Urban Microclimate and Thermal Comfort in Tropical Climate

Wei Yang,[1,2] **Yaolin Lin ⓘ,**[1] **and Chun-Qing Li ⓘ**[3]

[1]*School of Civil Engineering and Architecture, Wuhan University of Technology, Wuhan 430070, China*
[2]*College of Engineering and Science, Victoria University, Melbourne, VIC 8001, Australia*
[3]*School of Engineering, RMIT University, Melbourne, VIC 3000, Australia*

Correspondence should be addressed to Yaolin Lin; yaolinlin@gmail.com

Academic Editor: Andreas Matzarakis

A climate-responsive landscape design can create a more livable urban microclimate with adequate human comfortability. This paper aims to quantitatively investigate the effects of landscape design elements of pavement materials, greenery, and water bodies on urban microclimate and thermal comfort in a high-rise residential area in the tropic climate of Singapore. A comprehensive field measurement is undertaken to obtain real data on microclimate parameters for calibration of the microclimate-modeling software ENVI-met 4.0. With the calibrated ENVI-met, seven urban landscape scenarios are simulated and their effects on thermal comfort as measured by physiologically equivalent temperature (PET) are evaluated. It is found that the maximum improvement of PET reduction with suggested landscape designs is about 12°C, and high-albedo pavement materials and water bodies are not effective in reducing heat stress in hot and humid climate conditions. The combination of shade trees over grass is the most effective landscape strategy for cooling the microclimate. The findings from the paper can equip urban designers with knowledge and techniques to mitigate urban heat stress.

1. Introduction

The world is at its fastest pace of urbanization. Since 2008, more than half of the world's population live in urban areas. The trend in global population increase has led to an increase in housing demand. Singapore has gone from one of the worst housing shortages in the world in the 1960s to a country where 90% of its citizens now own their own home and homelessness is virtually eliminated—despite its population has tripled in the last 50 years. With success of housing policies, natural land has been replaced by artificial surfaces in Singapore with undesirable thermal effects. This issue, together with increasing industrialization, has caused a considerable deterioration of the urban environment. In tropical countries like Singapore, hot climate in terms of high temperature, high humidity, and high solar radiation often causes heat stress to residents, resulting in negative impact on public health and productivity. Climate-responsive urban design can create microclimates that people experience as feeling cooler than the prevailing climate, making urban spaces pleasant.

Thus, the effect of urban landscaping on microclimate and human thermal comfort is necessary to be considered in the urban design and planning process.

It is acknowledged that the transfer of climatic knowledge into planning practice is still lacking [1, 2]. Although many measures to reduce urban heat stress and/or improve outdoor thermal comfort have been proposed by various researchers and at different spatial scales [2–6], their effectiveness is a subject for debate. The main reason is that the dominant professions for urban design and planning, namely, architecture and engineering, so far focus on the influence of landscaping on air and surface temperatures and their subsequent effect on buildings [7]. However, the impact of countermeasures by urban design on urban thermal comfort cannot be described sufficiently by simple microclimate factors, such as surface or air temperature. There are seven factors (or parameters) that affect human thermal comfort in an outdoor environment. They are air temperature, air humidity, wind, solar radiation, terrestrial radiation, metabolic heat, and clothing insulation [8]. The first five parameters are affected by urban environments, while the latter two are related to individual choice. At the neighborhood or community scale, landscape elements can modify not only the wind and radiation but also the air temperature and humidity [2–9]. Therefore, it is necessary to study the effect of different landscape elements on different microclimate parameters and corresponding human thermal comfort.

In recent years, some researchers have realized that urban heat stress can be reduced through appropriate landscape design. Many field measurements and numerical simulations have been carried out to study the effect of landscape elements on urban microclimate and thermal comfort. For example, Ng et al. [5] conducted parametric studies in Hong Kong and found that proper greening may greatly improve the urban microclimate and lower the summer urban air temperature. Yahia and Johansson [10] explored how vegetation and landscape elements affect outdoor thermal comfort for detached buildings in the hot dry climate of Damascus, Syria, and found that PET (physiologically equivalent temperature) can be reduced by about 19°C for east-west street orientation through appropriate landscape design. Perini and Magliocco [11] investigated effects of vegetation, urban density, building height, and atmospheric conditions on local temperatures and thermal comfort in three different cities in Italy and found that vegetation has higher cooling effects with taller buildings. Lee et al. [12] studied the potential of urban green coverage to mitigate human heat stress using the ENVI-met model and found that trees are more effective in mitigating human heat stress than just grasslands. Yahia et al. [2] investigated the relationship between urban design, urban microclimate, and outdoor comfort in four built-up areas with different morphologies and found that the use of dense trees helps to reduce heat stress, but vegetation might negatively affect the wind ventilation.

Although the previous studies have added new knowledge and provided new insights, they have mainly focused on the street design like street orientation, street greenery, and street geometry [3–5, 10, 13]. Little research has been conducted in urban residential areas, particularly in those with high-rise residential areas. The microclimate quality of outdoor spaces in a residential area affects the quality of life of its residents. Therefore, the aim of this paper is to investigate how landscape elements affect urban microclimate and human thermal comfort in a high-rise residential area in Singapore by investigating different landscape design scenarios of pavement materials, greenery, and water bodies. Studying the relationship between landscaping and microclimate in cities like Singapore can provide valuable guidance, both for keeping Singapore residents cool and informing temperate-climate cities that would be much warmer in the future.

2. Materials and Methods

2.1. Study Area.
The study area is two residential quarters at Bedok in southeast Singapore as shown in Figure 1. Bedok is an urban residential zone for new development in Singapore. The two residential quarters are condominiums named the Clearwater and Aquarius By The Park near Bedok Reservoir.

Figure 1: Study area and field measurement points at Bedok.

The two residential quarters are in close proximity to each other with the Clearwater on the west side of Bedok Reservoir View Road and Aquarius By The Park on the east side of the road. Buildings in the studied residential quarters are of 4 to 18 storeys. An urban park is located in the vicinity of the two residential quarters on the north.

2.2. Field Measurements.
Field measurements were conducted at the study area from 13 April to 06 June 2012. The purpose of field measurements is to validate ENVI-met modeling (see below) results and also help define the initial conditions of the general model of ENVI-met.

Five measurement points were stationed as shown in Figure 1. The measurement points were selected to represent variations in urban geometry, ground thermal properties, and greenery as shown in Figure 2. Points 1 and 2 are in the urban park, and points 3, 4, and 5 are in a high-density apartment area. The sky view factor (SVF) ranges from highly shaded point 2 (SVF = 0.17) to less shaded point 5 (SVF = 0.67). The measured microclimatic parameters are air temperature, globe temperature, relative humidity, and wind speed, which were measured for 24 hours continuously and taken at 2.0 m above the ground level. Table 1 shows the measured microclimatic parameters and equipment used for the field measurements.

2.3. Microclimate Simulation.
For this study, the thermal characteristics of different urban design scenarios were investigated by ENVI-met 4.0 [14, 15]. This is a microclimate analysis program that simulates the thermal characteristics and energy fluxes in the built environment with high spatial and temporal resolution. The model generates a large amount of output data including necessary variables for calculation of thermal stress indices. It has been employed by many researchers to study the effects of different urban design options on microclimate and outdoor thermal comfort [1–4,10–13]. ENVI-met 4.0 allows users to employ the measured meteorological data as inputs by forcing the model to follow user's inputs during the simulation. In the previous versions of ENVI-met, only relatively simple weather profiles as prescribed by ENVI-met can be used as

FIGURE 2: Photos and fisheye photos for each location (measurements were taken at 2.0 m above the ground level). (a) Point 1 (SVF = 0.61). (b) Point 2 (SVF = 0.17). (c) Point 3 (SVF = 0.48). (d) Point 4 (SVF = 0.66). (e) Point 5 (SVF = 0.67).

inputs. The details of the ENVI-met model have been fully explained and presented on its website [15] and in many research papers [1, 4, 14, 16].

The weather data from the nearest station at the Changyi Airport were selected. It was found that the daily air temperature on April 30, 2012, was the highest during the study period. Therefore, the simulation study was conducted on that day. The hourly meteorological data from the weather station and from the on-site observation were used to generate the "forcing file" (as inputs) for the simulation. It was observed that the weather condition during the measurement period was characterized by high temperature, strong solar radiation, and light wind with a prevailing wind direction of southwest. The model was run for 18 h starting at 4 am and ending at 10 pm for each simulation of microclimate.

TABLE 1: Equipment used for field measurement.

Variable	Instrument	Accuracy
Air temperature/ relative humidity	HOBO U12-012 Temp/RH Data Logger	±0.35°C from 0°C–50°C to a maximum of ±3.5%
Globe temperature	HOBO Thermocouple Data Logger, U12-014 with Type-T Copper-Constantan thermocouple sensors and 40 mm diameter ping pong ball	±1.5°C
Wind speed	Onset Wind Speed Smart Sensor, S-WSA-M003	±1.1 m/s or ±4% of reading, whichever is greater
Short- and long-wave radiation	Kipp & Zonen, CNR 4 with integrated pyranometer, pyrgeometer, Pt-100, and thermistor	Pyranometer: <5% uncertainty (95% confidence level) Pyrgeometer: <10% uncertainty (95% confidence level) Pt-100/thermistor: ±0.7°C

2.4. Parametric Study and Urban Thermal Comfort Assessment. The parametric study consists of a base case and seven design scenarios. The base case was constructed according to the actual conditions of the study area. The model domain covers the entire area of the study area and is expanded to the surrounding buildings, streets, and an urban park. The spatial extent of the study area is $600 \times 392 \times 120$ m in the X, Y, and Z dimensions, respectively. The horizontal and vertical grid resolutions are both set at 4 m. The model domain of the base case for the study area is shown in Figure 3. The input data of the general model setting, the initial atmospheric/soil condition, and the building properties are summarized in Table 2.

The other scenarios to be investigated are designed based on changing different landscape elements such as pavement materials (brick, concrete, wood, and light-color granite) and amount of trees, grass, and water bodies as listed in Table 3. For the first 5 scenarios, only one parameter is changed at a time in order to determine the relative effect of each. The last two scenarios are a combination of two design elements to further investigate the effect of ground materials and tree shading.

For the assessment of urban thermal comfort, the PET (physiologically equivalent temperature) is selected as the thermal comfort index. PET has been calibrated against subjective thermal sensation evaluation by Yang et al. [17] in Singapore (Table 4), which makes it possible to compare different urban design proposals. Tropical residents are found to tolerate higher levels of PET than Western/Middle European residents due to thermal adaption to the local climate. PET is calculated using the RayMan model [18, 19]. It can be easily estimated with air temperature, relative humidity, wind speed, mean radiant temperature, clothing, and activity level of people. The thermal comfort map in terms of PET is generated in the paper for comparison.

3. Results and Discussion

3.1. The Base Case Scenario: Measurement and Simulation. The microclimatic parameters of air temperature, mean radiant temperature, wind speed, and relative humidity collected at measuring points 1–5 have been compared with the corresponding ENVI-met model outputs.

Figure 4 shows the comparison between measured and simulated air temperatures. It can be seen that the simulated and measured air temperatures have the same

FIGURE 3: Model domain for the study area.

TABLE 2: Boundary conditions and initial setting of the ENVI-met model.

Location	Singapore 103°51′E, 1°18′N
Climate	Tropical climate
Date/time simulated	From 04:00 to 22:00 (18 h) on 30 April 2012
Model domain	Bedok: $150 \times 98 \times 30$ grids $\Delta x = \Delta y = \Delta z = 4$m Note: vertical grid with the equidistant method
Meteorological inputs	Air temperature and relative humidity: hourly data from the measurement on-site Wind speed and direction: hourly data from the meteorological station Specific humidity (2500 m) = 7 g/kg
Initial soil temperature and relative humidity	Upper layer (0–20 cm): 305 K/30% Middle layer (20–50 cm): 307 K/40% Deeper layer (below 50 cm): 306 K/50%
Building conditions	Inside temperature = 293 K (constant) Heat transmission walls = 1.94 W/m²·K Heat transmission roofs = 6 W/m²·K Albedo walls = 0.2 Albedo roofs = 0.3
Plants	Trees: 10 m dense, leafless base Trees: 20 m dense, leafless base Grass: 20 cm average dense

trend for all five points with perhaps more smooth curves for the simulated ones. The air temperature pattern is clearly influenced by the sky view factor and surrounding urban environment. Point 2 has the lowest air temperature

TABLE 3: Different design scenarios for Bedok.

Design scenario	Pavement materials	Vegetation and water body
Base case	Red brick (ID: KK) and concrete pavement (ID: PP)	Sparse trees and grass Small area of water bodies (30 m^2)
Scenario 1	Wooden boards (ID: WD)	As base case
Scenario 2	Light-color granite (ID: G2)	As base case
Scenario 3	Grass surface	As base case
Scenario 4	As base case	Add more trees (increase by 200%)
Scenario 5	As base case	Add more water bodies (increase by 200%)
Scenario 6	Light-color granite (ID: G2)	Add more trees (increase by 200%)
Scenario 7	Grass surface	Add more trees (increase by 200%)

TABLE 4: Thermal sensations and PET classes for Singapore and Western/Middle Europe.

Thermal sensation	PET range for Singapore (°C)	PET range for Western/Middle Europe (°C)
Very cold	Not applicable	<4
Cold	Not applicable	4–8
Cool	Not applicable	8–13
Slightly cool	20–24	13–18
Neutral	24–30	18–23
Slightly warm	30–34	23–29
Warm	34–38	29–35
Hot	38–42	35–41
Very hot	>42	>41

Source: Yang et al. [17].

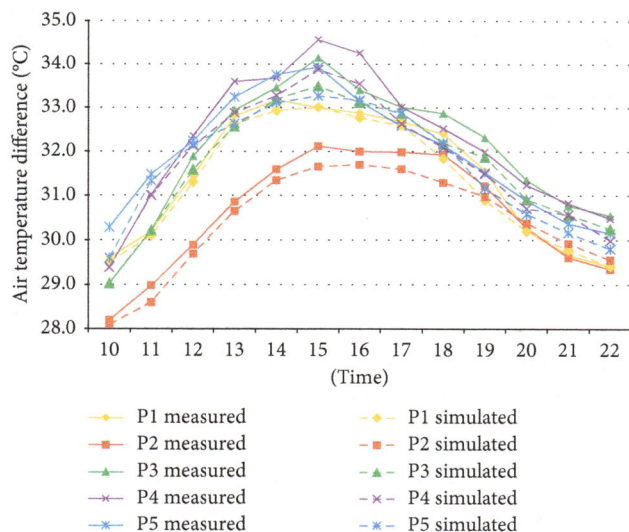

FIGURE 4: Comparison between simulated and measured air temperatures.

because it is located in the nearby park and has a low sky view factor (0.17).

Points 3, 4, and 5 have higher air temperatures than points 1 and 2 because these three points are located along high-density residential buildings. It can also be seen that ENVI-met underestimates the daytime air temperature by about 0.1–0.7°C. This is because ENVI-met calculates the urban climate at a microscale or a local scale and that larger regional

(mesoscale) effects are not taken into account [15]. During the night, the air temperature is underestimated by up to 0.5°C and overestimated by up to 0.3°C with ENVI-met in this study.

The mean radiant temperature comparison between simulated and measured results is shown in Figure 5. It can be seen that the simulated and measured mean radiant temperatures have the same trend for all the points. Points 3, 4, and 5 have a higher mean radiant temperature profile than points 1 and 2 during the day. This is because points 1 and 2 are located in the park and have lower sky view factors. It can also be found that the daytime mean radiant temperature is overestimated and nighttime mean radiant temperature is underestimated by ENVI-met. The daytime difference is about 0.1–6.7°C, and the nighttime difference is about 2.6–6.6°C. A number of other studies have also reported a mean radiant temperature difference of up to 7.97°C between the measured and simulated results [1, 4, 13, 20]. The discrepancies are due to that ENVI-met does not consider heat storage and transfer by buildings or anthropogenic heat production in an adequate manner [13, 21]. Therefore, studies on the effect of landscape design on nighttime outdoor thermal comfort and urban heat island need further investigation in the future due to limitations of ENVI-met modeling.

The results of the measured and simulated wind speed and relative humidity show little difference (less than 5%) for all the points. The input wind speed is less than 2 m/s in this study. It has also been reported that wind speeds predicted by ENVI-met are consistent with field data for input wind speeds below 2 m/s [22].

Tables 5 and 6 show the model fit between simulated and measured results for air temperature and mean radiant temperature, respectively. Very high overall agreement can be found for both air temperature (R^2 between 0.95 and 0.99) and mean radiant temperature (R^2 between 0.74 and 0.96). The relatively lower model fit of $R^2 = 0.74$ for point 2 as well as the 5°C difference between simulated and measured results in terms of mean radiant temperature can be partially explained by the error in the measurement; for example, solar radiation suddenly became very intense during that particular measurement time.

Therefore it is possible to say that although there are some discrepancies between the simulated and measured results, ENVI-met is able to present similar trends for microclimatic parameters compared with those from field measurement. Compared with a former study conducted in

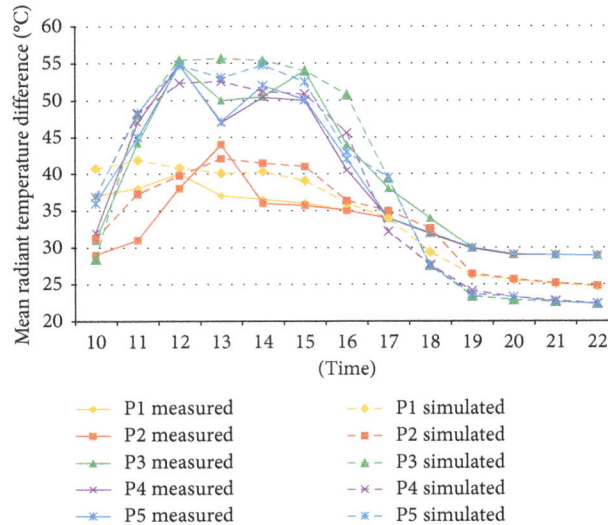

FIGURE 5: Comparison between simulated and measured mean radiant temperatures.

TABLE 5: Model fit between simulated and measured results for air temperature.

	Point 1	Point 2	Point 3	Point 4	Point 5
Minimum error (°C)	0	−0.1	0	0.01	0.03
Maximum error (°C)	−0.67	−0.63	−0.68	−0.72	−0.67
Mean error (°C)	−0.16	−0.19	−0.31	−0.41	−0.28
Standard deviation (°C)	0.21	0.26	0.20	0.23	0.30
R^2	0.98	0.97	0.99	0.99	0.95
RMSE (°C)	0.27	0.32	0.37	0.47	0.41

TABLE 6: Model fit between simulated and measured results for mean radiant temperature.

	Point 1	Point 2	Point 3	Point 4	Point 5
Minimum error (°C)	0	0.71	0.01	0.81	−0.07
Maximum error (°C)	−4.23	6.24	6.78	−6.25	−6.59
Mean error (°C)	0.08	1.10	0.88	−1.13	−0.6
Standard deviation (°C)	3.16	3.35	4.97	4.34	4.52
R^2	0.96	0.74	0.95	0.91	0.94
RMSE (°C)	3.16	3.56	5.05	4.49	4.56

Singapore [13], it can also be seen that the new ENVI-met model of version 4.0 shows much better performance than the previous version of ENVI-met 3.1. Since the thermal performance of different urban geometries and ground surface and their effect on mean radiant temperature can be modeled by ENVI-met, a relative comparison can be made for different design scenarios. In addition, the simulated results have been calibrated with field measurement data and then used as a benchmark for investigation of changes in design. Therefore, all the changes in design are consistent and relative to the simulated case, whereby the error from calibration has been effectively eliminated.

3.2. Microclimate Differences. It has been found from both the measurement and simulation results that the hottest time is at 3 pm on the simulation day. Therefore, the effects of different landscape design scenarios on microclimate and thermal comfort are compared based on results at 3 pm. Except for the surface temperature, the other microclimate parameters are compared at 2.0 m above the ground level.

3.2.1. Surface Temperature and Air Temperature. Figure 6 shows the surface temperature patterns for all design scenarios. The differences in surface temperatures are obvious. Pavement with light-color granite (Scenario 2) has the lowest surface temperature, with a maximum reduction of 12°C compared with the base case. Surface temperature reduction by grass surfacing (Scenario 3) and adding more trees (Scenario 4) is also obvious, with a reduction by up to 8°C for grass and 10°C for trees.

Surface temperature reduction by applying wood pavement (Scenario 1) can be up to 6°C. Not much difference in surface temperature can be found by adding more water bodies. Both Scenario 6 (combination of light-color granite and adding more trees) and Scenario 7 (combination of grass surfacing and adding more trees) resulted in a significant reduction of surface temperature. However, Scenario 6 is more effective in reducing the surface temperature than Scenario 7.

Figure 7 shows the air temperature patterns for all design scenarios. The differences in air temperature between different scenarios are not so obvious as those in the surface temperature. The air temperature is about 0.25–0.75°C lower for the scenarios with light-color granite compared with the base case. For scenarios with grass surfacing and more trees, the air temperature reduction is about 0.25–0.5°C. For wood scenario, the apparent reduction of 6°C of the temperature at the surface does not cause a significant reduction in local air temperature at 2.0 m above the ground level. However, the air temperature of areas under building shade for the wood scenario is about 0.25°C lower than that in the base case. Scenario 6 and Scenario 7 both cause an air temperature reduction by up to 0.75°C.

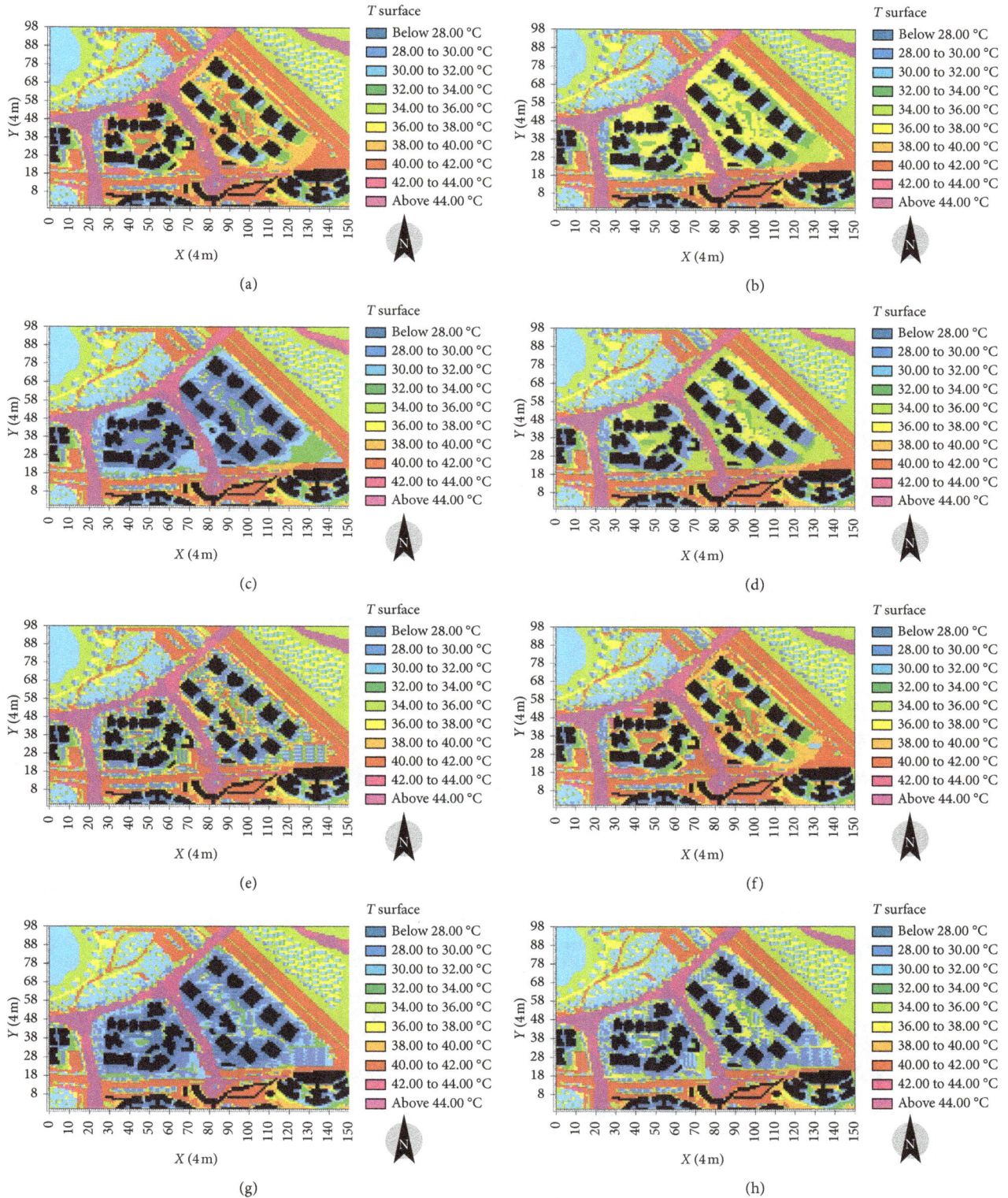

FIGURE 6: Simulated surface temperature for all design scenarios. (a) Base case. (b) Scenario 1 (wooden boards). (c) Scenario 2 (light-color granite). (d) Scenario 3 (grass surface). (e) Scenario 4 (more trees). (f) Scenario 5 (more water bodies). (g) Scenario 6 (light-color granite + more trees). (h) Scenario 7 (grass surface + more trees).

Not much difference in air temperature can be found for the scenario of adding more water bodies. Water bodies are found to be not effective in decreasing the air temperature in Singapore in the current study. This is in agreement with a field measurement study conducted by Wong et al. [23] who investigated the evaporative cooling performance of

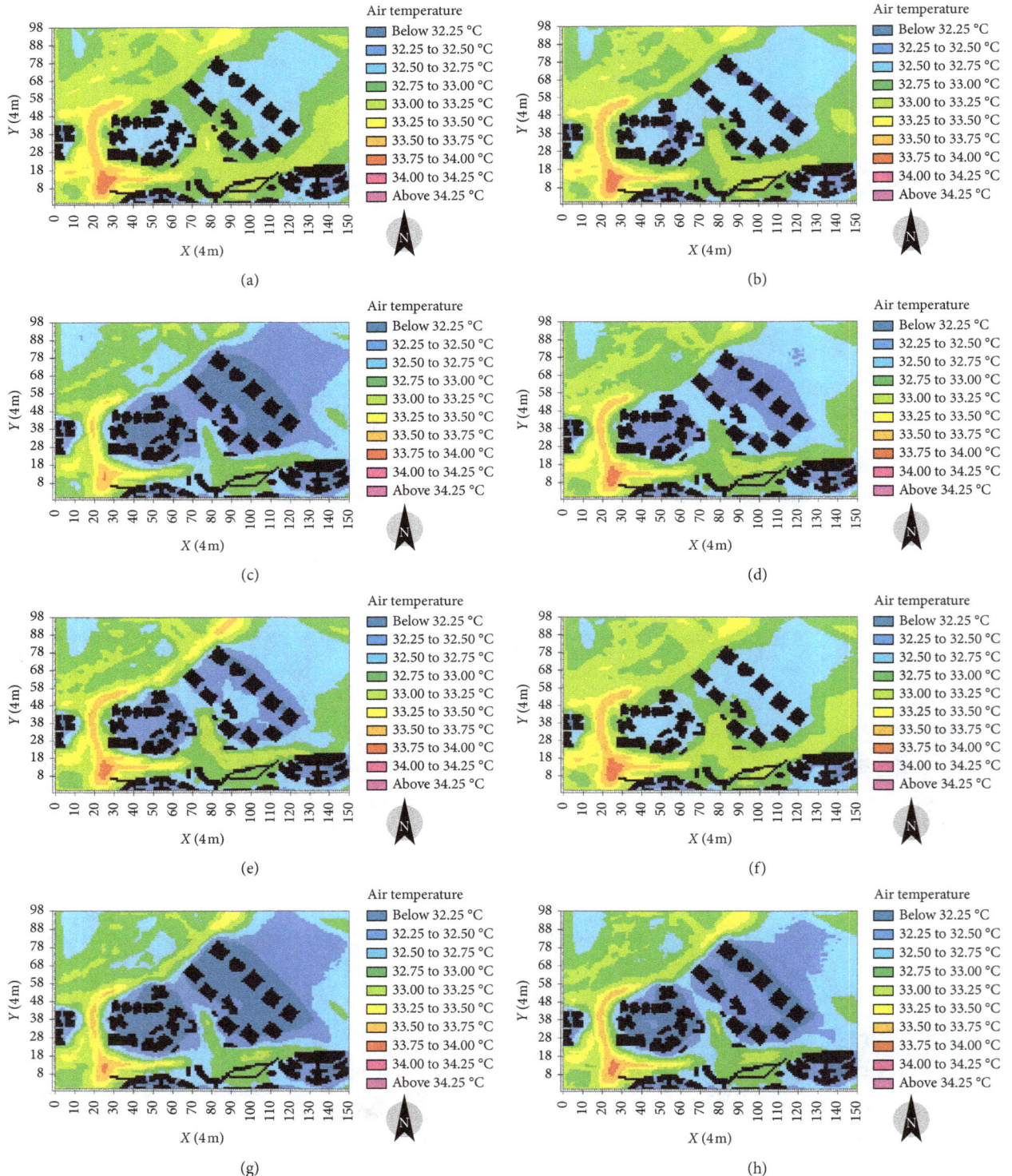

Figure 7: Simulated air temperature for all design scenarios. (a) Base case. (b) Scenario 1 (wooden boards). (c) Scenario 2 (light-color granite). (d) Scenario 3 (grass surface). (e) Scenario 4 (more trees). (f) Scenario 5 (more water bodies). (g) Scenario 6 (light-color granite + more trees). (h) Scenario 7 (grass surface + more trees).

a waterway in Singapore and found that the air temperature was merely reduced by 0.1°C on every 30 m away from the waterway. The high humidity climate and low wind condition might be one of the possible reasons for it.

3.2.2. *Mean Radiant Temperature*. The patterns of mean radiant temperature for all design scenarios are presented in Figure 8. For sunlit places, the mean radiant temperatures are 50–54°C for all scenarios except the grass surface

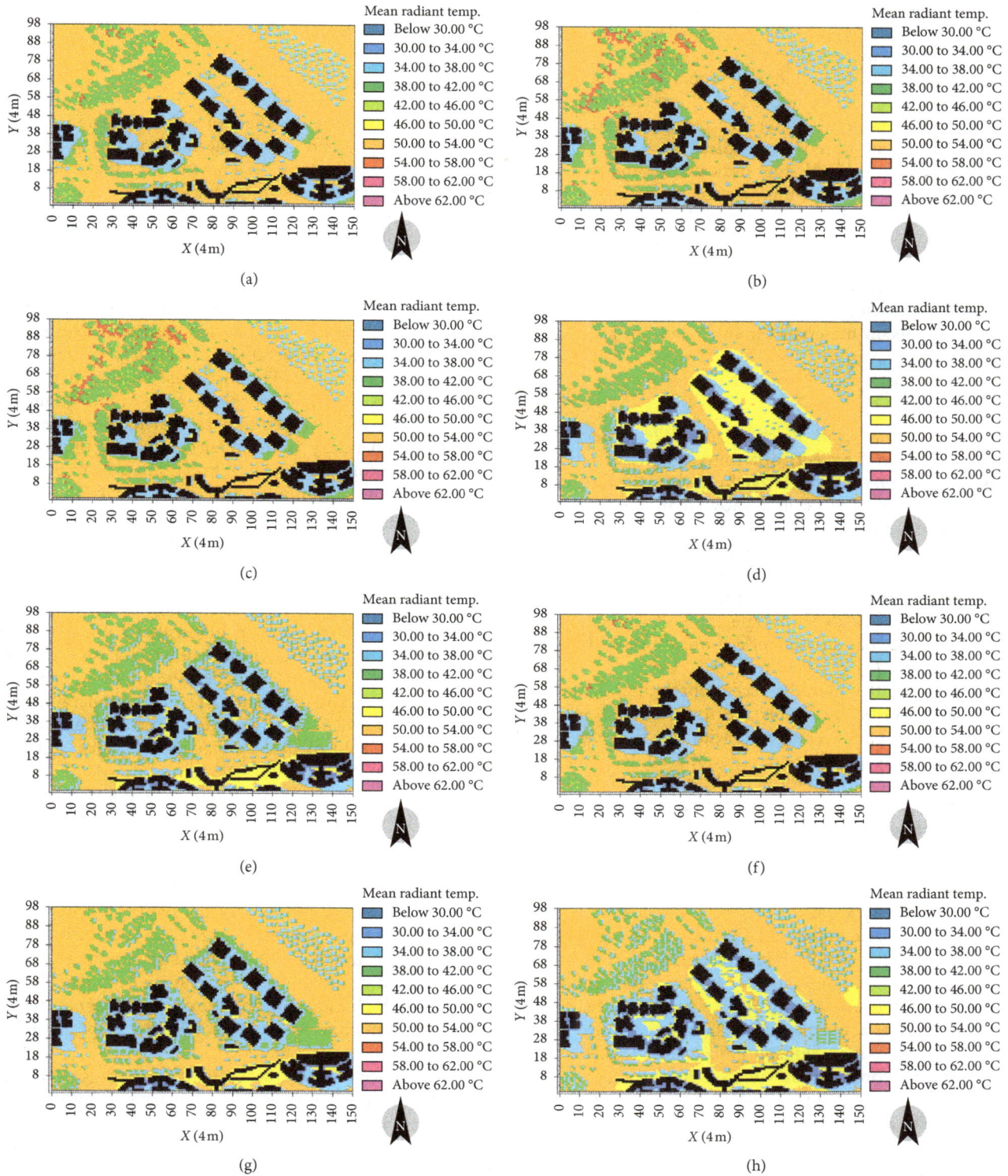

FIGURE 8: Simulated mean radiant temperature for all design scenarios. (a) Base case. (b) Scenario 1 (wooden boards). (c) Scenario 2 (light-color granite). (d) Scenario 3 (grass surface). (e) Scenario 4 (more trees). (f) Scenario 5 (more water bodies). (g) Scenario 6 (light-color granite + more trees). (h) Scenario 7 (grass surface + more trees).

scenarios, which have mean radiant temperatures 4–8°C lower than other scenarios. For places shaded by buildings, differences in mean radiant temperature are obvious. For both the wood and light-color granite scenarios, the mean radiant temperatures are 4–8°C higher than those of the base case. For the tree scenarios, there is a significant cooling effect and the mean radiant temperature under tree-shaded areas can be reduced by 12–16°C compared with sunlit areas. Scenario 7 (combination of grass surface and more trees) is the best with 4–8°C mean radiant temperature reduction for areas exposed to the sun and 12–16°C reduction for tree-shaded areas. Not much difference can be found for the scenario of adding more water bodies.

The results from ENVI-met indicate that the change of pavement materials has a minor effect on reducing mean radiant temperature for places exposed to high solar radiation. For places shaded by buildings, the mean radiant temperature is even increased by using high-albedo pavement materials. This is consistent with other studies which also found increases of mean radiant temperature by applying high-albedo materials [4, 16, 24] in hot and humid climates.

3.2.3. Wind Speed and Relative Humidity. Due to the small differences between different design scenarios in terms of wind speed and relative humidity, the figures are not shown here. The results show that wind speed is slightly reduced by 0.2 m/s with more trees planting. The differences in wind speed are not obvious for other design scenarios. This is because the layout of building blocks has been determined in the residential quarters for this study. Compared with landscape elements, the layout of building blocks has greater effect on air flow in urban spaces.

As to the relative humidity, scenarios with grass surface, more trees, and water bodies are more humid, with an increase of 4% to 6% compared with the base case. The change of landscape elements cannot lead to significant variation of relative humidity when the humidity is very high throughout the year. This is the climate in Singapore, and hence, the results make sense.

3.3. Thermal Comfort Maps of PET. Figure 9 shows the simulated thermal comfort (PET) maps for all the design scenarios at 3 pm. The PET values of sunlit places for all the design scenarios are dominated by extremely hot condition with the PET between 46 and 50°C, which is under severe heat stress and far above the comfortable temperature range (24–30°C) required for Singapore occupants (Table 4). Although thermal comfort is difficult to achieve under such hot climate conditions, some improvements can be made through landscape design.

The best thermal conditions are in the areas with shading, either shaded by buildings or trees, with a PET of 34–38°C, which corresponds to "warm" according to Table 4. The shade enhancement by trees or buildings has a clear positive effect on alleviating outdoor heat stress, as indicated by decreased PET.

Scenario 3 (grass surface) only leads to a PET reduction of 4–8°C for limited areas, and the heat stress conditions for most of the study areas are not improved. Scenarios with trees (4, 6, and 7) have the same PET patterns despite that each scenario has different pavement materials. Again adding more water bodies is found to have little effect on PET.

4. Discussion

Table 7 summarizes the effect of different design scenarios on microclimate and human thermal comfort (PET).

It can be seen that design strategies that can reduce surface temperature and air temperature may not necessarily reduce heat stress condition. Design strategies such as applying wooden boards and light-color granite have some extent of cooling effect, but heat stress is marginally reduced. Both the wooden board and light-color granite are high-albedo materials with an albedo of 0.8 in this study. While higher albedo reduces the surface temperatures, and consequently, the air temperature, it increases the amount of reflected short-wave radiation in the environment at the same time. As it is known, the increase of energy flux will result in the increase of mean radiant temperature. Mean radiant temperature is the main factor affecting outdoor thermal comfort in hot and humid climate as in Singapore [17]. Thus, the insignificant effect of high-albedo materials on reducing heat stress can be expected. However, the effectiveness of high-albedo covering on heat stress is disputed because PET does not take surface temperature into consideration. The decrease of surface temperature is not reflected in PET, which raises a question of whether surface temperature has an effect on urban thermal comfort. Different from the indoor environment which has uniform and relative lower surface temperatures, outdoor space has a large variation and fluctuation of surface temperatures. The evaluation of urban thermal comfort is a challenging topic in the research field of human bioclimate, which still needs further study.

Water can mitigate the urban heat island effect since more incoming heat can be transformed into latent heat rather than sensible heat. However, water bodies are found to be not effective in mitigating heat stress in hot and humid climate as studied in this paper. Adding more water bodies does not change any microclimate parameters except that humidity increases slightly. This may be because the area of water bodies in this study is not large enough to create a cooling effect for the surrounding environment. Besides, due to the high humidity conditions in Singapore, thermal comfort cannot benefit too much from the evaporation from water bodies.

It has been widely accepted that shading is the key strategy for promoting outdoor thermal comfort in hot climate. Interception of solar radiation is the most effective means in improving thermal comfort in outdoor areas in hot and dry climate [6]. The current study also vindicates this design principle because the scenarios shaded by more trees have the best thermal comfort condition, with the maximum PET reduced by 12°C. However, not much difference is found for scenarios with trees (Scenario 4, 6, and 7) in terms of urban heat stress even though each scenario has different pavement materials. Different pavement materials can lead to variations in surface temperature, air temperature, and mean radiant temperature in urban spaces, but these

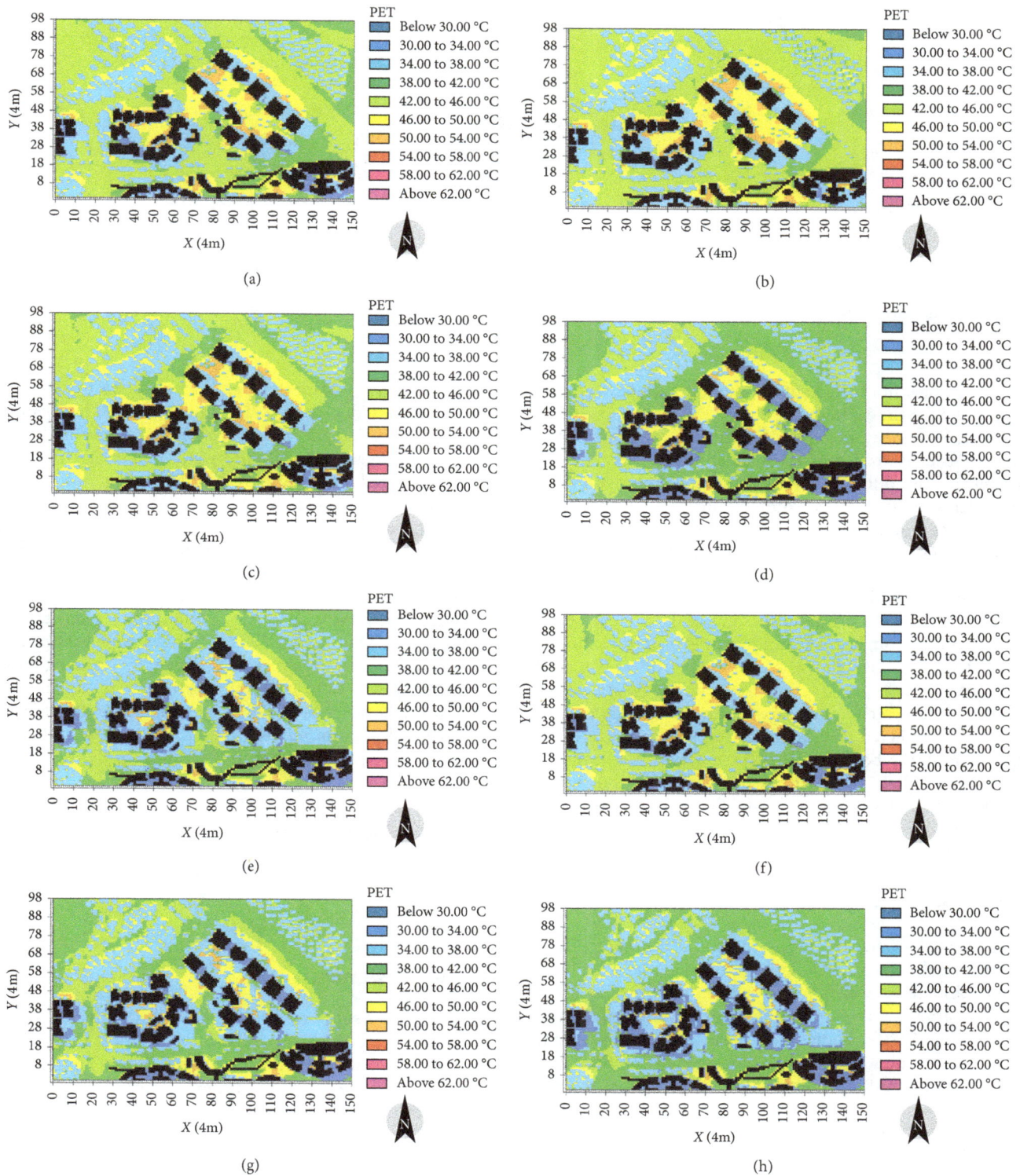

Figure 9: Simulated PET for all the design scenarios. (a) Base case. (b) Scenario 1 (wooden boards). (c) Scenario 2 (light-color granite). (d) Scenario 3 (grass surface). (e) Scenario 4 (more trees). (f) Scenario 5 (more water bodies). (g) Scenario 6 (light-color granite + more trees). (h) Scenario 7 (grass surface + more trees).

TABLE 7: Summary of the effect of different design scenarios on urban microclimate and thermal comfort (PET).

	Design scenarios	Surface temp. reduction	Air temp. reduction	Mean radiant temp. reduction	PET reduction
1	Wooden boards	2–6°C	0–0.25°C for building-shaded areas	−8 to −4°C for building-shaded areas	No change
2	Light-color granite	2–12°C	0.25–0.75°C	−8 to −4°C for building-shaded areas	No change
3	Grass surface	2–8°C	0.25–0.5°C	4–8°C for areas exposed to the sun	4–8°C for limited areas
4	More trees	2–10°C	0.25–0.5°C	12–16°C for tree-shaded areas	4–12°C
5	More water bodies	No change	No change	No change	No change
6	Light-color granite and more trees	2–12°C	0.25–0.75°C	−8 to −4°C for building-shaded areas 12–16°C for tree-shaded areas	4–12°C
7	Grass surface and more trees	2–10°C	0.25–0.75°C	4–8°C for areas exposed to the sun 12–16°C for tree-shaded areas	4–12°C

variations may not be effective enough to reduce heat stress during the daytime. However, during the night, the effect of different pavement materials on thermal comfort can be obvious because different materials have different thermal properties. In addition, air temperature is the main factor that affects urban thermal comfort during the nighttime. It needs to be noted that due to time constraints and limitations of ENVI-met modeling, nighttime thermal comfort is not investigated in this study.

Compared with grass surfacing, tree planting is a more effective strategy to promote shading, thus reducing urban heat stress. Although tree planting would lead to an increase of relative humidity and a decrease of wind speed, those negative effects are minor compared with the positive effects of reduction of air temperature and mean radiant temperature. As predicted, the combination of shade trees over grass is found to be the most effective landscape strategy in terms of cooling provided, with the maximum surface temperature reduced by 10°C, air temperature reduced by 0.75°C, mean radiant temperature reduced by 16°C, and PET reduced by 12°C.

5. Conclusions

The effects of urban landscape design on urban microclimate and thermal comfort in a high-rise residential area in the tropic climate of Singapore have been investigated in this paper. Various landscape elements of pavement materials, greenery, and water bodies have been studied. Real data on microclimate obtained from a comprehensive field measurement with multiple points have been presented and used to calibrate the new version of the microclimate-modeling software EVNI-met. With the calibrated ENVI-met, seven urban design scenarios of different surface albedo, greenery, and water bodies have been simulated with different microclimatic parameters, and their effects on human thermal comfort as measured by PET have been evaluated. It has been found that the maximum improvement of PET between the existing landscape (i.e., the base case) and suggested landscape design is about 12°C, and achieving thermal comfort during the hottest time of the day is impossible. It has also been found that the combination of shade trees over grass is the most effective landscape strategy for cooling with

the maximum surface temperature reduced by 10°C, air temperature reduced by 0.75°C, mean radiant temperature reduced by 16°C, and PET reduced by 12°C. Although high-albedo pavement materials and water bodies are found not effective in reducing heat stress in hot and humid climate conditions, the results are dubious since the evaluation of urban thermal comfort does not include the surface temperature. The evaluation of urban thermal comfort is a challenging topic in the research field of human bioclimate, which still needs further study. It can be concluded that the findings from the paper can equip urban planners and designers with knowledge and techniques when they plan for future urban areas/regions and replan for existing urban areas/regions so as to mitigate urban heat stress. However, due to the limitations of ENVI-met modeling, the effect of landscape design on nighttime thermal comfort and urban heat island requires further investigation in the future.

Conflicts of Interest

The authors declare that they have no conflicts of interest.

Acknowledgments

This work was supported by the NUS Research Scholarship from the National University of Singapore and Natural Science Foundation of Hubei Province, China, under Grant number 2015CFB510. The authors would like to express their sincere thanks to Professor Wong Nuyk Hien and his Ph.D. students from the National University of Singapore for their assistance in the field measurement work of this paper.

References

[1] F. Salata, I. Golasi, R. de Lieto Vollaro, and A. de Lieto Vollaro, "Urban microclimate and outdoor thermal comfort. A proper procedure to fit ENVI-met simulation outputs to experimental data," *Sustainable Cities and Society*, vol. 26, pp. 318–343, 2016.

[2] M. W. Yahia, E. Johansson, S. Thorsson, F. Lindberg, and M. I. Rasmussen, "Effect of urban design on microclimate and thermal comfort outdoors in warm-humid Dar es Salaam, Tanzania," *International Journal of Biometeorology*, vol. 62, no. 3, pp. 373–385, 2018.

[3] F. Ali-Toudert and H. Mayer, "Effects of asymmetry, galleries, overhanging facades and vegetation on thermal comfort in urban street canyons," *Solar Energy*, vol. 81, no. 6, pp. 742–754, 2007.

[4] R. Emmanuel, H. Rosenlund, and E. Johansson, "Urban shading—a design option for the tropics? A study in Colombo, Sri Lanka," *International Journal of Climatology*, vol. 27, no. 14, pp. 1995–2004, 2007.

[5] E. Ng, L. Chen, Y. Wang, and C. Yuan, "A study on the cooling effects of greening in a high-density city: an experience from Hong Kong," *Building and Environment*, vol. 47, pp. 256–271, 2012.

[6] N. Mazhar, R. D. Brown, N. Kenny, and S. Lenzholzer, "Thermal comfort of outdoor spaces in Lahore, Pakistan: lessons for bioclimatic urban design in the context of global climate change," *Landscape and Urban Planning*, vol. 138, pp. 110–117, 2015.

[7] A. M. Hunter, N. S. G. Williams, J. P. Rayner, L. Aye, D. Hes, and S. J. Livesley, "Quantifying the thermal performance of green facades: a critical review," *Ecological Engineering*, vol. 63, pp. 102–113, 2014.

[8] R. D. Brown and T. J. Gillespie, "Estimating outdoor thermal comfort using a cylindrical radiation thermometer and an energy budget model," *International Journal of Biometeorology*, vol. 30, no. 1, pp. 43–52, 1986.

[9] R. D. Brown, "Ameliorating the effects of climate change: modifying micro-climates through design," *Landscape and Urban Planning*, vol. 100, no. 4, pp. 372–374, 2011.

[10] M. W. Yahia and E. Johansson, "Landscape interventions in improving thermal comfort in the hot dry city of Damascus, Syria—the example of residential spaces with detached buildings," *Landscape and Urban Planning*, vol. 125, pp. 1–16, 2014.

[11] K. Perini and A. Magliocco, "Effects of vegetation, urban density, building height, and atmospheric conditions on local temperatures and thermal comfort," *Urban Forestry and Urban Greening*, vol. 13, no. 3, pp. 495–506, 2014.

[12] H. Lee, H. Mayer, and L. Chen, "Contribution of trees and grasslands to the mitigation of human heat stress in a residential district of Freiburg, Southwest Germany," *Landscape and Urban Planning*, vol. 148, pp. 37–50, 2016.

[13] W. Yang, N. H. Wong, and C. Q. Li, "Effect of street design on outdoor thermal comfort in an urban street in Singapore," *Journal of Urban Planning and Development*, vol. 142, no. 1, article 05015003, 2016.

[14] M. Bruse and H. Fleer, "Simulating surface-plant–air interactions inside urban environments with a three dimensional numerical model," *Environmental Modelling and Software*, vol. 13, no. 3-4, pp. 373–384, 1998.

[15] *ENVI-met 4.0 (Computer Software)*, Michael Bruse & Team, Bochum, Germany, http://www.envi-met.com/.

[16] F. Yang, S. S. Y. Lau, and F. Qian, "Thermal comfort effects of urban design strategies in high-rise urban environments in a sub-tropical climate," *Architecture Science Review*, vol. 54, no. 4, pp. 285–304, 2011.

[17] W. Yang, N. H. Wong, and G. Zhang, "A comparative analysis of human thermal conditions in outdoor urban spaces in summer season in Singapore and Changsha, China," *International Journal of Biometeorology*, vol. 57, no. 6, pp. 895–907, 2013.

[18] A. Matzarakis, F. Rutz, and H. Mayer, "Modeling radiation fluxes in simple and complex environments—application of the RayMan model," *International Journal of Biometeorology*, vol. 51, no. 4, pp. 323–334, 2007.

[19] A. Matzarakis, F. Rutz, and H. Mayer, "Modeling radiation fluxes in simple and complex environments: basics of the RayMan model," *International Journal of Biometeorology*, vol. 54, no. 2, pp. 131–139, 2010.

[20] A. N. Kakon, N. Mishima, and S. Kojima, "Simulation of the urban thermal comfort in a high density tropical city: analysis of the proposed urban construction rules for Dhaka, Bangladesh," *Building Simulation*, vol. 2, no. 4, pp. 291–305, 2009.

[21] C. Ketterer and A. Matzaraki, "Comparison of different methods for the assessment of the urban heat island in Stuttgart, Germany," *International Journal of Biometeorology*, vol. 59, no. 9, pp. 1299–1309, 2015.

[22] E. L. Krüger, F. O. Minella, and F. Rasia, "Impact of urban geometry on outdoor thermal comfort and air quality from field measurements in Curitiba, Brazil," *Building and Environment*, vol. 46, no. 3, pp. 621–634, 2011.

[23] N. H. Wong, C. Tan, A. Nindyani, S. Jusuf, and E. Tan, "Influence of water bodies on outdoor air temperature in hot and humid climate," in *Proceedings of International Conference on Sustainable Design and Construction (ICSDC 2011)*, Kansas City, MO, USA, March 2011.

[24] F. Salata, I. Golasi, A. de Lieto Vollaro, and R. de Lieto Vollaro, "How high albedo and traditional buildings' materials and vegetation affect the quality of urban microclimate. A case study," *Energy and Buildings*, vol. 99, pp. 32–49, 2015.

Spatiotemporal Variability and Trends in Extreme Temperature Events in Finland over the Recent Decades: Influence of Northern Hemisphere Teleconnection Patterns

Masoud Irannezhad (iD),[1,2] Hamid Moradkhani,[3] and Bjørn Kløve[1]

[1]Water Resources and Environmental Engineering Research Unit, Faculty of Technology, University of Oulu, 90014 Oulu, Finland
[2]School of Environmental Science and Engineering, Southern University of Science and Technology, Shenzhen 518055, China
[3]Center for Complex Hydrosystems Research, Department of Civil, Construction and Environmental Engineering,
  University of Alabama, Tuscaloosa, AL 35487, USA

Correspondence should be addressed to Masoud Irannezhad; masoud.irannezhad@oulu.fi

Academic Editor: Jorge E. Gonzalez

Fifteen temperature indices recommended by the ETCCDI (Expert Team on Climate Change Detection and Indices) were applied to evaluate spatiotemporal variability and trends in annual intensity, frequency, and duration of extreme temperature statistics in Finland during 1961–2011. Statistically significant relationships between these high-resolution (10 km) temperature indices and seven influential Northern Hemisphere teleconnection patterns (NHTPs) for the interannual climate variability were also identified. During the study period (1961–2011), warming trends in extreme temperatures were generally manifested by statistically significant increases in cold temperature extremes rather than in the warm temperature extremes. As expected, warm days and nights became more frequent, while fewer cold days and nights occurred. The frequency of frost and icing days also decreased. Finland experienced more (less) frequent warm (cold) temperature extremes over the past few decades. Interestingly, significant lengthening in cold spells was observed over the upper part of northern Finland, while no clear changes are found in warm spells. Interannual variations in the temperature indices were significantly associated with a number of NHTPs. In general, warm temperature extremes show significant correlations with the East Atlantic and the Scandinavia patterns and cold temperature extremes with the Arctic Oscillation and the North Atlantic Oscillation patterns.

1. Introduction

In recent decades, changes in climatic extremes have received considerable attention in international communities due to potential effects of floods, droughts, hurricanes, severe cyclonic storms, heat waves, and cold spells (e.g., [1–5]). Previous studies showed that changes in extreme temperatures are consistent with the climate warming at the global scale [6–8]. However, the intensity and spatial pattern of such changes are different around the world (e.g., [9, 10]). Hence, updated and systematic analysis of changes in the past extreme temperatures at the country scale plays a key role in defining national climate change adaptation strategies.

The most recent global analysis of trends in extreme temperatures shows decreases (increases) in cold (warm) extreme indices over majority of regions since 1900 (e.g., [7]). Similarly, Alexander et al. [6] reported that the annual occurrence of cold (warm) nights has decreased (increased) over 70% of the global land area during 1951–2003. For Europe, increases in warm days and nights, while decreases in cold days and nights, have also been determined for the last 50 years [11]. A recent analysis of trend detection in European extreme temperatures [2] confirms more warm days and nights as well as fewer cold days and nights over time during the 20th century. They also concluded increases (decreases) in the intensity and frequency of high- (low-) temperature extreme events. Besides, there has been a growing interest in identifying spatiotemporal changes in extreme temperatures at the country scale (e.g., [12–14]). For Finland, however, only few studies have investigated changes in some extreme temperature indices, all focusing on several important meteorological stations (e.g., [2, 9, 15]). Hence, a comprehensive evaluation of fine resolution spatial patterns of variability and trends in extreme temperatures over Finland is well motivated.

Atmospheric circulation is one of the key physical processes causing the variability and changes in extreme temperatures [16] and mainly follows a number of preferred modes [17]. Such atmospheric circulation modes are generally described by teleconnection patterns (e.g., Arctic Oscillation, AO), which are defined as persistent, recurrent, and large-scale modes of atmospheric pressure anomalies. The intensity of teleconnection patterns over a particular geographical area during a specific period of the year is often expressed by numerical indices of well-known atmospheric circulation modes. Previous studies have reported most influential teleconnection patterns and their connections to surface air temperature (SAT) variability on regional, national, and global scales (e.g., [18–21]). A number of studies have particularly focused on links between changes in extreme temperatures and teleconnection patterns in different parts of the world (e.g., [22–24]).

In a previous study, Irannezhad et al. [25] focused only on long-term variations and trends in annual, seasonal, and monthly mean temperatures in Finland and their connections to different Northern Hemisphere teleconnection patterns (NHTPs). However, the present study aims at characterizing the fine resolution spatial variability and trends in extreme temperatures in Finland during 1961–2011 and improving our knowledge on the roles of NHTPs in controlling such changes. Specific objectives of the present study are to (1) determine trends in a set of extreme temperature indices; (2) evaluate spatial variations and changes in these indices throughout the country; and (3) explore relationships between the temperature indices and a number of well-established NHTPs.

2. Materials and Methods

2.1. Study Area and Data Description. Finland is a northern European country, extended ~1320 km in the south-north direction between 60 and 70°N across the northwest of the Eurasian continent, close to the northern part of the Atlantic Ocean (Figure 1). The Arctic Ocean, the Atlantic Ocean, the Baltic Sea, the Scandinavian mountain range, latitudinal gradients, and continental Eurasia are the most important geographical factors influencing climate conditions over Finland. Temperature in Finland is generally moderate compared with the other geographical areas at the similar latitudes [26] and naturally decreases from south to north. The Köppen-Geiger climate classification system categorizes Finland as lying in the temperate of boreal zone, with no dry season (Df) [27].

For this study, daily minimum and maximum SAT time series spatially interpolated onto 3322 regular grid (10*10 km^2) points across Finland for the years 1961–2011 (Figure 1(c)) were obtained from PaITuli-Spatial Data for Research and Teaching at the CS-IT centre for Science Ltd. website: http://www.csc.fi/english. The Finnish Meteorological Institute (FMI) used daily minimum and maximum SAT measurements at 100–200 meteorological stations properly uniformly scattered throughout Finland (Figure 1(d)) as input to a spatial model [28] developed based on the kriging approach [29] to generate the Finnish daily gridded SAT datasets. As external forcing factors, the spatial model took into account the geographical coordinates, land elevation, the percentage of sea, and lakes at each grid cell [28, 30]. The effects of environmental changes (including urbanisation) on actual SAT records in densely populated areas (typically located in southern Finland) were also eliminated during the producing of gridded datasets [31, 32]. Interannual variations in the number of SAT measurement stations used by the FMI for creating the gridded minimum and maximum daily SAT time series for the period 1961–2011 are shown in Figure 1(e). The accuracy and homogeneity of the gridded daily minimum and maximum SAT data generated were qualified during the establishment of the PaITuli database. For more details of the gridding, see Venäläinen et al. [33]. The gridded time series of the PaITuli database have previously been applied in some studies, for example, Tietäväinen et al. [30], Aalto et al. [34], Irannezhad and Kløve [35], and Gao et al. [36].

The World Meteorological Organization (WMO) Expert Team on Climate Change Detection and Indices (ETCCDI) recommends a suite of indices derived from daily SAT for evaluation of changes in extreme temperature regimes (for the complete list and detailed descriptions of the extreme temperature indices, please check the ETCCDI webpage http://etccdi.pacificclimate.org/list_27_indices.shtml). Zhang et al. [37] also gives a review on such indices for detecting changes in extreme daily temperatures. The extreme events defined by the ETCCDI generally occur several times through a season or year. This yields more robust statistical features than other measures of extremes, which are far enough into the tails of distribution and may not be seen in quite some years [6]. Application of such predefined extreme temperature indices also allows comparability between observational and modelled SAT across different parts of the world [5]. Thus, this study uses fifteen extreme temperature indices (Table 1) recommended by the ETCCDI, with the base period of 1961–2011. Prior to calculating these fifteen indices at national scale, maximum (minimum) daily temperatures nationwide over Finland were computed as the average values of daily maximum (minimum) temperature at all 3322 grids across the country for each day throughout the period 1961–2011.

For describing the modes of atmospheric circulations, this study considers seven influential teleconnection patterns for climate variability in the Northern Hemisphere based on previous studies. These include the North Atlantic Oscillation (NAO), the Arctic Oscillation (AO), the East Atlantic (EA), the West Pacific (WP), the East Atlantic/West Russia (EA/WR), the Scandinavia (SCA), and the Polar/Eurasian (POL) patterns. The key characteristics of these seven NHTPs are summarized in Table 2. At the National Oceanic and Atmospheric Administration (NOAA) of the US, the Climate Prediction Centre (CPC) calculates standardized monthly values of these NHTPs since 1950 (available online at: http://www.cpc.ncep.noaa.gov/data/teledoc/telecontents.shtml). Corresponding to different extreme temperature indices, the present study calculates winter (Dec-Feb), summer (Jun-Aug), and annual (Jan-Dec) ACPs for the years between 1961 and 2011 as the mean of their standardized monthly values (Table 2).

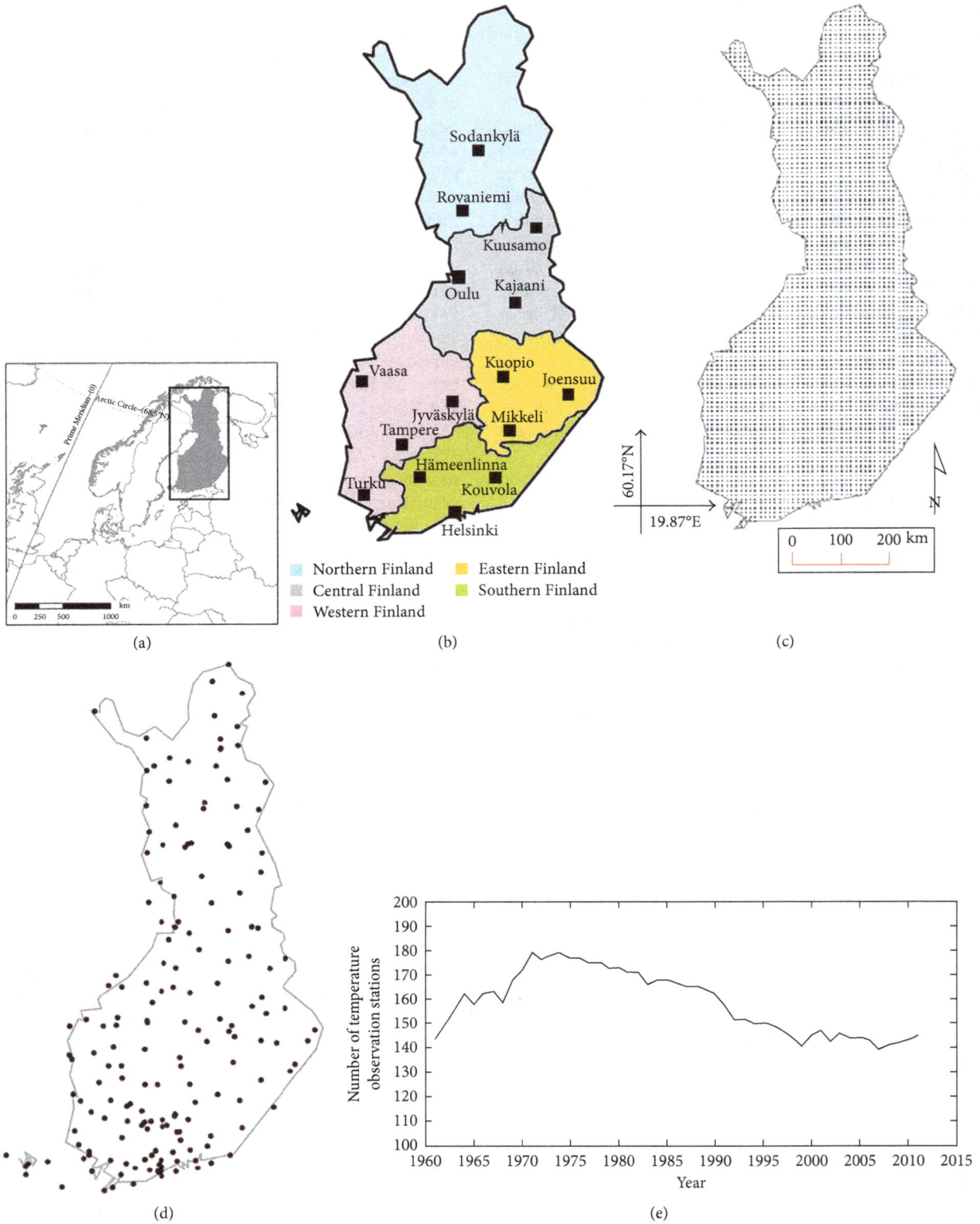

FIGURE 1: (a) Location of the study area, (b) different regions in Finland, (c) regular grid points (10×10 km^2) covering daily minimum and maximum temperature datasets across Finland obtained from PaiTuli, (d) daily temperature measurement stations in Finland used for calculation of the gridded time series, and (e) temporal variability in the number of daily temperature measurement stations in Finland over 1961–2011.

TABLE 1: Definitions of selected daily extreme temperature indices and their period of Northern Hemisphere teleconnection patterns (NHTPs) for this study.

ID number	ID	Indicator name	Definition	Units	Period for NHTPs
1	TXx	Maximum Tmax	Maximum value of daily maximum temperature (TX)	°C	Summer (Jun-Jul-Aug)
2	TXn	Minimum Tmax	Minimum value of daily maximum temperature	°C	Winter (Dec-Jan-Feb)
3	TNx	Maximum Tmin	Maximum value of daily minimum temperature (TN)	°C	Summer (Jun-Jul-Aug)
4	TNn	Minimum Tmin	Minimum value of daily minimum temperature	°C	Winter (Dec-Jan-Feb)
5	DTR	Diurnal temperature range	Annual mean difference between TX and TN	°C	Annual (Jan-Dec)
6	ETR	Extreme temperature range	Annual difference between TXx and TNn	°C	Annual (Jan-Dec)
7	FD	Frost days	Annual count of days when TN < 0°C	Days	Winter (Dec-Jan-Feb)
8	ID	Icing days	Annual count of days when TX < 0°C	Days	Winter (Dec-Jan-Feb)
9	SU25	Summer days	Annual count of days when TX (daily maximum) > 25°C	Days	Summer (Jun-Jul-Aug)
10	TX10p	Cold days	Percentage of days when TX < 10th percentile	%	Winter (Dec-Jan-Feb)
11	TX90p	Warm days	Percentage of days when TX > 90th percentile	%	Summer (Jun-Jul-Aug)
12	TN10p	Cold nights	Percentage of days when TN < 10th percentile	%	Winter (Dec-Jan-Feb)
13	TN90p	Warm nights	Percentage of days when TN > 90th percentile	%	Summer (Jun-Jul-Aug)
14	WSDI	Warm spell duration	Annual count of days with at least six consecutive days when TX > 90th percentile	Days	Summer (Jun-Jul-Aug)
15	CSDI	Cold spell duration	Annual count of days with at least six consecutive days when TN < 10th percentile	Days	Winter (Dec-Jan-Feb)

TABLE 2: Summary of the Northern Hemisphere teleconnection patterns (NHPs) affecting climate variability in Finland considered in this study.

NHTP	Centre(s) of circulation	Natural signature over Finland in positive phase	Reference
North Atlantic Oscillation (NAO)	Ponta Delagada (Azores) and Stykkisholmur (Iceland)	Strong westerly circulation bringing warmer and wetter weather than normal	Barnston and Livezey [38], Irannezhad et al. [35, 39]
Arctic Oscillation (AO)	A dipole between the polar cap area and the adjacent zonal ring centred along 45°N	Low pressure in the Arctic and high pressure at midlatitudes, leading to warmer and wetter weather than normal	Thompson and Wallace [40], CPC [11], Irannezhad et al. [35, 39]
East Atlantic (EA)	North-south dipoles across the North Atlantic	Intensive westerly circulation, causing the weather to be warmer and wetter than normal	Barnston and Livezey [38], CPC [41], Irannezhad et al. [35, 39]
West Pacific (WP)	Kamchatka (Russia) and a centre between the western, North Pacific, and South Asia and East Asia	North-south dipole anomalies causing the weather to be wetter than normal in February and colder than normal in spring and summer	Irannezhad et al. [35, 39]
East Atlantic/ West Russia (EA/WR)	Western Europe, northwest Europe, and Portugal in spring and autumn; Caspian Sea in winter and Russia	Northerly and northwesterly circulation across the Baltic Sea, resulting in milder and drier weather than normal	Barnston and Livezey [38], Lim and Kim [42], Irannezhad et al. [35, 39]
Scandinavia (SCA)	West of Europe, Mongolia, and Scandinavia	High pressure over Scandinavia, bringing milder and drier weather than normal	Barnston and Livezey [38], Bueh and Nakamura [43], Irannezhad et al. [35, 39]
Polar/Eurasia (POL)	North Pole, Europe, and northeastern China,	Strong polar vortex resulting in milder and drier weather than normal	CPC [41], Irannezhad et al. [35, 39]

2.2. Statistical Analyses. The Mann–Kendall (MK) nonparametric test [44, 45] was applied for detecting statistically significant ($p < 0.05$) trends in extreme temperatures. To identify relationships between extreme temperature indices in Finland and NHTPs during 1961–2011, Spearman's rank correlation (ρ) was used. However, for autocorrelated time series, the trend free prewhitening (TFPW) method [46] to detect significant ($p < 0.05$) trends and the residual bootstrap (RB) approach [47] with 5000 independent replications to estimate the standard deviation of the ρ values were applied. Using the Sen method [48], the present study calculated the slope of detected significant trends as an estimate of the trend.

To assess the uncertainties associated with the estimated trends, their 95% confidence intervals were calculated [49].

3. Results

3.1. Country Scale Assessment of Extreme Temperatures. Number of warm days (TX90p) and nights (TN90p) on the country scale significantly increased by 0.8 ± 0.4 and 1.4 ± 0.7 (% per decade, $p < 0.05$) during 1961–2011, but cold days (TX10p) and nights (TN10p) decreased by 1.2 ± 0.7 and 1.3 ± 0.7 (% per decade, $p < 0.05$), respectively (Figures 2(a)–2(d)). The maximum number of events during a year

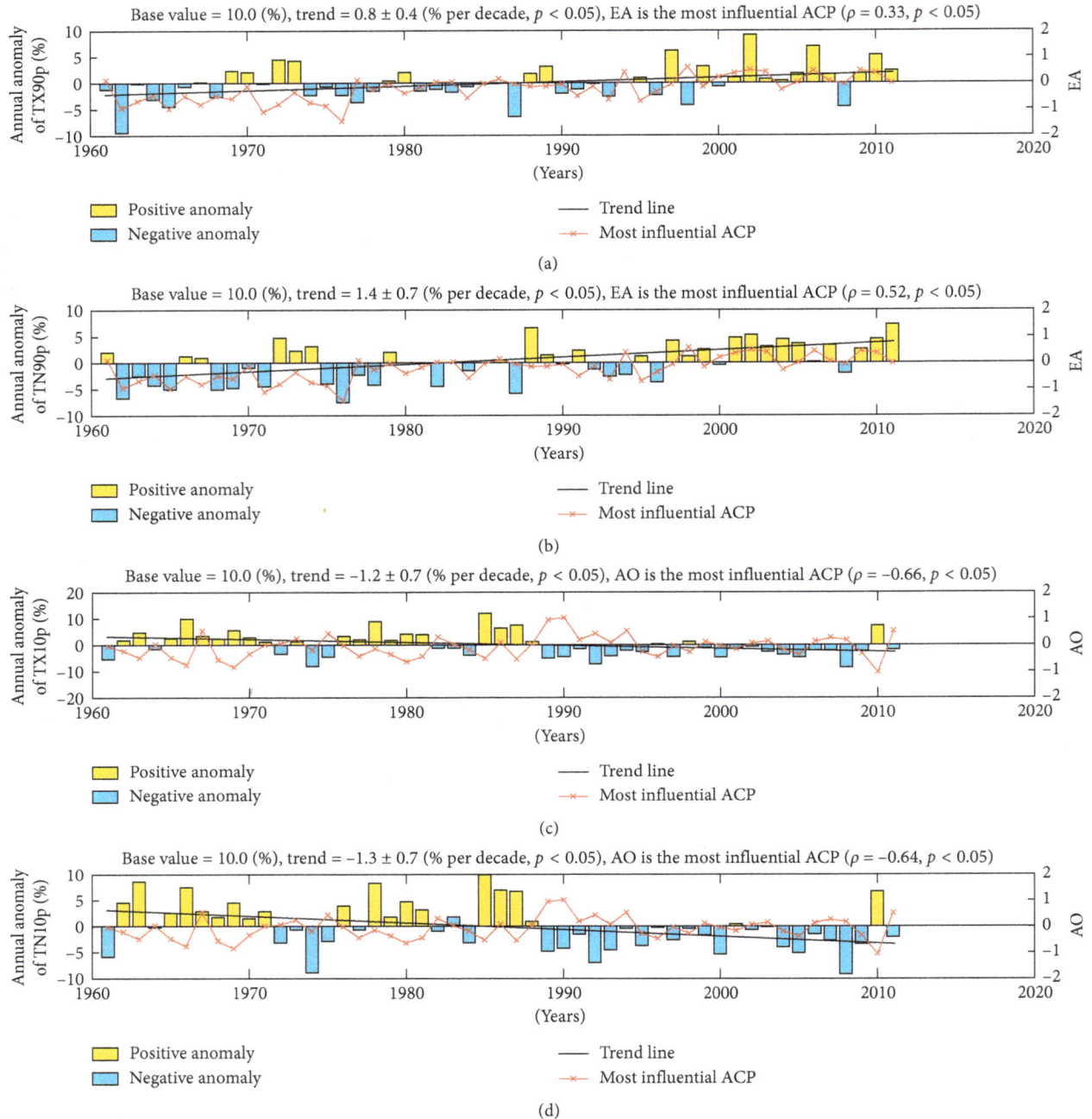

FIGURE 2: Time series with trend line and most influential NHTP for (a) TX90p, (b) TN90p, (c) TX10p, and (d) TN10p, on the country scale in Finland during 1961–2011.

for warm days was about 70 days (19.2%) in 2007 (Figure 2(a)); for warm nights was almost 63 days (17.3%) in 2011 (Figure 2(b)); and for cold days and nights was correspondingly equal to 80 (21.9%) and 73 (20%) days, both in 1985 (Figures 2(c) and 2(d)). The EA pattern was the most influential NHTP for variations in both warm days ($\rho = 0.33$) and nights ($\rho = 0.52$) and the AO for variability in both cold days ($\rho = -0.66$) and nights ($\rho = -0.64$) (Table 3).

On country scale, significant decreasing trends were detected in frost (FD) and icing (ID) days during 1961–2011, at the rates of -4.7 ± 2.7 and -3.6 ± 3.2 (days/decade, $p < 0.05$), respectively (Figures 3(a) and 3(b)). The base value was 200.1 days for the annual FD and about 117 days for the annual ID (Table 3). The annual FD varied from 172 days in 2000 to 230 days in 1968 (Figure 3(a)), and the annual ID ranged between 77 days in 1975 and 148 days in 1980 (Figure 3(b)). Such variations in the annual FD and ID were most significantly correlated with the NAO ($\rho = -0.28$) and the AO ($\rho = -0.49$) patterns, in turn (Table 3).

TABLE 3: Base value (average during 1961–2011), statistically significant ($p < 0.05$) trends, and Spearman rank correlation (ρ) with NHTPs for daily extreme temperature indices given in Table 1.

ID number	ID	Unit	Base value	Trend (/decade, $p < 0.05$)	NAO	EA	WP	EA/WR	SCA	POL	AO
1	TXx	°C	26.7	—	−0.05	0.21	−0.06	−0.02	**0.39**	**0.28**	0.01
2	TXn	°C	−20.0	—	**0.35**	−0.21	−0.15	0.01	−0.22	−0.16	**0.32**
3	TNx	°C	14.9	0.3 ± 0.2	0.01	**0.55**	−0.22	**−0.37**	0.11	−0.02	0.18
4	TNn	°C	−29.3	—	**0.47**	−0.12	−0.04	0.10	−0.19	−0.22	**0.41**
5	DTR	°C	8.2	−0.09 ± 0.08	**−0.35**	−0.23	0.07	**0.33**	0.17	0.19	**−0.28**
6	ETR	°C	56.0	—	−0.10	0.07	0.21	0.14	0.06	0.06	−0.13
7	FD	days	200.1	−4.7 ± 2.7	**−0.28**	−0.04	−0.21	0.04	0.08	0.22	−0.11
8	ID	days	117.0	−3.6 ± 3.2	**−0.39**	−0.05	−0.12	−0.18	0.10	−0.07	**−0.49**
9	SU25	days	5.3	—	0.01	**0.27**	−0.13	−0.07	**0.39**	0.21	0.07
10	TX10p	%	10.0	−1.2 ± 0.7	**−0.64**	−0.01	0.09	−0.23	**0.28**	0.11	**−0.66**
11	TX90p	%	10.0	0.8 ± 0.4	0.07	**0.33**	−0.28	−0.17	**0.27**	0.20	0.21
12	TN10p	%	10.0	−1.3 ± 0.7	**−0.62**	−0.03	−0.14	−0.16	0.23	0.17	**−0.64**
13	TN90p	%	10.0	1.4 ± 0.7	−0.05	**0.52**	**−0.42**	**−0.48**	0.07	−0.10	0.11
14	WSDI	days	42.0	—	0.17	**−0.34**	0.24	0.16	0.14	−0.11	−0.13
15	CSDI	days	286.5	—	**0.42**	0.07	−0.13	−0.08	−0.23	−0.02	**0.35**

If significant ($p < 0.05$), ρ is given in bold.

Diurnal temperature range (DTR) has slightly decreased by 0.09 ± 0.08 (°C/decade, $p < 0.05$), while maximum Tmin (TNx) has increased by 0.30 ± 0.20 (°C/decade, $p < 0.05$) (Figures 3(c) and 3(d)). On average, the annual DTR was about 8.2°C and ranged from 7.3°C in 2008 and 9.4°C in 1963 (Figure 3(c)). The NAO was the strongest ($p < 0.05$) NHTP affecting the variability of DTR, with $\rho = -0.35$ (Table 3). With the base value of 14.9°C, the annual TNx varied between 11.7°C in 1962 and 18.5°C in 2003 and showed the strongest relationships with the EA pattern ($\rho = 0.55$, $p < 0.05$) (Figure 3(d)). Other indices related to the extreme temperature intensity (TXx, TXn, TNn, and ETR in Table 1) and frequency (SU25 in Table 1) showed no clear trends (Table 3). For such extreme temperatures, wintertime-dependent indices (TXn and TNn) were most significantly associated with the NAO index and summertime indices (TXx and SU25) with the SCA pattern (Table 3). However, no clear connections between ETR and NHTPs were found (Table 3).

On average, annual warm (WSDI) and cold (CSDI) spells were 42.0 and 286.5 days, respectively (Table 3). The longest WSDI (CSDI) was 77 (358) days in 1963 (1974), and the shortest was 10 (19) days in 1986 (2008). The EA pattern was the most influential NHTP for the WSDI variability ($\rho = -0.34$), while the CSDI showed strongest relationships with the NAO ($\rho = 0.42$) (Table 3). Besides, trend analysis indicates no statistically significant changes in warm and cold spells (Table 3).

3.2. Spatial Distribution of Warm and Cold Temperature Indices. In general, TXx was warmer (ranging 28.0–30.0°C) over the southern, eastern, and western Finland (Figure 4(a)), while TNx was warmer (ranging 17.3–18.5°C) mostly in the east of country (Figure 4(c)). The warmest range for both TXn (from −14.0 to −10.0°C) and TNn (from −23.0 to −20.0°C) was seen over the southwest coastal areas of Finland (Figures 4(b) and 4(d)). The highest range for DTR was 8.3–9.5°C largely observed in the north, centre,

west, and south (Figure 4(e)), and for ETR was between 60.8 and 66.0°C seen over the northern, central, and eastern parts (Figure 4(f)). Both FD and ID were more frequent at the higher latitudes, with the highest range of 225–250 (Figure 5(a)) and 165–205 (Figure 5(b)) days, respectively. In contrast, the highest frequency range (15–20 events) of summer days (SU25 in Table 1) was found in the south (Figure 5(c)). All TX10p, TX90p, TN10p, and TN90p obviously ranged between 9.5 and 10.0% of days during a year (~35–37 days) scattered over Finland (not shown). The longest warm spell ranging from 52 to 57 days was observed in the southern areas of Finland (Figure 5(d)), and the longest cold spell ranging between 285 and 305 days was located in the northern and central parts (Figure 5(e)).

Significant ($p < 0.05$) spatial trends determined in TXx, TXn, TNx, and TNn during 1961–2011 were all positive (Figures 6(a)–6(d)). Such warming trends in TXx were observed over small areas in southern and western Finland (Figure 6(a)); in TXn across the large parts of the northern and western Finland (Figure 6(b)); in TNx mostly throughout the entire country except central parts (Figure 6(c)); and in TNn mainly over the southern, western, and northern Finland (Figure 6(d)). For both diurnal (Figure 6(e)) and extreme (Figure 7(a)) temperature ranges (DTR and ETR), significant decreases are primarily observed over the northern parts and southwestern coastal areas of Finland. During 1961–2011, the range of such trends was from −0.50 to −0.25 (°C/decade) for DTR (Figure 6(e)) and from −3 to −1 (°C/decade) for ETR (Figure 7(a)). The highest rate of decreasing trends in FD (5.3–10.6 days/decade) was seen over some areas in the southern, eastern, western, and central Finland (Figure 7(b)). For ID, significant decreasing trends, ranging from −1.2 to −7.2 (days/decade), were largely observed over the west and north of the country (Figure 7(c)). All significant trends in SU25 found mainly in the southern Finland were positive, ranging between 1.6 and 3.2 (days/decade) (Figure 7(d)). Both TX10p (Figure 7(e)) and TN10p (Figure 8(b)) showed decreases over most parts

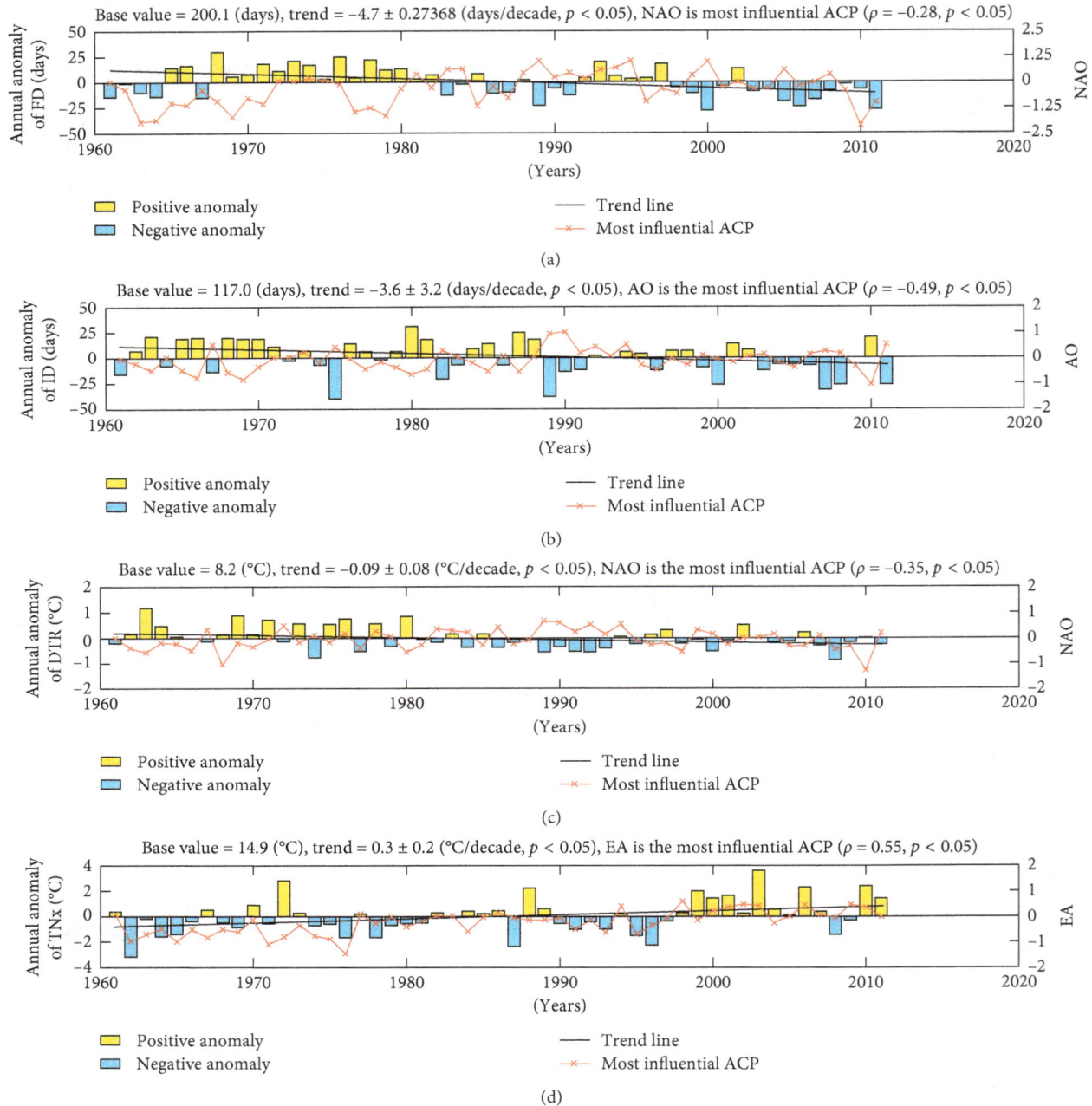

FIGURE 3: Time series with trend line and most influential NHTP for (a) FD, (b) ID, (c) DTR, and (d) TNx, on the country scale in Finland during 1961–2011.

of Finland, but both TX90p (Figure 8(a)) and TN90p (Figure 8(c)) revealed increasing trends. WSDI significantly increased at a few grids in the western Finland (Figure 8(d)), while increases in CSDI were mainly found in the northern Finland (Figure 8(e)).

TXx showed significant positive correlations with the SCA pattern in the north, centre, east, and upper west of Finland, with the EA pattern over the lower parts of western Finland and with the POL pattern in the south of Finland (Figure 9(a)). The EA and SCA patterns were also the most influential NHTPs for TNx and SU25 in most parts of

Finland, respectively (Figures 9(b) and 9(c)). Besides, TX90p showed strong positive correlations with the EA pattern in the northern and centre of Finland, while with the SCA pattern in other parts of the country (Figure 9(d)). Both TN90p and WSDI in the northern Finland were significantly associated with the EA pattern (Figures 9(e) and 9(f)). However, the EA/WR pattern was the most significant NHTPs affecting TN90p in most parts of the central, eastern, and western Finland (Figure 9(e)). On the contrary, the POL pattern was the most influential NHTP for variations in SU25 and WSDI in the south of Finland (Figures 9(c) and 9(f)).

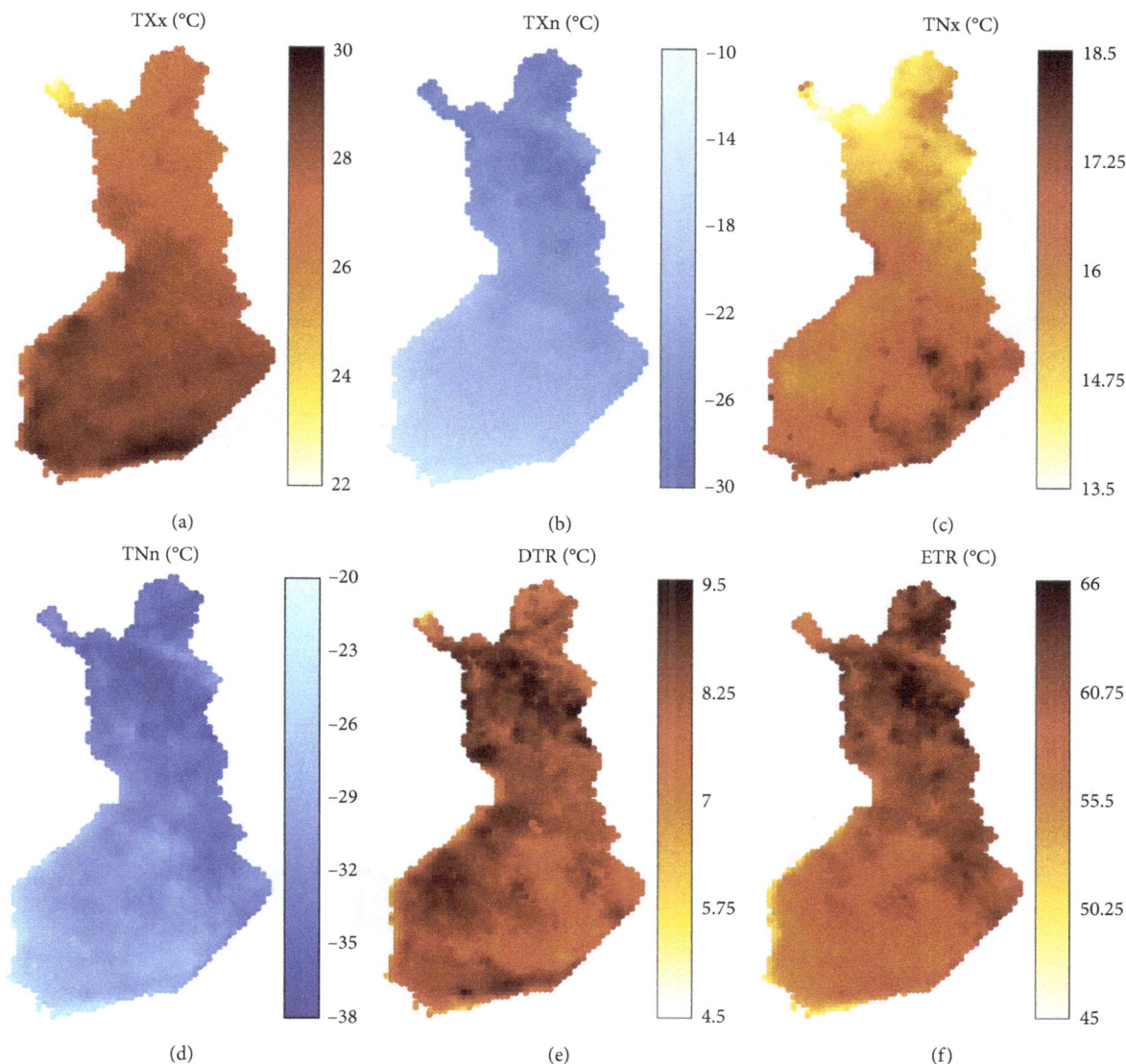

FIGURE 4: Spatial distribution maps of base values for (a) TXx, (b) TXn, (c) TNx, (d) TNn, (e) DTR, and (f) ETR in Finland during 1961–2011.

In general, variations in cold temperature extreme indices (TXn, TNn, FD, ID, TX10p, TN10p, and CSDI) in Finland were significantly correlated with the NAO and AO (Figure 10). The AO showed negative associations with ID and TX10p over most parts of Finland (Figures 10(d) and 10(e)). It (AO) also influenced TN10p in the country except the eastern parts of the northern and central Finland, where the NAO was the most influential NHTP (Figure 10(f)). Likewise, the NAO was the only strong NHTP affecting FD in the southern, western, and eastern areas (Figure 10(c)) as well as CSDI in the northern, central, and some parts of eastern and western Finland (Figure 10(g)). This NHTP (NAO), besides, was in strong associations with TXn in most parts of eastern, western, northern, and southern Finland, while TXn in central areas were significantly linked to the AO index (Figure 10(a)). Similarly, TNn strongly correlated with the AO index in small areas mainly scattered over the eastern, western, southern, central, and

western parts of northern Finland, while the NAO was the most influential NHTP for TNn throughout other areas of the country (Figure 10(b)).

DTR was negatively correlated with the NAO mostly in the northern and central Finland, while positively with the EA/WR pattern in the western, southern, and eastern areas (Figure 11(a)). On the contrary, ETR showed positive correlations with the WP pattern in the northern, western, and southern Finland; with the EA pattern in central and western parts; and the EA/WR pattern in small areas throughout the north of the country (Figure 11(b)).

4. Discussion

4.1. Changes in Extreme Temperatures. Study of changes in intensity, frequency, and duration of extreme temperatures in response to climate warming has received a substantial attention around the world in recent years (e.g. [2, 6, 7]). Using

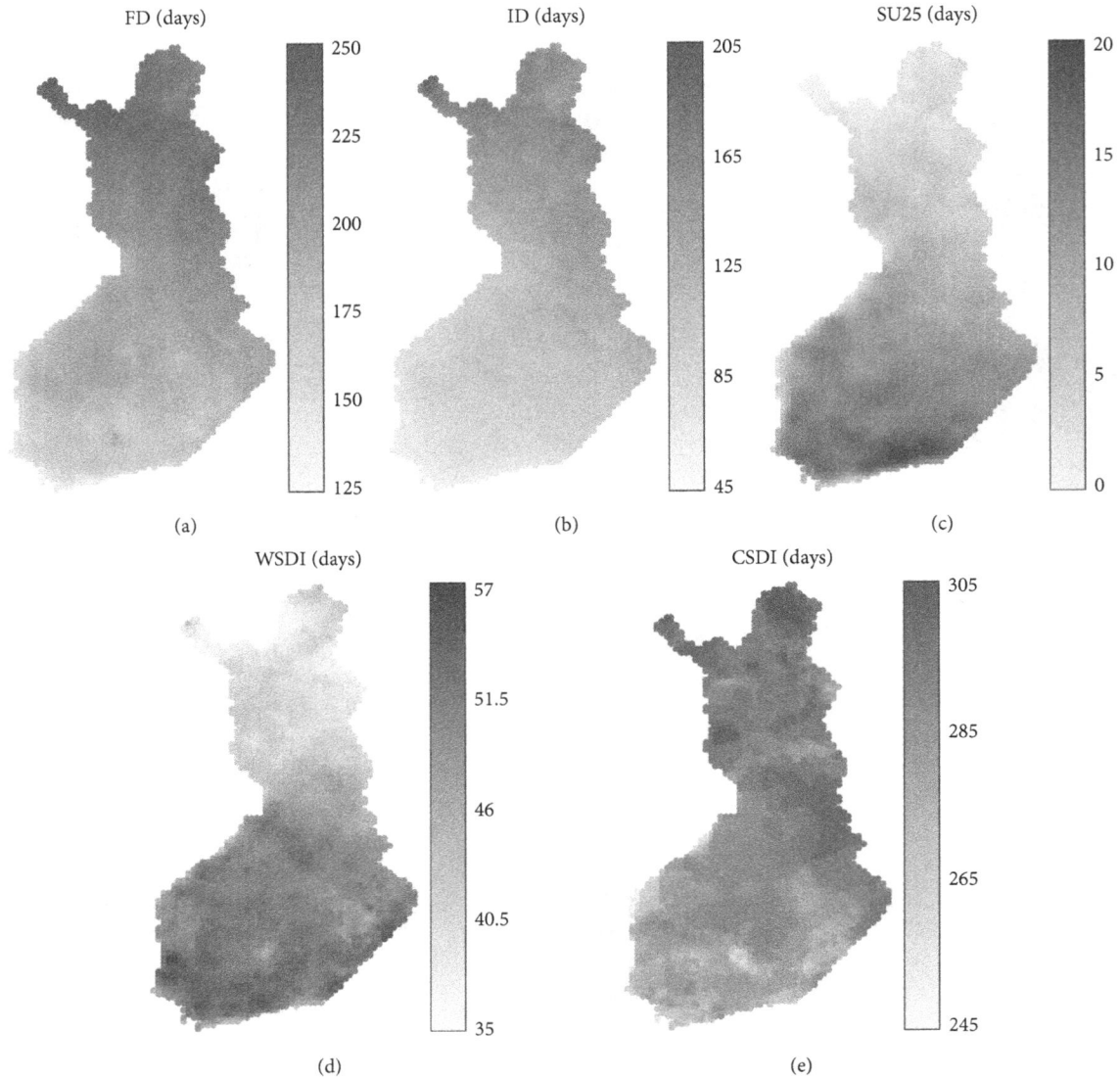

FIGURE 5: Spatial distribution maps of base values for (a) FD, (b) ID, (c) SU25, (d) WSDI, and (e) CSDI in Finland during 1961–2011.

daily extreme temperature indices recommended by ETCCDI, this paper evaluated interannual variability and trends in warm and cold extreme events in Finland during 1961–2011. Intensification of warm extreme temperatures in Finland was generally manifested by significant increasing trends detected in TNx over most parts of the country. For cold extreme temperatures, warming trends revealed by the TXn and TNn indices were largely seen over the southwest and north of Finland. Difference in the rate of such increasing trends in warm and cold extreme temperatures caused DTR and ETR indices to decline, particularly over the southwestern and northern Finland. Such decreases in DTR and ETR are generally referred to increases in the cold temperature extreme indices. In fact, extreme temperature warming in Finland results primarily from rises in the cold temperature extreme indices than in the warm ones. Similarly, Alexander et al. [6] reported warming trends in minimum temperature

extremes were greater than in maximum temperature extremes over many parts of the world. Chen et al. [2] also concluded that the intensity of extreme temperatures has increased over the northern Europe associated with mean temperature rise over the 20th century. Likewise, previous studies reported statistically significant warming trends over Finland during the last century (e.g., [25, 50]).

This study determined that warm temperature extremes have become more frequent in Finland over 1961–2011, reflected by the statistically significant increasing trends in warm days (TX90p) and nights (TN90p). Also, lower frequency of cold temperature extremes were manifested by statistically significant decreasing trends in frost (FD) and icing (ID) days as well as cold days (TX10p) and nights (TN10p). Similar to these findings, Alexander et al. [6] concluded more (less) frequent warm (cold) days and nights at the global scale. For

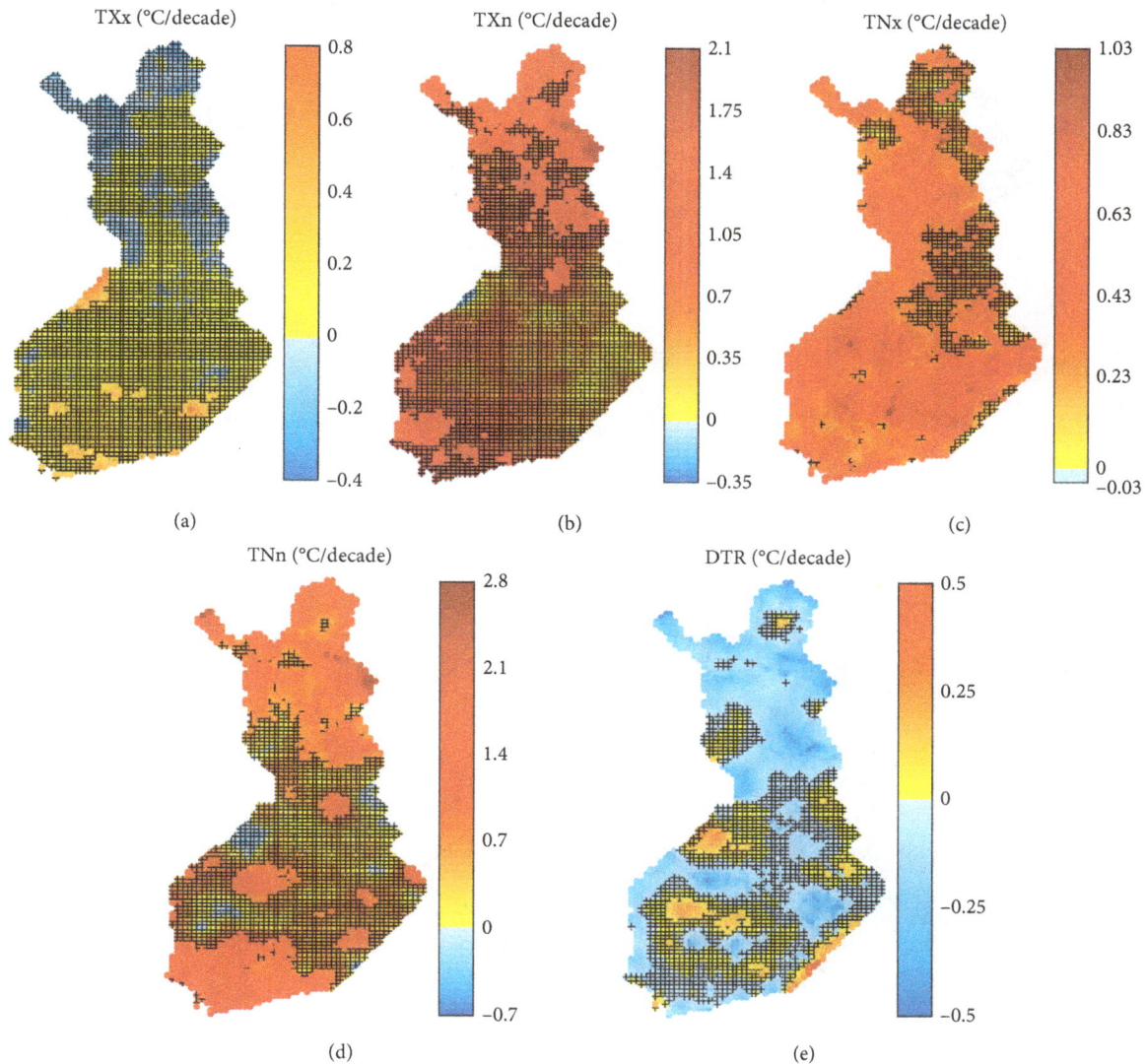

FIGURE 6: Spatial distribution maps of trends (/decade) in (a) TXx, (b) TXn, (c) TNx, (d) TNn, and (e) DTR in Finland during 1961–2011. The stippling indicates areas where the trends are statistically insignificant ($p > 0.05$).

the north of Chen et al. [2] also determined increases (decreases) in the frequency of high- (low-) temperature extreme events during 1901–2000.

On national scale, there were no clear changes in warm and cold spells during 1961–2011. However, spatial trend analysis indicates significant increases in cold spells in the upper parts of the northern Finland. On the contrary, Alexander et al. [6] reported that cold (warm) spells over Finland have slightly shortened (lengthened) over the period 1951–2003. For the northern Finland, Chen et al. [2] concluded significant lengthening in the warm spells only during the winter season (December–February) over the years 1901–2000, while shortening in the cold spells in the summer (June–August) and autumn (September–November) seasons. Such differentiated changes identified by these studies confirm that extreme temperature events are spatially and temporally inhomogeneous.

4.2. Influence of Northern Hemisphere Teleconnection Patterns. The present study indicates that warm temperature extreme events (TXx, TNx, SU25, TX90p, TN90p, and WSDI) were mostly controlled by the EA and SCA patterns. Similarly, Irannezhad et al. [25] concluded that the EA and SCA patterns are the first and second most significant NHTPs positively influencing variability in temperature over summer season in Finland. The EA pattern is originally defined by Wallace and Gutzler [51] as a teleconnection with four different action centres: two low pressures in the west of the British Islands and over central Serbia and two high pressures in the southwest of the Canary Islands and between the Caspian and Black Seas, respectively [52]. It principally describes the airflow coming from the Biscayan towards the centre of Europe. Hence, the EA positive phase results in a negative pressure anomaly over the western Ireland and a positive pressure anomaly from the western to

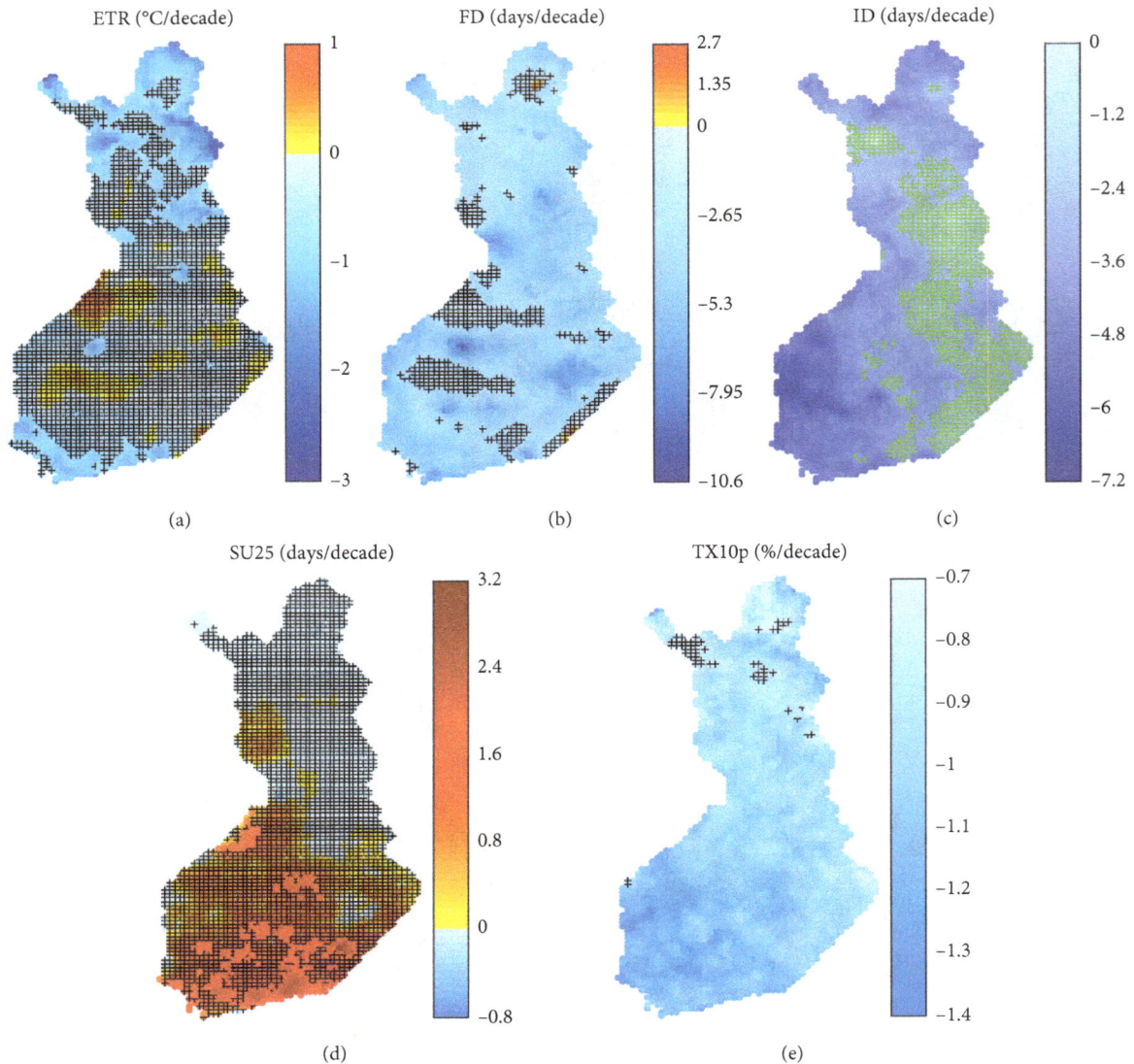

FIGURE 7: Spatial distribution maps of trends (/decade) in (a) ETR, (b) FD, (c) ID, (d) SU25, and (e) TX10p in Finland during 1961–2011. The stippling indicates areas where the trends are statistically insignificant ($p > 0.05$).

the eastern Atlantic, which brings warmer airflow than normal to Finland. Irannezhad et al. [25] also showed that the predominant surface wind over Finland during summer under the positive EA is from the south direction, which brings warm air from the southern Europe to Finland and thereby controlling warm extreme temperature events over the country. For the SCA pattern, the main action centre is located over Scandinavia and large parts of the Arctic Ocean in the north of Siberia, while two other centres are across the west of Europe and the west of China. The positive (negative) SCA refers to high- (low-) pressure system throughout Greenland, Norwegian Sea, and Scandinavia [43] resulting in warmer (colder) climate in, particularly during warm months of the year, Finland [25].

Cold temperature extreme indices (including TXn, TNn, FD, ID, TX10p, TN10p, and CSDI) showed most significant correlations with the AO and the NAO patterns. Similar to this study, Irannezhad et al. [25] identified the AO and the

NAO pattern as the most influential NHTPs for variability in the mean temperature over Finland, particularly during winter. The AO describes the power of circumpolar vortex [40], and the NAO index expresses the intensity of westerly airflow from the North Atlantic to the Atlantic European sector [53]. Their positive (negative) phase corresponds to the strengthening (weakening) of westerly circulation and prevailing of mild maritime (cold) airflow over the northern Europe, particularly during wintertime (e.g., [20, 54, 55]). Over Eurasia, Thompson et al. and Wallace et al. [40, 51] showed temperature is more strongly connected to the AO than to the NAO. Serreze et al. [56] concluded that the NAO could be considered as a major component of the AO. Previous studies reported shift in both AO and NAO from the negative to the positive phase in the early 1970s (e.g., [54, 57, 58]). Wang et al. [59] also found statistically significant increasing trends in both AO (0.26 per decade) and NAO (0.20 per decade) during recent decades.

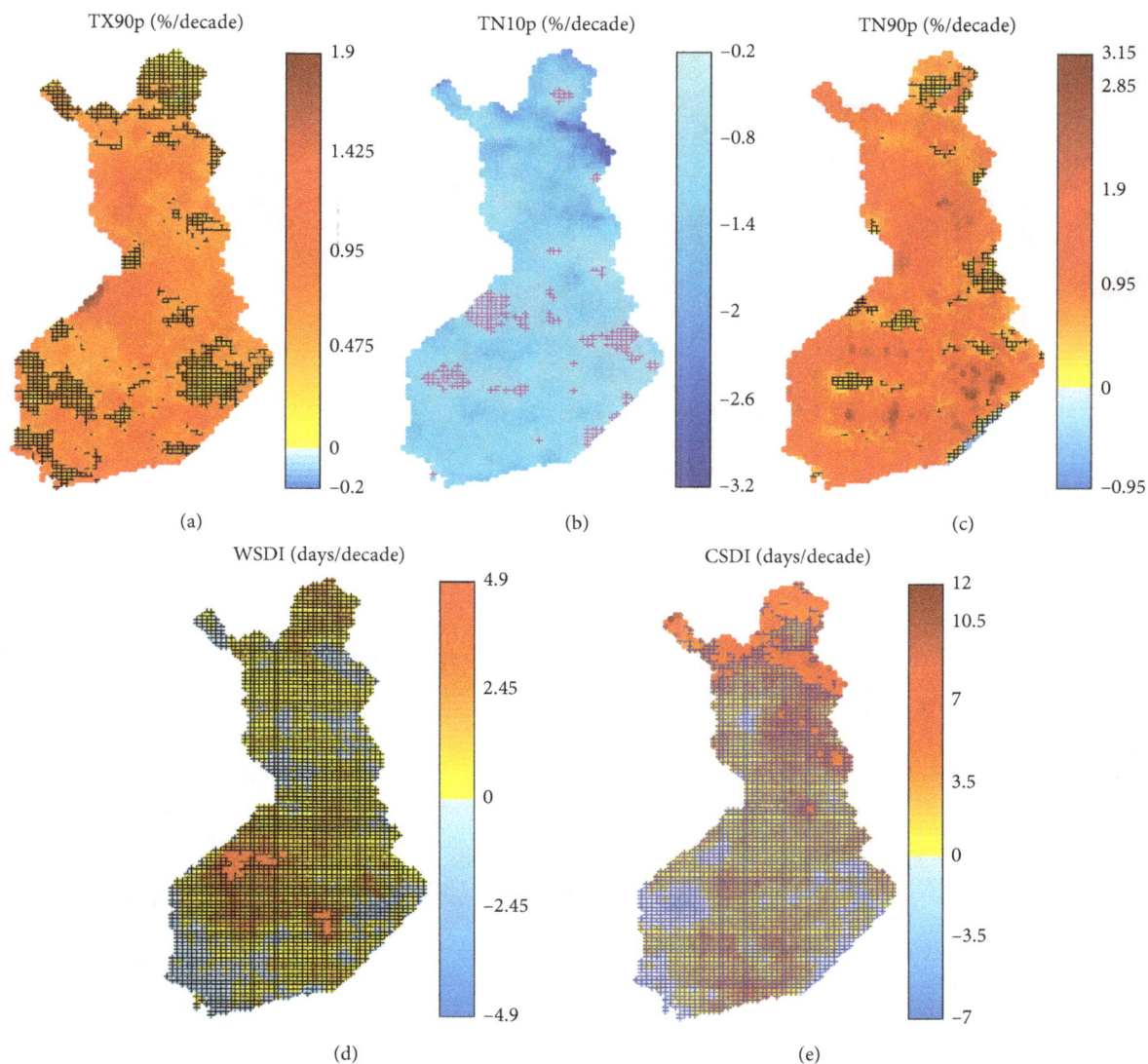

FIGURE 8: Spatial distribution maps of trends (/decade) in (a) TX90p, (b) TN10p, (c) TN90p, (d) WSDI, and (e) CSDI in Finland during 1961–2011. The stippling indicates areas where the trends are statistically insignificant ($p > 0.05$).

Strengthening of westerly circulation, such increases in AO, and NAO indices are generally responsible for recent winter temperature warming and consequently increases in the cold temperature extremes over the northern Europe.

This study found that EA/WR pattern negatively affected variations in DTR largely over the southern, eastern, and western Finland. Irannezhad et al. [25] concluded that the EA/WR pattern significantly influenced temperature in Finland during all seasons except summer. This NHTP (EA/WR) describes the meridional circulation for Finland that generally declines with the strengthening of westerly airflow. The positive EA/WR value is accompanied by the anomalous northerly and northwesterly airflow, while its negative value corresponds to the anomalous southerly and southeasterly circulation. Hence, the positive EA/WR phase associated with cold temperature in large parts of western Russia, northeast Africa, and the Arctic area, while warm temperature in east Asia (e.g., [38, 42]). Besides, observed

significant increasing trend in EA/WR pattern during recent decade [60] can play a critical role in the climate variability over northern Europe throughout a year, which has received little attention in research communities.

WP pattern was the most influential NHTP for ETR variability mostly in the northern, southern, and eastern Finland. For this NHTP (WP), the positive phase results in warmer than normal climate at midlatitudes of the western North Pacific during summer and winter but colder climate in the east of Siberia in all seasons [51]. Irannezhad et al. [25] reported that the positive WP is usually associated with anomalous northerly airflow over Finland, causing negative temperature anomalies in the country during spring. However, the colder climate in the east of Siberia during the negative phase of the WP pattern seems to play a role in spring temperature variability in Finland, but the mechanisms through which this is identified remain to be determined.

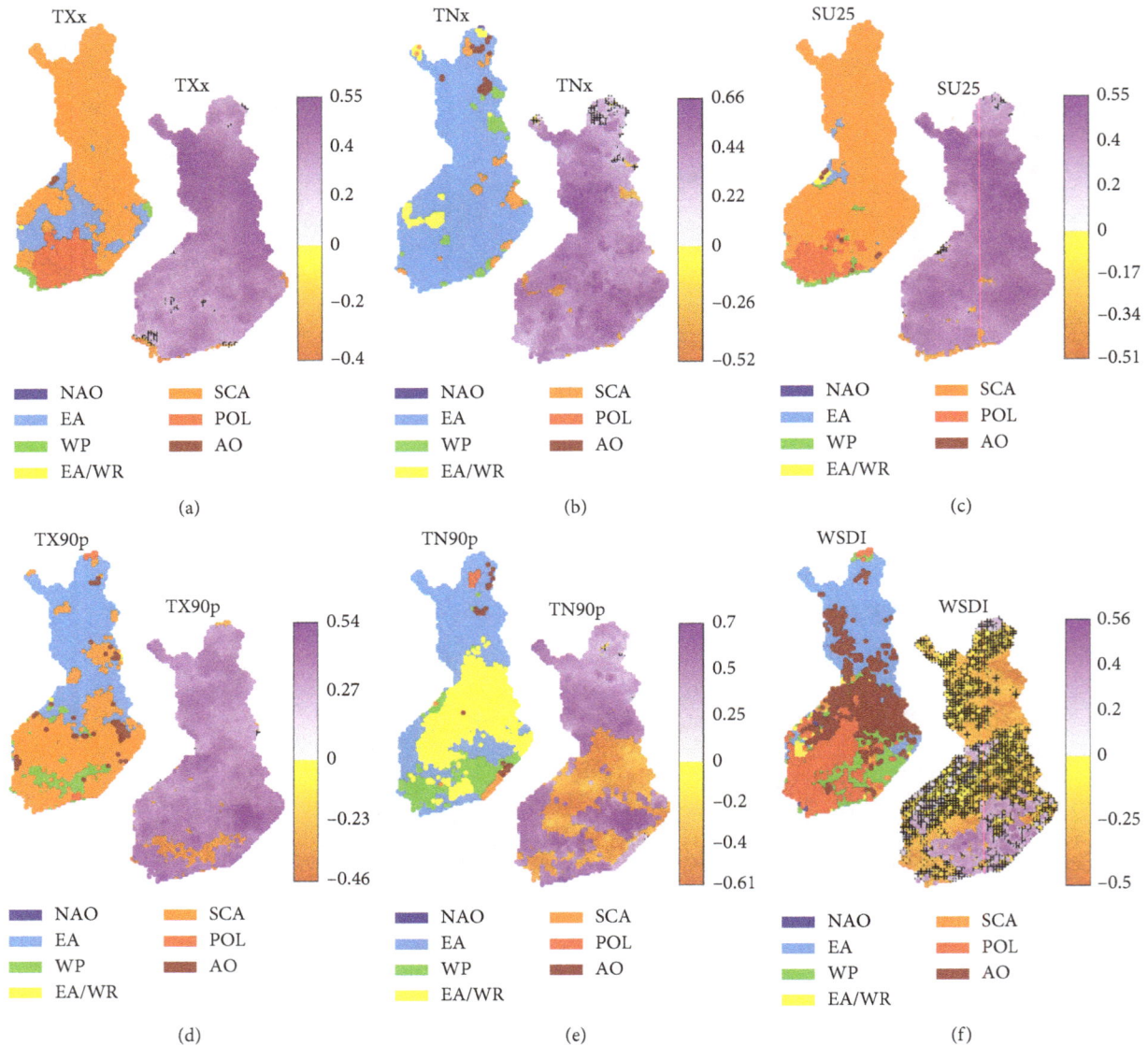

FIGURE 9: Spatial distribution maps of most influential NHTP (left) and its corresponding Spearman rank correlation (right) for warm extreme temperature indices including (a) TXx, (b) TNx, (c) SU25, (d) TX90p, (e) TN90p, and (f) WSDI in Finland during 1961–2011. The stippling indicates areas where the correlations are statistically insignificant ($p > 0.05$).

5. Conclusions

The present study evaluated the impacts of seven well-known large-scale atmospheric circulation patterns, often referred to as Northern Hemisphere teleconnection patterns (NHTPs), on temperature statistics in Finland with a focus on extreme events in recent decades. Using daily gridded maximum and minimum temperature time series across Finland for the period 1961–2011, variations and trends in the fifteen temperature indices recommended by the ETCCDI were evaluated. The connections of these indices to the seven NHTPs were also identified. Major conclusions are drawn as follows:

(1) Significant increases in extreme temperatures were mainly revealed by warming trends in the TNx, TNn,

and TXn indices. Different rates of the increasing trends in warm and cold extreme temperatures principally caused both diurnal (DTR) and extreme (ETR) temperature ranges to decline over the southwest and north of Finland. In general, all extreme temperature warming trends were associated with significant increases in the mean temperature over the country. However, warming trends in extreme temperatures were largely attributed to increases in cold temperature extremes rather than in warm temperature extremes.

(2) Significant increasing trends were determined in the numbers of warm days and nights, but the number of cold days and nights showed significant decreasing trends. Both the numbers of frost and icing days have

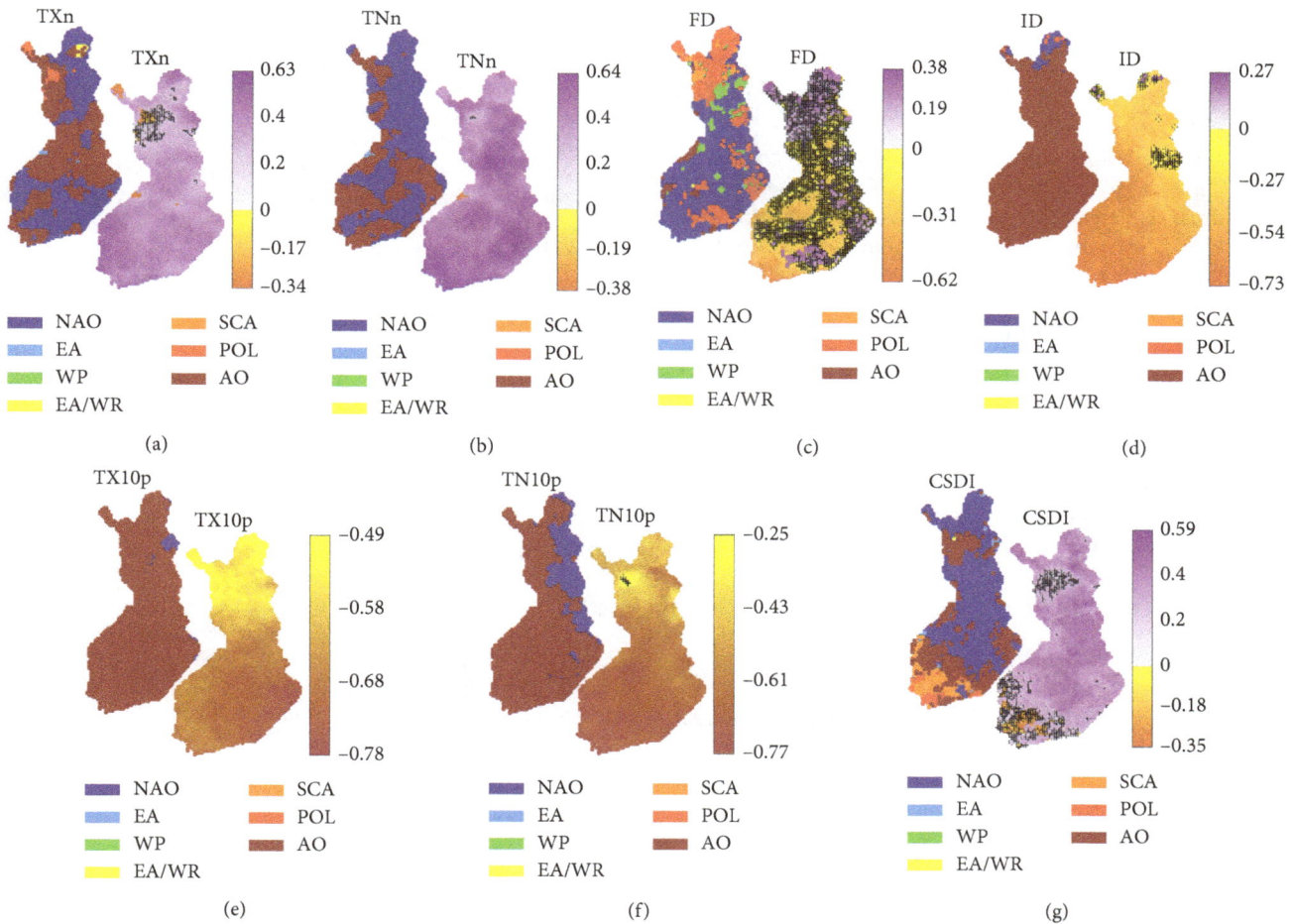

Figure 10: Spatial distribution maps of most influential NHTP (left) and its corresponding Spearman rank correlation (right) for cold temperature extreme indices including (a) TXn, (b) TNn, (c) FD, (d) ID, (e) TX10p, (f) TN10p, and (g) CSDI in Finland during 1961–2011. The stippling indicates areas where the correlations are statistically insignificant ($p > 0.05$).

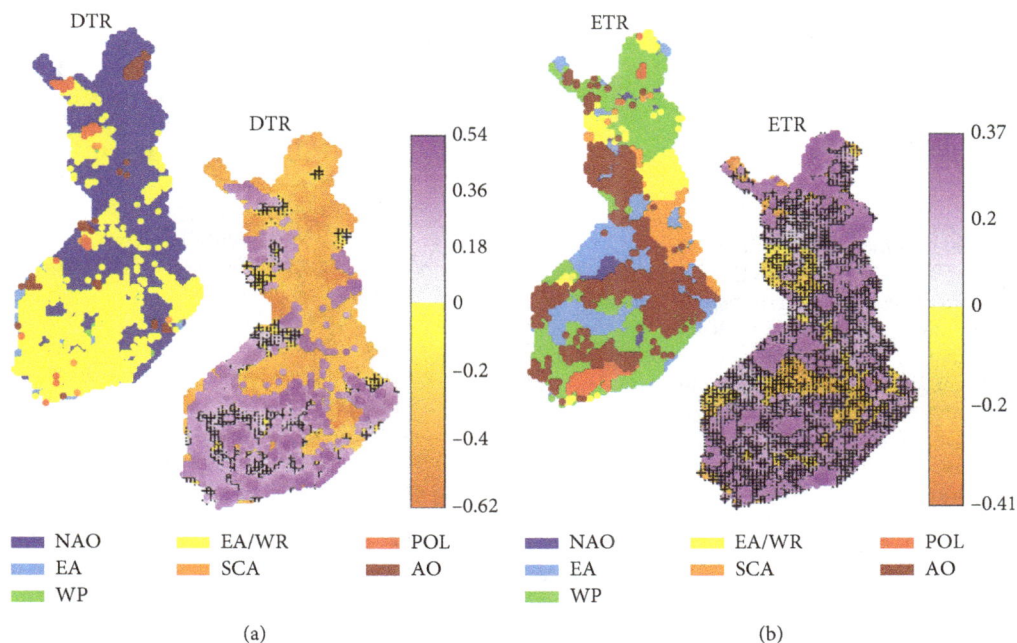

Figure 11: Spatial distribution maps of most influential NHTP (left) and its corresponding Spearman rank correlation (right) for (a) DTR and (b) ETR in Finland during 1961–2011. The stippling indicates areas where the correlations are statistically insignificant ($p > 0.05$).

also decreased. Hence, warm temperature extreme events have become more frequent in Finland in recent decades, while the opposite occurred for cold temperature extreme events.

(3) For the whole country, both warm and cold spells showed no clear changes. Similarly, spatial trend analysis determined no significant changes in warm spells over different parts of Finland. In contrast, lengthening in cold spells has been observed over the upper areas of the northern Finland, which indicates the nonlinearity of the climate dynamics and demonstrates the complexity of extreme climate events.

(4) Temperature extremes over Finland were influenced by a number of NHTPs such as the EA, SCA, AO, WP, and POL patterns. In general, warm temperature extremes were associated with the EA and SCA patterns, while cold temperature extremes with the AO and NAO patterns.

Conflicts of Interest

The authors declare that they have no conflicts of interest.

Acknowledgments

The research presented in this paper was financially supported by Maa-Ja Vesitekniikan Tuki R.Y. and the Finnish Cultural Foundation. The authors gratefully thank the CSC-IT Centre for Science Ltd. for providing gridded daily minimum and maximum temperature time series for Finland and the Climate Prediction Centre (CPC) at the National Oceanic and Atmospheric Administration (NOAA) of the USA for making the standardized monthly values of ACPs accessible online.

References

[1] A. Ahmadalipour, H. Moradkhani, and M. Svoboda, "Centennial drought outlook over the CONUS using NASA-NEX downscaled climate ensemble," *International Journal of Climatology*, vol. 35, no. 5, pp. 2477–2491, 2016.

[2] D. Chen, A. Walther, A. Moberg, P. D. Jones, J. Jacobeit, and D. Lister, *European Trend Atlas of Extreme Temperature and Precipitation*, Springer, Dordrecht, Netherlands, 2015.

[3] S. Christoph and J. Gerd, "Hot news from summer 2003," *Nature*, vol. 432, pp. 559-560, 2004.

[4] U. S. De, R. K. Dube, and G. S. Prakasa Rao, "Extreme weather events over India in the last 100 years," *Journal of Indian Geophysical Union*, vol. 9, no. 3, pp. 173–187, 2005.

[5] IPCC, *Managing the Risks of Extreme Events and Disasters to Advance Climate Change Adaptation: A Special Report of Working Groups I and II of the Intergovernmental Panel on Climate Change*, Cambridge University Press, Cambridge, UK, 2012.

[6] L. V. Alexander, X. Zhang, T. C. Peterson et al., "Global observed changes in daily climate extremes of temperature and precipitation," *Journal of Geophysical Research*, vol. 111, no. D5, article D05109, 2006.

[7] M. G. Donat, L. V. Alexander, H. Yang et al., "Updated analyses of temperature and precipitation extreme indices since the beginning of the twentieth century, the HadEx2 dataset," *Journal of Geophysical Research: Atmospheres*, vol. 118, no. 5, pp. 2098–2118, 2013.

[8] P. Frich, L. V. Alexander, P. Della-Marta et al., "Observed coherent changes in climatic extremes during the second half of the 20th Century," *Climate Research*, vol. 19, pp. 193–212, 2002.

[9] A. M. G. Klein Tank, T. C. Peterson, D. A. Quadir et al., "Changes in daily temperature and precipitation extremes in Central and South Asia," *Journal of Geophysical Research*, vol. 111, no. D16, article D16105, 2006.

[10] L. A. Vincent and E. Mekis, "Changes in daily and extreme temperature and precipitation indices for Canada over the twentieth century," *Atmosphere-Ocean*, vol. 44, no. 2, pp. 177–193, 2006.

[11] EEA, "Climate change, impacts and vulnerability in Europe 2012—an indicator based report," European Environment Agency Report No. 12/2012, 2012, ISSN 1725-9177.

[12] A. M. Araghi, M. Mousavi-Baygi, and J. Adamowski, "Detection of trends in days with extreme temperatures in Iran from 1961 to 2010," *Theoretical and Applied Climatology*, vol. 125, no. 1, pp. 213–225, 2016.

[13] M. I. P. de Lima, F. E. Santo, A. M. Ramos, and J. L. M. P. de Lima, "Recent changes in daily precipitation and surface air temperature extremes in mainland Portugal, in the period 1941–2007," *Atmospheric Research*, vol. 127, pp. 195–209, 2013.

[14] G. Fioravanti, E. Piervitali, and F. Desiato, "Recent changes of temperature extremes over Italy: an index-based analysis," *Atmospheric Research*, vol. 123, no. 3-4, pp. 473–486, 2016.

[15] A. Moberg, P. D. Jones, D. Lister et al., "Indices for daily temperature and precipitation extremes in Europe analyzed for the period 1901–2000," *Journal of Geophysical Research*, vol. 111, no. D22, article D22106, 2006.

[16] P. A. O'Gorman and T. Schneider, "The physical basis for increases in precipitation extremes in simulations of 21st-century climate change," *Proceedings of the National Academy of Sciences*, vol. 106, no. 35, pp. 14773–14777, 2009.

[17] IPCC, *Climate Change 2013: the Physical Science Basis: Working Group I Contribution to the Intergovernmental Panel on Climate Change Fifth Assessment Report (AR5)–Changes to the Underlying Scientific/Technical Assessment*, Cambridge University Press, Cambridge, UK, 2013.

[18] M. H. Glantz, R. W. Katz, and N. Nicholls, *Teleconnections Linking Worldwide Climate Anomalies: Scientific Basis and Societal Impact*, Cambridge University Press, New York, NY, USA, 2009.

[19] A. Hoy, M. Sepp, and J. Matschullat, "Large-scale atmospheric circulation forms and their impact on air temperature in Europe and northern Asia," *Theoretical and Applied Climatology*, vol. 113, no. 3-4, pp. 643–658, 2013.

[20] A. Omstedt, C. Pettersen, J. Rodhe, and P. Winsor, "Baltic Sea climate: 200 yr of data on air temperature, sea level variation, ice cover, and atmospheric circulation," *Climate Research*, vol. 25, pp. 205–216, 2004.

[21] R. M. Trigo, T. J. Osborn, and J. M. Corte-Real, "The North Atlantic Oscillation influence on Europe: climate impacts and

associated physical mechanisms," *Climate Research*, vol. 20, pp. 9–17, 2002.

[22] S. Filahi, M. Tanarhte, L. Mouhir, M. El Morhit, and Y. Tramblay, "Trends in indices of daily temperature and precipitations extremes in Morocco," *Theoretical and Applied Climatology*, vol. 124, no. 3-4, pp. 959–972, 2016.

[23] S. Malinovic-Milicevic, M. M. Radovanovic, G. Stanojevic, and B. Milpvanovic, "Recent changes in Serbian climate extreme indices from 1961–2011," *Theoretical and Applied Climatology*, vol. 124, no. 3-4, pp. 1089–1098, 2016.

[24] W. Sun, X. Mu, X. Song, D. Wu, A. Cheng, and B. Qiu, "Changes in extreme temperature and precipitation events in the Loess Plateau (China) during 1960-2013 under global warming," *Atmospheric Research*, vol. 168, pp. 33–48, 2016.

[25] M. Irannezhad, D. Chen, and B. Kløve, "Interannual variations and trends in surface air temperature in Finland in relation to atmospheric circulation patterns, 1961–2011," *International Journal of Climatology*, vol. 35, no. 10, pp. 3078–3092, 2015.

[26] M. Tikkanen, *The Physical Geography of Fennoscandia*, M. Seppälä, Ed., Oxford University Press, Oxford, UK, 2005.

[27] D. Chen and H. W. Chen, "Using the Köppen classification to quantify climate variation and change: an example for 1901–2010," *Environmental Development*, vol. 6, pp. 69–79, 2013.

[28] H. Henttonen, *Kriging in Interpolating July Mean Temperatures and Precipitation Sums. Reports from the Department of Statistics*, University of Jyvaskyla, Jyvaskyla, Finland, 1991.

[29] B. D. Ripley, *Spatial Statistic*, Wiley, New York, NY, USA, 1981.

[30] H. Tietäväinen, H. Tuomenvirta, and A. Venäläinen, "Annual and seasonal mean temperatures in Finland during the last 160 years based on gridded temperature data," *International Journal of Climatology*, vol. 30, no. 15, pp. 2247–2256, 2010.

[31] H. Tuomenvirta, "Homogeneity adjustments of temperature and precipitation series—Finnish and Nordic data," *International Journal of Climatology*, vol. 21, no. 4, pp. 495–506, 2001.

[32] H. Tuomenvirta, *Reliable Estimation of Climatic Variations in Finland*, Finnish Meteorological Institute Contributions, Helsinki, Finland, 2004.

[33] A. Venäläinen, H. Tuomenvirta, P. Pirinen, and A. Drebs, "A basic Finnish climate data set 1961-2000—description and illustrations," Reports No. 2005:5, Finnish Metrological Institute, Helsinki, Finland, 2005.

[34] J. Aalto, P. Pirinen, J. Heikkinen, and A. Venäläinen, "Spatial interpolation of monthly climate data for Finland: comparing the performance of kriging and generalized additive models," *Theoretical and Applied Climatology*, vol. 112, no. 1-2, pp. 99–111, 2013.

[35] M. Irannezhad and B. Kløve, "Do atmospheric teleconnection patterns explain variations and trends in thermal growing season parameters in Finland?," *International Journal of Climatology*, vol. 35, no. 15, pp. 4619–4630, 2015.

[36] Y. Gao, T. Markkanen, T. Thum et al., "Assessing various drought indicators in representing summer drought in boreal forests in Finland," *Hydrology and Earth System Sciences*, vol. 20, no. 1, pp. 175–191, 2016.

[37] X. Zhang, L. Alexander, G. C. Hergel et al., "Indices for monitoring changes in extremes based on daily temperature and precipitation data," *Wiley Interdisciplinary Reviews: Climate Change*, vol. 2, no. 6, pp. 851–870, 2011.

[38] A. G. Barnston and R. E. Livezey, "Classification, seasonality and persistence of low-frequency atmospheric circulation patterns," *Monthly Weather Review*, vol. 115, no. 6, pp. 1083–1126, 1987.

[39] M. Irannezhad, H. Marttila, and B. Kløve, "Long-term variations and trends in precipitation in Finland," *International Journal of Climatology*, vol. 34, no. 10, pp. 3139–3153, 2014.

[40] D. W. J. Thompson and J. M. Wallace, "The Arctic Oscillation signature in the wintertime geopotential height and temperature fields," *Geophysical Research Letters*, vol. 25, no. 9, pp. 1297–1300, 1998.

[41] CPC, http://www.cpc.ncep.noaa.gov/data/teledoc/telecontents.shtml, 2011.

[42] Y. K. Lim and H. D. Kim, "Impact of the dominant large-scale teleconnections on winter temperature variability over East Asia," *Journal of Geophysical Research: Atmospheres*, vol. 118, no. 14, pp. 7835–7848, 2013.

[43] C. Bueh and H. Nakamura, "Scandinavian pattern and its climatic impact," *Quarterly Journal of the Royal Meteorological Society*, vol. 133, no. 627, pp. 2117–2131, 2007.

[44] M. G. Kendall, *Rank Correlation Methods*, Griffin, London, UK, 1975.

[45] H. B. Mann, "Nonparametric tests against trend," *Econometrica*, vol. 13, no. 3, pp. 245–259, 1945.

[46] S. Yue, P. Pilon, R. Phinney, and G. Cavadias, "The influence of autocorrelation on the ability to detect trend in hydrological series," *Hydrological Processes*, vol. 16, no. 9, pp. 1807–1829, 2002.

[47] E. Park and Y. J. Lee, "Estimates of standard deviation of Spearman's rank correlation coefficients with dependent observations," *Communications in Statistics-Simulation and Computation*, vol. 30, no. 1, pp. 129–142, 2001.

[48] P. K. Sen, "Estimates of the regression coefficient based on Kendall's tau," *International Journal of Statistics & Economics*, vol. 63, no. 324, pp. 1379–1389, 1968.

[49] D. R. Helsel and R. M. Hirsch, *Statistical Methods in Water Resources*, Studies in Environmental Science, Amsterdam, Netherlands, 1992.

[50] S. Mikkonen, M. Laine, H. M. Mäkelä et al., "Trends in the average temperature in Finland, 1847–2013," *Stochastic Environmental Research and Risk Assessment*, vol. 29, no. 6, pp. 1521–1529, 2015.

[51] J. M. Wallace and D. S. Gutzler, "Teleconnections in the geopotential height field during the Northern Hemisphere winter," *Monthly Weather Review*, vol. 109, no. 4, pp. 784–812, 1981.

[52] F. Panagiotopoulos, M. Shahgedanova, and D. B. Stephenson, "A review of Northern Hemisphere winter-time teleconnection patterns," *Journal of Physics*, vol. 12, no. 10, pp. 27–47, 2002.

[53] J. W. Hurrell, "Decadal trends in the North Atlantic Oscillation: regional temperatures and precipitation," *Science*, vol. 269, no. 5224, pp. 676–679, 1995.

[54] A. K. Gormsen, A. Hense, T. B. Toldam-Andersen, and P. Braun, "Large-scale climate variability and its effects on mean temperature and flowering time of Prunus and Betula in Denmark," *Theoretical and Applied Climatology*, vol. 82, no. 1-2, pp. 41–50, 2005.

[55] J. Jaagus, "Climatic changes in Estonia during the second half of the 20th century in relationship with changes in large-scale atmospheric circulation," *Theoretical and Applied Climatology*, vol. 83, no. 1-4, pp. 77–88, 2006.

[56] M. C. Serreze, J. E. Walsh, F. S. Chapin et al., "Observational evidence of recent change in the northern high-latitude environment," *Climatic Change*, vol. 46, no. 1-2, pp. 159–207, 2000.

[57] A. Bukantis and G. Bartkeviciene, "Thermal effects of the North Atlantic Oscillation on the cold period of the year in Lithuania," *Climate Research*, vol. 28, pp. 221–228, 2005.

[58] S. Järvenoja, *Arctic Oscillation and its impact on Finland's climate: XXII Geophysical Days, May 19–20, Helsinki*, 2005.

[59] D. Wang, C. Wang, X. Yang, and J. Lu, "Winter Northern Hemisphere surface air temperature variability associated with the Arctic Oscillation and North Atlantic Oscillation," *Geophysical Research Letters*, vol. 32, article L16706, 2005.

[60] S. O. Krichak, P. Kishcha, and P. Alpert, "Decadal trends in of main Eurasian oscillations and the Mediterranean precipitation," *Theoretical and Applied Climatology*, vol. 72, no. 3-4, pp. 209–220, 2002.

Assessment of Climate Change Impacts on Extreme Precipitation Events: Applications of CMIP5 Climate Projections Statistically Downscaled over South Korea

Jang Hyun Sung,[1] Hyung-Il Eum,[2] Junehyeong Park ⓘ,[3] and Jaepil Cho[4]

[1]Ministry of Environment, Han River Flood Control Office, Seoul, Republic of Korea
[2]Alberta Environment and Parks, Calgary, Canada
[3]Civil, Construction, and Environmental Engineering, University of Alabama, Tuscaloosa, AL, USA
[4]Climate Services and Research Department, APEC Climate Center, Busan, Republic of Korea

Correspondence should be addressed to Junehyeong Park; sai0259@gmail.com

Academic Editor: Stefano Dietrich

Climate change may accelerate the water cycle at a global scale, resulting in more frequent extreme climate events. This study analyzed changes in extreme precipitation events employing climate projections statistically downscaled at a station-space scale in South Korea. Among the CMIP5 climate projections, based on spatial resolution, this study selected 26 climate projections that provide daily precipitation under the representative concentration pathway (RCP) 4.5. The results show that a 20-year return period of precipitation event during a reference period (1980~2005) corresponds to a 16.6 yr for 2011 to 2040, 14.1 yr for 2041 to 2070, and 12.8 yr for 2071 to 2100, indicating more frequent extreme maximum daily precipitation may occur in the future. In addition, we found that the probability density functions of the future periods are located out of the 10% confidence interval of the PDF for the reference period. The result indicates that the design standard under the reference climate is not managed to cope with climate change, and accordingly the revision of the design standard is required to improve sustainability in infrastructures.

1. Introduction

Changes in the water cycle caused by climate change lead to temporal and spatial alteration in hydrological and ecological systems. Although climate models provide essential information to assess climate change impacts at a global scale, direct applications of climate projections have inherent problems due to a coarse resolution (~100 km) which induce a difficulty to capture climatic characteristics at regional or local scales. Therefore, an application of downscaling technique is a prerequisite to accurately complete climate change studies at a local scale. Since there is, in addition, considerable uncertainty in climate projections, caused by different dynamic systems, grid size, and parameterization of physics processes, many studies have paid attention to quantify uncertainty of climate change scenarios in climate change impact assessment [1–4].

An ensemble approach has been applied to deal with the uncertainty in climate scenarios because a specific scenario cannot represent all possible future climate conditions [5, 6]. However, it is still questionable in the climate change impact assessment which scenarios need to be included to capture future climate variability. Most studies have selected appropriate scenarios based on the performance in reproducing historical climate. However, it has the limitation that performance during a historical period cannot guarantee consistent performance during a future period [7]. The IPCC (Intergovernmental Panel of Climate Change) report suggested the use of as many climate scenarios as possible in climate change assessment [8]. In other words, employing multiple scenarios in climate change impact assessment may take into account the uncertainty. Therefore, the use of multimodel ensemble (MME) has been increasing to capture possible climate changes projected by multiple models [9–11].

Previous studies have assessed the changes in extreme precipitation and suggested that South Korea is expected to become more vulnerable to flood hazards due to an increase in the probability of severe extreme events in the future [12, 13]. Sung et al. [13] found overall increase in frequency of extreme precipitation over South Korea in association with climate change. According to [13]; particularly, daily extreme precipitation with 20-year return period during the reference climate from 1980 to 2005 is likely to happen about every 4.3 and 3.4 yr by the end of 21st century (2070~2099) employing HadGEM3-RA based on the RCP 4.5 and 8.5. Ahn et al. [14] suggested that multi-RCMs can be used to reduce uncertainties and assess the future change of extreme precipitation more reliably. According to [14], 50-year return value will change from −32.69% to 72.7% and from −31.6% to 96.32% in the mid-21st century and from −31.97% to 86.25% and from −19.45% to 134.88% in the late-21st century under RCP 4.5 and 8.5 scenarios. Im et al. [15] suggested that changes in return levels of annual maximum precipitation in a regional climate model indicate an increased frequency of present day in 20- and 50-year extreme precipitation events. Previous studies have projected future extreme precipitations with one or several models produced by dynamic downscaling.

In general, evaluating changes in extreme events in the future requires high-resolution climate change scenarios which are produced by dynamic or statistical downscaling methods. Especially in South Korea, spatial downscaling should be implemented because the climate of South Korea is highly dependent on topography, due to the large portion of mountainous area. The dynamic downscaling takes advantage of considering interaction between climatic systems, the nonstationarity of climate change, and the temporal and spatial correlation between variables which can be interpreted as a physics process of the Earth system [13, 16–19]. However, the dynamic downscaling techniques require a huge computing facility and highly skilled experts. On the other hand, statistical downscaling directly incorporates correlations between climate models' simulations and observational data into algorithms. Therefore, statistical downscaling techniques are inexpensive to apply to convert low-resolution into high resolution. Although statistical downscaling has the limitation that it assumes the stationarity of the climate processes over time, statistical downscaling methods have been actively applied to produce a high-resolution regional climate projections [20–25]. The recent NEX-GDDP (NASA Earth Exchange Global Daily Downscaled Projections) and DCHP (Downscaled CMIP3 and CMIP5 Climate and Hydrology Projections) applied bias-correction/spatial disaggregation (BCSD) [25] for climate change impact assessment at local scales. Abatzoglou and Brown [20] suggested multivariate adaptive constructed analogs (MACA) to improve coincidence of climate events. Burger et al. [21] proposed detrended quantile mapping (DQM) and [24] applied spatial disaggregation/quantile delta mapping (SD-QDM) to preserve long-term trends driven by climate models.

There are limitations in establishing adaption and response to climate change due to different information produced by each climate change model, physical processes, and resolutions. Furthermore, extreme values are associated with large uncertainty so that it is unwise to use the result of statistical frequency analysis using climate change scenarios naively. Therefore, in this study, we applied the downscaling preserving the long-term trend of the climate model to 26 RCP 4.5 scenarios and projected the change in extreme precipitation—the 20-year return value of annual maximum daily precipitation—over South Korea and estimated the uncertainty with the confidence intervals.

2. Data and Method

2.1. Procedure. In this study, we used MME, which combines multiple model results, to project changes in extreme precipitation in South Korea (Figure 1). Employing the daily precipitation of 26 CMIP5 climate projections downscaled by three statistical downscaling methods, we collected the annual maximum daily precipitation. Then, we estimated the frequency and magnitude of extreme precipitation using the generalized extreme value (GEV) distribution for the reference period (1980~2005) and the three 30-year future periods (Future1: 2011~2040, Future2: 2041~2070, and Future3: 2071~2100).

2.2. Climate Change Scenarios. The uncertainties among scenarios should be taken into consideration, since adaptation strategies were different depending on whether a specific scenario was selected [26]. Climate change scenarios have different simulations due to GCMs with different dynamics and grid sizes and parameterization processes. Therefore, the use of a single model is likely to lead to a bias, so it is necessary to consider the uncertainties using GCM results. As an alternative to this, the use of MME has been increasing. The Coupled Model Intercomparison Project (CMIP) started in 1995 to compare various climate models. In CMIP Phase 5 (CMIP5), 4 representative concentration pathways (RCP) were proposed considering economic growth rate, industrialization and restoration technology. In this study, we used the RCP 4.5 scenario, in which the greenhouse gas mitigation policy is quite substantial. RCP 4.5 is a scenario that stabilizes a radiative forcing of $4.5 \, W \cdot m^{-2}$ in 2100 without exceeding this value. Table 1 shows the 26 GCMs selected in the previous climate change studies [24, 27] in descending order of spatial resolution, where grid points are collected within E119°–135° and N29°–43°. Using climate projections at the grid points of each GCM, we applied three statistical downscaling methods to downscale to the automated synoptic observing system (ASOS) stations in Figure 2.

2.3. Statistical Downscaling. In this research, three statistical downscaling methods are used. The first method used in this research is spatial disaggregation/quantile mapping (SDQM), which is equal to the daily BCSD [20]. This method is suggested to overcome the shortcoming of existing typical method, BCSD, that it may lose the climate characteristics in climate model driven daily sequencing. Moreover, spatial disaggregation/detrended quantile mapping (SD-DQM) and spatial disaggregation/quantile delta mapping (SD-QDM) are also applied to preserve GCM-driven long-term trends. More specific explanations and citations are shown below.

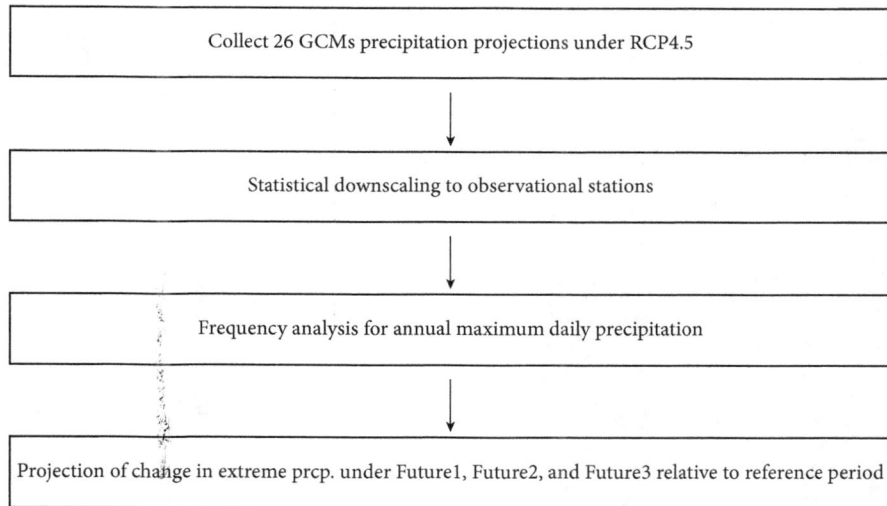

FIGURE 1: Procedure of this study.

TABLE 1: 26 GCMs of CMIP5 for this study.

No.	GCMs	Resolution (degree)	Grid points	Institution
1	CMCC-CM	0.750×0.748	22×18	Centro Euro-Mediterraneo per I Cambiamenti Climatici
2	CCSM4	1.250×0.942	13×15	
3	CESM1-BGC	1.250×0.942	13×15	National Center for Atmospheric Research
4	CESM1-CAM5	1.250×0.942	13×15	
5	BCC-CSM1-1-M	1.125×1.122	15×12	Beijing Climate Center, China Meteorological Administration
6	MRI-CGCM3	1.125×1.122	15×12	Meteorological Research Institute
7	CNRM-CM5	1.406×1.401	12×12	Centre National de Recherches Meteorologiques
8	MIROC5	1.406×1.401	12×10	Atmosphere and Ocean Research Institute (The University of Tokyo)
9	HadGEM2-AO	1.875×1.250	9×11	
10	HadGEM2-CC	1.875×1.250	9×11	Met Office Hadley Centre
11	HadGEM2-ES	1.875×1.250	9×11	
12	INM-CM4	2.000×1.500	8×10	Institute for Numerical Mathematics
13	IPSL-CM5A-MR	1.875×1.865	7×11	Institut Pierre-Simon Laplace
14	CMCC-CMS	1.875×1.865	9×7	Centro Euro-Mediterraneo per I Cambiamenti Climatici
15	MPI-ESM-LR	1.875×1.865	9×7	Max Planck Institute for Meteorology (MPI-M)
16	MPI-ESM-MR	1.875×1.865	9×7	
17	FGOALS-s2	2.813×1.659	6×9	LASG, Institute of Atmospheric Physics, Chinese Academy of Sciences
18	NorESM1-M	2.500×1.895	7×8	Norwegian Climate Centre
19	GFDL-ESM2G	2.500×2.023	6×7	Geophysical Fluid Dynamics Laboratory
20	GFDL-ESM2M	2.500×2.023	6×7	
21	IPSL-CM5A-LR	3.750×1.895	5×8	Institute Pierre-Simon Laplace
22	IPSL-CM5B-LR	3.750×1.895	5×8	
23	BCC-CSM1-1	2.813×2.791	6×5	Beijing Climate Center, China Meteorological Administration
24	CanESM2	2.813×2.791	6×5	Canadian Centre for Climate Modelling and Analysis
25	MIROC-ESM-CHEM	2.813×2.791	6×5	Japan Agency for Marine-Earth Science and Technology
26	MIROC-ESM	2.813×2.791	6×5	

2.3.1. Spatial Disaggregation/Quantile Mapping (SDQM). BCSD has been originally developed to downscale GCM's information to regional climate data at the monthly scale, and a temporal disaggregation technique is applied to generate bias-corrected daily climate data [25]. Therefore, it may lose the climate characteristics in climate model driven daily sequencing. Therefore, a daily BCSD has been applied not only to avoid the temporal disaggregation process but also to preserve climate model driven daily sequencing by incorporating cumulative density function (CDFs) of daily climate data [20, 28]. In the daily BCSD, the spatial disaggregation of the climate information is first performed by the inverse distance weighted interpolation [29] in which the squared distance between the GCM grid and the observation

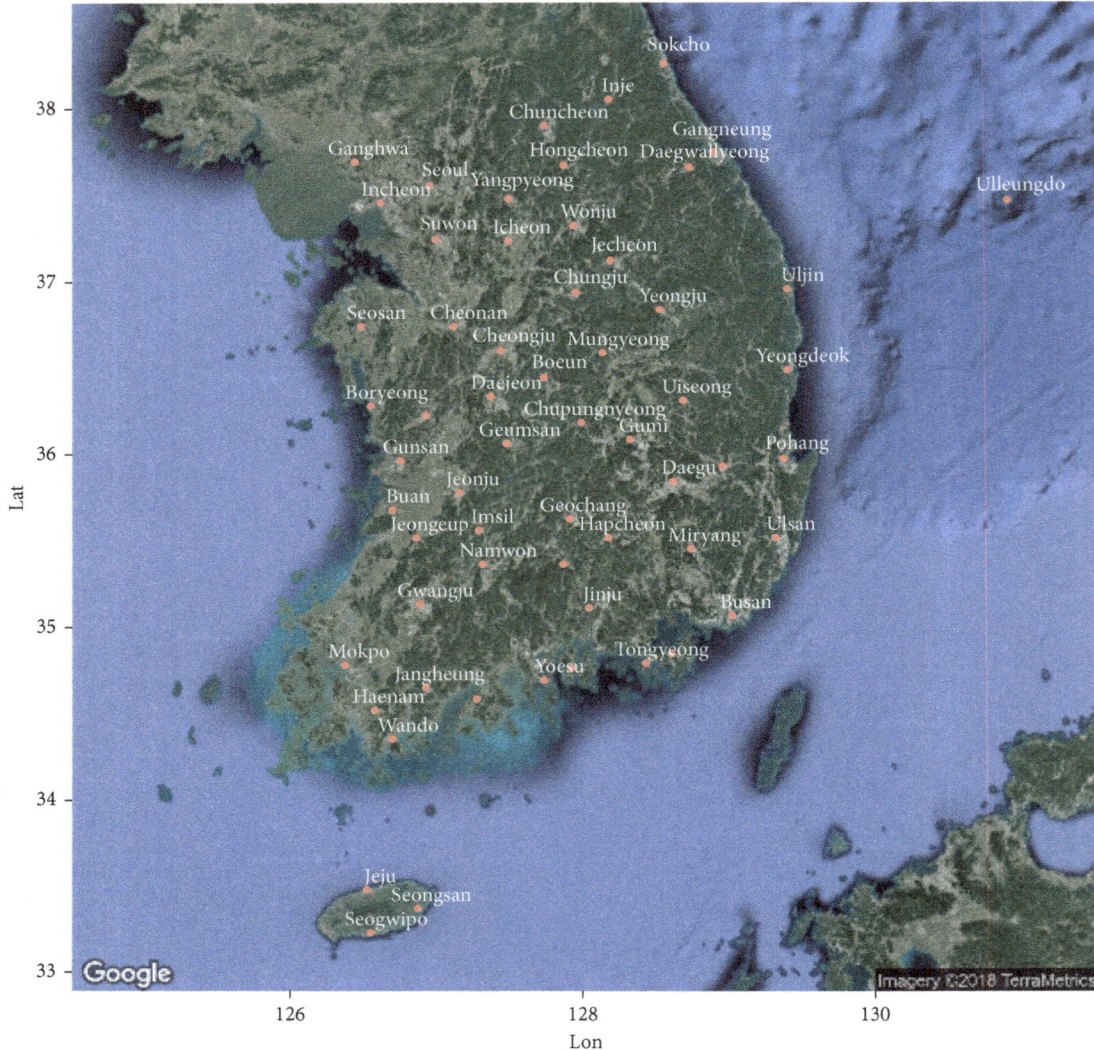

FIGURE 2: ASOS 60 weather stations in South Korea.

point is inversely weighted. Then, the bias correction is performed using the quantile delta algorithm shown in Equation (1) using empirical CDFs formulated with daily climate data within a moving window to reflect the seasonality of the area. Eum and Cannon [24] tested different half-widths of moving window, for example, 15, 30, 90, and 180 days, to investigate the impacts of moving window selection on extreme climate indices. The study showed that a 15-day moving window reflected better seasonality on the climate index for South Korea. Therefore, this study also employed the same moving window (15 d) for the quantile mapping method. For example, when downscaling for January 1, the density function is formed by observational and GCM data from December 15 to January 16, and this is applied to Equation (1) to bias-correct. In this study, daily BCSD was applied and will be referred to as SDQM.

$$\widehat{x}_{m,p}(t) = F_{o,h}^{-1}\big[F_{m,h}\{x_{m,p}(t)\}\big], \qquad (1)$$

where $x_{m,p}(t)$ and $\widehat{x}_{m,p}(t)$ represent the values before and after the bias correction at time t, respectively, $F_{m,h}(t)$

represents the cumulative density function of the past period generated by the global model, and $F_{o,h}^{-1}$ represents the inverse function of the cumulative density function of the observed data. For Equation (1), p is the projection, h is the past period, m is the global model, and o is the observation.

2.3.2. Long-Term Trend Preserving Downscaling Method. Extreme values outside of a range of historical data need to be extrapolated in SDQM. Eum and Cannon [24] estimated bias-corrected values using the Gumbel distribution which has been used to evaluate extreme flood events for South Korea due to simplicity in estimating parameters. This study also used the same method to extrapolate values outside of the historical range. For climate data with GCM-driven long-term (increasing or decreasing) trends, such as temperature, more extreme events may occur, and accordingly more frequent extrapolation has to be implemented, which may induce a substantial distortion of GCM-driven climate signals by the variance inflation [24, 30, 31]. Burger et al. [21] proposed the detrended quantile mapping (DQM) method that removes a long-term trend in climate projections to minimize the frequency of extrapolation. While DQM can directly consider the long-term trend of the monthly average, it cannot consider the long-term trend of the extreme values. Cannon et al. [22] proposed quantile delta mapping (QDM) designed to preserve absolute or relative changes in all of quantiles. As in [24], this study applied SD-QDM that combined the daily BCSD and QDM to produce downscaled climate projections over South Korea. More details on SD-QDM can be found in [24]. The CMIP5 historical run was simulated until 2005, and future climate scenarios were simulated from 2006 to 2100 forced by RCPs. Therefore, this study set the reference period from 1976 to 2005. For the future period, SD-QDM was applied for a total of 30 yr from 15 yr before to 14 yr after a certain year to consider gradual changes in quantiles between the reference and future periods. For example, when applying QDM for 2006, the reference period was from 1976 to 2005, and the future period was from 1991 to 2020.

2.4. Frequency Analysis.

Generally, if a certain precipitation is equaled or exceeded once during an average T year, it is said to have a return period T. The inverse of the return period T is an exceedance probability (P) of the event occurring in a certain year:

$$T = \frac{1}{P}. \quad (2)$$

The GEV distribution function is generally used to estimate the nonexceedance probability of extreme events because the upper tail of the GEV distribution is suitable to represent the extreme events. The GEV distribution has been used to describe the extreme probability in observed or GCM simulated hydrometeorological variables [32–34], because there is much evidence that the distributions of hydrologic variables have heavy tails. Although it can be difficult to determine from only a single site unless the record is relatively long, the distribution of annual maximum precipitation amount appears consistently to have a heavy tail [35–37]. Because GEV has shape parameters and is very useful for expressing heavy tail, we used the GEV distribution in this study. The cumulative distribution function, which estimates the nonexceedance probability of the GEV distribution, is given by Equation (3) and its solution is estimated using (4) [38]:

$$F(x) = \begin{cases} \exp\left[-\left(1 - \kappa \frac{x-\xi}{\alpha}\right)^{1/\kappa}\right], & \kappa \neq 0, \\ \exp\left[-\exp\left(-\frac{x-\xi}{\alpha}\right)\right], & \kappa = 0, \end{cases} \quad (3)$$

$$x = \begin{cases} \xi + \frac{\alpha}{\kappa}\{1 - [-\log(P)]^{\kappa}\}, & \kappa \neq 0, \\ \xi - \alpha \log[-\log(P)], & \kappa = 0, \end{cases} \quad (4)$$

where ξ, α, and κ are parameters related to location, scale, and shape, respectively. The GEV distribution can be divided into type I, type II, and type III distributions according to the sign of κ. Among them, type I is called Gumbel and type III is called Weibull distribution. In Equations (3) and (4) the range of the variable x depends on κ, which is $\xi + \alpha/\kappa \leq x < \infty$ for $\kappa < 0$, $-\infty < x < \infty$ for $\kappa = 0$, and $-\infty < x \leq \xi + \alpha/\kappa$ for $\kappa > 0$ [38]. The GEV distribution varies in the thickness of the upper tail depending on the κ. In other words, κ is getting smaller when the tail area of distribution function enlarges, which means that the occurrence probability of the extreme value increases. Also, most of annual maximum 24-hour precipitations are located in the interval where the κ are negative [39]. For $\kappa < 0$, the distribution has a thicker right-hand tail. We used the GEV type II distribution.

The estimation methods of the parameters of the GEV distribution include the method of moments, the method of maximum likelihood estimates, the method of probability weighted moments, and the method of L-moments. The method of moments is simple to calculate, but the higher the moments, the more inaccurate estimates are obtained. The maximum likelihood method is effective when the number of sample data is large enough, but the solution process is complicated and sometimes the solution cannot be obtained because it does not converge. Since the probability weighted moment method and the L-moment method are based on the same theory, the same result is obtained. They are not sensitive to the number of sample data because they use the order statistics of the observation data. In addition, even with distorted data, relatively stable results can be obtained [39]. In this study, the method of L-moment was applied to estimate the parameters of the GEV distribution. 20-year return value of annual maximum daily precipitation, which is referred to as 20-year precipitation in this study, is selected for target of this study because length of each projection period is 30 years so that we need extrapolation for events with return periods exceeding 30 years and Klein Tank et al. [40] suggested 20-year event for evaluating the intensity and frequency of rare events that lie far in the tails of the probability distribution of weather variables.

3. Results

3.1. Future Projection.

The spatial distributions of the 20-year precipitation averaged over 26 climate projections are shown in Figure 3 to identify regional variation of extreme precipitation during the reference and future periods. Spatial downscaling methods applied in Figure 3 were SDQM, SD-DQM, and SD-QDM. The average of 20-year precipitation for the 26 CMIP5 GCMs changed gradually from the reference period to the Future1, Future2, and Future3 periods. Therefore, Figure 3 only includes the reference period as a start point, the Future3 period as an end point, and the difference between these two periods. At the northeastern mountainous region and southern coastal regions, large extreme precipitation was projected due to their regional characteristics. These regions receive more precipitation by the effects of mountains, urbanization, and ocean climates [41].

In order to quantitatively examine the difference in results by the method of downscaling shown in Figure 3, the spatially averaged precipitation of South Korea was calculated for each downscaling method (Table 2). The table

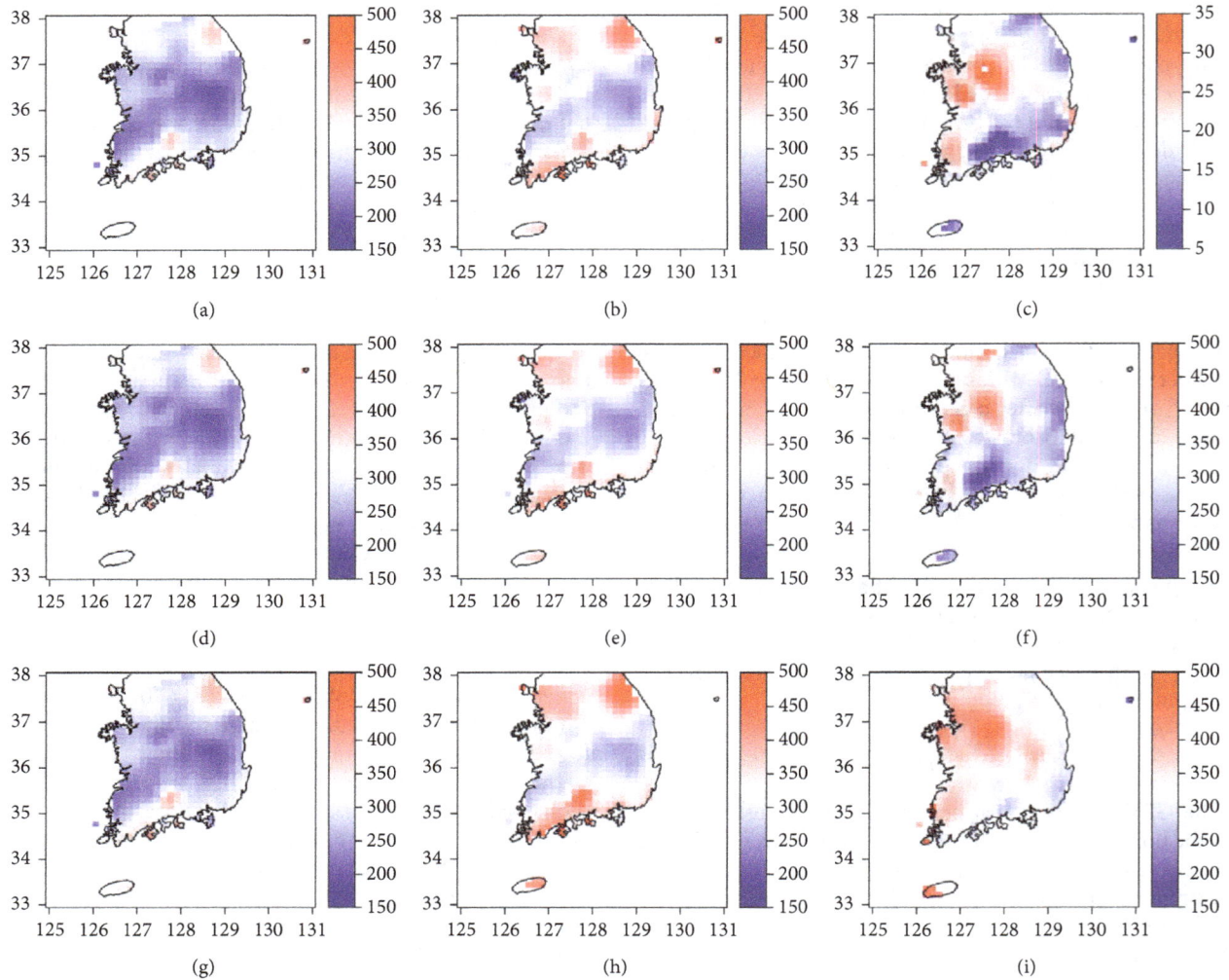

FIGURE 3: Average of 20-year return value of annual maximum daily precipitation for CMIP5 26 GCMs by SDQM ((a), (b), and (c)), SD-DQM ((d), (e), and (f)) and SD-QDM ((g), (h), and (i)) for the reference ((a), (d), and (g)), Future3 period ((b), (e), and (h)), and the difference between reference and Future3 period in percentage ((c), (f), and (i)).

TABLE 2: Average of 20-year precipitation under reference, Future1, Future2, and Future3 period for South Korea.

	20-year return value of annual maximum daily precipitation (mm·day^{-1})											
	Reference			Future1			Future2			Future3		
	SDQM	SD-DQM	SD-QDM	SDQM	SD-DQM	SD-QDM	SDQM	SD-DQM	SD-QDM	SDQM	SD-DQM	SD-QDM
Avg.	280.4	280.4	288.0	307.2	300.7	316.2	331.0	328.2	344.2	330.7	330.0	350.0
Med.	275.2	275.2	280.5	303.7	296.3	303.8	326.5	327.0	327.6	327.2	321.7	335.2

shows that the amount of precipitation downscaled with SD-QDM in the reference and each future period is generally larger than for the other methods. Comparing the mean and median of all the downscaling methods, including SD-QDM, the mean was greater than the median. Therefore, the right tail of the probability density function was thick. Overall, the results in Table 2 mean that 20-year precipitation in the reference period can occur more frequently in the future. Regarding the difference between the statistical methods, the amount of precipitation downscaled with SD-QDM was larger than other methods by better preserving GCM-driven relative change as shown in Figure 3(i).

The coefficient of variation (CV) was calculated to compare the distribution of 20-year precipitation derived from 26 GCMs (Figure 4). The variance of GCMs in the reference climate was small and differences among sites were relatively small (see Figure 4(a), 4(d), and 4(g)). There was a slight nonstationarity in the spatial distribution of the CV. The area of large CV was broader in the Future2 and the Future3 than Future1, indicating that the uncertainty in climate projections may considerably contribute to the distribution of 20-year precipitation in the future. The CV of southeastern (Future1), northeastern (Future2), and southeastern (Future3) region was large. Because the coefficients of variation in the reference

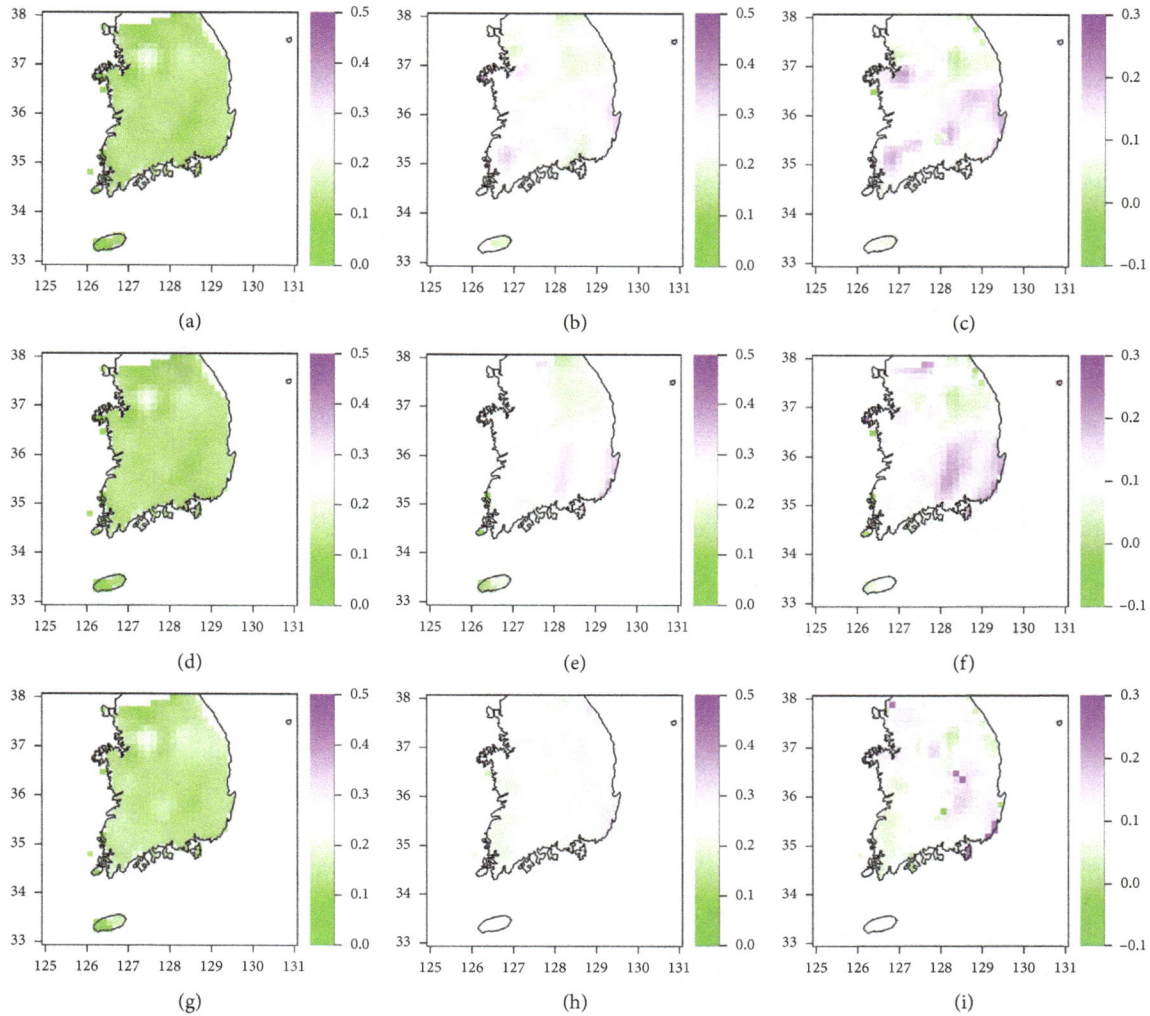

FIGURE 4: Coefficient of variation of 20-year return value of annual maximum daily precipitation for CMIP5 26 GCMs corrected by SDQM ((a), (b), and (c)), SD-DQM ((d), (e), and (f)), and SD-QDM ((g), (h), and (i)) for the reference ((a), (d), and (g)), Future3 period ((b), (e), and (h)), and the difference between reference and Future3 period in percentage ((c), (f), and (i)).

TABLE 3: Coefficient of variation of 20-year precipitation under reference, Future1, Future2, and Future3 period for South Korea.

	Reference			Future1			Future2			Future3		
	SDQM	SD-DQM	SD-QDM	SDQM	SD-DQM	SD-QDM	SDQM	SD-DQM	SD-QDM	SDQM	SD-DQM	SD-QDM
Avg.	0.132	0.132	0.148	0.235	0.228	0.216	0.235	0.249	0.237	0.234	0.240	0.243
Med.	0.128	0.128	0.141	0.226	0.222	0.204	0.234	0.240	0.232	0.235	0.236	0.238

period were all small and similar spatially, so the difference map (Figure 4(c), 4(f), and 4(i)) followed the results of Future3 period (Figure 4(b), 4(e), and 4(h)).

The uncertainty was investigated by calculating the CV of climate change scenarios according to the downscaling methods. Table 3 shows that the coefficient of variation in future climate is larger than in the reference climate which may be mainly due to internal variability. In the future climate, the variation according to downscaling methods was not clear. As shown in Figure 4, although the difference in the magnitude of the CV between the methods is not clear, large values of CV become widespread with time (Table 3).

3.2. Probability Density Function. The averages of 20-year precipitation increased over time, and the region where the CV was increasing in Figure 4 was larger during the further future. This trend was also confirmed by analyzing the changes in the probability distribution. Figure 5 shows the parameters and return periods of the GEV distribution for the extreme precipitation of the 26 climate models during the reference and future periods. Each subfigure in Figure 5 represents the parameter of GEV distribution of extreme precipitation downscaled with SDQM, SD-DQM and SD-QDM, respectively, and a change projection of 20-year precipitation in the reference period.

(a)

(b)

(c)

(d)

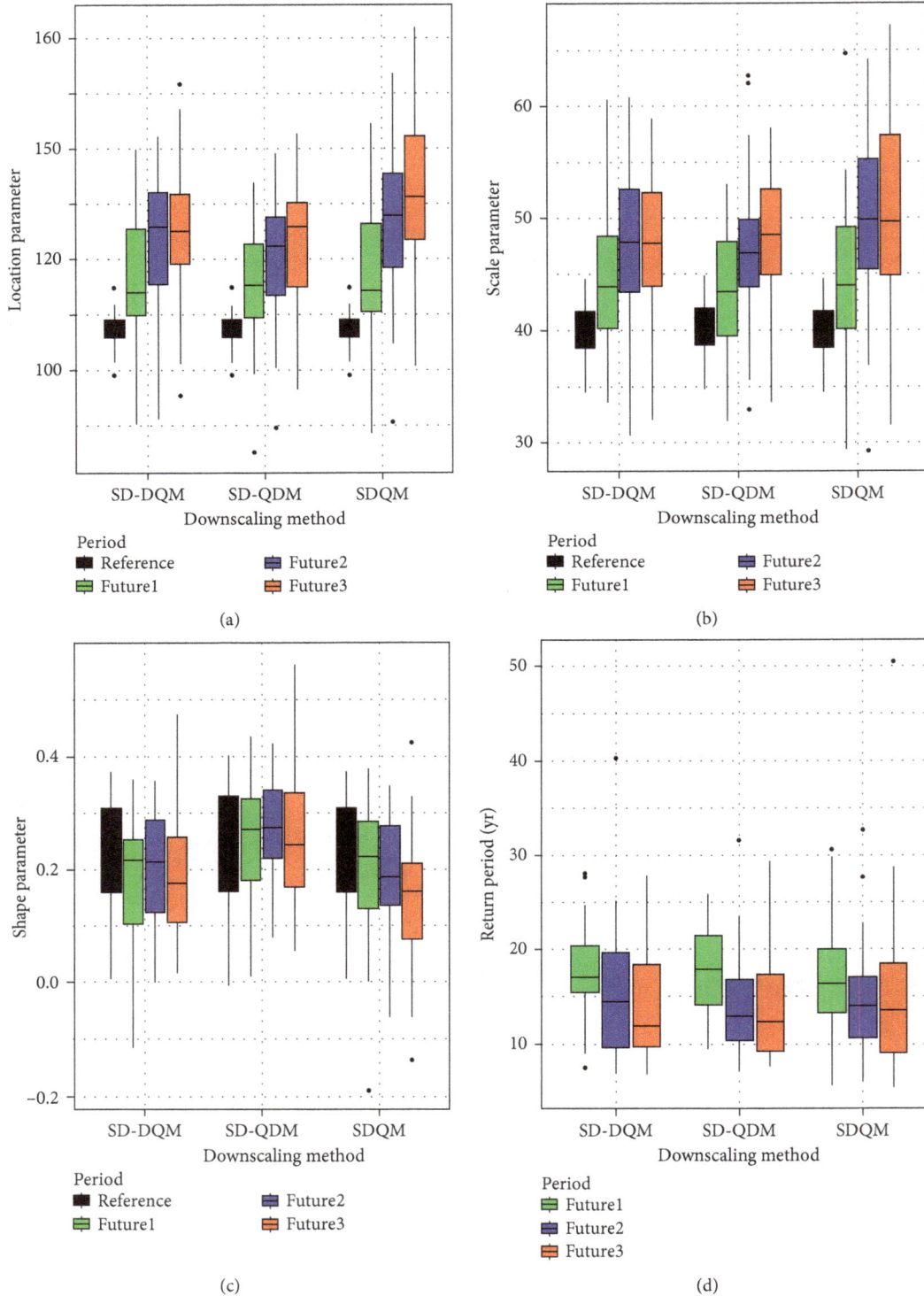

FIGURE 5: Box-plot of (a) location, (b) scale, (c) shape parameter of GEV distribution for CMIP5 precipitation, and (d) return period of future for 20-year precipitation of reference climate.

TABLE 4: Return periods of Future1, Future2, and Future3 using 20-year precipitation of reference climate as shown in Figure 5(d).

	Future1			Future2			Future3		
	QM	DQM	QDM	QM	DQM	QDM	QM	DQM	QDM
Avg.	16.93	17.39	17.47	14.18	14.85	14.80	15.04	14.71	14.58
Med.	16.30	16.63	16.76	13.90	14.21	14.16	13.49	12.68	12.28

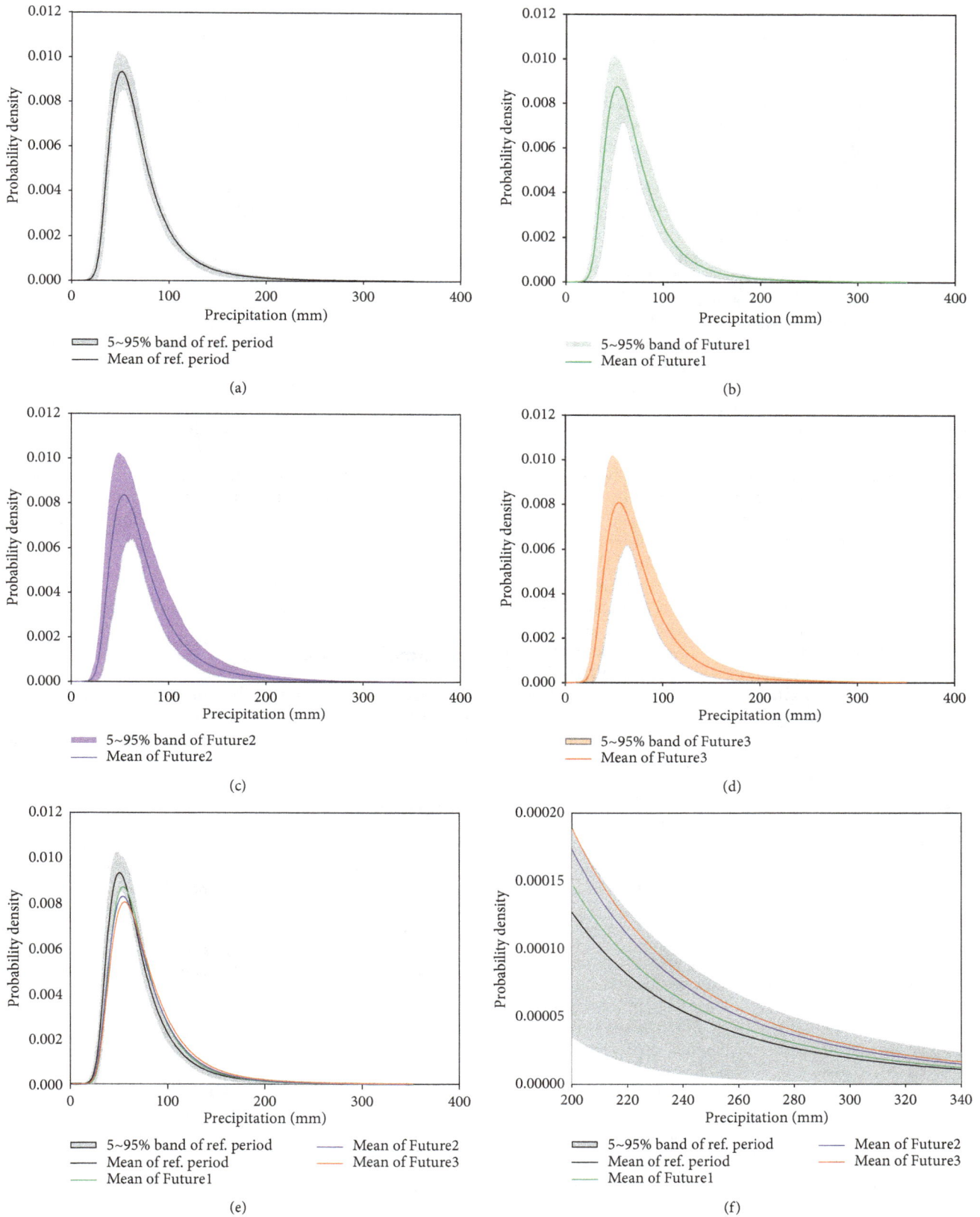

FIGURE 6: GEV probability density functions considering all downscaling methods for (a) reference, (b) Future1, (c) Future2, (d) Future3 period, (e) all together, and (f) zoomed Figure 6(e) for extreme values.

Location and scale parameters increased monotonously as shown in Figure 5. They were increasing in future climate compared to reference climate, but there was no big difference between Future2 and Future3. There was no difference in shape parameters over the periods. In recent studies on the nonstationary probability distribution model, the shape parameter has become smaller overall, indicating that the right tail of the GEV probability density function becomes thicker and the occurrence of extreme precipitation becomes more frequent in the future [42]. The median of the return period estimated by the inverse of the GEV distribution is expected to be decrease from the reference 20-year to the Future1: 16.56 yr; Future2: 14.09 yr; and Future3: 12.82 yr (Table 4), which means that 20-year precipitation of the reference climate would be relatively frequent in the future. However, it should be noted that this result is only the result of using RCP 4.5. This study used a single scenario to investigate the changes in future prospects closely. However, since we cannot be certain of which of scenarios will be realized, the actual uncertainty is larger than that shown in Figure 6, particularly in the later future.

Changes in GEV-PDFs of reference and future climate were analyzed in Figure 6. The black lines and gray bands are the mean and confidence intervals (5~95%) of the PDFs of the reference climate. The confidence interval corresponds to the upper 95% and 5% of the 26 climate models. The green solid line and band mean the PDFs for Future1, the blue solid line and band for Future2, and the red solid line and band for Future3 (Figure 6(a)~6(d)). Figures 6(e) and 6(f) illustrate the change in GEV distribution in the future climate relative to the reference climate, and the PDFs shifted to the right and the right tail thickened over time. In addition, we found that the PDFs of the future periods are located out of the 10% confidence interval of the PDF for the reference period, which presents a significant difference of PDFs between the reference and future periods. The result indicates that the design standard under the reference climate is not managed to cope with climate change, and accordingly the revision of the design standard is required to improve sustainability in infrastructures.

4. Conclusion and Summary

This study investigated changes of extreme precipitation over time using the 26 CMIP5 climate models under RCP 4.5, which were statistically downscaled by three methods frequently used in climate change impacts assessment. The ensemble average 20-year return value of precipitation was projected to increase while the coefficient of variation of this return value within the ensemble expanded in the future. In other words, the uncertainty of the models increased over time and the extreme precipitation events may be more severe in the future as a whole. For the case of using the SD-QDM downscaling method, 20-year precipitation in the period 2071–2100 was estimated to increase to 350 mm, 121.5% compared to the reference climate, and especially in the south, east coast and metropolitan areas. Because the recent summer precipitation of South Korea has been dominated by unstable convective activity, typhoon, and low

pressure, there was much precipitation in the middle and coast of the Youngdong, the southern coast, the eastern region of Jeju, and the northern region of Gyeonggi-do [43]. Our results are consistent with these trends.

Although the confidence interval of the PDF of GEV widened over time, which means the uncertainty also increased over time. However, due to the increasing trend of location and scale parameter and the decreasing trend of shape parameter, the PDF moved to the right, and the upper tail was expected to be thicker. However, the downscaled extreme precipitation showed considerable variability between the models and between the downscaling methods in this study. This means that simulation of extreme precipitation is dependent on each resolution and parameterization. This leads to the limit in the quantitative projection of extreme precipitation based on climate change scenarios. Nevertheless, the future PDF is located outside of the 10% confidence interval of the reference PDF, so it was confirmed that the revisit of the design standard was required. There is limitation to not provide acceptable methodology for design standard revision. Sen et al. [44] provided methodology for design standard revision. In the future, we will examine the performance of engineering structure based on including the climate change factor in the risk calculation formulation.

Conflicts of Interest

The authors declare that they have no conflicts of interest.

Acknowledgments

This work was supported by the Korea Agency for Infrastructure Technology Advancement (KAIA) grant funded by the Ministry of Land, Infrastructure and Transport (Grant 18AWMP-B083066-05).

References

[1] A. L. Kay, H. N. Davies, V. A. Bell, and R. G. Jones, "Comparison of uncertainty sources for climate change impacts: flood frequency in England," *Climatic Change*, vol. 92, no. 1-2, pp. 41–63, 2009.

[2] C. Prudhomme and H. Davies, "Assessing uncertainties in climate change impact analyses on the river flow regimes in the UK. Part 2: future climate," *Climatic Change*, vol. 93, no. 1-2, pp. 197–222, 2009.

[3] R. L. Wilby and I. Harris, "A framework for assessing uncertainties in climate change impacts: low-flow scenarios for the River Thames, U.K.," *Water Resources Research*, vol. 42, no. 2, article W02419, 2006.

[4] D. M. Wolock and G. J. McCabe, "Estimates of runoff using water-balance and atmospheric general circulation models," *Journal of the American Water Resources Association*, vol. 35, no. 6, pp. 1341–1350, 1999.

[5] F. J. Doblas-Reyes, R. Hagendorn, and T. N. Palmer, "The rationale behind the success of multi-model ensembles in seasonal forecasting-II. Calibration and combination," *Tellus A: Dynamic Meteorology and Oceanography*, vol. 57, no. 3, pp. 234–252, 2005.

[6] F. Giorgi and L. O. Mearns, "Calculation of average, uncertainty range, and reliability of regional climate changes

from AOGCM simulations via the "Reliability Ensemble Averaging" (REA) method," *Journal of Climate*, vol. 5, pp. 1141–1158, 2002.

[7] J. K. Lee, Y.-O. Kim, and Y. Kim, "A new uncertainty analysis in the climate change impact assessment," *International Journal of Climatology*, vol. 37, no. 10, pp. 3837–3846, 2016.

[8] IPCC, *Climate Change 2014: Mitigation of Climate Change. Working Group III Contribution to the Fifth Assessment Report of the Intergovernmental Panel on Climate Change*, Cambridge University Press, Cambridge, UK, 2014.

[9] A. E. Raftery, T. Gneiting, F. Balabdaoui, and M. Polakowski, "Using Bayesian model averaging to calibrate forecast ensemble," *Monthly Weather Review*, vol. 133, pp. 1155–1174, 2005.

[10] J. Raisanen and T. N. Palmer, "A probability and decision-model analysis of a multimodel ensemble of climate change simulation," *Journal of Climate*, vol. 14, pp. 3212–3226, 2001.

[11] B. Rajagopalan, U. Lall, and S. E. Zebiak, "Categorical climate forecasts through regularization and optimal combination of multiple GCM ensembles," *Monthly Weather Review*, vol. 130, pp. 1792–1811, 2002.

[12] E. S. Im, B. J. Lee, J. H. Kwon, S. R. In, and H. O. Han, "Potential increase of flood hazards in Korea due to global warming from a high-resolution regional climate simulation," *Asia-Pacific Journal of the Atmospheric Sciences*, vol. 48, pp. 107–113, 2012.

[13] J. H. Sung, H. S. Kang, S. Park, C. Cho, D. H. Bae, and Y.-O. Kim, "Projection of extreme precipitation at the end of 21st century over South Korea based on Representative Concentration Pathways (RCP)," *Atmosphere*, vol. 22, no. 2, pp. 221–231, 2012, in Korean.

[14] J. B. Ahn, S.-R. Jo, M.-S. Suh, and K.-M. Shim, "Changes of precipitation extremes over South Korea projected by the 5 RCMs under RCP scenarios," *Asia-Pacific Journal of the Atmospheric Sciences*, vol. 52, no. 2, pp. 223–236, 2016.

[15] E. S. Im, J. B. Ahn, and S. R. Jo, "Regional climate projection over South Korea simulated by the HadGEM2-AO and WRF model chain under RCP emission scenarios," *Climate Research*, vol. 63, pp. 249–266, 2015.

[16] D. H. Cha and D. K. Lee, "Reduction of systematic errors in regional climate simulations of the summer monsoon over East Asia and the western North Pacific by applying the spectral nudging technique," *Journal of Geophysical Research*, vol. 114, article D14108, 2009.

[17] F. Giorgi and L. O. Mearns, "Introduction tospecial section: regional climate modeling revisited," *Journal of Geophysical Research*, vol. 104, pp. 6335–6352, 1999.

[18] D. K. Lee and M. S. Suh, "Ten-year east Asian summer monsoon simulation using a regional climate model (RegCM2)," *Journal of Geophysical Research*, vol. 105, no. 22, pp. 29565–29577, 2000.

[19] M. S. Suh, S. G. Oh, D. K. Lee et al., "Development of new ensemble methods based on the performance skills of regional climate models over South Korea," *Journal of Climate*, vol. 25, pp. 7067–7082, 2012.

[20] J. T. Abatzoglou and T. J. Brown, "A comparison of statistical downscaling methods suited for wildfire applications," *International Journal of Climatology*, vol. 32, no. 5, pp. 772–780, 2012.

[21] G. Burger, S. R. Sobie, A. J. Cannon, A. T. Werner, and T. Q. Murdock, "Downscaling extremes: an Intercomparison of multiple methods for future climate," *Journal of Climate*, vol. 26, pp. 3429–3449, 2013.

[22] A. J. Cannon, S. R. Sobie, and T. Q. Murdock, "Bias correction of GCM precipitation by quantile mapping: how well do methods preserve changes in quantiles and extremes?," *Journal of Climate*, vol. 28, pp. 6938–6959, 2015.

[23] I. Dabanlı and Z. Şen, "Precipitation projections under GCMs perspective and Turkish Water Foundation (TWF) statistical downscaling model procedures," *Theoretical and Applied Climatology*, vol. 132, no. 1-2, pp. 153–166, 2017.

[24] H.-I. Eum and A. J. Cannon, "Intercomparison of projected changes in climate extremes for South Korea: application of trend preserving statistical downscaling methods to the CMIP5 ensemble," *International of Journal of Climatology*, vol. 37, no. 8, pp. 3381–3397, 2017.

[25] A. W. Wood, L. R. Leung, V. Sridhar, and D. P. Lettenmaier, "Hydrologic implications of dynamical and statistical approaches to downscaling climate model outputs," *Climatic Change*, vol. 62, pp. 189–216, 2004.

[26] D. G. Groves and R. J. Lempert, "A new analytic method for finding policy-relevant scenarios," *Global Environmental Change*, vol. 17, pp. 73–85, 2007.

[27] M.-J. Shin, H.-I. Eum, C.-S. Kim, and I.-W. Jung, "Alteration of hydrologic indicators for Korean catchments under CMIP5 climate projections: alteration of hydrologic indicators for the seven Korean catchments," *Hydrological Processes*, vol. 30, no. 24, pp. 4517–4542, 2016.

[28] S. Hwang and W. D. Graham, "Development and comparative evaluation of a stochastic analog method to downscale daily GCM precipitation," *Hydrology and Earth System Sciences*, vol. 17, no. 11, pp. 4481–4502, 2013.

[29] D. Shepard, "A two-dimensional interpolation function for irregularly-spaced data," in *Proceedings of the 1968 ACM National Conference*, New York, USA, 1968.

[30] D. Maraun, "Bias correction, quantile mapping, and downscaling revisiting the inflation issue," *Journal of Climate*, vol. 26, no. 6, pp. 2137–2143, 2013.

[31] E. P. Maurer and D. W. Pierce, "Bias correction can modify climate model simulated precipitation changes without adverse effect on the ensemble mean," *Hydrology and Earth System Sciences*, vol. 18, no. 3, pp. 915–925, 2014.

[32] R. W. Katz, M. B. Parlange, and P. Naveau, "Statistics of extremes in hydrology," *Advances in Water Resources*, vol. 25, pp. 1287–1304, 2002.

[33] V. V. Kharin and F. W. Zwiers, "Estimating extremes in transient climate change simulations," *Journal of Climate*, vol. 18, no. 8, pp. 1156–1173, 2005.

[34] V. V. Kharin, F. W. Zwiers, and X. Zhang, "Intercomparison of near surface temperature and precipitation extremes in AMIP-2 simulations, reanalyses, and observations," *Journal of Climate*, vol. 18, no. 24, pp. 5201–5223, 2005.

[35] J. J. Egozcue and C. Ramis, "Bayesian hazard analysis of heavy precipitation in eastern Spain," *International Journal of Climatology*, vol. 21, no. 10, pp. 1263–1279, 2001.

[36] R. L. Smith, *Trends in Rainfall Extremes*, University of North Carolina, Chapel Hill, NC, USA, 1999, In press.

[37] R. L. Smith, "Extreme value statistics in meteorology and environment," *Environmental Statistics*, Chapter 8, pp. 300–357, 2001.

[38] J. R. Stedinger, R. M. Vogel, and E. Foufoula-Georgiou, *Frequency analysis of extreme events, Handbook of Hydrology*, Chapter 18, D. R. Maidment, Ed., McGraw-Hill, New York, NY, USA, 1993.

[39] E. S. Martin and J. R. Stedinger, "Generalized maximum likelihood GEV quantile estimator for hydrologic data," *Water Resources Research*, vol. 28, no. 11, pp. 3001–3010, 2000.

[40] A. M. G. Klein Tank, F. W. Zwiers, and X. Zhang, "Guidelines on analysis of extremes in a changing climate in support of informed decisions for adaptation," *Climate Data and Monitoring*, WCDMP-72, WMO-TD/No. 1500, p. 56, 2009.

[41] S. R. In, S. O. Han, E. S. Im, K. H. Kim, and J. Shim, "Study on temporal and spatial characteristics of summertime precipitation over Korean peninsula," *Atmosphere*, vol. 24, no. 2, pp. 159–171, 2014, in Korean.

[42] J. Park, J. H. Sung, Y.-J. Lim, and H.-S. Kang, "Introduction and application of non-stationary standardized precipitation index considering probability distribution function and return period," *Theoretical and Applied*, 2018.

[43] C. Park, J. Moon, E.-J. Cha, W.-T. Yun, and Y. Choiz, "Recent changes in summer precipitation characteristics over South Korea," *Journal of the Korean Geographical Society*, vol. 43, no. 3, pp. 324–336, 2008, in Korean.

[44] Z. Sen, M. A. Mohorji, and M. Almazroui, "Engineering risk assessment on water structures under climate change effects," *Arabian Journal of Geosciences*, vol. 10, p. 517, 2017.

4

Temporal Trends and Spatial Patterns of Temperature and its Extremes over the Beijing-Tianjin Sand Source Region (1960–2014), China

Wei Wei [ID],[1] Baitian Wang [ID],[1] Kebin Zhang [ID],[1] Zhongjie Shi [ID],[2] Genbatu Ge,[2] and Xiaohui Yang[2]

[1]College of Water and Soil Conservation, Beijing Forestry University, Beijing 100083, China
[2]Institute of Desertification Studies, Chinese Academy of Forestry, Beijing 100091, China

Correspondence should be addressed to Baitian Wang; wbaitian@bjfu.edu.cn and Kebin Zhang; ctccd@126.com

Academic Editor: Harry D. Kambezidis

In order to examine temperature changes and extremes in the Beijing-Tianjin Sand Source Region (BTSSR), ten extreme temperature indices were selected, categorized, and calculated spanning the period 1960–2014, and the spatiotemporal variability and trends of temperature and extremes on multitimescales in the BTSSR were investigated using the Mann-Kendall (M-K) test, Sen's slope estimator, and linear regression. Results show that mean temperatures have increased and extreme temperature events have become more frequent. Annual temperature has recorded a significant increasing trend over the BTSSR, in which 51 stations exhibited significant increasing trends ($p < 0.05$); winter temperature recorded the most significant increasing trend in the northwest subregion. All extreme temperature indices showed warming trends at most stations; a higher warming slope in extreme temperature mainly occurred along the northeast border and northwest border and in the central-southern mountain area. As extreme low temperature events decrease, vegetation damage due to freezing temperatures will reduce and low cold-tolerant plants may expand their distribution range northward to revegetate barren areas in the BTSSR. However, in water-limited areas of the BTSSR, increasing temperatures in the growing season may exacerbate stress associated with plants relying on precipitation due to higher temperatures combining with decreasing precipitation.

1. Introduction

As highlighted by the IPCC, climate change is the most important environmental problem and the greatest challenge facing mankind [1]. From 1951 to 2012, global average surface temperatures are reported to have risen by 0.72 degrees [1]; as temperatures have increased, extreme climatic events, which can be sudden and destructive, have also increased in frequently, and their occurrence directly or indirectly has resulted in prominent social problems and natural disasters [2], especially in ecologically fragile areas. Recent global temperature extremes [3] indicate that the number of cold days, cold nights, and frost days has decreased while the number of warm days and warm nights has increased [3], these changes having obvious regional characteristics [4–7].

Global regional-scale temperature changes are currently well documented. For example, Rehman and Al-Hadhrami [8] showed that the annual average maximum temperature on the west coast of Saudi Arabia has increased more than the average minimum temperature, while Khomsi et al. [9] recorded that hot and very hot events during the summer in Morocco (Safi and Marrakech) showed statistically significant decreasing trends. The number of very cold nights recorded annually and during the winter has significantly decreased across the Basilicata region (southern Italy) over the period 1951–2010, while the frequency of cold days has shown a weak increase [10]. Extreme temperature indices have indicated steady warming trends for arid and semiarid areas of southeastern Kenya [11], while days with 1-day extreme minimum or maximum apparent temperature have exhibited an increasing trend across many stations in the United States [12]. Temporal and spatial distribution of extreme temperatures across China shows that hot extremes indicate an upward trend, while cold extremes show downward trends; between 1961 and 2011, more frequent extreme temperature events were recorded in the southern area of the major grain producing region in China [13] and some trends of extreme temperature indices in the Songhua River Basin were stronger than those recorded in the Yangtze River Basin, southwestern China, and the Tibetan Plateau. Zhong et al. [14] also found the number of warm nights, warm days, and summer days would be more significant at higher latitudes. Sun et al. [15], investigating extreme temperature events in the Loess Plateau, found that the trend magnitudes of cold extremes were greater than those of warm extremes and that the growing season increased and diurnal temperature ranges declined in the region.

Surface temperature changes have been mainly considered to be due to anthropogenic activities, especially due to changes in concentrations of greenhouse gases (GHG) [16, 17]. In areas surrounding Beijing and Tianjin, a large amount of land was reclaimed in the 1960s which destroyed the surface vegetation and caused mobile and semimobile sand dunes to expand [18]. Since the implementation of the reform and development policy (1979), environmental degradation has continued to intensify in these areas due to over grazing and excessive reclamation of grasslands [19, 20]; as a result, the Beijing, Tianjin, and Hebei areas experienced huge dust storms during March to April in the year 2000. To improve the environment of these areas, the Chinese government established the sand source control project (phase I) in the Beijing-Tianjin Sand Source Region (BTSSR) in the year 2000 [21, 22]. Following the implementation of this project, more than half of the land in this region has experienced an increase in vegetation productivity during 2000–2010 [23]. Studies on the characteristics of climate change and the response of vegetation cover to weather changes in this region have found that NDVI has recorded a slight increasing trend in the growing season [24, 25]. In order to consolidate the results of phase I of the Blown-Sand Control Project, and to further reduce sandstorm hazards and construct the northern ecological barrier, phase II has been implemented, spanning the period 2013–2022. However, with continued increasing temperatures, decreasing precipitation, and enlarging of the drought area, it is very difficult to consolidate the achievements of phase I and to implement and improve the construction of phase II at the same time [23, 26]. Evaluation of extreme climate changes in the BTSSR can identify factors causing regional environmental and social problems and provide reference for the construction of phase II to cope with extreme climatic events and to improve disaster prevention and mitigation work in this area.

The objectives of this study, therefore, are (1) to undertake trend analysis of temperature and extremes in the BTSSR; (2) to investigate the spatial patterns of trends of temperature and extremes; (3) and to discuss the possible impact temperature and extremes have on vegetation.

2. Study Area

The study area, located at the scope of the phase II project in the BTSSR, includes six provinces with 138 counties. This area includes the Mu Us Sandland, situated between Yulin city (Shaanxi province) and Erdos city (Inner Mongolia), and the Kubuqi Desert, situated in the northern Ordos Plateau. The region covers an area of $71.05 \times 10^4 \, km^2$ (approximately $36° \sim 46°N$, $107° \sim 119°E$; Figure 1). The area of land desertification covers an area of $22.69 \times 10^4 \, km^2$, exceeding one-third of the total area of phase II [22].

The climate in the study region is complex, including temperate semihumid, semiarid, arid, and extremely arid climate types [27]. Annual precipitation across the area varies from 105 to 743 mm, with the majority of precipitation occurring from June to September. Average temperature is $6.2°C$ with significant regional differences, such as a maximum temperature of $12.7°C$ in Beijing and a minimum temperature of $-2.6°C$ in Arxan, Inner Mongolia. The average sunshine duration is 7.9 h/d, and the regional mean wind speed is 4.5 m/s. The average growing season length is 206 days. In general, the climate of the BTSSR is dry, windy, and has a low temperature, with climatic characteristics varying significantly across the region.

3. Data and Methods

3.1. Data Source, Quality Control, and Homogeneity Test. Daily temperature data from the BTSSR was provided by the China Meteorological Data Network (http://data.cma.cn/). When more than 10% of daily temperature data was missing, or more than three months contained more than 20% missing days, annual temperature records were deemed as missing. To ensure data integrity and consistency, meteorological stations lacking more than 75% of a year's temperature records were omitted; missing days were not superseded by estimated daily temperature values [28, 29]. A total of 53 stations evenly distributed across the BTSSR were selected for this study; data from these stations spanned the period from 1960 to 2014 (Figure 1).

Quality control of the data was undertaken using the RClimDex package developed by Zhang and Yang [30], a control package that has been previously applied to test climate data [31]. The RClimDex package enabled errors in the data to be eliminated, including errors in manual keying, daily maximum temperatures lower than daily minimum temperatures, and any identified outliers. Daily temperature outliers were identified by manually examining visual data graphs and histograms; suspicious outliers were identified using statistical tests, local knowledge, and comparisons with adjacent days or the same day at neighboring stations. Data that was clearly flawed were adjusted or removed.

Due to relocation of meteorological stations, instrument changes, or observing procedures, step changes were identified at specific stations which resulted in observed meteorological data to not be fully comparable. To eliminate problems associated with step changes, daily temperature values were used to calculate the monthly temperatures; log transformations were subsequently used to test data

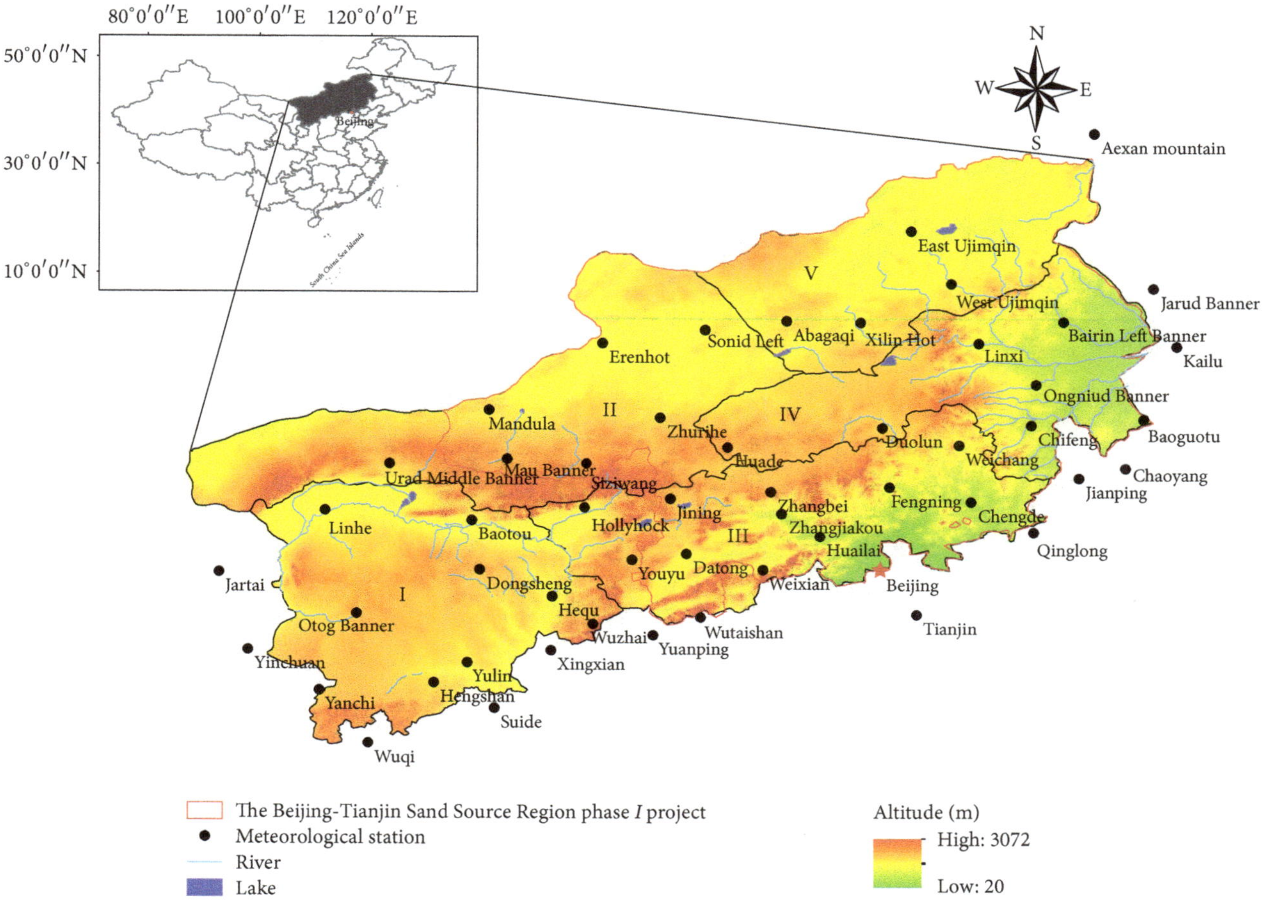

FIGURE 1: Location of the study area and meteorological stations. Greek capital numbers indicate the subregion division: I is the southwest subregion (SR-I), II is the northwest subregion (SR-II), III is the south-central subregion (SR-III), IV is the east-central subregion (SR-IV), and V is the northeast subregion (SR-V). Source: [27].

homogeneity using the RHtest V4 [32]. Any step changes identified using RHtest V4 were analyzed using metadata to evaluate the changes. Results from these tests showed that all data from the 53 meteorological stations were acceptable for use in this investigation.

After quality control and homogeneity tests, monthly temperature values were used to calculate mean annual temperatures, the standard climatological seasons (spring: March–May; summer: June–August; autumn: September–November; Winter: December–February) and the growing season (GS: April–October).

3.2. Calculation of Mean Temperature and Indices. To identify spatial changes in average temperature and extremes, the study region was divided into five subregions: the southeast subregion (SR-I), the northwest subregion (SR-II), the south-central subregion (SR-III), the east-central subregion (SR-IV), and the northeast subregion (SR-V), as per the investigation of Lu and Wu [27] (Figure 1). Temporally, two time cut points were determined as the starting year of reform and opening policy (ROP) (1979) and the starting year of the Blown-Sand Control Project Phase I in the BTSSR (hereafter

referred to as project phase I) (2000). To identify the impact of land use/land cover on temperature and extremes, the study period (S: 1960–2014) was divided into three time stages: 1960–1978 (S_1) (prior to ROP); 1979–1999 (S_2) (ROP to before project phase I); and 2000–2014 (S_3) (Project phase I).

Ten core indices for extreme temperature from ETC-CDMI, indices which have been widely applied to evaluate extreme temperature shifts [33–40], were selected and classified into three categories: cold-related indices, warm-related indices, and variability indices (Table 1). All calculations of these indices were undertaken using RClimDex 1.0 software [30].

3.3. Trend Analysis of Mean Temperature and Indices. Trends in mean temperature and indices were analyzed for the three time stages and for the duration of the study period. Trends for individual station data and regional series were estimated; the significance of the trends was determined using the Mann-Kendall test [41, 42]. This test does not assume that data are normally distributed, and it robustly responds to the effects of outliers in the series; this test has been previously

TABLE 1: Definitions of temperature and extreme indices used in this study.

Type	Index	Descriptive name	Definition	Units
Mean temperature	TA	Annual temperature	Annual mean temperature	°C
	T_{spring}	Spring temperature	Mean temperature in spring	°C
	T_{summer}	Summer temperature	Mean temperature in summer	°C
	T_{autumn}	Autumn temperature	Mean temperature in autumn	°C
	T_{winter}	Winter temperature	Mean temperature in winter	°C
	TG	Growing season temperature	Mean temperature in growing season	°C
Cold-related indices	TN10p	Cold nights	Percentage of days when TN < 10th percentile	Days
	TX10p	Cold days	Percentage of days when TX < 10th percentile	Days
	CSDI	Cold spell duration indicator	Annual count of days with at least 6 consecutive days when TN < 10th percentile	Days
	TNn	Min Tmin	Monthly minimum value of daily minimum temperature	°C
Warm-related indices	TN90p	Warm nights	Percentage of days when TN > 90th percentile	Days
	TX90p	Warm days	Percentage of days when TX > 90th percentile	Days
	WSDI	Warm spell duration indicator	Annual count of days with at least 6 consecutive days when TX > 90th percentile	Days
	TXx	Max Tmax	Monthly maximum value of daily maximum temperature	°C
Variability indices	DTR	Diurnal temperature range	Monthly mean difference between TX and TN	°C
	GSL	Growing season length	Annual (1st Jan to 31st Dec in NH, 1st July to 30th June in SH) count between first span of at least 6 days with TG > 5°C and first span after July 1 (January 1 in SH) of 6 days with TG < 5°C	Days

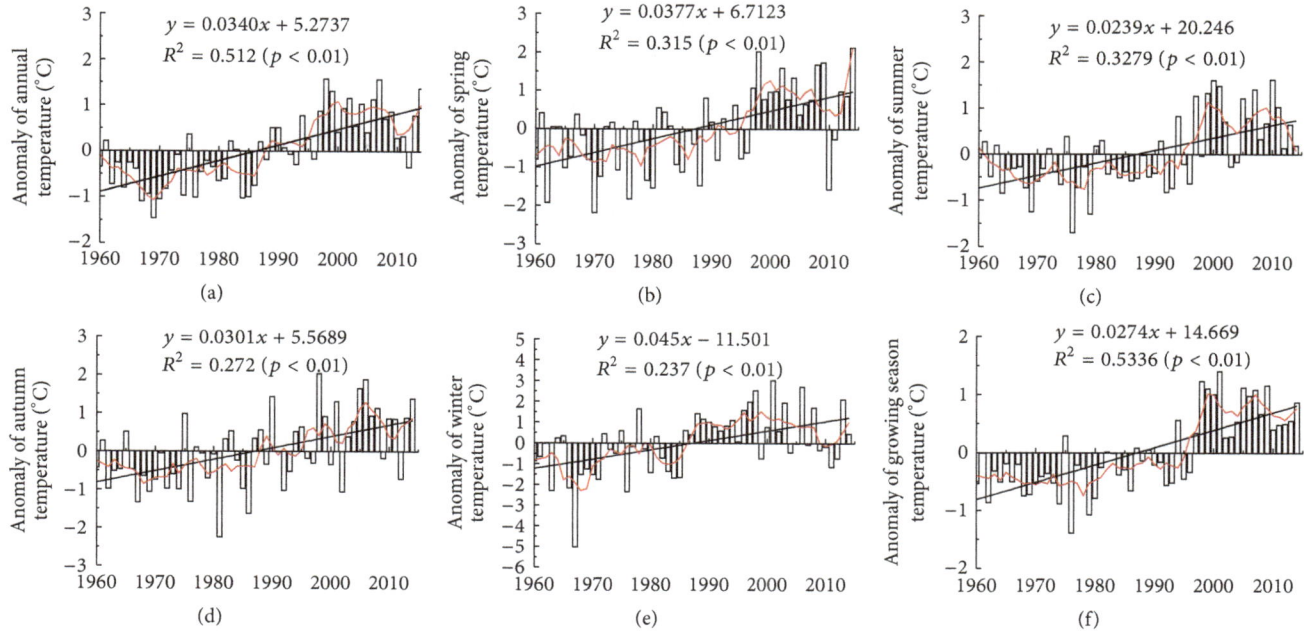

FIGURE 2: Anomaly of regional average temperature curves over the BTSSR, 1960–2014. The *blackline* represents the linear trends for (a) annual, (b) spring, (c) summer, (d) autumn, (e) winter, and (f) growing season. The *redline* represents the 5-year moving average.

used to test trends in hydrological and meteorological data [26, 43, 44].

The Mann-Kendall test, enabling data with missing values and values below a detection limit to be analyzed, is one of the most important and commonly used statistical methods to detect trends as nonparametric tests in hydroclimatic time series data. The null hypothesis that Z_c is not statistically significant or has no significant trend will be accepted if $-Z_{(1-\alpha)/2} \le Z_c \le Z_{(1-\alpha)/2}$, where $\pm Z_{(1-\alpha)/2}$ are the standard normal deviates and α is the significance level for the test [45].

Kendall's statistic S assessing the monotonic trend was calculated as

$$S = \sum_{i=1}^{n-1} \sum_{k=i+1}^{n} \text{sgn}\left(x_k - x_i\right),$$

$$\text{var}\left(S\right) = \frac{n\left(n-1\right)\left(2n+5\right) - \sum_{i=1}^{m} e_i\left(e_i - 1\right)\left(2e_i + 5\right)}{18},$$

(1)

where x_k and x_i are sequential data values representing the annual values in years k and i, respectively; n is the time length of the dataset; $\text{sgn}(\theta) = -1$ if $\theta < 0$, $\text{sgn}(\theta) = 0$ if $\theta = 0$, and $\text{sgn}(\theta) = 1$ if $\theta > 0$; e_i is the number of ties of the ith value and m is the number of tied values. The Mann-Kendall test statistic Z_c was calculated as

$$Z_c = \frac{S-1}{\sqrt{\text{var}\left(S\right)}}, \quad \text{if } S > 0,$$

$$Z_c = 0, \quad \text{if } S = 0,$$

(2)

$$Z_c = \frac{S+1}{\sqrt{\text{var}\left(S\right)}}, \quad \text{if } S < 0,$$

where Z_c is a standard normal variable. A positive Z_c indicates an upward trend while a negative Z_c indicates a downward trend [46]. A two-sided test at the significance level of 5% was undertaken in this study, and Sen's slope estimator was used to estimate the true slope of an existing trend (the change per year) [47]. All of the trend calculations for these indices were undertaken using MAKESENS 1.0 [48]. Prior to performing the Mann-Kendall test, the trend-free, prewhitening method of Yue et al. [49] was used to limit the effect of serial correlations on the Mann-Kendall test [50, 51].

3.4. Spatial Interpolation of Mean Temperature and Extremes Trends. Mean temperature and indices trends were spatially interpolated using the spline method. This is characterized by fitting a smooth and continuous surface with the observation points, and it does not require a preliminary estimate for the structure of a temporal variance and statistical hypothesis [52]. In comparison with other methods, previous studies, for example, those using Kriging [53, 54] and those analyzing climatic and meteorological change [55–58], have shown the spline method to be more advantageous around topographic features. Spatial interpolation with the spline method was completed using ArcGIS software (Version 10.2).

4. Results and Discussion

4.1. Mean Temperature Spatiotemporal Trends

4.1.1. Annual Temperature. MK test results across the study period for the BTSSR show that annual temperature (TA) significantly increased ($p < 0.001$). The TA slope was 0.34°C/decade (Figures 2(a) and 3(a), Table 2), a result which

TABLE 2: Trend slopes for average temperature and extreme temperature indices in the BTSSR during the different stages.

Index	S_1	S_2	S_3	S		Increasing (%)		Decreasing (%)	Units
T_A	−0.01	0.77**	−0.14	0.34***	52	**96.23%**	1	**0.00%**	1.89%	°C/decade
T_{spring}	−0.01	0.76*	0.00	0.38***	52	**98.11%**	1	**0.00%**	1.89%	°C/decade
T_{summer}	−0.24	0.35	−0.32	0.24***	52	**84.91%**	1	**0.00%**	1.89%	°C/decade
T_{autumn}	−0.03	0.64	0.29	0.30***	51	**83.02%**	2	**3.77%**	0.00%	°C/decade
T_{winter}	0.64	1.28**	−1.10	0.45***	53	**79.25%**	0	**0.00%**	0.00%	°C/decade
T_G	−0.02	0.47*	−0.15	0.27***	52	**94.34%**	1	**1.89%**	0.00%	°C/decade
TN10p	−0.98	−3.40**	−0.35	−2.10***	3	0.00%	50	**81.13%**	13.21%	days/decade
TX10p	1.38	−2.19*	−0.73	−0.96*	0	0.00%	53	**56.60%**	43.40%	days/decade
CSDI	−0.22	−0.73*	−0.81*	−0.80**	4	0.00%	49	**49.06%**	43.40%	days/decade
TNn	−0.17	1.61*	0.50	0.47*	48	**52.83%**	5	**1.89%**	7.55%	°C/decade
TN90p	−0.17	3.49*	−0.12	1.74***	51	**88.68%**	2	**3.77%**	7.55%	days/decade
TX90p	−1.06	2.77	−1.37	1.13**	52	**54.72%**	1	**0.00%**	43.40%	days/decade
WSDI	−0.27	3.52**	−1.02	1.11**	53	**79.25%**	0	**0.00%**	20.75%	days/decade
TXx	−0.23	0.36	−0.95	0.16	48	18.87%	5	**0.00%**	71.70%	°C/decade
DTR	−0.43*	−0.11	−0.12	−0.17***	10	13.21%	43	**66.04%**	5.66%	°C/decade
GSL	1.33	4.41	0.62	3.20***	52	**79.25%**	1	**0.00%**	18.87%	days/decade

S_1 is 1960–1978, S_2 is 1979–2000, and S_3 is 2000–2014; ∗ ∗ ∗ is 0.001 level of significance, ∗∗ is 0.01 level of significance, and ∗ is 0.05 level of significance; percentage of stations with a significant ($p < 0.05$) trend (**bold**), and percentage of stations with an insignificant ($p > 0.05$) trend (*italics*).

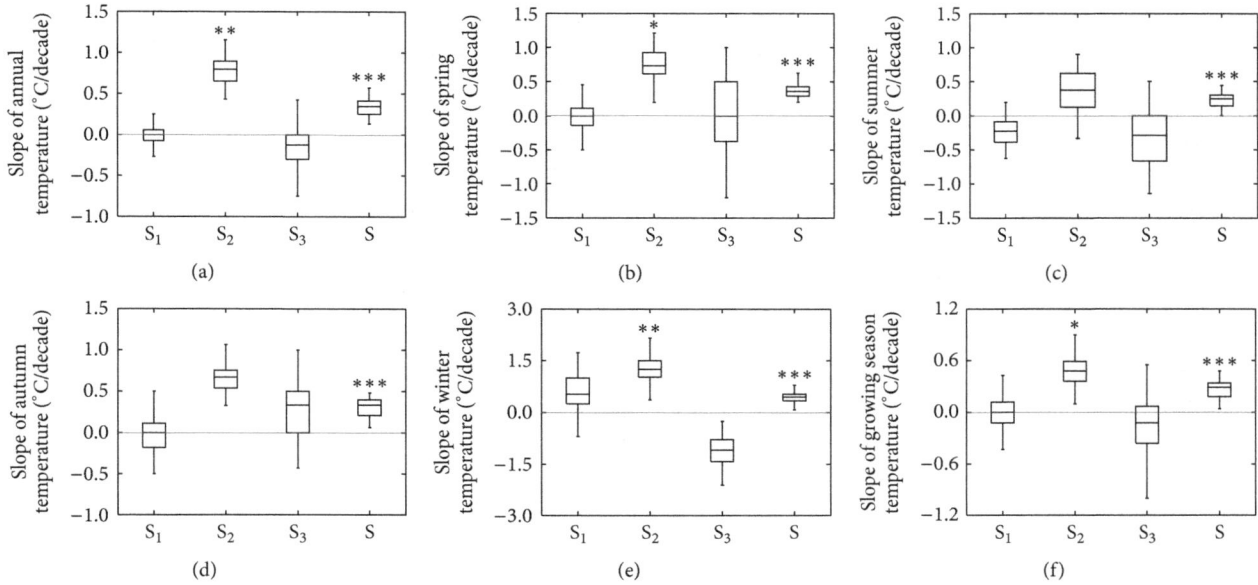

FIGURE 3: Box-Plot of Slopes for regional average temperature over the BTSSR in different stages. (a) Annual, (b) spring, (c) summer, (d) autumn, (e) winter, and (f) growing season. S_1 is 1960–1978, S_2 is 1979–2000, S_3 is 2000–2014, and S is 1960–2014. $***$ is 0.001 level of significance, $**$ is 0.01 level of significance, and $*$ is 0.05 level of significance.

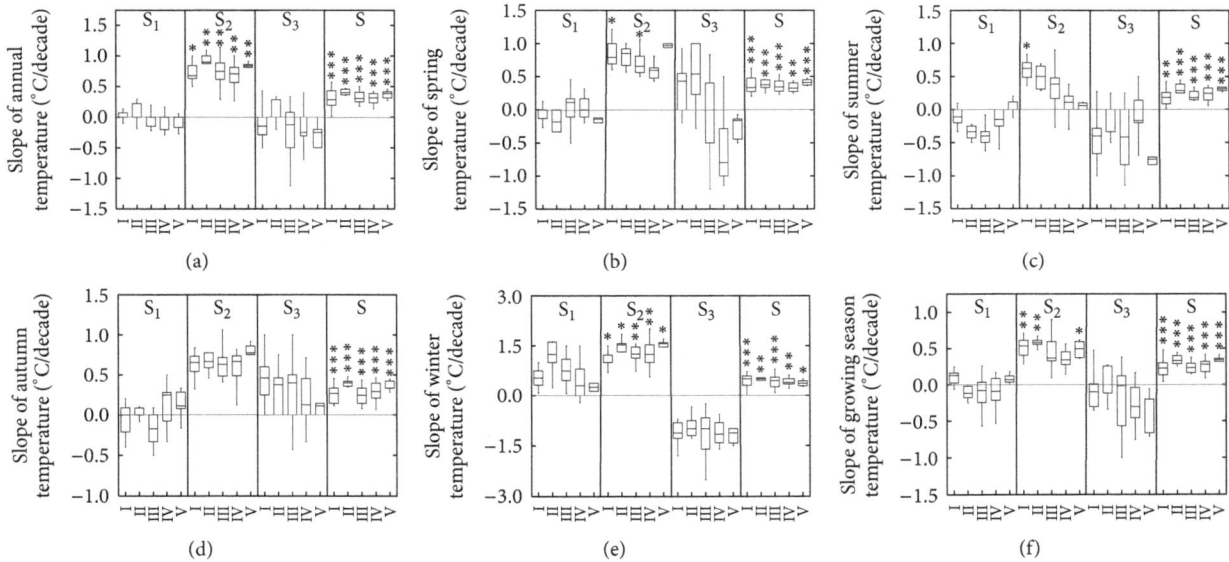

FIGURE 4: Box-Plot of Slopes for regional average temperature over the five subregions in different stages. (a) Annual, (b) spring, (c) summer, (d) autumn, (e) winter, and (f) growing season. S_1 is 1960–1978, S_2 is 1979–2000, S_3 is 2000–2014, and S is 1960–2014. I is the southwest subregion (SR-I), II is the northwest subregion (SR-II), III is the south-central subregion (SR-III), IV is the east-central subregion (SR-IV), and V is the northeast subregion (SR-V). $***$ is 0.001 level of significance, $**$ is 0.01 level of significance, and $*$ is 0.05 level of significance.

coincides with TA change trends identified in the China territory [59, 60], in regions such as the Yellow River basin, the Yangtze River Basin [61], arid and semiarid regions in northwest China [62], and those in the monsoon area of eastern China [63]. TA results for the different stages showed some differences to those recorded for the entire study period. TA recorded a slightly downward trend in S_1, a significantly upward trend in S_2 (with a slope of 0.77°C/decade), and an

insignificant downward trend in S_3 (Figure 3(a), Table 2), results which may be due to alternating surface albedo due to large areas of desertification in S_2 and rapid revegetation activities in S_3 [23, 37, 64–70].

The variations in T_A trends for the different stages in the five subregions are shown in Figure 4(a) and Table 3. Results show that T_A trends in these subregions significantly increased from 1960 to 2014 ($p < 0.001$); the highest T_A

TABLE 3: Trend slopes (°C/decade) of mean temperature and extreme temperature indices in the five subregions during the different stages.

Stage	S_1					S_2					S_3					S					Units
Index	I	II	III	IV	V	I	II	III	IV	V	I	II	III	IV	V	I	II	III	IV	V	
T_A	0.00	0.08	−0.02	−0.02	−0.08	0.74*	0.92**	0.77**	0.68**	0.85***	−0.09	0.08	−0.22	−0.18	−0.29	0.31***	0.41***	0.36***	0.30***	0.37***	°C/decade
T_{spring}	−0.01	−0.18	0.05	0.03	−0.11	0.85*	0.82	0.73*	0.56	0.91	0.36	0.51	−0.04	−0.58	−0.27	0.36***	0.38***	0.40***	0.34**	0.41***	°C/decade
T_{summer}	−0.09	−0.35	−0.42	−0.20	0.04	0.61*	0.49	0.38	0.08	−0.01	−0.40	−0.12	−0.36	−0.09	−0.75	0.19*	0.31**	0.25***	0.22***	0.31***	°C/decade
T_{autumn}	−0.10	0.03	−0.16	0.15	0.13	0.64	0.71	0.62	0.56	0.75	0.43	0.33	0.28	0.19	0.08	0.24***	0.40***	0.29***	0.29***	0.38***	°C/decade
T_{winter}	0.54	1.17	0.77	0.42	0.25	1.07*	1.37*	1.34**	1.28**	1.50*	−1.13	−0.94	−1.09	−1.12	−1.21	0.45***	0.46***	0.50***	0.41**	0.37***	°C/decade
T_G	0.11	−0.12	−0.07	−0.11	0.08	0.53**	0.58**	0.45	0.34	0.49*	−0.04	0.02	−0.15	−0.28	−0.46	0.23***	0.34***	0.27***	0.26***	0.35***	°C/decade
TN10p	−0.93	−1.62	−0.23	−1.55	−2.07	−3.32	−3.04*	−4.72**	−2.65**	−2.64***	−0.20	−0.64	0.39	−1.04	−0.53	−1.72***	−2.30***	−2.30***	−2.14***	−2.28***	days/decade
TX10p	1.10	1.13	1.74	1.29	1.52	−2.38	−2.59*	−2.20*	−1.80	−1.95*	−1.33	−1.20	−0.71	−0.07	0.09	−0.97**	−0.86*	−1.13**	−0.82*	−0.81	days/decade
CSDI	0.74	−2.64	0.29	−0.35	−0.92	−0.45	−0.66	−0.89*	−0.83	−0.92*	−0.61	−0.53	0.18	−2.54*	−1.14**	−0.72*	−1.38**	−0.54**	−0.81**	−1.04**	days/decade
TNn	−0.07	0.98	0.09	0.05	0.19	1.03	1.26	1.78**	2.10**	2.05**	0.81	0.90	0.52	0.17	−0.20	0.25	0.55**	0.60***	0.49**	0.49*	°C/decade
TN90p	−0.33	0.20	−0.58	0.11	0.48	3.96*	4.01**	3.34*	2.96	3.09*	1.17	1.96	−1.31	−1.18	−0.48	1.60***	2.07***	1.80***	1.63***	1.70***	days/decade
TX90p	−0.55	−1.44*	−1.44	−1.07	−0.71	3.29	3.69	2.69	1.98	2.03	−0.56	0.45	−3.16	−1.53	−2.13	1.31	1.07*	1.17***	0.87	1.13**	days/decade
WSDI	0.58**	−1.57**	−0.32	−0.28	−0.46	5.42**	3.29**	4.31**	0.57*	2.45**	1.20	2.16	4.31	−2.76**	−2.92	1.34**	1.00**	1.40**	0.57**	0.88**	days/decade
TXx	0.33	−0.03	−0.78	−0.42	0.11	0.54	1.02	0.33	−0.16	0.22	−1.22	−0.13	−1.08	−1.12	−0.54	0.11	0.23	0.23	0.05	0.24	°C/decade
DTR	−0.31	−0.60**	−0.33	−0.54**	−0.58**	−0.04	−0.06	−0.18	−0.10	−0.21	−0.09	−0.01	−0.07	−0.29	−0.14	−0.09	−0.23***	−0.17***	−0.20***	−0.21***	°C/decade
GSL	−4.26	5.92	3.56	1.11	3.90	3.39	5.17	3.03	5.64	7.95	0.14	3.52	−0.84	0.35	3.23	2.92**	3.42***	3.40***	3.08***	3.28***	days/decade

S_1 is 1960–1978, S_2 is 1979–2000, and S_3 is 2000–2014; * * * is 0.001 level of significance, * * is 0.01 level of significance, and * is 0.05 level of significance.

slope was 0.41°C/decade in the northwest subregion (SR-II) while the east-central subregion (SR-IV) recorded the lowest slope (0.30°C/decade) (Figures 3(a) and 4(a), Table 3). Results for all five subregions recorded a slight fluctuation during S_1 and a significant increasing trend in S_2, especially in the northwest subregion (SR-II; 0.92°C/decade), a result which may be attributed to serious steppe degradation in this area [71, 72]. During the last decade, T_A recorded slightly negative trends, except for the northwest subregion (SR-II).

Results for the spatial distribution of T_A trends (Figure 5(a)) show that most areas of the BTSSR had positive slopes, with the largest occurring in the central-southern mountain area; however, negative slopes were identified in the southeastern region in the Luanhe River Basin. Based on data from 51 stations (96.23% of all stations), the spatial distribution of T_A showed a significant increase in the BTSSR ($p < 0.01$) (Table 2), corresponding to the average slope of 0.35°C/decade. A strong increasing trend was observed at the Wutaishan station and a weak negative trend was found at the Chengde station, both meteorological stations being situated in the south-central subregion (SR-III).

In general, an increasing T_A trend in the BTSSR can accelerate plant growth by altering plant photosynthesis processes [73, 74] and lengthening the growing season [75, 76]. However, increasing T_A trends can also increase regional evapotranspiration [26] which, in conjunction with a decreasing trend in precipitation in the BTSSR [77], will increase the risk of drought [78]. The tradeoff of change trends of these meteorological factors is currently unclear.

4.1.2. Seasonal Temperature. MK test results indicate that the seasonal temperature of the BTSSR significantly increased ($p < 0.05$) during the study period; the lowest (0.24°C/decade) and highest (0.45°C/decade) rate of temperature rise were recorded in summer and winter, respectively (Figures 2(b)–2(e) and 3(b)–3(e), Table 2). This result was similar to that recorded by Wang et al. [79] for seasonal temperature changes across the arid region of northwestern China. Results for the different stages of the BTSSR showed differing changes. Results during S_1 recorded seasonal temperature to have a slightly negative trend, except for winter temperature (T_{winter}). Seasonal temperature in S_2 increased, with T_{winter} recording a significant increase (having a slope of 1.28°C/decade) and T_{summer} having an insignificant increase (with a slope of 0.35°C/decade), results which were in agreement with seasonal temperature changes across northeastern China [80]. Seasonal temperatures during S_3 recorded a mixture of slightly negative and positive trends (Table 2).

Results for seasonal temperature trends across the five subregions (Figures 4(b)–4(e), Table 3) recorded a significant increase ($p < 0.05$) during the study period, especially for T_{winter} in SR-III (having a slope of 0.50°C/decade) (Figures 5(b)–5(e)). During S_1, insignificant increasing trends for T_{winter} were recorded in all subregions, while other seasonal temperature series showed a mixture of slightly negative and positive trends across the five subregions. During S_2, T_{winter} showed a significant increasing trend in all subregions, especially in SR-V (with a slope of 1.50°C/decade), and other

seasonal temperature series showed insignificant positive trends for most subregions. Insignificant decreasing trends were recorded in the five subregions during the winter months and summer months, while spring temperatures showed a mixture of slightly negative and positive trends, with more negative than positive trends being recorded, and slightly increasing trends were found in all five subregions during the autumn months over the last decade (Figures 4(b)–4(e), Table 3).

The spatial distribution of seasonal temperature trends (Figures 5(b)–5(e)) shows that a positive temperature slope was identified for almost the entire BTSSR in the four seasons; the southeastern area of SR-III and the eastern area of SR-I were exceptions, these areas recording negative T_{spring}, T_{summer}, and T_{autumn} slopes and a negative T_{autum} slope, respectively (Figures 5(b)–5(e)). Most stations exhibited increasing trends, with the lowest winter slope of 0.01°C/decade (Hequ station) and the highest winter slope of 1.51°C/decade (Wutaishan station) being located in SR-I and SR-III, respectively. In spring, data from 98.11% of stations recorded a significant increasing trend (Figure 5(b), Table 2). In summer and autumn, the number of stations recording significant positive trends was similar, the trends mainly occurring in SR-I, SR-III, and SR-IV. Similarly, the number of stations recording negative trends was few (Figures 5(c) and 5(d), Table 2). In the winter, 42 stations recorded significant positive trends (79.25% of all stations), these being mainly located in SR-I, SR-III, and SR-IV, with no station recording a negative trend (Figure 5(e), Table 2). This result implies that the southern and east-central areas of the BTSSR experienced a warmer climate each year.

The winter warming trends identified in our study across the BTSSR are similar to those previously identified in Europe [81], North Asia and North America [82], subarctic [83], China [84, 85], and other areas across the globe [86]. Winter warming will undoubtedly be beneficial in alleviating vegetation damage due to freezing temperatures, to both native and planted vegetation, and it will increase the possibility of using low cold-tolerant plants to revegetate the BTSSR.

4.1.3. Growing Season Temperature. MK test results for growing season temperature (T_G) from 1960 to 2014 across the BTSSR recorded a significant increase ($p < 0.001$), with a slope of 0.27°C/decade (Figures 2(f) and 3(f), Table 2). Results during the different stages showed T_G to record a slight decrease in S_1, a significant increase in S_2 (with a slope of 0.47°C/decade), and an insignificant decline in S_3 (Figure 3(f), Table 2). The overall trend of T_G was similar with that of T_A.

Results for the variations in T_G trends for the five subregions in the different stages recorded a significant increase ($p < 0.001$; Figure 4(f), Table 3) from 1960 to 2014, especially in SR-V, with a slope of 0.35°C/decade (Figures 5(f) and 4(f), Table 3). As temperatures increased, the date of initiation of the growing season advanced and earlier growth resulted in vegetation covering the land earlier, thus reducing the occurrence of dust storms [87]. However, earlier vegetation growth could increase water stress in the reproductive growth period of water-limited vegetation in the BTSSR [24]. T_G

(a)

(b)

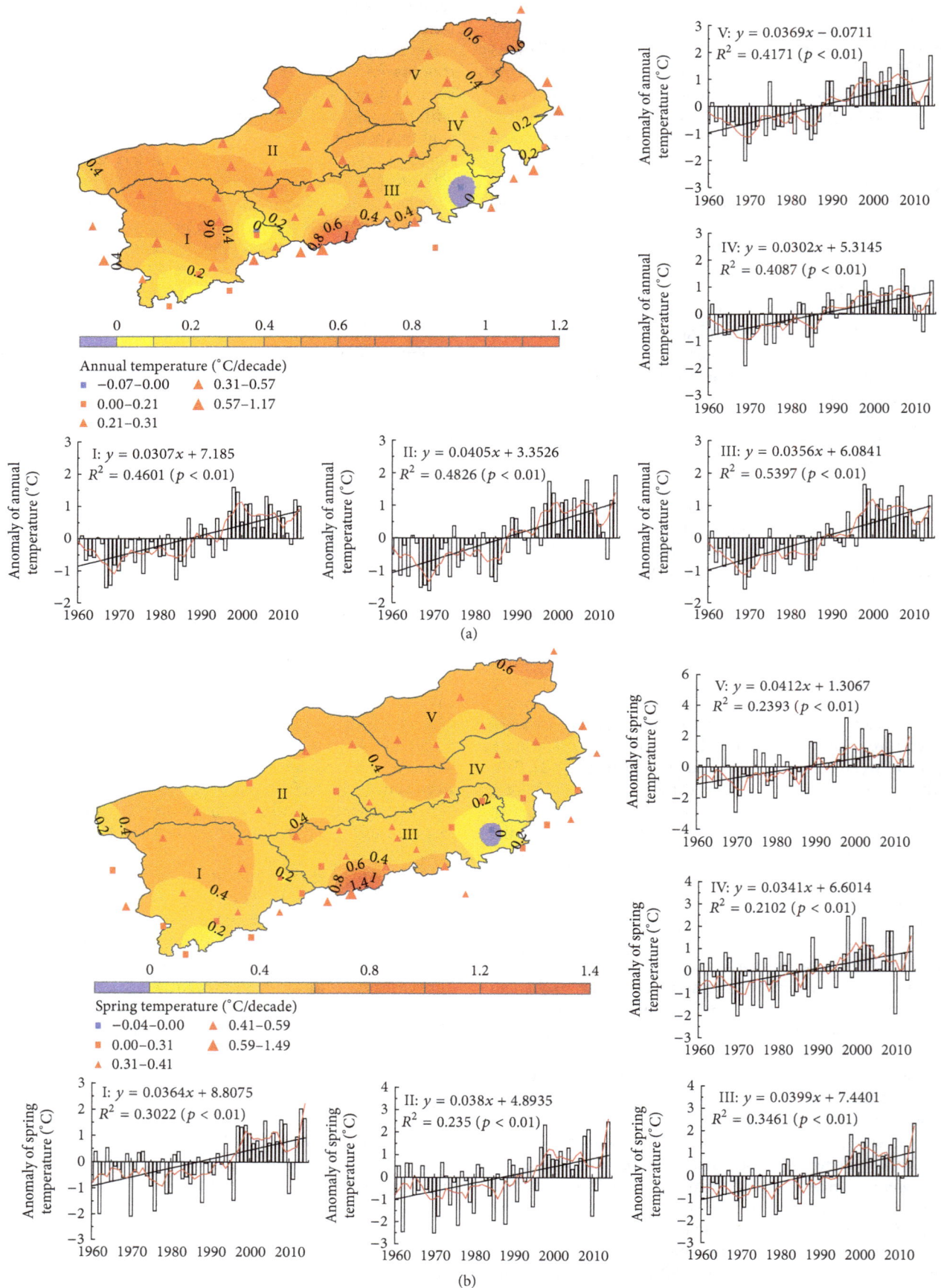

FIGURE 5: Continued.

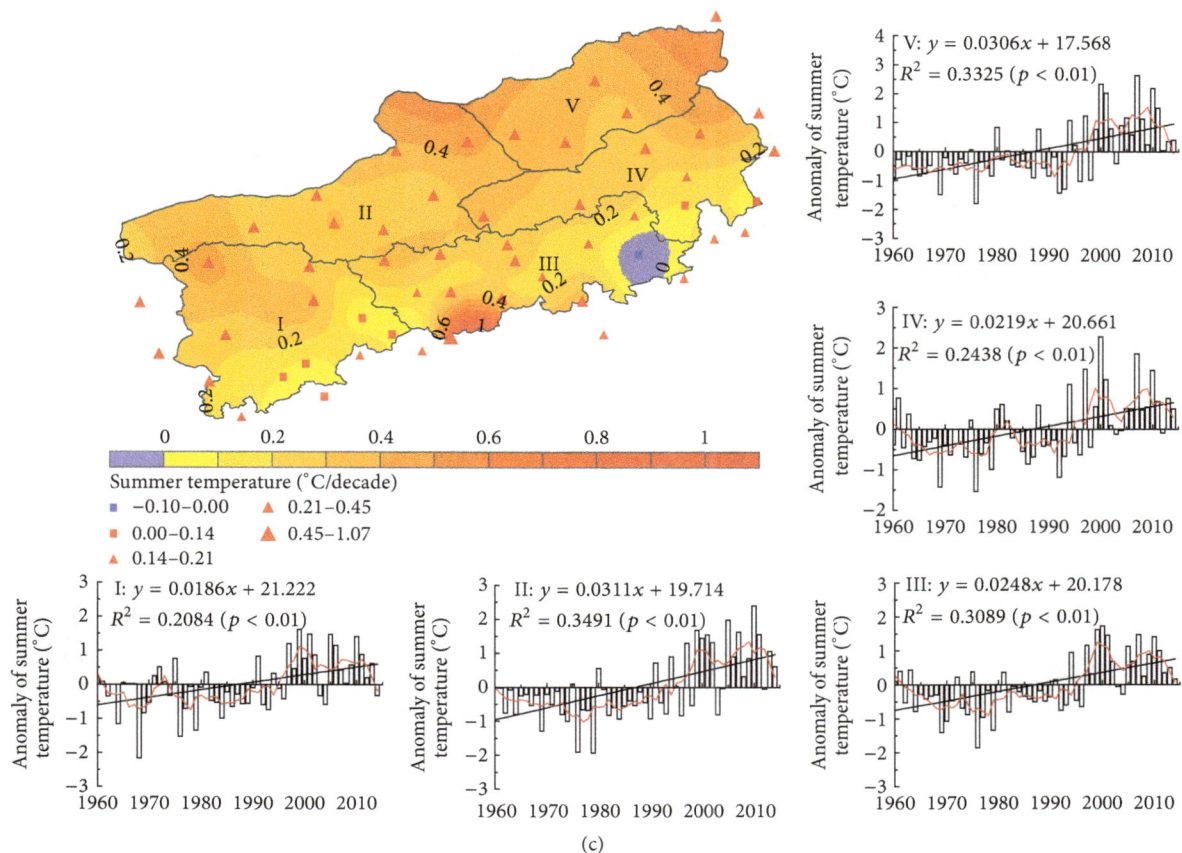

(c)

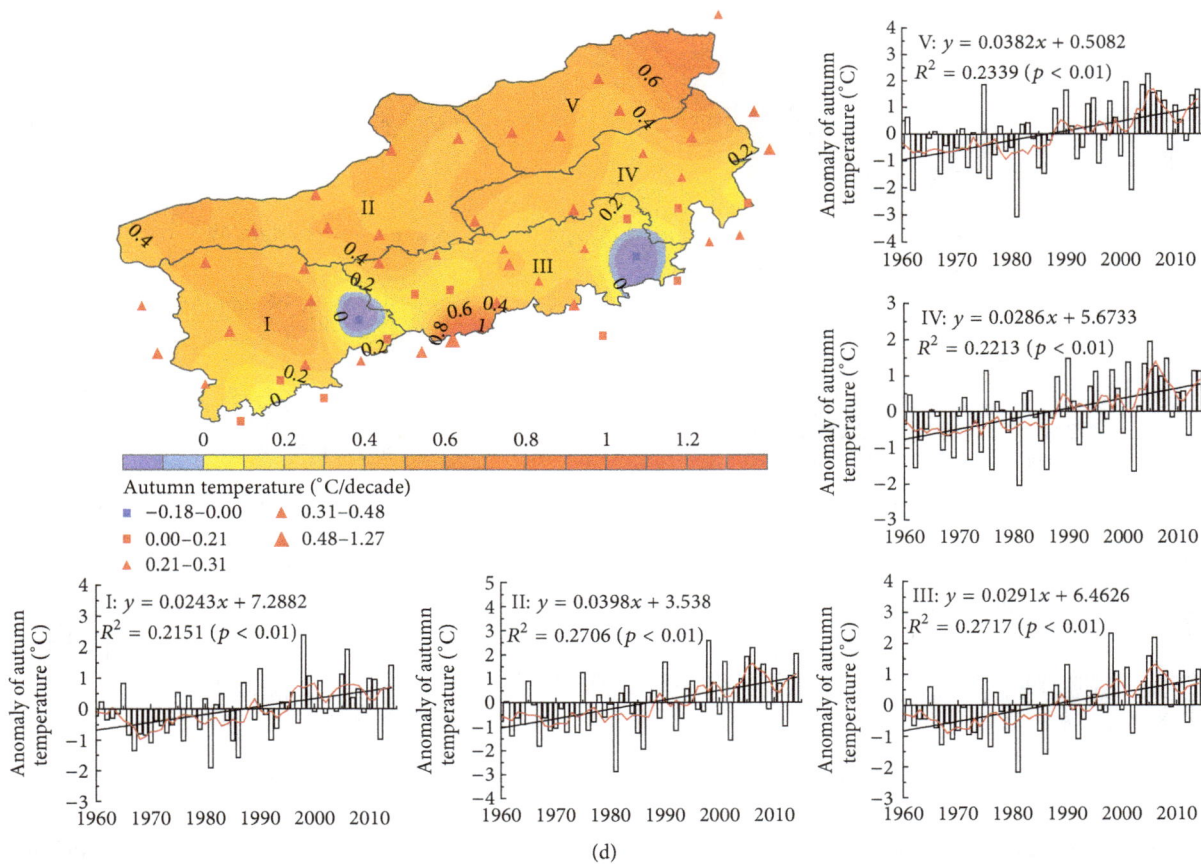

(d)

FIGURE 5: Continued.

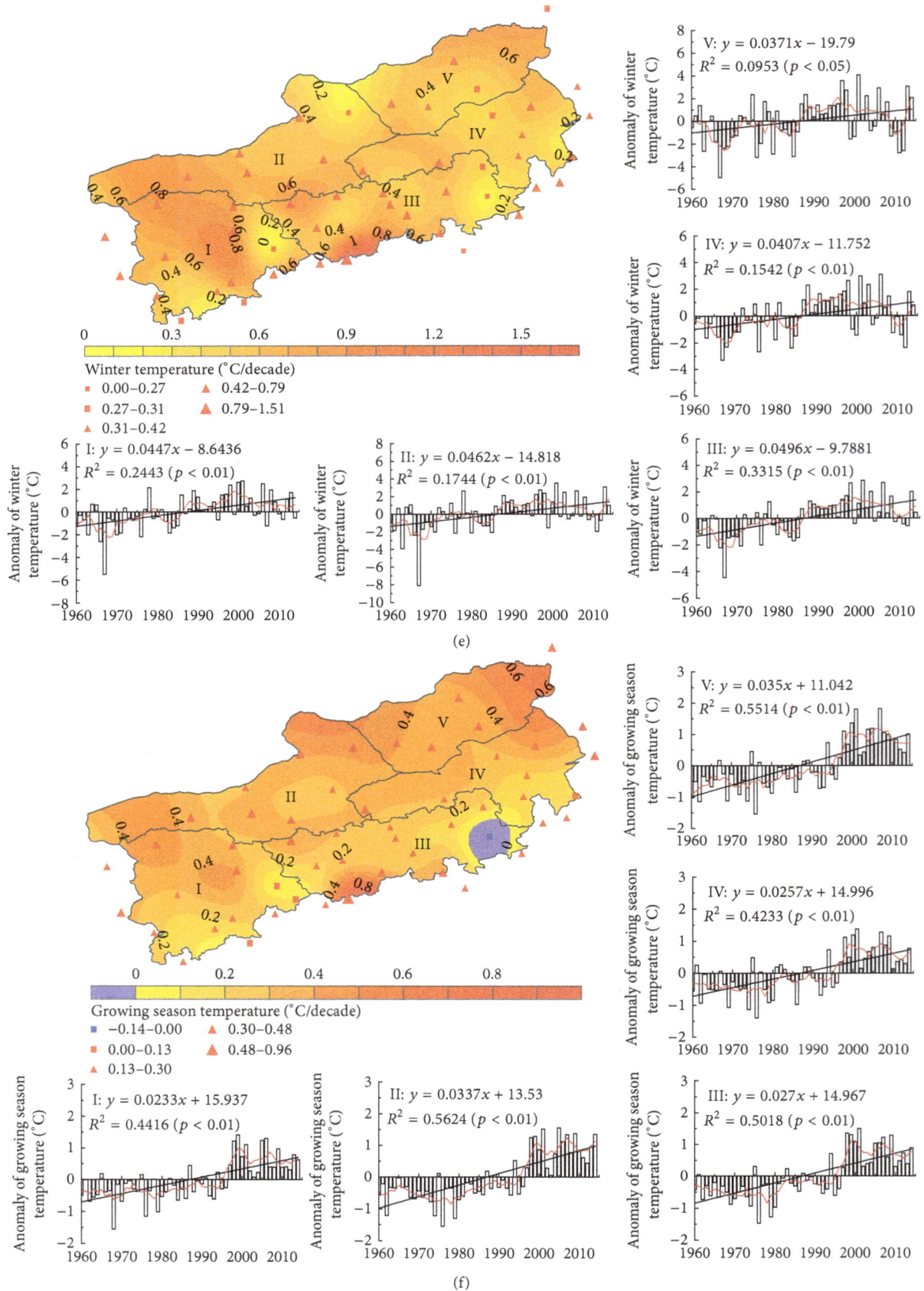

FIGURE 5: Spatial patterns of the trend slopes of mean temperature (1960–2014) over the BTSSR and anomaly of subregional indices curves. The *blackline* represents linear trends for (a) annual, (b) spring, (c) summer, (d) autumn, (e) winter, and (f) growing season. The *redline* represents the 5-year moving average.

trends for all five subregions recorded slight fluctuations during S_1. During S_2, a slope of $0.58°C$/decade was recorded, indicating a significant increasing trend, especially in SR-II. This result was consistent with previous findings showing T_G to have significantly increased across most regions of China since 1980 [88]. During the last decade (S_3), T_G showed slightly negative trends, except in SR-II (Figure 4(f), Table 3).

Results for the T_G slope showed a similar spatial pattern with those for T_A (Figure 5(f)). A larger positive slope was recorded in the central-southern mountain area and the northeast region, and a negative slope occurred in the southeastern area of SR-III. Based on data from 50 stations (94.34% of all stations; Table 2), the spatial distribution of T_G showed a significant increase over the BTSSR ($p < 0.05$) (Figure 5(f)), corresponding to an average slope of $0.29°C$/decade. A strong increasing trend was observed at the Wutaishan station and a significant negative trend was only found at the Chengde station, both stations being located in SR-III (Figure 5(f)).

Our results showed that a trend of increasing T_G can enhance plant photosynthetic activities and further productivity without causing water stress [89, 90]. However, in water-limited areas of the BTSSR, increasing temperature in the growing season may exacerbate rain-fed plant stress due to high temperatures combined with decreasing precipitation [77, 91], especially for vegetation planted (shrubs or trees) as part of the degraded land restoration project in the BTSSR.

4.2. Spatiotemporal Trends of Indices

4.2.1. Cold-Related Indices. Results for temporal variation of cold nights (TN10p), cold days (TX10p), and cold spell duration indicator (CSDI), representing the frequency and duration of extreme low temperature for the BTSSR, are shown in Figures 6(a)–6(c). In general, from 1960 to 2014, the fluctuations of these indices continuously significantly decreased ($p < 0.05$), especially TN10p (having a slope of -2.10 days/decade; Figure 7(a), Table 2). Changes observed for TN10p are more significant than those of TX10p across the BTSSR (Figures 7(a) and 7(b), Table 2); a reduction in TN10p will result in a reduction in frost damage during the night [92, 93]. The results of our study were in accordance with those over the arid region of northwestern China [79] and in Iran [94]. Results for the different stages showed that, during S_1, the frequency and duration of extreme low temperature recorded a slightly negative trend, except for TX10p. During S_2, the frequency and duration of extreme low temperatures significantly decreased, especially TN10p (with a slope of -3.40 days/decade), and results during S_3 showed CSDI to significantly decrease (a slope of -0.81 days/decade), while TN10p and TX10p insignificantly decreased (Figures 7(a)–7(c), Table 2). Figure 6(d) shows the temporal variation of Min Tmin (TNn) across the BTSSR. From 1960 to 2014, the fluctuations of TNn continuously significantly increased, having a slope of $0.47°C$/decade ($p < 0.01$). During the different stages, TNn showed a slightly positive trend during S_1, a significant increase in S_2 (with a slope of $1.61°C$/decade), and an insignificant increased in S_3 (Figure 7(d), Table 2).

Variations in the trends for the extreme cold indices at the different stages in the five subregions (Figures 8(a)–8(d), Table 3) show that their frequency and duration for the majority of the subregions significantly decreased from 1960 to 2014 ($p < 0.05$). TN10p results in SR-III recording a slope of -2.30 days/decade (Figures 8(a) and 9(a), Table 3) and, for the majority of subregions, TNn significantly increased during the study period ($p < 0.05$), notably in SR-III (with a slope of $0.60°C$/decade, Figures 8(d) and 9(d), Table 3). During the period S_1, all subregions showed a mixture of negative and positive trends, except for TN10p and TX10p. These fluctuating trends stabilized during S_2 as a significant warming trend occurred, especially for TN10p in SR-III (with a slope of -4.72 days/decade), and for TNn in SR-IV (with a slope of $2.10°C$/decade), these results being in accordance with observations over China during 1979–1999 [95]. During S_3, TN10p, TX10p, and CSDI showed negative trends for most subregions while TNn showed a positive trend for most subregions, especially in SR-II (with a slope of $-0.90°C$/decade).

The slope distribution of TN10p showed a spatial disparity across the BTSSR (Figure 9(a)). A positive slope was identified in the eastern area of SR-I, the northern area of SR-II, and the southeastern area of SR-III, this being the area around the Chengde station; a negative slope prevailed across the rest of the region where higher negative slopes occurred along the northeast border and northwest border and in the south-central part. Results for the spatial distribution of TN10p, based on data from 43 stations (81.13% of all stations), showed a significant decrease across the BTSSR ($p < 0.05$) (Table 2), corresponding to an average slope of -5.05 days/decade. A strong decreasing trend was observed at the Wutaishan station in SR-III and a weak positive trend was identified at the Sonid Left station (SR-II; Figure 9(a)).

A negative slope for TX10p was identified across the entire region, with the largest slope occurring in the central-southern mountain area and in the east Siramulen River Basin (Figure 9(b)). Data from 56.60% of stations recorded a significant decreasing trend ($p < 0.05$) with no stations recording an increasing trend (Table 2). The Wutaishan station in SR-III recorded the strongest decreasing trend in the BTSSR (Figure 9(b)).

Results for CSDI also recorded a negative slope across the majority of the region, an exception being in the eastern parts of SR-I and SR-III where positive slopes occurred (Figure 9(c)). Data from 49.06% of stations recorded a significant decreasing trend, concentrated in the northern part of the study area (Figure 9(c), Table 2); a strong decreasing trend was observed at the Linhe station (SR-I) and a weak positive trend was identified at the Yuanping station (SR-III; Figure 9(c)).

The slopes for TNn and TN10p recorded similar patterns (Figure 9(d)), with larger positive slopes being recorded along the northeast-northwest border and in the south-central part, and negative slopes in the eastern and southern areas of SR-I, the northern area of SR-II, and the eastern area of SR-III. Based on data from 28 stations (52.83% of all stations; Table 2), the spatial distribution of TNn showed a significant increase across the BTSSR ($p < 0.05$), corresponding to an

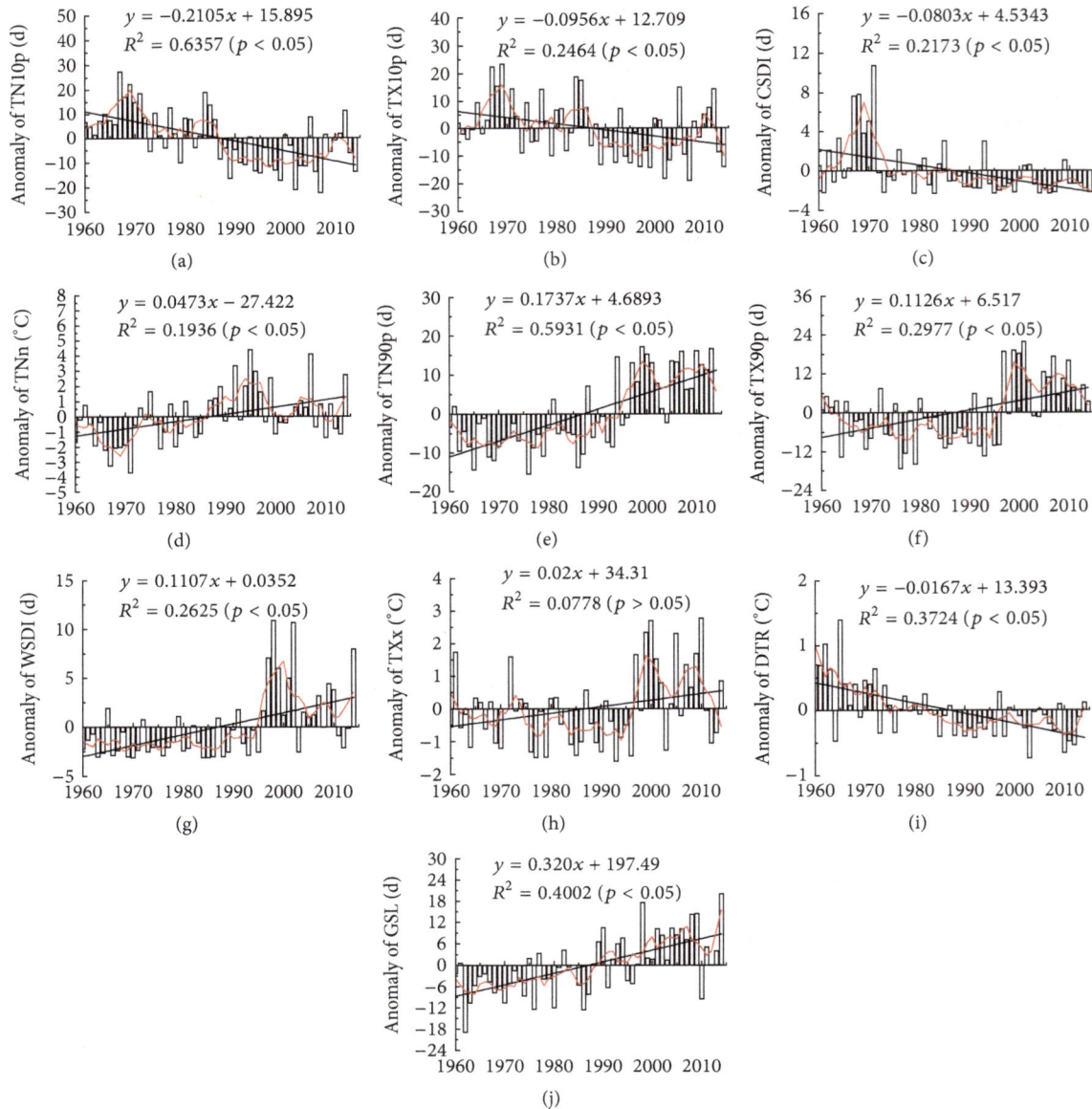

FIGURE 6: Anomaly of extreme temperature indices curves across the BTSSR (1960–2014). The *black line* represents the linear trends for (a) TN10p, (b) TX10p, (c) CSDI, (d) TNn, (e) TN90p, (f) TX90p, (g) WSDI, (h) TXx, (i) DTR, and (j) GSL. The *redline* represents the 5-year moving average.

average slope of $0.75°C$/decade. A strong increasing trend was observed at the Wutaishan station (SR-III) and a significant negative trend was identified at the Hequ station (SR-I; Figure 9(d)).

Results for the study period across the BTSSR showed a general trend of warming for the cold-related indices, trends which were also identified in northern Mongolia [96], South Africa [97], northeast China [98], and globally [99]. As extreme low temperature events decreased in the BTSSR, vegetation damage due to freezing conditions was mitigated, thus resulting in an expansion of range northwards of low, cold-tolerant plants which can result in the enrichment of plant species in this area.

4.2.2. Warm-Related Indices. The frequency and duration of extreme high temperature events across the BTSSR are shown by the results for the temporal variation of warm nights (TN90p), warm days (TX90p), and the warm spell duration indicator (WSDI) (Figures 6(e)–6(g)). In general, these indices fluctuated continuously, having a significant increasing trend during the study period ($p < 0.01$). The results for TN90p (with a slope of 1.74 days/decade; Figures 7(e)–7(g), Table 2) were noticeable, these being in accordance with previous findings across China [100]. Results for the different stages showed that the frequency and duration of extreme high temperature recorded an insignificant negative trend during S_1. Results during S_2 only recorded significant

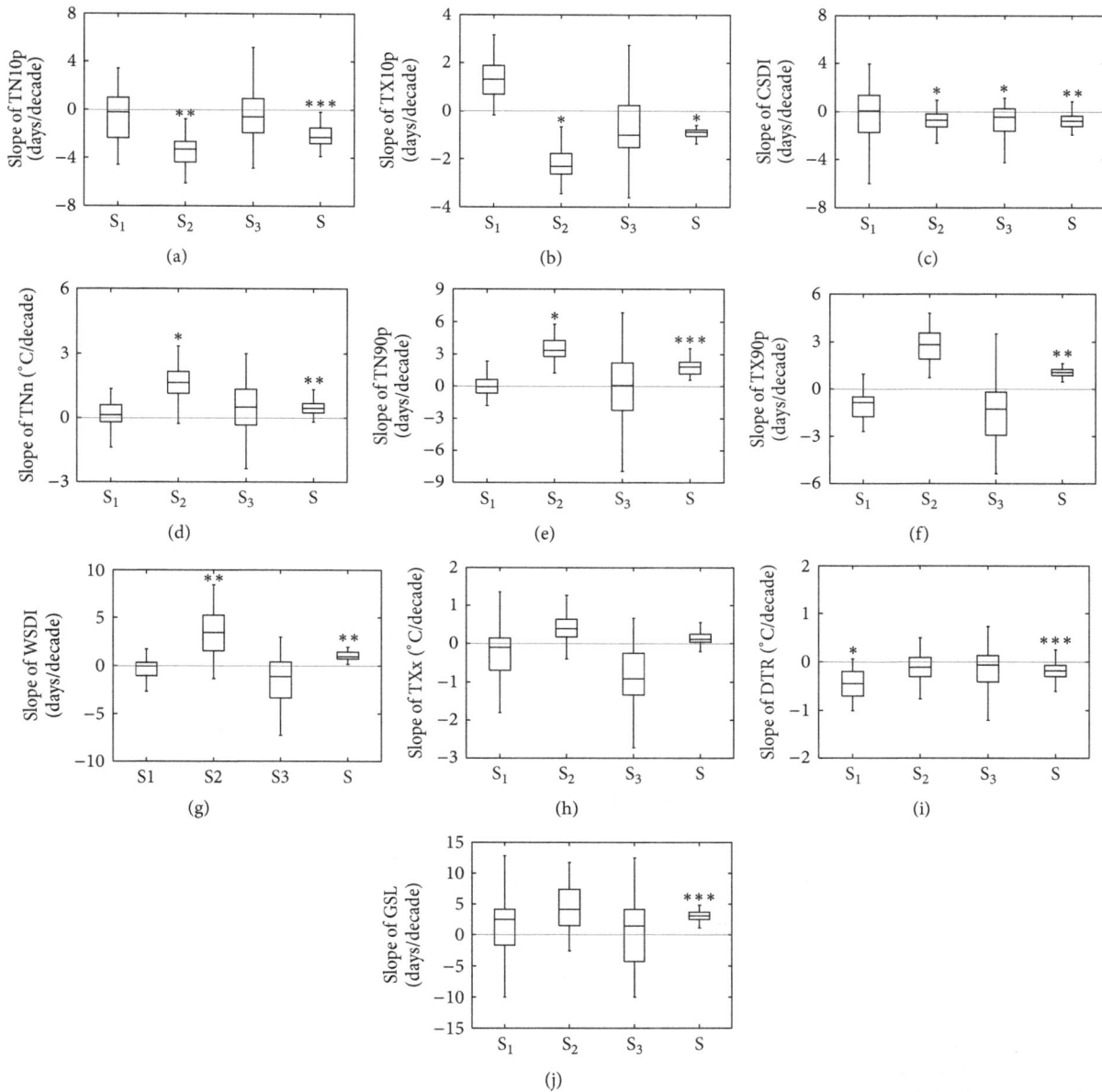

FIGURE 7: Box-Plot of Slopes for extreme temperature indices across the BTSSR in different stages. (a) TN10p, (b) TX10p, (c) CSDI, (d) TNn, (e) TN90p, (f) TX90p, (g) WSDI, (h) TXx, (i) DTR, and (j) GSL. S_1 is 1960–1978, S_2 is 1979–2000, S_3 is 2000–2014, and S is 1960–2014. $***$ is 0.001 level of significance, $**$ is 0.01 level of significance, and $*$ is 0.05 level of significance.

increases for TN90p and WSDI; TN90p had the most noticeable increase with a slope of 3.49 days/decade; during S_3, the frequency and duration of extreme high temperature decreased (Figures 7(e)–7(g), Table 2). The temporal variation of Max Tmax (TXx) across the BTSSR (Figure 6(h)) from 1960 to 2014 recorded fluctuations to insignificantly increase, having a slope of 0.16°C/decade. TXx results during the different stages recorded a slightly negative trend in S_1, an insignificant increase in S_2, and an insignificant decrease in S_3 (Figure 7(h), Table 2).

Variations in extreme warm indices in the five subregions (Figures 8(e)–8(h), Table 3) showed that frequency

and duration of extreme high temperature recorded significant increasing trends (1960–2014) for the majority of the subregions, especially TN90p (Figures 10(a)–10(c)). The change of TN90p is more pronounced than that of TX90p, a finding which is in accordance with previous studies [79, 94]. The temporal variation of TXx for the five subregions (Figure 10(d)) shows an increasing trend for all subregions, especially SR-V (with a slope of 0.24°C/decade). As daily maximum temperatures increase, transpiration and plant water stress will be exacerbated which may lead to an increase in water demand for vegetation production [101]. All five subregions showed that there was a combination of negative

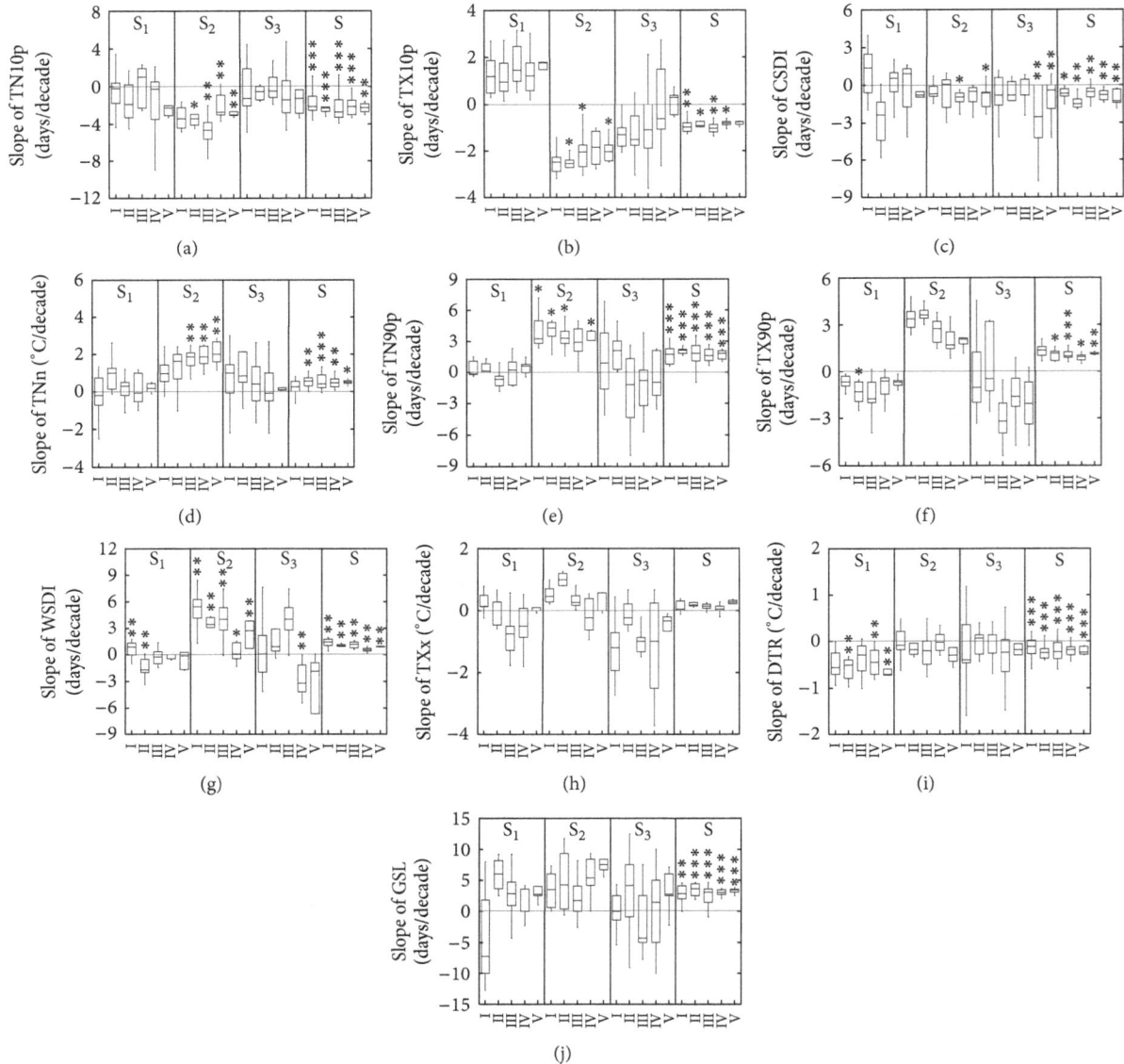

FIGURE 8: Box-Plot of Slopes for extreme temperature indices across the five subregions in different stages. (a) TN10p, (b) TX10p, (c) CSDI, (d) TNn, (e) TN90p, (f) TX90p, (g) WSDI, (h) TXx, (i) DTR, and (j) GSL. S_1 is 1960–1978, S_2 is 1979–2000, S_3 is 2000–2014, S is 1960–2014. * * * is 0.001 level of significance, ** is 0.01 level of significance, and * is 0.05 level of significance.

and positive trends during S_1, with more negative trends being identified. During S_2, an increasing trend occurred for the majority of the subregions, with more insignificant than significant trends, especially for TN90p in SR-III (with a slope of 3.34 days/decade). These findings are also in accordance with changes of temperature extremes in China as a whole [95]. During the last decade (S_3), the five subregions showed a mixture of negative and positive trends, with more negative trends recorded than positive (Figures 8(e)–8(h), Table 3).

Results across the BTSSR showed that positive slopes in TN90p occurred across the majority of the region, with the largest trend occurring in the central-southern mountain area. However, negative slopes were identified in the Luanhe River Basin of the southeastern region and the east area

of the southwest subregion (SR-I) (Figure 10(a)). Based on data from 47 stations (88.68% of all stations) (Table 2), the spatial distribution of TN90p showed a significant increase for the BTSSR ($p < 0.05$), corresponding to an average slope of 4.58 days/decade. Significant decreasing trends (at the 99% confidence level) were identified at the Hequ and Chengde stations (Figure 10(a)), accounting for 3.77% of the total stations, and a strong increasing trend was observed at the Wutaishan station (SR-III).

A positive slope for TX90p was identified across almost the entire BTSSR, the only exception being the southeastern area of SR-I where a negative slope was recorded. The larger positive slopes were recorded in the central-southern mountain area, and the number of stations recording positive

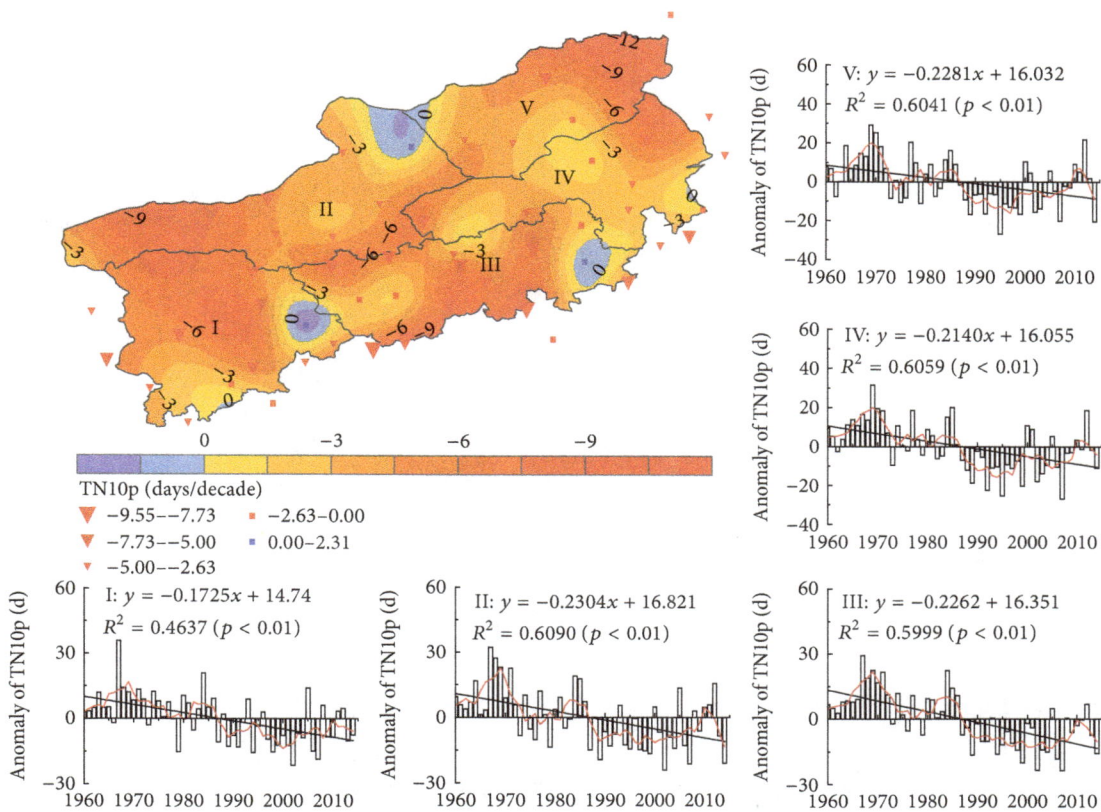

(a)

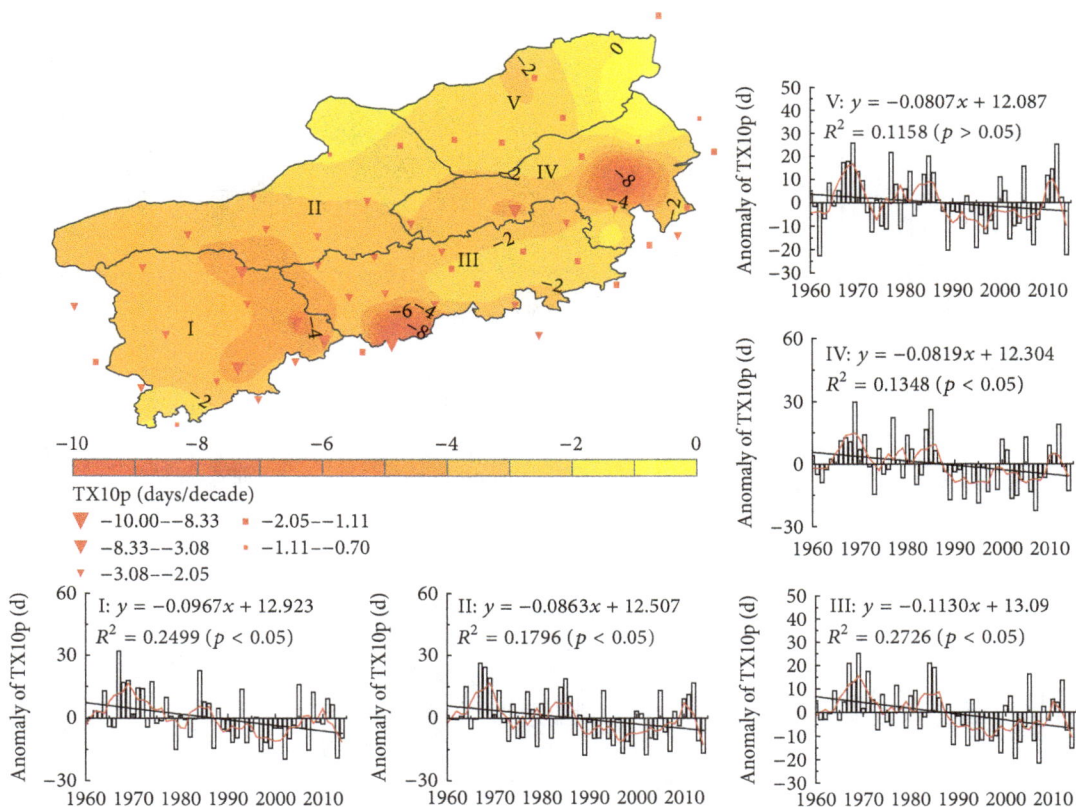

(b)

FIGURE 9: Continued.

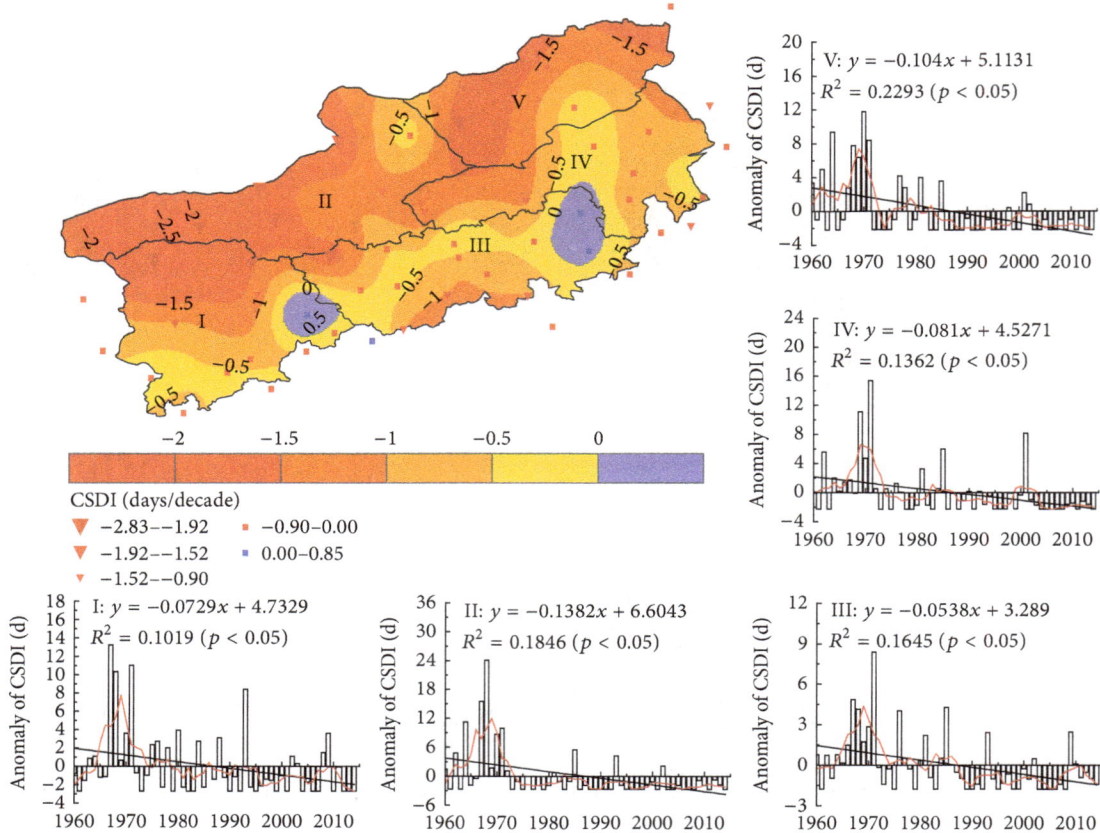

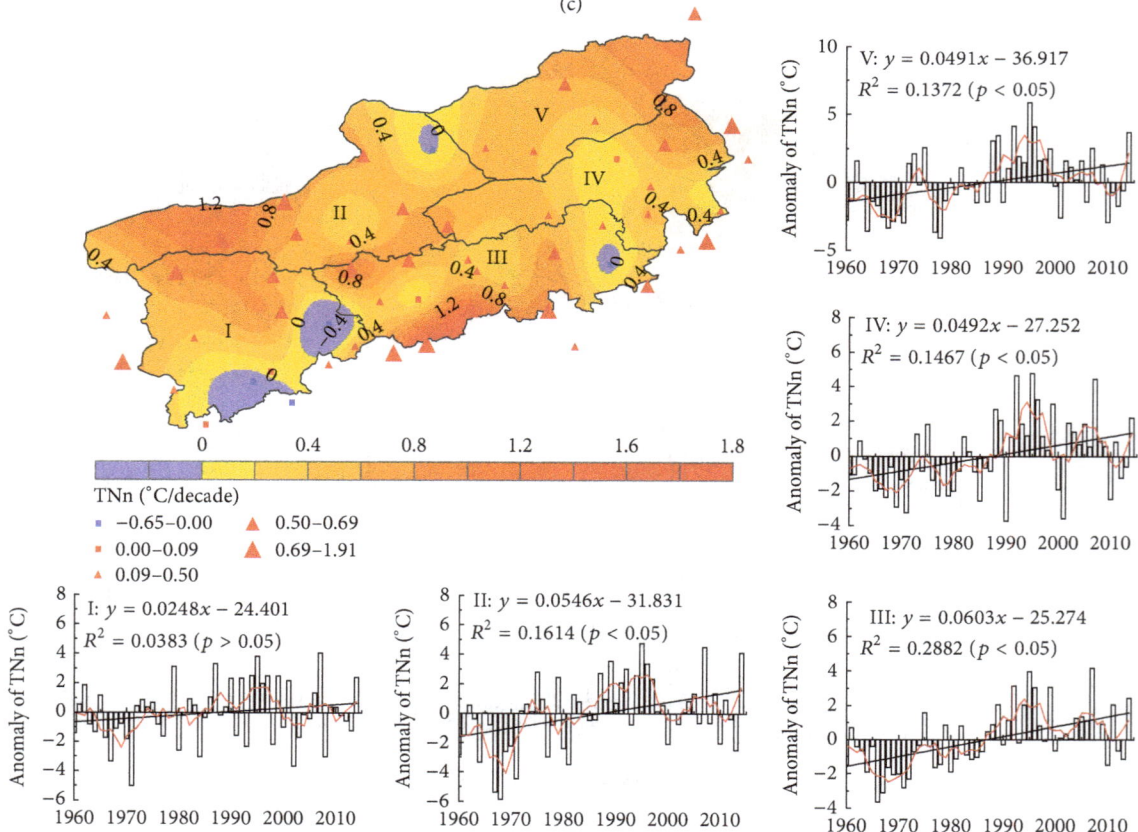

FIGURE 9: Spatial patterns of the trend slopes of cold-related indices across the BTSSR and anomaly of subregional indices curves, 1960–2014. The *black line* represents the linear trends for (a) TN10p, (b) TX10p, (c) CSDI, and (d) TNn. The *redline* represents the 5-year moving average.

(a)

(b)

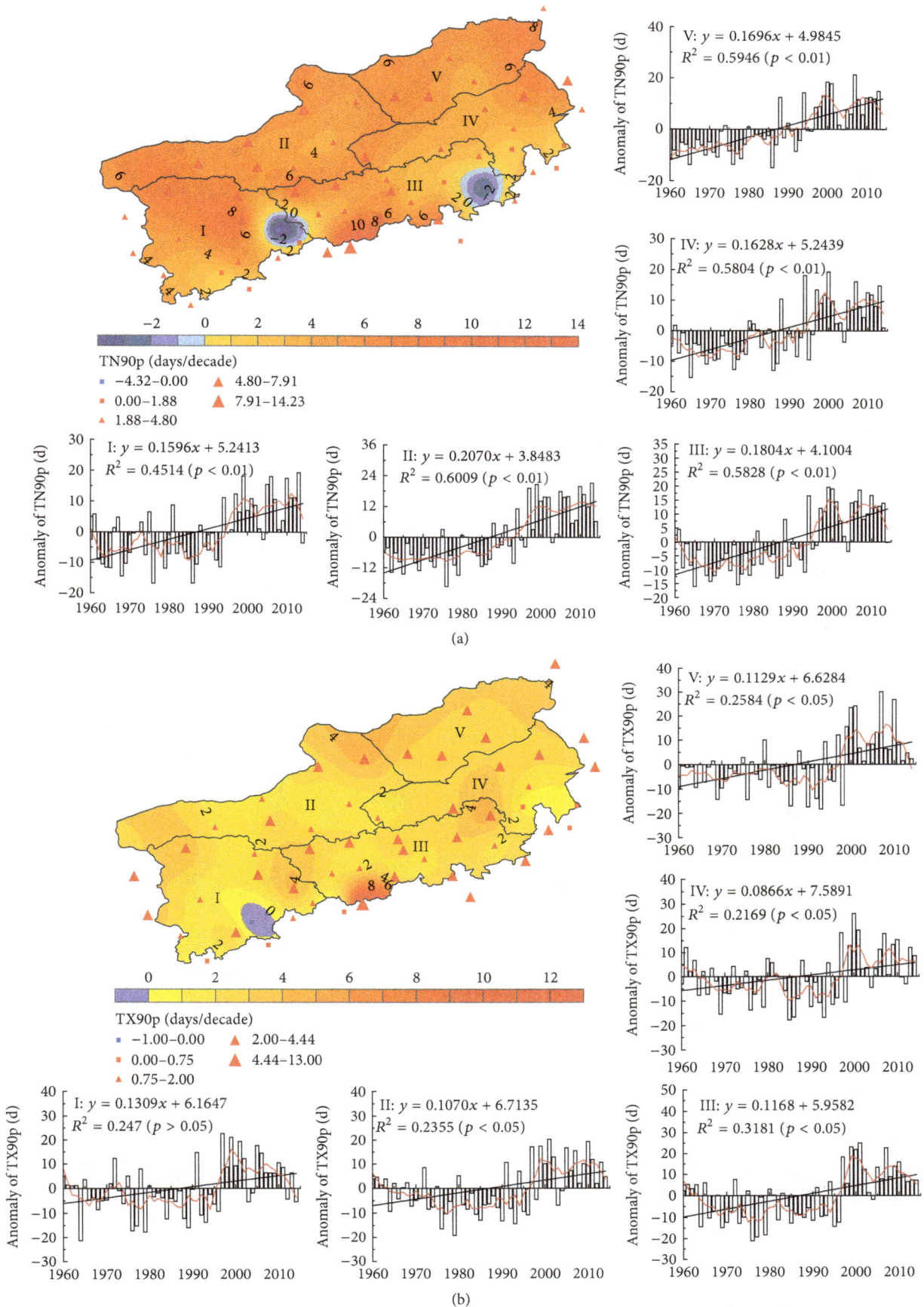

FIGURE 10: Continued.

(c)

(d)

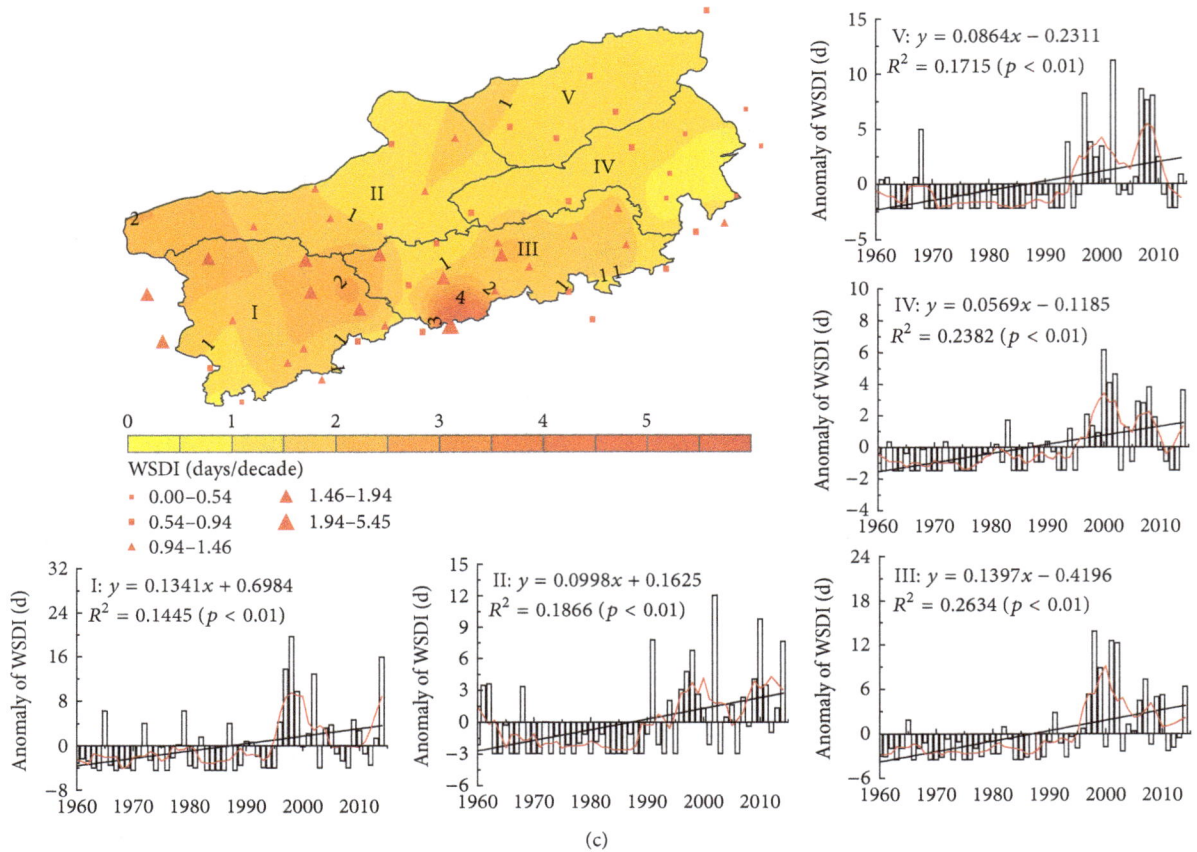

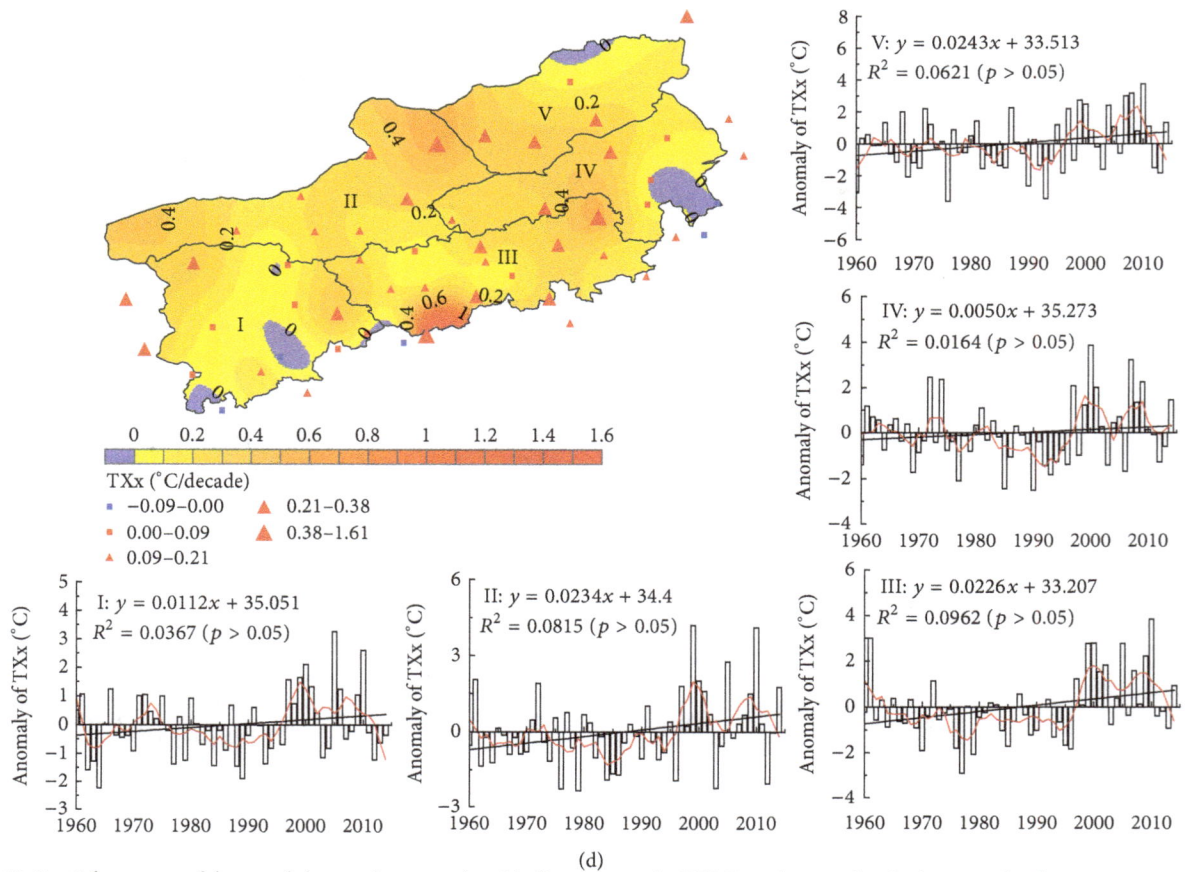

FIGURE 10: Spatial patterns of the trend slopes of warm-related indices across the BTSSR and anomaly of subregional indices curves, 1960–2014. The *black line* represents the linear trends for (a) TN90p, (b) TX90p, (c) WSDI, and (d) TXx. The *redline* represents the 5-year moving average.

trends was much larger than those recording negative trends (Figure 10(b)). Twenty-nine stations recorded significant increases (54.72% of the total stations), with the Wutaishan station in SR-III recording the strongest increasing trend, and only one station recorded a decrease (Table 2; Figure 10(b)).

WSDI results across the region recorded positive slopes, with the largest being in the central-southern mountain area (Figure 10(c)). Positive trends were significant at 42 stations (79.25%), these being evenly distributed across the region; there were no negative trends recorded for this variable (Figure 10(c), Table 2). The strongest increasing trend was observed at the Wutaishan station in SR-III (Figure 10(c)).

TXx slope distribution results across the BTSSR showed a spatial disparity. Negative slopes were identified in the eastern and southern areas of SR-I, in southeastern areas of SR-IV, and in the northern area of SR-V. Positive slopes were present in the other areas of the study region, with the highest positive slopes occurring in the central-southern mountain area (Figure 10(d)). The number of stations recording positive trends was much greater than those recording negative trends; 10 stations recorded significant increases (18.87%), with no significant decreases being recorded at any station (Table 2). The strongest increasing trend was recorded at the Wutaishan station (SR-III; Figure 10(d)).

Differences in results between the minimum nighttime temperature (TNn) and the maximum daytime temperature (TXx) (Figures 7(d) and 7(h), Table 2) indicated that temperatures across the BTSSR showed a general warming trend, a result which is consistent with global increases [102]. The rate of warming of TXx was lower than that of TNn over all of the subregions (Figures 8(d) and 8(h), Table 3), an occurrence which is not conducive to daytime vegetation growth; this warming will have the opposite effect on nighttime vegetation growth by affecting carbon absorption and consumption [103, 104]. Moreover, the increase of maximum daily temperature may inhibit the growth of grass across the BTSSR; an increase of the minimum temperature may promote grassland growth via the compensatory effect of increasing nocturnal respiration [105], as well as reducing desert vegetation productivity by accelerating plant respiration [106], increasing the rate of nutrient metabolism, and shortening the duration of grain filling [107, 108].

4.2.3. Variability Indices. The temporal variation of the diurnal temperature range (DTR) across the BTSSR showed DTR fluctuations to continuously significantly decrease ($p < 0.001$), with a slope of $-0.17°C$/decade (Figures 6(i) and 7(i), Table 2). During the different stages, DTR showed a significant negative trend during S_1 (with a slope of $-0.43°C$/decade; $p < 0.05$), and insignificant decreases during S_2 and S_3 (Figure 7(i), Table 2). DTR trend variations for the different stages for the majority of the five subregions (Figure 8(i), Table 3) showed a significant decreasing trend across the study period, especially in SR-II which had a slope of $-0.23°C$/decade (Figures 8(i) and 11(a), Table 3). All five subregions showed negative trends during S_1; SR-II recorded a more significant trend, having a slope of $-0.60°C$/decade. During S_2, insignificant decreasing trends were recorded for all subregions, and during the last decade (S_3) the five

subregions showed negative trends (Figure 8(i), Table 3). Higher DTR is propitious to the accumulation of biomass [93, 109], therefore, in the BTSSR, as DTR decreased the quality of vegetation may decline.

The slope distribution of DTR shows a spatial disparity in the BTSSR (Figure 11(a)). A positive slope was recorded in the north, east, and southwestern areas of the region, and a negative slope prevailed in the other areas where higher negative slopes occurred along the northeast-northwest border and the central-southern mountain area. Based on data from 35 stations (66.04% of all stations), the spatial distribution of DTR showed a significant decrease across the BTSSR ($p < 0.05$) (Table 2), corresponding to an average slope of $-0.29°C$/decade. Seven stations (13.21% of total stations) had a significant increasing trend at the 95% confidence level. A strong decreasing trend was observed at the Weixian station (SR-III) and a strong increasing trend was observed at the Hequ station (SR-I; Figure 11(a)).

The temporal variation of the growing season length (GSL) across the BTSSR during the study period continuously increased significantly by 3.20 days/decade from 1960 to 2014 ($p < 0.001$) (Figures 6(j) and 7(j), Table 2). Results for the three stages showed that GSL across the BTSSR showed an insignificant positive trend during S_1, S_2, and S_3 (Figure 7(j), Table 2). The variations in GSL trends for the different stages in the five subregions showed significant increasing trends for the majority of the subregions during the study period, especially for GSL results in SR-II (with a slope of 3.42 days/decade; Table 3, Figures 8(j) and 11(b)). As the length of the growing season increased, the potential number of harvests and seasonal yields for perennial forage crops may be promoted at high latitudes [110]. All five subregions showed a mixture of negative and positive trends, with more positive trends recorded during S_1. During S_2, insignificant increasing trends occurred for all subregions and, during the last decade (S_3), the five subregions recorded a mixture of negative and positive trends, with more positive than negative trends recorded (Figure 8(j), Table 3).

The slope distribution of GSL shows a spatial disparity across the BTSSR (Figure 11(b)). Negative slopes were recorded in the eastern areas of SR-III and positive slopes were recorded for all of the other areas in the region; the highest positive slopes occurred in the central-southern mountain area. The number of stations recording positive trends was much greater than those recording negative trends; 42 stations recorded significant increases (79.25%) and no station recorded a significant decrease (Table 2). A strong increasing trend was observed at the Wutaishan station (SR-III) and a negative trend was only found at the Chengde station (SR-III; Figure 11(b)).

5. Conclusions

Based on daily temperature data from 53 meteorological stations across the Beijing-Tianjin Sand Source Region (BTSSR), spanning 1960–2014, annual and seasonal temperature temporal and spatial trends were analyzed, as well as extreme temperature, using the Mann-Kendall test, Sen's slope estimator, and linear regression.

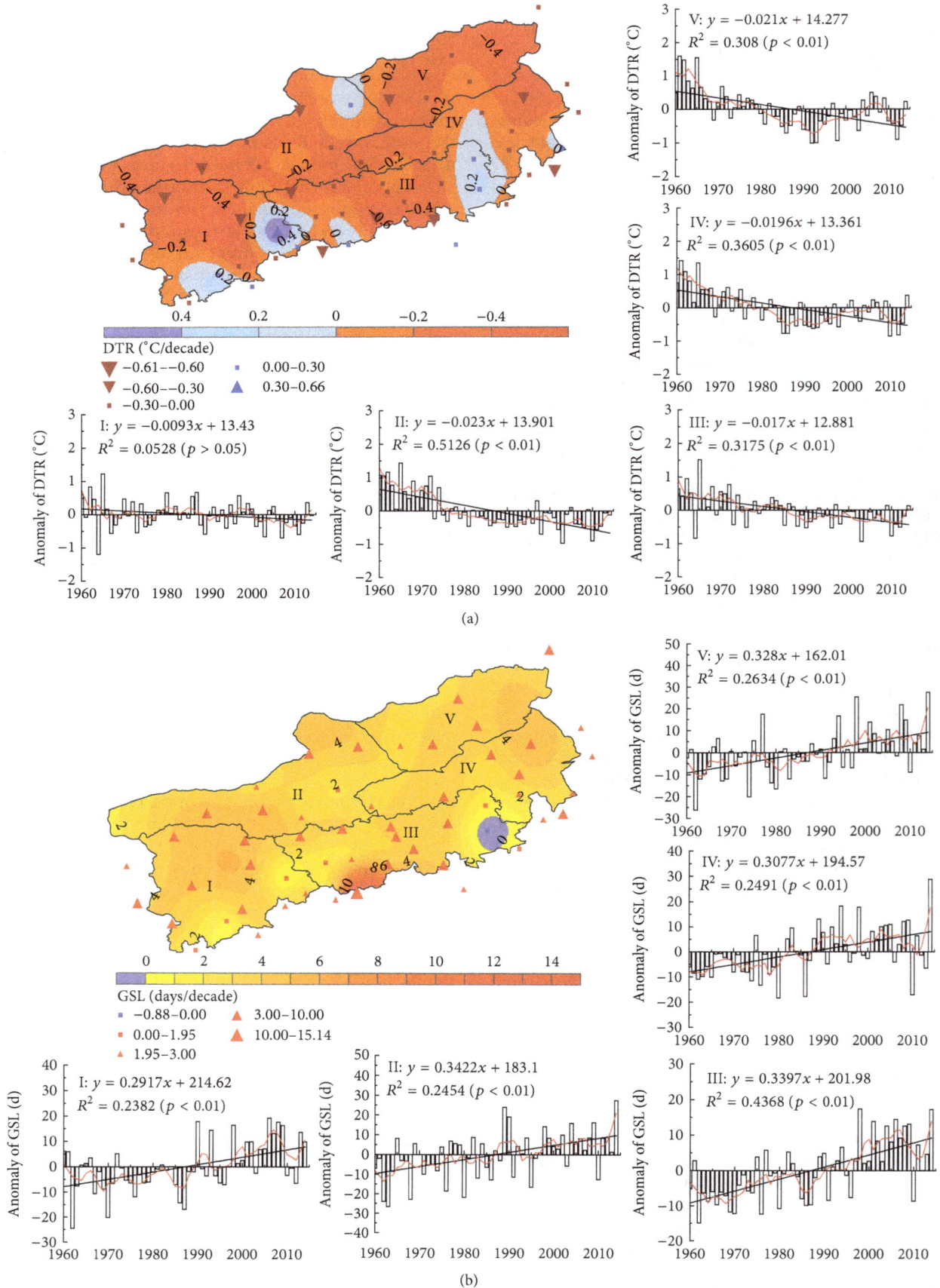

FIGURE 11: Spatial patterns of the trend slopes of variability indices over the BTSSR and anomaly of subregional indices curves, 1960–2014. The *black line* represents the linear trends for (a) DTR, (b) GSL. The *redline* represents 5-year moving average.

Results from 1960 to 2014 showed that a long-term warming trend is evident in this region which is affecting weather conditions. Across the whole area, annual temperature recorded a significant increase of 0.34°C/decade, with the trend test for annual temperature showing that the majority of meteorological stations (96.23% of all stations) exhibited an increasing trend at the 95% confidence level. The highest upward trends were recorded during the winter for the five subregions, with winter temperature in SR-III exhibiting the most significant increasing trend (with a value of 0.50°C/decade).

Significant upward trends were also recorded in warm-related extreme temperature indices (TN90p, TX90p, and WSDI), a cold extreme index (TNn) and growing season length (GSL) during the study period, while significant downward trends were identified in cold-related extreme temperature indices (TN10p, TX10p, and CSDI) and for the diurnal temperature range (DTR). In addition, cold nights (TN10p) in SR-III and warm nights (TN90p) in SR-II exhibited higher warming trends with values of −2.30 and 2.07 days/decade, respectively. A higher positive slope for extreme temperature indices was recorded in the west, south-central, and northeastern regions; higher negative slopes occurred in the northwest, south-central, and eastern regions. A higher warming slope was also recorded in extreme temperature along the northeast border, northwest border, and in the central-south mountain part. The trend test in extreme temperature showed that most stations (>79.25% of all stations) exhibited a warming trend.

A warming trend for temperature and extremes can result in accelerated plant growth, alleviate plant damage due to freezing, and increase the possibility of using low cold-tolerant plants to revegetate in the BTSSR. As the length of the growing season increases, the potential number of harvests and seasonal yields for perennial forage crops may also be promoted at high latitudes. However, in water-limited areas of the BTSSR, increasing temperatures during the growing season may exacerbate rain-fed plant stress due to higher temperatures and a decrease in precipitation, especially vegetation (shrubs or trees) used for restoration of degraded areas. The increase of the minimum temperature may also reduce productivity of desert vegetation by accelerating plant respiration across the BTSSR.

The accurate estimation of changes of temperature and extremes across the BTSSR may be useful for designing and operating appropriate irrigation systems, as well as revegetation strategies, to protect vulnerable arid and semiarid ecosystems against degradation. These findings may be very important for the implementation of future phases of the regeneration project in this area.

Conflicts of Interest

The authors declare that they have no conflicts of interest.

Acknowledgments

This research was supported by the International Science & Technology Cooperation Program of China (2015DFR31130), the National Key Research and Development Program of China (2016YFC0500801; 2016YFC0500804; 2016YFC0500908), the Fundamental Research Funds of CAF (CAFYBB2017ZA006), and the National Natural Science Foundation of China (31670715; 41471029; 41271033; 41371500; 31200350; 41701249).

References

[1] IPCC, *Climate Change 2013—The Physical Science Basis. Contribution of Working Group I to the Fifth Assessment Report of the Intergovernmental Panel on Climate Change*, Cambridge University Press, Cambridge, UK, 2013.

[2] S. J. Asl, A. M. Khorshiddoust, Y. Dinpashoh, and F. Sarafrouzeh, "Frequency analysis of climate extreme events in Zanjan, Iran," *Stochastic Environmental Research and Risk Assessment*, vol. 27, no. 7, pp. 1637–1650, 2013.

[3] J. Caesar, L. Alexander, and R. Vose, "Large-scale changes in observe daily maximum and minimum temperatures: creation and analysis of a new gridded data set," *Journal of Geophysical Research: Atmospheres*, vol. 111, no. 5, Article ID D05101, 2006.

[4] J. A. Marengo, M. Rusticucci, O. Penalba, and M. Renom, "An intercomparison of observed and simulated extreme rainfall and temperature events during the last half of the twentieth century: Part 2: Historical trends," *Climatic Change*, vol. 98, no. 3, pp. 509–529, 2010.

[5] M. R. Haylock, T. C. Peterson, L. M. Alves et al., "Trends in total and extreme South American rainfall in 1960–2000 and links with sea surface temperature," *Journal of Climate*, vol. 19, no. 8, pp. 1490–1512, 2006.

[6] M. Rusticucci and B. Tencer, "Observed changes in return values of annual temperature extremes over Argentina," *Journal of Climate*, vol. 21, no. 21, pp. 5455–5467, 2008.

[7] L. A. Vincent, T. C. Peterson, V. R. Barros et al., "Observed trends in indices of daily temperature extremes in South America, 1960–2002," *Journal of Climate*, vol. 18, no. 23, pp. 5011–5023, 2005.

[8] S. Rehman and L. M. Al-Hadhrami, "Extreme temperature trends on the West Coast of Saudi Arabia," *Atmospheric and Climate Sciences*, vol. 2, no. 3, pp. 351–361, 2012.

[9] K. Khomsi, G. Mahe, Y. Tramblay, M. Sinan, and M. Snoussi, "Regional impacts of global change: Seasonal trends in extreme rainfall, run-off and temperature in two contrasting regions of Morocco," *Natural Hazards and Earth System Sciences*, vol. 16, no. 5, pp. 1079–1090, 2016.

[10] M. Piccarreta, M. Lazzari, and A. Pasini, "Trends in daily temperature extremes over the Basilicata region (southern Italy) from 1951 to 2010 in a Mediterranean climatic context," *International Journal of Climatology*, vol. 35, no. 8, pp. 1964–1975, 2015.

[11] S. N. Marigi, A. K. Njogu, and W. N. Githungo, "Trends of extreme temperature and rainfall indices for arid and semi-arid lands of South Eastern Kenya," *Journal of Geoscience and Environment Protection*, vol. 04, no. 12, pp. 158–171, 2016.

[12] A. Grundstein and J. Dowd, "Trends in extreme apparent temperatures over the United States, 1949-2010," *Journal of Applied Meteorology and Climatology*, vol. 50, no. 8, pp. 1650–1653, 2011.

[13] J. Tian, J. Liu, J. Wang, C. Li, H. Nie, and F. Yu, "Trend analysis of temperature and precipitation extremes in major grain producing area of China," *International Journal of Climatology*, vol. 37, no. 2, pp. 672–687, 2017.

[14] K. Zhong, F. Zheng, H. Wu, C. Qin, and X. Xu, "Dynamic changes in temperature extremes and their association with atmospheric circulation patterns in the Songhua River Basin, China," *Atmospheric Research*, vol. 190, pp. 77–88, 2017.

[15] W. Sun, X. Mu, X. Song, D. Wu, A. Cheng, and B. Qiu, "Changes in extreme temperature and precipitation events in the Loess Plateau (China) during 1960-2013 under global warming," *Atmospheric Research*, vol. 168, pp. 33–48, 2016.

[16] N. Brooks, M. Legrand, S. J. McLaren et al., *Linking Climate Change to Land Surface Change*, Springer, Netherlands, 2000.

[17] S. Fall, D. Niyogi, A. Gluhovsky, R. A. Pielke, E. Kalnay, and G. Rochon, "Impacts of land use land cover on temperature trends over the continental United States: assessment using the North American Regional Reanalysis," *International Journal of Climatology*, vol. 30, no. 13, pp. 1980–1993, 2010.

[18] T. Jie and L. Nianfeng, "Some problems of ecological environmental geology in arid and semiarid areas of China," *Environmental Geology*, vol. 26, no. 1, pp. 64–67, 1995.

[19] J. G. Han, Y. J. Zhang, C. J. Wang et al., "Rangeland degradation and restoration management in China," *The Rangeland Journal*, vol. 30, no. 2, pp. 233–239, 2008.

[20] J. Liu and J. Diamond, "China's environment in a globalizing world," *Nature*, vol. 435, no. 7046, pp. 1179–1186, 2005.

[21] J. Wu, L. Zhao, Y. Zheng, and A. Lü, "Regional differences in the relationship between climatic factors, vegetation, land surface conditions, and dust weather in China's Beijing-Tianjin Sand Source Region," *Natural Hazards*, vol. 62, no. 1, pp. 31–44, 2012.

[22] Z. T. Wu, J. J. Wu, J. H. Liu, B. He, T. Lei, and Q. F. Wang, "Increasing terrestrial vegetation activity of ecological restoration program in the Beijing-Tianjin Sand Source Region of China," *Ecological Engineering*, vol. 52, pp. 37–50, 2013.

[23] X. Li, H. Wang, S. Zhou, B. Sun, and Z. Gao, "Did ecological engineering projects have a significant effect on large-scale vegetation restoration in Beijing-Tianjin Sand Source Region, China? A remote sensing approach," *Chinese Geographical Science*, vol. 26, no. 2, pp. 216–228, 2016.

[24] B. Qu, W. Zhu, S. Jia, and A. Lv, "Spatio-temporal changes in vegetation activity and its driving factors during the growing season in China from 1982 to 2011," *Remote Sensing*, vol. 7, no. 10, pp. 13729–13752, 2015.

[25] J. Liu, J. Wu, Z. Wu, and M. Liu, "Response of NDVI dynamics to precipitation in the Beijing-Tianjin sandstorm source region," *International Journal of Remote Sensing*, vol. 34, no. 15, pp. 5331–5350, 2013.

[26] N. Shan, Z. Shi, X. Yang, J. Gao, and D. Cai, "Spatiotemporal trends of reference evapotranspiration and its driving factors in the Beijing-Tianjin Sand Source Control Project Region, China," *Agricultural and Forest Meteorology*, vol. 200, pp. 322–333, 2015.

[27] Q. Lu and B. Wu, *The Beijing-Tianjin Sand Source Region Phase II Project Team, 'Planning Research about The Beijing-Tianjin Sand Source Region Phase II Project'*, China Forestry press, Beijing, China, 2013.

[28] P. Frich, L. V. Alexander, P. Della-Marta et al., "Observed coherent changes in climatic extremes during the second half of the twentieth century," *Climate Research*, vol. 19, no. 3, pp. 193–212, 2002.

[29] E. L. J. Booth, J. M. Byrne, and D. L. Johnson, "Climatic changes in western North America, 1950-2005," *International Journal of Climatology*, vol. 32, no. 15, pp. 2283–2300, 2012.

[30] X. B. Zhang and F. Yang, *RClimDex (1.0) User Manual*, Climate Research Branch of Meteorological Service of Canada, 2004.

[31] S. Wang, M. Zhang, B. Wang, M. Sun, and X. Li, "Recent changes in daily extremes of temperature and precipitation over the western Tibetan Plateau, 1973-2011," *Quaternary International*, vol. 313-314, pp. 110–117, 2013.

[32] X. L. Wang and Y. Feng, *RHtestsV4 User Manual*, Climate Research Division, Atmospheric Science and Technology Directorate, Science and Technology Branch, Environment Canada, 2013.

[33] M. J. Manton, P. M. Della-Marta, M. R. Haylock et al., "Trends in extreme daily rainfall and temperature in southeast Asia and the south Pacific: 1961–1998," *International Journal of Climatology*, vol. 21, no. 3, pp. 269–284, 2001.

[34] T. C. Peterson, M. A. Taylor, R. Demeritte et al., "Recent changes in climate extremes in the caribbean region," *Journal of Geophysical Research: Atmospheres*, vol. 107, no. D21, pp. ACL 16-1–ACL 16-9, 1984.

[35] Z. Yan, P. D. Jones, T. D. Davies et al., "Trends of extreme temperatures in Europe and China based on daily observations," *Climatic Change*, vol. 53, no. 1, pp. 355–392, 2002.

[36] E. Aguilar, T. C. Peterson, P. R. Obando et al., "Changes in precipitation and temperature extremes in Central America and northern South America, 1961–2003," *Journal of Geophysical Research: Atmospheres*, vol. 110, no. 23, Article ID D23107, pp. 1–15, 2005.

[37] X. Zhang, E. Aguilar, S. Sensoy et al., "Trends in Middle East climate extreme indices from 1950 to 2003," *Journal of Geophysical Research: Atmospheres*, vol. 110, no. 22, Article ID D22104, 2005.

[38] A. M. G. Klein Tank, T. C. Peterson, D. A. Quadir et al., "Changes in daily temperature and precipitation extremes in central and south Asia," *Journal of Geophysical Research: Atmospheres*, vol. 111, no. 16, Article ID D16105, 2006.

[39] M. New, B. Hewitson, D. B. Stephenson et al., "Evidence of trends in daily climate extremes over southern and west Africa," *Journal of Geophysical Research: Atmospheres*, vol. 111, no. 14, 2006.

[40] L. A. Vincent, E. Aguilar, M. Saindou et al., "Observed trends in indices of daily and extreme temperature and precipitation for the countries of the western Indian Ocean, 1961–2008," *Journal of Geophysical Research: Atmospheres*, vol. 116, no. 10, Article ID D10108, 2011.

[41] H. B. Mann, "Nonparametric tests against trend," *Econometrica*, vol. 13, pp. 245–259, 1945.

[42] M. G. Kendall, *Rank Correlation Methods*, Charles Griffin, London, England, 4th edition, 1975.

[43] Z. Shi, N. Shan, L. Xu et al., "Spatiotemporal variation of temperature, precipitation and wind trends in a desertification prone region of China from 1960 to 2013," *International Journal of Climatology*, vol. 36, no. 13, pp. 4327–4337, 2016.

[44] Z. Shi, L. Xu, X. Yang et al., "Trends in reference evapotranspiration and its attribution over the past 50 years in the Loess Plateau, China: implications for ecological projects and agricultural production," *Stochastic Environmental Research and Risk Assessment*, vol. 31, no. 1, pp. 257–273, 2017.

[45] Z. X. Xu, K. Takeuchi, H. Ishidaira, and J. Y. Li, "Long-term trend analysis for precipitation in Asian Pacific FRIEND river basins," *Hydrological Processes*, vol. 19, no. 18, pp. 3517–3532, 2005.

[46] R. M. Hirsch, J. R. Slack, and R. A. Smith, "Techniques of trend analysis for monthly water quality data." *Water Resources Research*, vol. 18, no. 1, pp. 107–121, 1982.

[47] P. K. Sen, "Estimates of the regression coefficient based on Kendall's tau," *Journal of the American Statistical Association*, vol. 63, pp. 1379–1389, 1968.

[48] T. Salmi, A. Määttä, P. Anttila et al., "Detecting trends of annual values of atmospheric pollutants by the Mann–Kendal test and Sen's slope estimates the excel template application MAKESENS," *Air Quality*, no. 31, 2002.

[49] S. Yue, P. Pilon, B. Phinney, and G. Cavadias, "The influence of autocorrelation on the ability to detect trend in hydrological series," *Hydrological Processes*, vol. 16, no. 9, pp. 1807–1829, 2002.

[50] D. Jhajharia, Y. Dinpashoh, E. Kahya, R. R. Choudhary, and V. P. Singh, "Trends in temperature over Godavari River basin in Southern Peninsular India," *International Journal of Climatology*, vol. 34, no. 5, pp. 1369–1384, 2014.

[51] D. Jhajharia, Y. Dinpashoh, E. Kahya, V. P. Singh, and A. Fakheri-Fard, "Trends in reference evapotranspiration in the humid region of northeast India," *Hydrological Processes*, vol. 26, no. 3, pp. 421–435, 2012.

[52] G. Zhu, Y. He, T. Pu et al., "Spatial distribution and temporal trends in potential evapotranspiration over Hengduan Mountains region from 1960 to 2009," *Journal of Geographical Sciences*, vol. 22, no. 1, pp. 71–85, 2012.

[53] D. A. Jones, W. Wang, and R. Fawcett, "High-quality spatial climate data-sets for Australia," *Australian Meteorological and Oceanographic Journal*, vol. 58, no. 4, pp. 233–248, 2009.

[54] D. T. Price, D. W. McKenney, I. A. Nalder, M. F. Hutchinson, and J. L. Kesteven, "A comparison of two statistical methods for spatial interpolation of Canadian monthly mean climate data," *Agricultural and Forest Meteorology*, vol. 101, no. 2-3, pp. 81–94, 2000.

[55] P. A. Hancock and M. F. Hutchinson, "Spatial interpolation of large climate data sets using bivariate thin plate smoothing splines," *Environmental Modelling & Software*, vol. 21, no. 12, pp. 1684–1694, 2006.

[56] D. Zuo, Z. Xu, H. Yang, and X. Liu, "Spatiotemporal variations and abrupt changes of potential evapotranspiration and its sensitivity to key meteorological variables in the Wei River basin, China," *Hydrological Processes*, vol. 26, no. 8, pp. 1149–1160, 2012.

[57] A. P. Cuervo-Robayo, O. Téllez-Valdés, M. A. Gómez-Albores, C. S. Venegas-Barrera, J. Manjarrez, and E. Martínez-Meyer, "An update of high-resolution monthly climate surfaces for Mexico," *International Journal of Climatology*, vol. 34, no. 7, pp. 2427–2437, 2014.

[58] Z. Ye and Z. Li, "Spatiotemporal Variability and Trends of Extreme Precipitation in the Huaihe River Basin, a Climatic Transitional Zone in East China," *Advances in Meteorology*, vol. 2017, Article ID 3197435, 15 pages, 2017.

[59] H. Yang, D. Yang, Q. Hu, and H. Lv, "Spatial variability of the trends in climatic variables across China during 1961–2010," *Theoretical and Applied Climatology*, vol. 120, no. 3-4, pp. 773–783, 2015.

[60] Q. X. Li, W. J. Dong, W. Li et al., "Assessment of the uncertainties in temperature change in China during the last century," *Chinese Science Bulletin*, vol. 55, no. 19, pp. 1974–1982, 2010.

[61] Q. Tian, M. Prange, and U. Merkel, "Precipitation and temperature changes in the major Chinese river basins during 1957-2013 and links to sea surface temperature," *Journal of Hydrology*, vol. 536, pp. 208–221, 2016.

[62] S. Y. Chen, Y. Y. Shi, Y. Z. Guo, and Y. X. Zheng, "Temporal and spatial variation of annual mean air temperature in arid and semiarid region in northwest China over a recent 46 year period," *Journal of Arid Land*, vol. 2, no. 2, pp. 87–97, 2010.

[63] S. Chen, J. Wang, Y. Shi, and Z. Guo, "The change of annual mean temperature in Monsoon Area of East China," *Resources Science*, vol. 3, 2009.

[64] R. D. Jackson, S. B. Idso, and J. Otterman, "Surface albedo and desertification," *Science*, vol. 189, no. 4207, pp. 1012–1015, 1975.

[65] F. M. Vukovich, D. L. Toll, and R. E. Murphy, "Surface temperature and albedo relationships in Senegal derived from NOAA-7 satellite data," *Remote Sensing of Environment*, vol. 22, no. 3, pp. 413–421, 1987.

[66] R. Bintanja and J. Oerlemans, "The influence of the albedo-temperature feed-back on climate sensitivity," *Annals of Glaciology*, vol. 21, pp. 353–360, 1995.

[67] T. Wang, W. Wu, X. Xue, Z. Han, W. Zhang, and Q. Sun, "Spatial-temporal changes of Sandy desertified land during last 5 decades in Northern China," *Acta Geographica Sinica*, vol. 59, no. 2, pp. 203–212, 2004.

[68] N. Zeng and J. Yoon, "Expansion of the world's deserts due to vegetation-albedo feedback under global warming," *Geophysical Research Letters*, vol. 36, no. 17, Article ID L17401, 2009.

[69] H. Pelgrum, T. Schmugge, A. Rango, J. Ritchie, and B. Kustas, "Length-scale analysis of surface albedo, temperature, and normalized difference vegetation index in desert grassland," *Water Resources Research*, vol. 36, no. 7, pp. 1757–1765, 2000.

[70] Z. Xuezhen, W. W. Chyung, F. Xiuqi, Y. Yu, and Z. Jingyun, "Agriculture development-induced surface albedo changes and climatic implications across northeastern china," *Chinese Geographical Science*, vol. 22, no. 3, pp. 264–277, 2012.

[71] C. Tong, J. Wu, S. Yong, J. Yang, and W. Yong, "A landscape-scale assessment of steppe degradation in the Xilin River Basin, Inner Mongolia, China," *Journal of Arid Environments*, vol. 59, no. 1, pp. 133–149, 2004.

[72] K. Kawada, W. W., and T. Nakamura, "Land degradation of abandoned croplands in the Xilingol steppe region, Inner Mongolia, China," *Grassland Science*, vol. 57, no. 1, pp. 58–64, 2011.

[73] J. Liang, J. Xia, L. Liu, and S. Wan, "Global patterns of the responses of leaf-level photosynthesis and respiration in terrestrial plants to experimental warming," *Journal of Plant Ecology*, vol. 6, no. 6, pp. 437–447, 2013.

[74] Z. Xu, H. Shimizu, S. Ito et al., "Effects of elevated CO2, warming and precipitation change on plant growth, photosynthesis and peroxidation in dominant species from North China grassland.," *Planta*, vol. 239, no. 2, pp. 421–435, 2014.

[75] M. Reyes-Fox, H. Steltzer, M. J. Trlica et al., "Elevated CO2 further lengthens growing season under warming conditions," *Nature*, vol. 510, no. 7504, pp. 259–262, 2014.

[76] J. S. Clark, J. Melillo, J. Mohan, and C. Salk, "The seasonal timing of warming that controls onset of the growing season," *GCB Bioenergy*, vol. 20, no. 4, pp. 1136–1145, 2014.

[77] W. Wei, Z. Shi, X. Yang et al., "Recent trends of extreme precipitation and their teleconnection with atmospheric circulation in the Beijing-Tianjin Sand source region, China, 1960-2014," *Atmosphere*, vol. 8, no. 5, article no. 83, 2017.

[78] Z. Chen and G. Yang, "Analysis of drought hazards in North China: distribution and interpretation," *Natural Hazards*, vol. 65, no. 1, pp. 279–294, 2013.

[79] Y. Wang, B. Zhou, D. Qin, J. Wu, R. Gao, and L. Song, "Changes in mean and extreme temperature and precipitation over the arid region of northwestern China: Observation and

projection," *Advances in Atmospheric Sciences*, vol. 34, no. 3, pp. 289–305, 2017.

[80] W. He, R. Bu, Y. Hu, Z. Xiong, and M. Liu, "Analysis of temporal and spatial characteristic of temperature change over the last 45 years in Northeastern China," *Advanced Materials Research*, vol. 518-523, pp. 1367–1370, 2012.

[81] J. Otterman, R. Atlas, S.-H. Chou et al., "Are stronger North-Atlantic southwesterlies the forcing to the late-winter warming in Europe?" *International Journal of Climatology*, vol. 22, no. 6, pp. 743–750, 2002.

[82] G. J. Zhang, M. Cai, and A. Hu, "Energy consumption and the unexplained winter warming over northern Asia and North America," *Nature Climate Change*, vol. 3, no. 5, pp. 466–470, 2013.

[83] S. Bokhorst, J. W. Bjerke, F. W. Bowles, J. Melillo, T. V. Callaghan, and G. K. Phoenix, "Impacts of extreme winter warming in the sub-Arctic: Growing season responses of dwarf shrub heathland," *GCB Bioenergy*, vol. 14, no. 11, pp. 2603–2612, 2008.

[84] J. Liu, J. Xie, T. Gong, H. Wang, and Y. Xie, "Impacts of winter warming and permafrost degradation on water variability, upper Lhasa River, Tibet," *Quaternary International*, vol. 244, no. 2, pp. 178–184, 2011.

[85] Z. Jiang, T. Ma, and Z. Wu, "China coldwave duration in a warming winter: Change of the leading mode," *Theoretical and Applied Climatology*, vol. 110, no. 1-2, pp. 65–75, 2012.

[86] J. M. Wallace, I. M. Held, D. W. J. Thompson, K. E. Trenberth, and J. E. Walsh, "Global warming and winter weather," *Science*, vol. 343, no. 6172, pp. 729–730, 2014.

[87] B. Fan, L. Guo, N. Li et al., "Earlier vegetation green-up has reduced spring dust storms," *Scientific Reports*, vol. 4, article no. 6749, 2014.

[88] W. Xiong, I. P. Holman, L. You, J. Yang, and W. Wu, "Impacts of observed growing-season warming trends since 1980 on crop yields in China," *Regional Environmental Change*, vol. 14, no. 1, pp. 7–16, 2014.

[89] L. You, M. W. Rosegrant, S. Wood, and D. Sun, "Impact of growing season temperature on wheat productivity in China," *Agricultural and Forest Meteorology*, vol. 149, no. 6-7, pp. 1009–1014, 2009.

[90] K. Mix, V. L. Lopes, and W. Rast, "Growing season expansion and related changes in monthly temperature and growing degree days in the Inter-Montane Desert of the San Luis Valley, Colorado," *Climatic Change*, vol. 114, no. 3-4, pp. 723–744, 2012.

[91] J. Larkindale, M. Mishkind, and E. Vierling, "Plant responses to high temperature," in *Plant Abiotic Stress*, pp. 100–144, Blackwell Publishing, Hoboken, NJ, USA, 2005.

[92] G. Søgaard, A. Granhus, and Ø. Johnsen, "Effect of frost nights and day and night temperature during dormancy induction on frost hardiness, tolerance to cold storage and bud burst in seedlings of Norway spruce," *Trees - Structure and Function*, vol. 23, no. 6, pp. 1295–1307, 2009.

[93] D. B. Lobell, "Changes in diurnal temperature range and national cereal yields," *Agricultural and Forest Meteorology*, vol. 145, no. 3-4, pp. 229–238, 2007.

[94] M. Darand, A. Masoodian, H. Nazaripour, and M. R. Mansouri Daneshvar, "Spatial and temporal trend analysis of temperature extremes based on Iranian climatic database (1962–2004)," *Arabian Journal of Geosciences*, vol. 8, no. 10, pp. 8469–8480, 2015.

[95] P. Zhai and X. Pan, "Trends in temperature extremes during 1951–1999 in China," *Geophysical Research Letters*, vol. 30, no. 17, 2003.

[96] B. Nandintsetseg, J. S. Greene, and C. E. Goulden, "Trends in extreme daily precipitation and temperature near Lake Hövsgöl, Mongolia," *International Journal of Climatology*, vol. 27, no. 3, pp. 341–347, 2007.

[97] A. C. Kruger and S. S. Sekele, "Trends in extreme temperature indices in South Africa: 1962-2009," *International Journal of Climatology*, vol. 33, no. 3, pp. 661–676, 2013.

[98] L. Wang, Z. Wu, F. Wang, H. Du, and S. Zong, "Comparative analysis of the extreme temperature event change over Northeast China and Hokkaido, Japan from 1951 to 2011," *Theoretical and Applied Climatology*, vol. 124, no. 1-2, pp. 375–384, 2016.

[99] T. R. Karl, N. Nicholls, and A. Ghazi, "Clivar/GCOS/WMO workshop on indices and indicators for climate extremes workshop summary," *Climatic Change*, vol. 42, no. 1, pp. 3–7, 1999.

[100] D. Huang, Y. Qian, and J. Zhu, "Trends of temperature extremes in China and their relationship with global temperature anomalies," *Advances in Atmospheric Sciences*, vol. 27, no. 4, pp. 937–946, 2010.

[101] X. Wu, Z. Wang, X. Zhou, C. Lai, and X. Chen, "Trends in temperature extremes over nine integrated agricultural regions in China, 1961–2011," *Theoretical and Applied Climatology*, vol. 129, no. 3-4, pp. 1279–1294, 2017.

[102] S. Solomon, D. Qin, M. Manning et al., *Climate Change 2007: The Physical Science Basis. Contribution of Working Group I to The Fourth Assessment Report of The Intergovernmental Panel on Climate Change*, Cambridge University Press, Cambridge, UK, 2007.

[103] O. K. Atkin, M. H. Turnbull, J. Zaragoza-Castells et al., "Light inhibition of leaf respiration as soil fertility declines along a post-glacial chronosequence in New Zealand: An analysis using the Kok method," *Plant and Soil*, vol. 367, no. 1-2, pp. 163–182, 2013.

[104] S. Peng, S. Piao, P. Ciais et al., "Asymmetric effects of daytime and night-time warming on Northern Hemisphere vegetation," *Nature*, vol. 501, no. 7465, pp. 88–92, 2013.

[105] S. Q. Wan, J. Y. Xia, W. X. Liu, and S. Niu, "Photosynthetic overcompensation under nocturnal warming enhances grassland carbon sequestration," *Ecology*, vol. 90, no. 10, pp. 2700–2710, 2009.

[106] C. L. Liang, Q. Z. Yu, Y. J. Liu et al., "Effects of air temperature circadian on the NDVI of Nansi Lake Wetland Vegetation," *Tropical Geography*, vol. 35, no. 3, pp. 422–426, 2015 (Chinese).

[107] J. R. Welch, J. R. Vincent, M. Auffhammer, P. F. Moya, A. Dobermann, and D. Dawe, "Rice yields in tropical/subtropical Asia exhibit large but opposing sensitivities to minimum and maximum temperatures," *Proceedings of the National Acadamy of Sciences of the United States of America*, vol. 107, no. 33, pp. 14562–14567, 2010.

[108] J. Tan, S. Piao, A. Chen et al., "Seasonally different response of photosynthetic activity to daytime and night-time warming in the Northern Hemisphere," *GCB Bioenergy*, vol. 21, no. 1, pp. 377–387, 2015.

[109] H. Yu, J. Hammond, S. Ling, S. Zhou, P. E. Mortimer, and J. Xu, "Greater diurnal temperature difference, an overlooked but important climatic driver of rubber yield," *Industrial Crops and Products*, vol. 62, pp. 14–21, 2014.

[110] J. Berner, T. V. Callaghan, H. Huntington et al., *Impacts of a Warming Arctic: Arctic Climate Impact Assessment*, Cambridge University Press, Cambridge, UK, 2004.

5

Impacts of Water Consumption in the Haihe Plain on the Climate of the Taihang Mountains, North China

Jing Zou [ID],[1] Chesheng Zhan [ID],[2] Ruxin Zhao,[3] Peihua Qin [ID],[4] Tong Hu,[1] and Feiyu Wang[5]

[1]Institute of Oceanographic Instrumentation, Qilu University of Technology (Shandong Academy of Sciences), Qingdao 266001, China
[2]Institute of Geographic Sciences and Natural Resources Research, Chinese Academy of Sciences, Beijing 100101, China
[3]Beijing Normal University, Beijing 100875, China
[4]Institute of Atmospheric Physics, Chinese Academy of Sciences, Beijing 100029, China
[5]School of Geography and Tourism, Shaanxi Normal University, Xi'an 710119, China

Correspondence should be addressed to Chesheng Zhan; zhancs@igsnrr.ac.cn

Academic Editor: Theodore Karacostas

In this study, the RegCM4 regional climate model was employed to investigate the impacts of water consumption in the Haihe Plain on the local climate in the nearby Taihang Mountains. Four simulation tests of twelve years' duration were conducted with various schemes of water consumption by residents, industries, and agriculture. The results indicate that water exploitation and consumption in the Haihe Plain causes wetting and cooling of the local land surface and rapid increases in the depth of the groundwater table. These wetting and cooling effects increase atmospheric moisture, which is transported to surrounding areas, including the Taihang Mountains to the west. In a simulation where water consumption in the Haihe Plain was doubled, the wetting and cooling effects in the Taihang Mountains were enhanced but at less than double the amount, because a cooler land surface does not enhance atmospheric convective activities. The impacts of water consumption activities in the Haihe Plain were more obvious during the irrigation seasons (primarily spring and summer). In addition, the land surface variables in the Taihang Mountains, e.g., sensible and latent heat fluxes, were less sensitive to the climatic impacts due to the water consumption activities in the Haihe Plain because they were strongly affected by local surface energy balance.

1. Introduction

The Taihang Mountains, which extend over 400 km from north to south in North China, form a geographic boundary between the Loess Plateau (to the west) and the North China Plain (to the east). Mountains and intermontane valleys lie to the west and north, and plains lie to the east and south (Figure 1). A number of reservoirs in the Taihang Mountains are important water sources for North China. However, due to long-term human exploitation and climatic changes, the Taihang Mountains suffer from soil erosion and vegetation degradation [1, 2].

During the past fifty years, significant drying and warming trends have been detected in North China [3, 4]. The mean precipitation in the Taihang Mountains was found to decrease by 10.2 mm per decade, which is slightly less than the trend observed on the piedmont plain of these mountains, of 12.5 mm per decade [5].

Numerous studies have investigated climatic changes in North China and have mainly attributed these changes to human activity [6–9]. For example, Jia et al. attributed the observed changes in water resources from 1961 to 2000 to various factors and demonstrated that local human activity accounted for about 60% of the observed changes [10].

For the Taihang Mountains, the most intense human activities occur in its eastern plains area, Haihe Plain, which is in the northern part of the North China Plain. The Haihe Plain is an important agricultural area in China and supports rapidly developing industries. According to the *Urban*

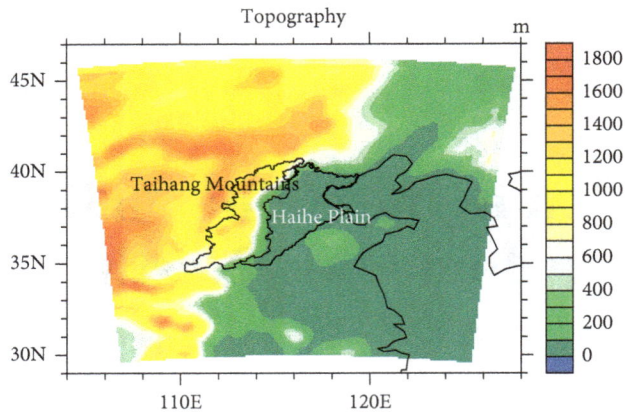

Figure 1: Topography of the simulation domain. The Taihang Mountains and Haihe Plain are outlined in black.

Statistical Yearbook of China 2000, the population density and gross domestic product (GDP) in 2000 in the Taihang Mountains were 322 people per km and 239×10^4 yuan·km^{-2}, respectively. This compares with 687 people per km and 530×10^4 yuan·km^{-2} in the Haihe Plain [11]. The dense population and industry in the Haihe Plain have induced a serious imbalance between water demands and surface water supplies [12–14]. Large amounts of groundwater have been exploited for crop cultivation and industrial development, leading to a rapidly dropping groundwater table and a series of eco-environmental crises [15–17].

Water consumption activities have been shown to enhance evapotranspiration and reduce local temperature, and moreover, increasing local water vapor from land may lead to more convection and further changes in atmospheric circulation [18–22]. Numerous studies have investigated the regional impacts of agricultural irrigation, which constitutes a major proportion of water consumption. For example, Saeed et al. applied the Max Planck Institute's Regional Model (REMO) and found that the irrigation over India caused increased evapotranspiration and less westerlies entering into land from the Arabian Sea [23]. DeAngelis et al. analyzed the precipitation observations in the Great Plains of the United States and found the increased evapotranspiration due to irrigation contributed to more downwind precipitation [24].

For the Haihe Plain, Chen and Xie investigated the climatic effects of large-scale irrigation due to interbasin water transfer and demonstrated that such irrigation causes increased local precipitation and decreased temperature at the land surface [25]. Leng et al. revealed that the changes in topsoil moisture and subsurface water flux induced by groundwater irrigation exceeded the potential change that could be attributed to various climate projections for the Haihe Plain [26]. These studies have made valuable progress in investigating the relationship between local climate changes and water consumption activities. However, the extent of the effect of water consumption within the Haihe Plain on its western water source area, the Taihang Mountains, remains unclear. Further discussion should address the way in which high-level water consumption in

the eastern Haihe Plain affects the climate of the western Taihang Mountains.

Therefore, in this study, a series of water exploitation and consumption simulations were conducted by the regional climate model RegCM4, using data from 1996 to 2007. Through the comparison of these simulations, we investigated the effect of water consumption processes in the Haihe Plain on the spatiotemporal variability of the local climate of the Taihang Mountains.

2. Methodology

2.1. Model Description. The regional climate model used in this study, RegCM4, was developed by the International Center for Theoretical Physics (ICTP) in Italy [27]. RegCM4 is a hydrostatic climate model with a sigma vertical coordinate. Three convective precipitation schemes, Kuo, Grell and Emanuel, and a large-scale precipitation scheme, SUBEX, were available as simulation options. The land surface models BATS (biosphere-atmosphere transfer scheme) and CLM (community land model) were available for use as land modules in RegCM4.

In this study, version 3.5 of CLM was used. The CLM model divides grid cells into multiple land units (glacier, wetland, vegetation, etc.). Vegetation units can be further divided into 17 plant function types (PFTs) [28]. The hydrological processes in CLM3.5 were obtained from a runoff parameterization scheme with a simple groundwater model developed by Niu et al. [29, 30].

The RegCM4 model has been implemented around the globe and demonstrated excellent climate simulation abilities. For example, Wang et al. evaluated the monthly precipitation simulation of RegCM4 over the Tibetan Plateau using the station observation and TRMM (Tropical Rainfall Measurement Mission) data and found RegCM4/CLM3.5 performed better than RegCM4/BATS in the statistical indices [31]. Ashfaq et al. used RegCM4 to dynamically downscale the historical and near-term future outputs from 11 global climate models, in order to present high-resolution ensemble projections of climatic changes over the continental United States [32]. Mbienda et al. compared the simulations performed by different convective schemes of RegCM4 over Central Africa and found the Emanuel-MIT convective scheme showed better indices in temperature and the Grell scheme with Arakawa–Schubert closure assumption was better to downscale precipitation and surface wind [33].

2.2. Water Exploitation Scheme Description. The scheme of water exploitation and consumption used in this study was developed based on the CLM3.5 approach used in our previous studies [34, 35]. The scheme relies on preset water demand and is composed of exploitation and consumption components. As shown in Figure 2, the total water demand D_t is supplied by water resources pumped from rivers (Q_s) and wells (Q_g). The total water demand D_t is consumed by three consumption sections—irrigation consumption (D_a), domestic consumption (D_d), and industrial consumption (D_i).

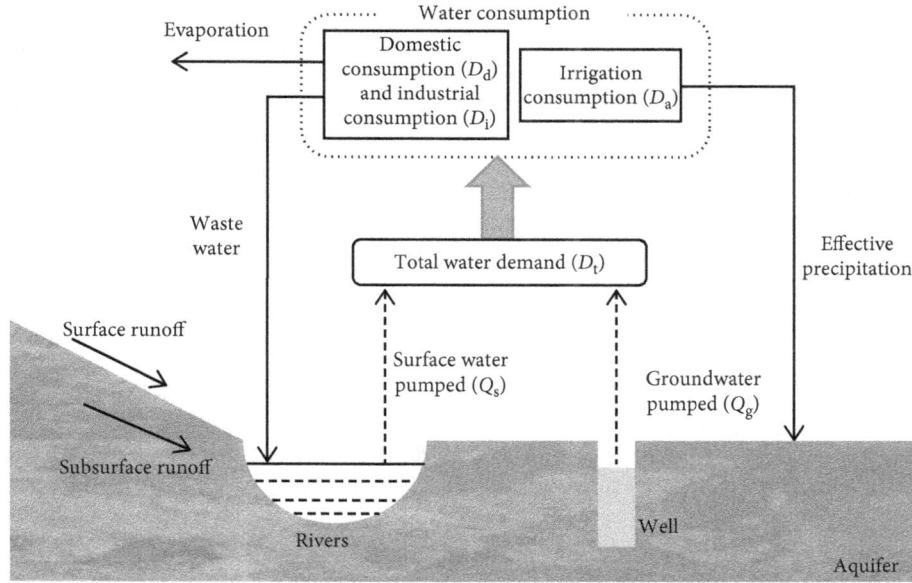

FIGURE 2: Framework of the water exploitation and consumption scheme.

The irrigation water is treated as effective rainfall reaching the topsoil, as per the approach in previous studies [23, 25, 36, 37]. Furthermore, according to the descriptions of urban water use by Shiklomanov and the approach of specific hydrological modeling studies, the water for domestic and industrial consumption is highly simplified in the scheme to partly increase local evaporation and partly return to river channels [36–38].

The scheme is based on the balance of water demand and supply, which can be described as

$$D_t = Q_s + Q_g. \tag{1}$$

The surface water supply Q_s is composed of the total runoff R_t and stream discharge R_{str} in each model grid cell, and it has a higher priority than the groundwater supply Q_g for satisfying demand D_t. When the surface water supply cannot meet the total demand, the groundwater supply Q_g is subtracted from the groundwater storage and can be calculated as

$$Q_g = \max[(D_t - R_t - R_{str}), 0]. \tag{2}$$

In the consumption section, the effective rainfall in the model increases by D_a due to irrigation, which will further participate in the processes of runoff generation and infiltration. The increased evaporation in the model is defined as $\varepsilon(D_d + D_i)$ and the remaining water of domestic and industrial consumption, $(1 - \varepsilon)(D_d + D_i)$, is regarded as wastewater recharge returning to river channels.

In this study, domestic and industrial water consumption occurred year-round, while irrigation consumption only occurred during the main crop growing periods. In North China, the main crops are winter wheat and summer maize. According to their known phenology, irrigation first occurred from March 10 (when wheat turns green) to June 18 (wheat harvest time). Irrigation also occurred from June

22 to August 28 (summer maize planting and harvesting times, respectively) [39–42]. The evaporation rate of domestic and industrial consumption ε was set as 0.26 based on the statistical value of water consumption rate for industrial and urban domestic consumption in the *Water Resource Bulletin of China* for the year 2000 (http://www.mwr.gov.cn/sj/tjgb/szygb/201612/t20161222_776035.html).

2.3. Water Demand Estimation. Due to the lack of spatial distribution data, the water demand in each grid cell was indirectly estimated from relevant eco-social data, as other studies have done [36]. In this study, the total demand D_t was divided into three components—domestic, industrial, and irrigation demand—which were estimated based on data of population, GDP, and irrigated cropland areas in China for the year 2000. The estimation equation can be expressed as

$$\begin{aligned} D_{t,2000} &= D_{d,2000} + D_{i,2000} + D_{a,2000} \\ &= \rho\lambda^{-1}S^{-1}\left(\gamma_1 A_{pop} + \beta\gamma_2 A_{GDP} + \alpha\gamma_3 A_{agr}\right), \end{aligned} \tag{3}$$

where $D_{t,2000}$ is the total water demand in the year 2000 ($kg\cdot m^{-2}\cdot s^{-1}$); ρ is water density ($1 \times 10^3\, kg\cdot m^{-3}$); λ is the number of seconds in a year ($31{,}536{,}000\, s\cdot yr^{-1}$); S is the grid cell area (m^2); γ_1 is per capita annual domestic water consumption ($m^3\cdot person^{-1}\cdot yr^{-1}$); A_{pop} is the population in each grid cell (persons); β is the conversion ratio between GDP and industrial output; A_{GDP} is GDP in each grid cell (yuan); γ_2 is industrial water consumption for unit industrial output ($m^3\cdot yuan^{-1}\cdot yr^{-1}$); α is the ratio between the irrigated and total cropland areas; γ_3 is annual irrigation water consumption per hectare ($m^3\cdot ha^{-1}\cdot yr^{-1}$); and A_{agr} is the cropland area in each grid cell (ha).

Population and GDP data for the year 2000 were obtained from the Data Center for Resources and Environment of Sciences, Chinese Academy of Sciences (http://www.resdc. cn). Cropland area data for 2000 were obtained from the China land cover classification dataset available at the Science Data Center for Cold and Arid Regions (http://westdc.westgis. ac.cn) [43]. The three eco-social datasets had a resolution of 1 km × 1 km and were then interpolated to suit the RegCM4 model's resolution of 20 km × 20 km for the subsequent simulation tests. The parameters of water consumption were collected from the *Water Resource Bulletin of China* for the year 2000, where the provincial mean values were available. The other two parameters, β and α, were collected from the *China Statistical Yearbook 2000* and were also provincial mean values [44].

The spatial distribution of estimated water demand $D_{t,2000}$ per unit area is shown in Figure 3(a), and the demands for the Taihang Mountains (Figure 3(b)) and Haihe Plain (Figure 3(c)) were then cropped from Figure 3(a). As shown in Figure 3(b), more water is consumed in the eastern piedmont areas and south-central valleys of the Taihang Mountains. For the Haihe Plain (Figure 3(c)), and areas near Beijing (116.5°E, 40.0°N) and Tianjin (117.0°E, 39.0°N), there were higher water demands due to intense industrial activity and high population densities. The central areas of the plain are the main agricultural areas, and water demands there are also high.

Table 1 provides a comparison of the estimated and actual total water demands in 2000 for China and the Haihe River Basin. The estimation and actual values are basically consistent with each other, although they do not strictly coincide due to the use of different data sources. For the whole of China, the estimated irrigation demand provides most of the error, because of inconsistencies in cropland area between the remote sensing data and *China Statistical Yearbook* data used in this study. In addition, a comparison for the Haihe River Basin is also presented, because the whole Haihe Plain and most areas of the Taihang Mountains lie in the basin, and few data for the Taihang Mountains are available in water resource bulletins.

To obtain the demands during the simulation period (1996–2007), the mean annual water use for the Haihe River Basin was derived from the *Water Resource Bulletin of Haihe River Basin* (http://www.hwcc.gov.cn/hwcc/wwgj/xxgb/). Using these statistics, the estimated water demands in 2000 for the Taihang Mountains and Haihe Plain were scaled up or down, keeping the three water-consuming sectors unchanged, to approximate interannual variations. The estimated annual demands from 1996 to 2007 are shown in Table 2.

2.4. Experimental Design. The simulation domain of RegCM4 is shown in Figure 1, where the domains of Taihang Mountains and Haihe Plain are outlined by black lines. The central projection of RegCM4 was located at 116°E, 38°N, and it used a spatial resolution of 20 × 20 km. The Grell scheme was used as the convective precipitation scheme, and ERA-Interim reanalysis data from 1992 to 2007 were used as the lateral boundary forcing. The time steps were 30 seconds for the atmosphere module and 30 minutes for the land

surface module of CLM3.5. Before the simulations were conducted, a test simulation using the original RegCM4 with ERA40 reanalysis data from 1961 to 1991 was conducted to obtain the balanced groundwater depth. A spin-up simulation from 1992 to 1995 was then conducted using the original RegCM4 without exploitation. Based on the final status of the spin-up simulation, four simulation tests were conducted from 1996 to 2007. A control test (CTL) continued to simulate the natural state without exploitation; Exploitation Test 1 (T1) used estimated water demand data for the Haihe Plain, and Exploitation Test 2 (T2) used double this demand. In addition, Exploitation Test 3 (T3) used combined demand data of the Taihang Mountains and Haihe Plain. By comparing differences between the exploitation and control tests, this study aimed to investigate the effects of water consumption activities in the Haihe Plain on local climate changes in the Taihang Mountains.

3. Result

3.1. Validation of the Control Test. Station observed precipitation, and 2 m air temperature data during the simulation period was derived from the China National Meteorological Information Center (http://data.cma.cn/en) and then was interpolated as gridded data with a resolution of 0.5° × 0.5°, by using the inverse distance square weighting method as preformed in previous studies [45, 46]. The spatial distributions of annual mean precipitation and 2 m air temperature from 1996 to 2007 are shown in Figure 4. The CTL test run in RegCM4 simulated the regional climatology over the study domain well. Specifically, the simulated precipitation and temperature increased from northwest to southeast, which is consistent with observations. However, slight low biases were detected in the Taihang Mountains, of −25.1 mm/year for precipitation and −0.64 K for 2 m air temperature. Also, a dry and warm bias, of −115.3 mm/year for precipitation and 0.88 K for 2 m air temperature, was detected in the Haihe Plain. Additionally, the statistical indices of monthly series in the Taihang Mountains and Haihe Plain are listed in Table 3, including the temporal correlation coefficient, root-mean-square error, and standard deviation. These statistics indicate that the simulated series corresponds with observations; hence, the RegCM4 model is suitable for simulating temporal variations in the local climate.

3.2. Spatial Distribution of Climatic Differences. Figure 5 shows the spatial distribution of differences between the three exploitation tests and control test for groundwater depth, 10 cm depth soil moisture, and precipitation. Regarding the groundwater withdrawn continuously in the exploitation tests, Figures 5(a)–5(c) show the differences in groundwater depth in December 2007 (final status) and January 1996 (initial status). The other two climatic variables in Figures 5(d)–5(i) use the 12-year mean differences from 1996 to 2007.

As shown in Figure 5(a), the 12-year exploitation process in the Haihe Plain led to a significant drop of local

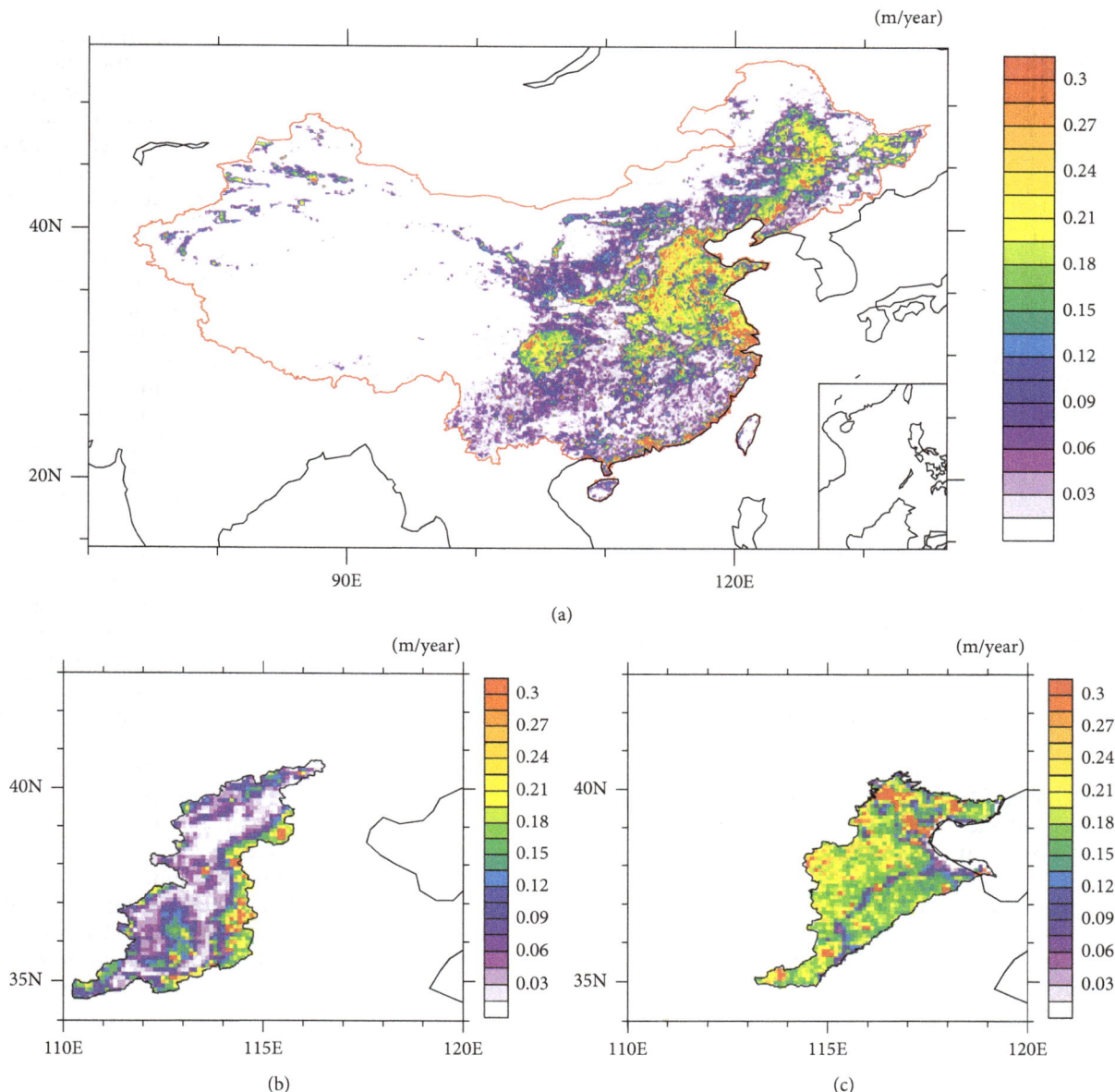

FIGURE 3: Spatial distributions of estimated water demand per unit area in 2000 over (a) all of China, (b) the Taihang Mountains, and (c) Haihe Plain.

TABLE 1: Estimated water demands and actual statistical values in 2000 (unit: $10^8 \, \mathrm{m}^3/\mathrm{year}$).

Domain	Data source	Domestic	Industrial	Irrigation	Total
China	Actual	574.9	1139.1	3783.6	5497.6
	Estimated	611.1	1035.7	3366.6	5013.4
Haihe river basin	Actual	51.8	65.7	280.9	398.4
	Estimated	50.8	80.6	247.7	379.1

groundwater table, of a mean of 13.86 m. The groundwater table dropped in the areas near Beijing (116.5°E, 40.0°N), Tianjin (117.0°E, 39.0°N), and Shijiazhuang (114.5°E, 38.0°N), corresponding with their high water demands. The

differences in the areas outside of the Haihe Plain were almost negligible. The groundwater depth only rose by a mean of 0.06 m in the Taihang Mountains, indicating that the local processes of water exploitation and consumption in the Haihe Plain do not cause obvious changes to groundwater resources in the Taihang Mountains. For the T2 test (Figure 5(b)), when the water demands in the Haihe Plain were assumed to be double those of the T1 test, the groundwater depth dropped by more than 20 m in most areas of the plain, with a mean of 25.49 m. Also, the groundwater table in the Taihang Mountains rose by 0.10 m due to slightly increased precipitation in the Haihe Plain. For the T3 test (Figure 5(c)), which considers the total demands in the Taihang Mountains and Haihe Plain, the mean

TABLE 2: Actual statistical water use over the Haihe River Basin and estimated water demands over the Taihang Mountains and Haihe Plain from 1996 to 2007 (unit: $10^8 \, \text{m}^3$/year).

Year	Haihe river basin	Taihang mountains	Haihe Plain
1996	413.1	140.1	244.1
1997	432.8	146.8	255.8
1998	432.0	146.5	255.3
1999	431.5	146.4	255.0
2000	398.4	135.1	235.5
2001	392.0	133.0	231.7
2002	399.83	135.6	236.3
2003	377.0	127.9	222.8
2004	368.01	124.8	217.5
2005	379.1	128.6	224.1
2006	391.0	132.6	231.1
2007	384.48	130.4	227.2

groundwater depth difference in the Haihe Plain was 13.82 m, which is approximately the same as the difference observed in the T1 test. Due to local groundwater exploitation, the groundwater table in the Taihang Mountains dropped by 3.78 m on average. The greatest drops appeared in the eastern piedmont areas and the south-central valleys, corresponding with the higher water demand there, as shown in Figure 3(b).

The processes of water exploitation and consumption withdraw groundwater and consume it at the surface, which leads to increased upper soil moisture and humidity of the lower atmosphere. Increased moisture at the land surface may further enhance regional wetting effects via vertical convective activity and horizontal transport in the upper troposphere.

The differences in soil moisture at 10 cm depth between the exploitation and control tests are shown in Figures 5(d)–5(f). The changes in soil moisture at 10 cm depth were highly dependent on changes in precipitation and irrigation water reaching the soil surface. As shown in Figure 5(d), the soil moisture increased by a mean of $0.015 \, \text{m}^3/\text{m}^3$ in the Haihe Plain, and a large increase in soil moisture appeared in the central areas, where irrigation demands occupy a high proportion of the total demand. Wetting effects in the Haihe Plain also led to an increase in soil moisture in the Taihang Mountains; however, this increase was only $0.002 \, \text{m}^3/\text{m}^3$, because the increase was entirely from increased local precipitation. For the T2 test with doubled water demand (Figure 5(e)), soil moisture increased by $0.040 \, \text{m}^3/\text{m}^3$ in the Haihe Plain and $0.003 \, \text{m}^3/\text{m}^3$ in the Taihang Mountains. Although the water consumption processes in the Haihe Plain led to increased soil moisture in the Taihang Mountains, the extent was weaker than that of the wetting effects caused by local water consumption processes. Considering the local water consumption in the Taihang Mountains (Figure 5(f)), the soil moisture increased by $0.016 \, \text{m}^3/\text{m}^3$ in the Haihe Plain and $0.008 \, \text{m}^3/\text{m}^3$ in the Taihang Mountains.

Local wetting changes at the land surface in the Haihe Plain led to changes in precipitation, not only in the local plain areas, but also in the surrounding areas (including the Taihang Mountains) via atmospheric moisture transfer.

As shown in Figure 5(g), increased precipitation was detected in most areas of the Haihe Plain, with a mean value of 0.046 mm/d. The mean precipitation also increased by 0.020 mm/d in the Taihang Mountains, and the spatial distribution of increased precipitation was in accordance with the distribution of soil moisture (Figure 5(d)). The distribution of precipitation differences in Figure 5(h) is similar to that in Figure 5(g) but with approximately double values, specifically, 0.030 mm/d in the Taihang Mountains and 0.089 mm/d in the Haihe Plain. For the T3 test (Figure 5(i)), when the water consumption processes in the Taihang Mountains were considered in conjunction with the processes in the Haihe Plain, the precipitation in the Haihe Plain was slightly enhanced to 0.050 mm/day due to more land water over a larger scale being transferred into the atmosphere. In addition, precipitation increased by 0.028 mm/day in the Taihang Mountains, which is slightly higher than the mean difference in the T1 test, due to local increased evaporation.

Accompanying the regional wetting effects, the water consumption processes also led to cooling effects via changes in the heat fluxes emitted from the land surface. As shown in Figure 6(a), the mean 2 m air temperature decreased by 0.613 K in the Haihe Plain due to the local water consumption of the T1 test. Weak cooling effects were also detected in the Taihang Mountains, with a mean temperature decrease of 0.049 K. For the T2 test with doubled demand (Figure 6(b)), the changes in temperature were slightly less than double the changes observed in the T1 test. There were decreases of 1.121 K in the Haihe Plain and 0.089 K in the Taihang Mountains. For the T3 test (Figure 6(c)), the cooling effects were slightly enhanced in the Haihe Plain. The mean temperature decrease was 0.646 K in the Haihe Plain and 0.201 K in the Taihang Mountains.

In CLM, changes in the upward heat fluxes emitted from the land surface are reliant on differences in temperature and humidity between the land surface and the lower atmosphere. Thus, locally wetter and cooler air, due to water consumption in the Haihe Plain, was transferred to the Taihang Mountains via atmospheric activity. Subsequently, changes in land-atmosphere differences in temperature and humidity further affected the heat fluxes in the Taihang Mountains. Therefore, the changes in the heat fluxes of the Taihang Mountains, which were indirectly affected by the wetter and cooler atmosphere of the Haihe Plain, were far less than the changes in the Haihe Plain. For the T1 test (Figure 6(d)), the sensible heat flux decreased by 7.682 W/m^2, on average, in the Haihe Plain, yet only decreased by 0.098 W/m^2 in the Taihang Mountains. For the T2 test (Figure 6(e)), the mean decrease was 13.884 W/m^2 in the Haihe Plain and 0.168 W/m^2 in the Taihang Mountains. If local water consumption is considered, the heat fluxes will be affected significantly due to the increased evaporation associated with water consumption processes. Thus, the sensible heat flux in the T3 test (Figure 6(f)) decreased by 2.837 W/m^2 in the Taihang Mountains and by 7.793 W/m^2 in the Haihe Plain.

The changes of latent heat flux shown in Figures 6(g)–6(i) were similar to the changes of sensible heat flux except for their larger magnitudes. In the Haihe Plain, the latent heat

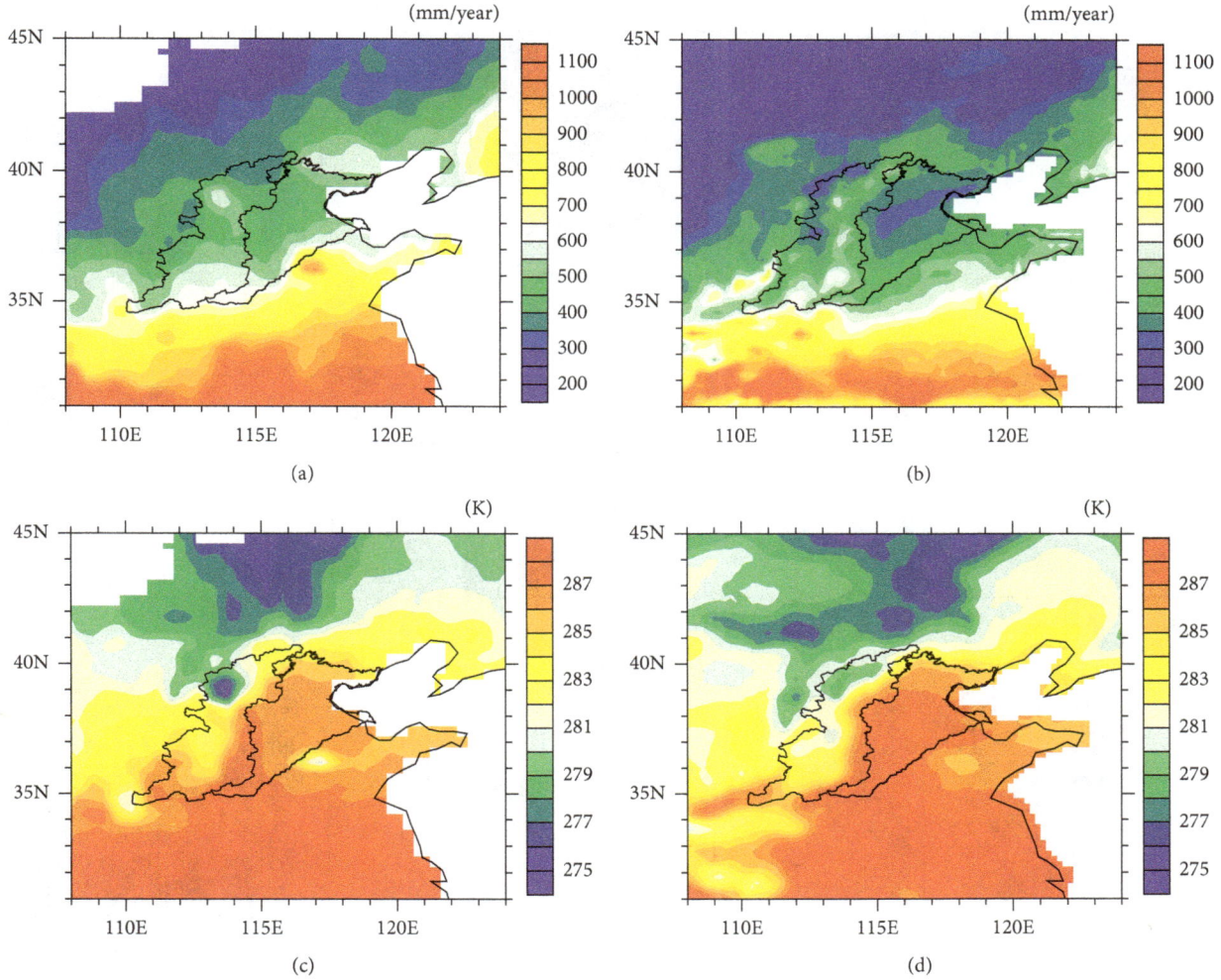

FIGURE 4: Spatial distributions of mean (a) observed precipitation, (b) precipitation simulated by the control test, (c) observed 2 m air temperature, and (d) 2 m air temperature simulated by the control test.

TABLE 3: Statistics of control test in the Taihang Mountains and Haihe Plain.

Domain	Variable	COR*	RMSE	STD	
				Observation	Simulation
Taihang mountains	Precipitation	0.931	0.519 mm/day	1.380 mm/day	1.418 mm/day
	Temperature	0.998	1.032 K	10.163 K	9.935 K
Haihe Plain	Precipitation	0.936	0.732 mm/day	1.691 mm/day	1.312 mm/day
	Temperature	0.997	1.186 K	10.296 K	10.640 K

*COR: correlation coefficient, $\text{COR} = (\sum_{i=1}^{n}(x_{mi} - \overline{x_m})(x_{oi} - \overline{x_o}))/(\sqrt{\sum_{i=1}^{n}(x_{mi} - \overline{x_m})^2 \cdot \sum_{i=1}^{n}(x_{oi} - \overline{x_o})^2})$, x_{mi} and x_{oi} are the simulated and observed monthly mean precipitation or 2 m air temperature in the month i, and n is the number of months; RMSE: root-mean-square error, $\text{RMSE} = \sqrt{(1/n)\sum_{i=1}^{n}(x_{mi} - x_{oi})^2}$; STD: standard deviation, $\text{STD} = \sqrt{(1/n)\sum_{i=1}^{n}(x_i - \overline{x})^2}$.

flux increased by 11.683 W/m², 21.412 W/m², and 11.994 W/m² for the T1, T2, and T3 tests, respectively. In the Taihang Mountains, the fluxes increased by 0.413 W/m², 0.673 W/m², and 4.644 W/m², respectively. The sum of sensible and latent heat flux was positive in most areas of the Taihang Mountains, indicating that the total heat flux emitted into the atmosphere from the land surface increased and the temperature of the soil layers would subsequently change.

3.3. Vertical Profiles in Soil and Atmosphere Layers.

To investigate changes in vertical profiles, Figure 7 provides profiles of mean soil moisture and temperature differences in the Taihang Mountains. The profiles of soil moisture differences (Figure 7(a)) indicate that soil moisture does not change greatly with depth, except for the wetter upper layers in the T1 and T2 tests, because changes in soil moisture in the Taihang Mountains are reliant on increased precipitation

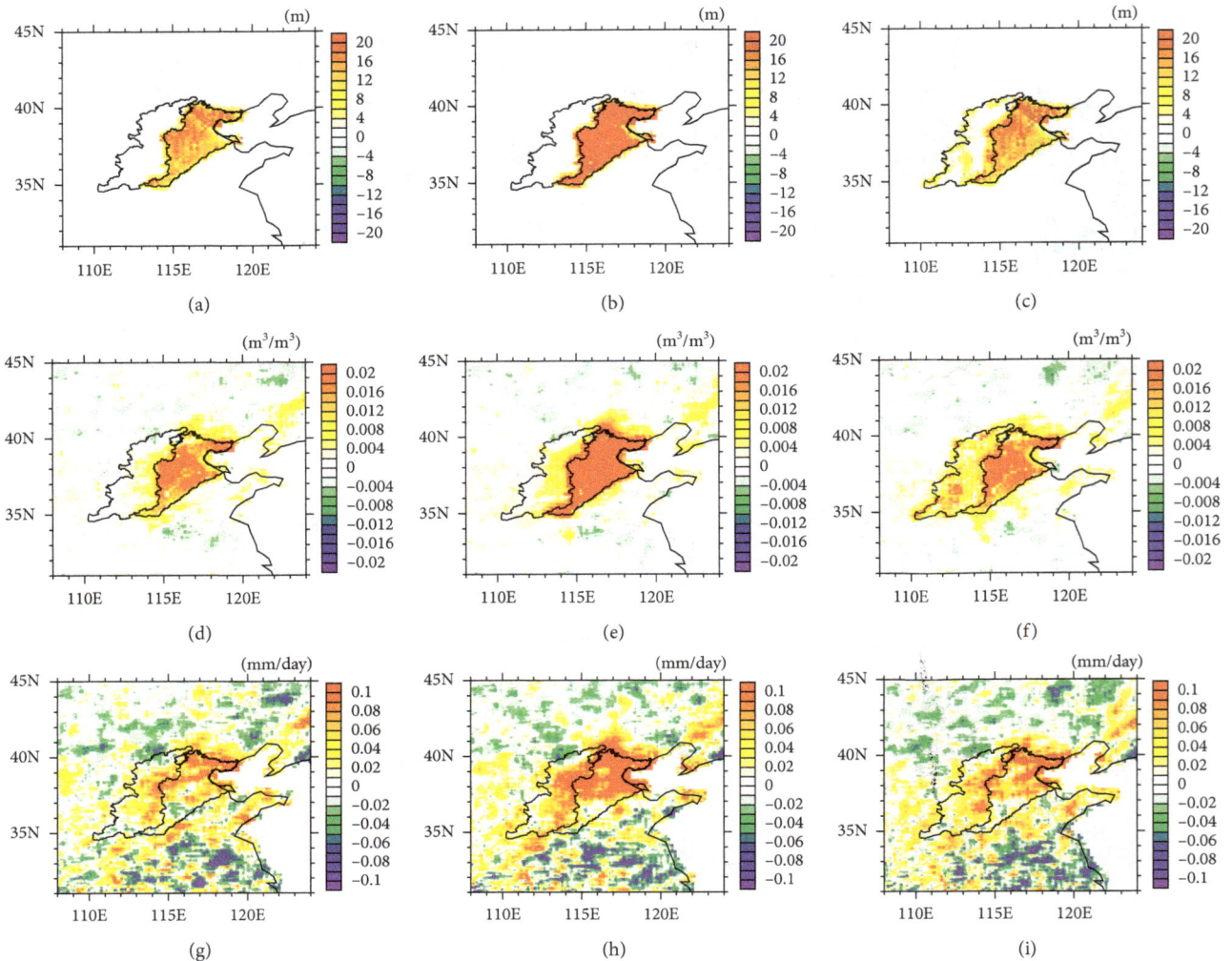

FIGURE 5: Spatial distributions of groundwater depth difference between tests (a) T1–CTL, (b) T2–CTL, and (c) T3–CTL. Mean soil moisture difference at 10 cm depth for (d) T1–CTL, (e) T2–CTL, and (f) T3–CTL. Mean precipitation difference for (g) T1–CTL, (h) T2–CTL, and (i) T3–CTL.

reaching the topsoil layer. The differences between the T1 and T2 tests in each soil layer basically remain constant. For the T3 test, which considered local water exploitation and consumption, soil moisture increased significantly in the upper layers due to increased water fluxes (e.g., irrigation and precipitation) into the top layer. In CLM, drainage from soil layers to aquifers is reliant on a water potential gradient between the soil layers and aquifers. This drainage will gradually approach the maximum infiltration capacity with increasing groundwater table depth. Therefore, in the lower soil layers, soil moisture was detected to decrease with depth.

Unlike the characterization of soil moisture, the soil bottom is regarded as being heat-insulated in CLM, and soil temperature is only dependent on the energy balance at the surface. As shown in Figure 7(b), soil temperature differences basically remained constant with depth in the T1 and T2 tests, and the differences in the T2 test were slightly less than double the differences in the T1 test. For the T3 test, due to the continuous water consumption process, the

temperature decrease in the upper soil layers was greater than that in the lower layers, and the decrease tends to be constant with depth.

While domain-averaged profile curves are provided in Figure 7, latitude-averaged contours of atmospheric temperature and humidity between 35°N and 40°N are provided in Figure 8 to investigate the transfer of cooling and wetting effects into the atmosphere. As shown in Figure 8(a), increased air humidity over the Haihe Plain (approximately 114.5°E–119°E) below a pressure of 800 hPa was detected, and at a pressure of around 600 hPa, a center of dry values was detected because the convergent flow from the lower atmosphere becomes divergent at 600 hPa. For the T2 test (Figure 8(b)), the wetting effects in the lower atmosphere were enhanced, and more moisture was transferred to the atmosphere over the Taihang Mountains (approximately 111°E–114.5°E). The wetting effects in the south-central valleys of the Taihang Mountains (about 112.5°E) were even stronger in the T3 test (Figure 8(c)) than the wetting

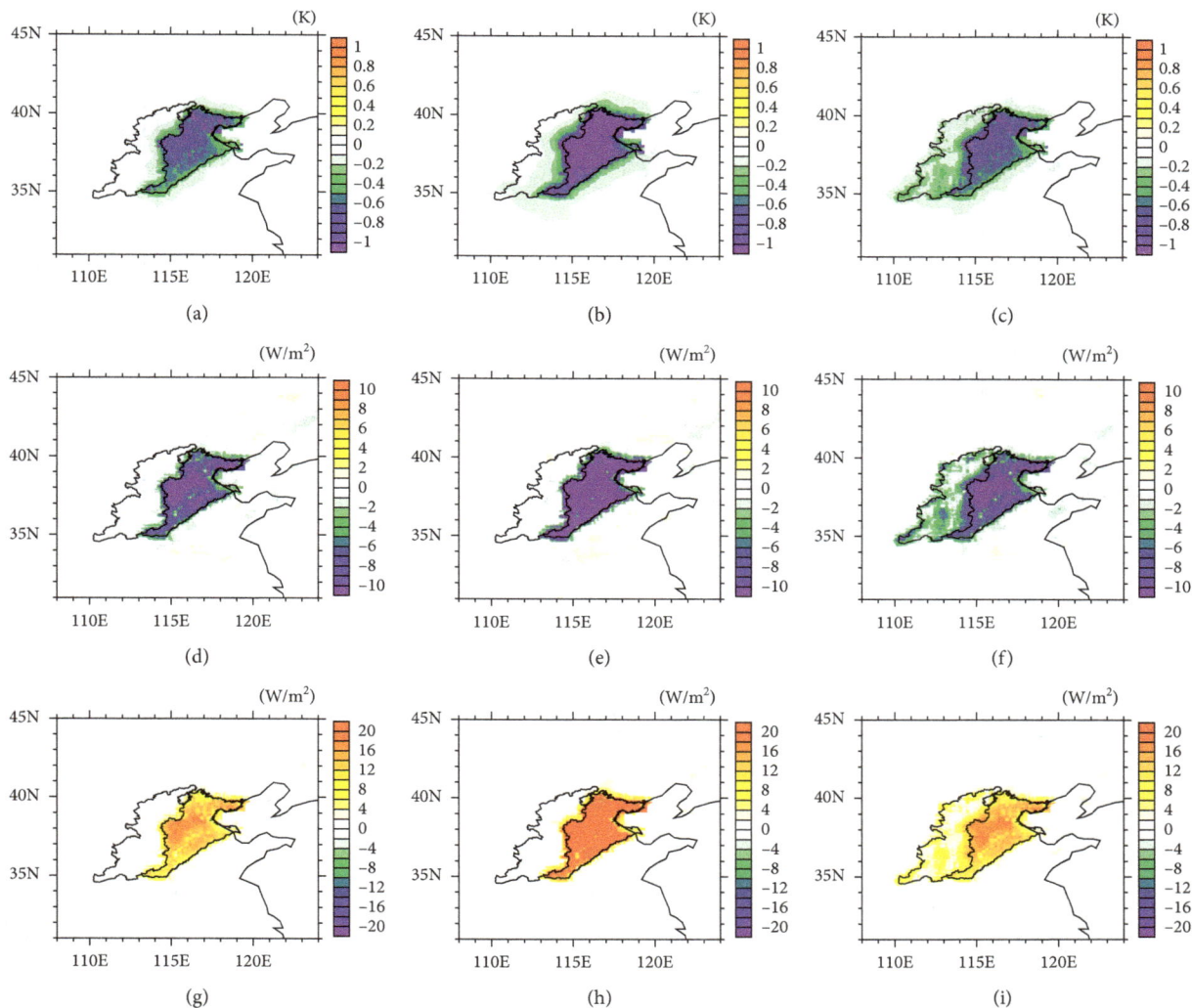

FIGURE 6: Spatial distributions of mean 2 m air temperature difference between tests (a) T1–CTL, (b) T2–CTL, and (c) T3–CTL. Sensible heat flux difference for (d) T1–CTL, (e) T2–CTL, and (f) T3–CTL. Latent heat flux difference for (g) T1–CTL, (h) T2–CTL, and (i) T3–CTL.

effects in the Haihe Plain. When compared with the changes in the T1 test (Figure 8(a)), the atmospheric wetting effects were enhanced over the Haihe Plain, and this slight enhancement was in accordance with the mean differences in climatic variables shown in Figures 5 and 6.

The profiles of air temperature differences (Figures 8(d)–8(f)) were similar to the air humidity profiles; that is, water consumption processes in the Haihe Plain lead to regional cooling effects in the atmosphere; not only over the Haihe Plain, but also over its surrounding areas, including the western Taihang Mountains. The cooling effects were basically below the 800 hPa level and were enhanced as the water volume consumed at the surface increased. Also, the cooling effects over the Haihe Plain were enhanced slightly due to water consumption over a larger-scale domain.

3.4. Annual and Seasonal Variability. Figure 9 shows the mean annual series of differences for the Taihang

Mountains. The series in Figure 9(a) indicates that the increased groundwater depths in the T1 and T2 tests were almost negligible and the groundwater depth dropped continuously due to local groundwater exploitation in the T3 test. For the other series of variables, the differences between the T1 and T2 tests indicate that doubled water consumption in the Haihe Plain did not cause doubled changes in the Taihang Mountains in most years. A negative feedback mechanism exists between increased water demand and subsequent climatic changes.

Meanwhile, the differences between the T1 and T3 tests indicate that changes in land surface variables due to water consumption are usually far greater when the water is consumed locally rather than in neighboring areas. The differences between the T1 and T3 tests also differed between variables. For the atmospheric variables, these differences were the lowest, because cooling and wetting effects occurring in the Haihe Plain can be transferred to neighboring areas via atmospheric activity, including to the Taihang

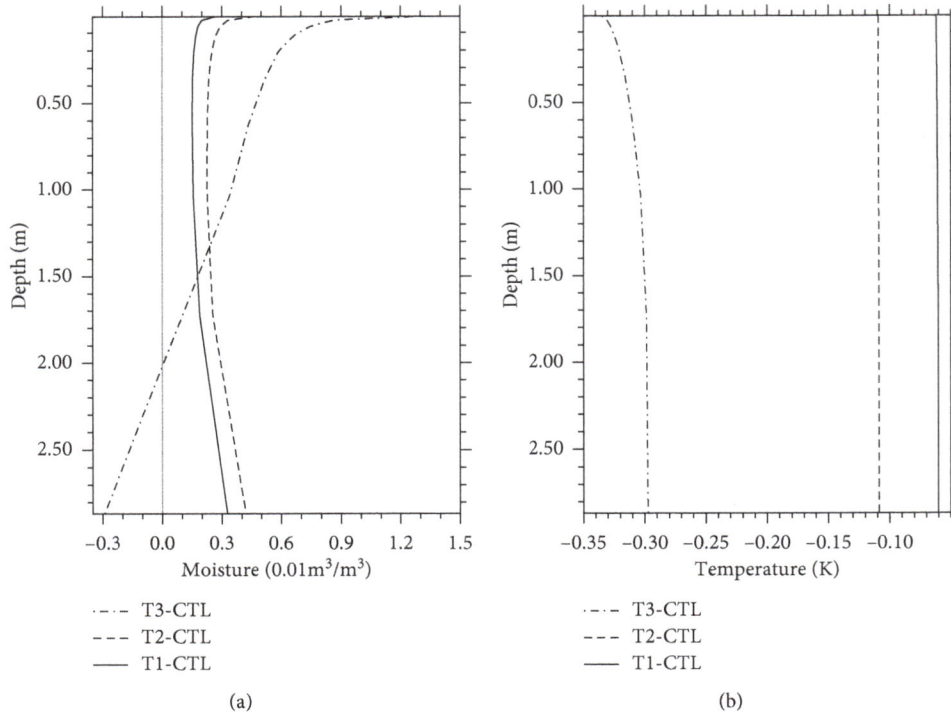

FIGURE 7: Mean profiles of (a) soil moisture difference and (b) soil temperature difference in the Taihang Mountains.

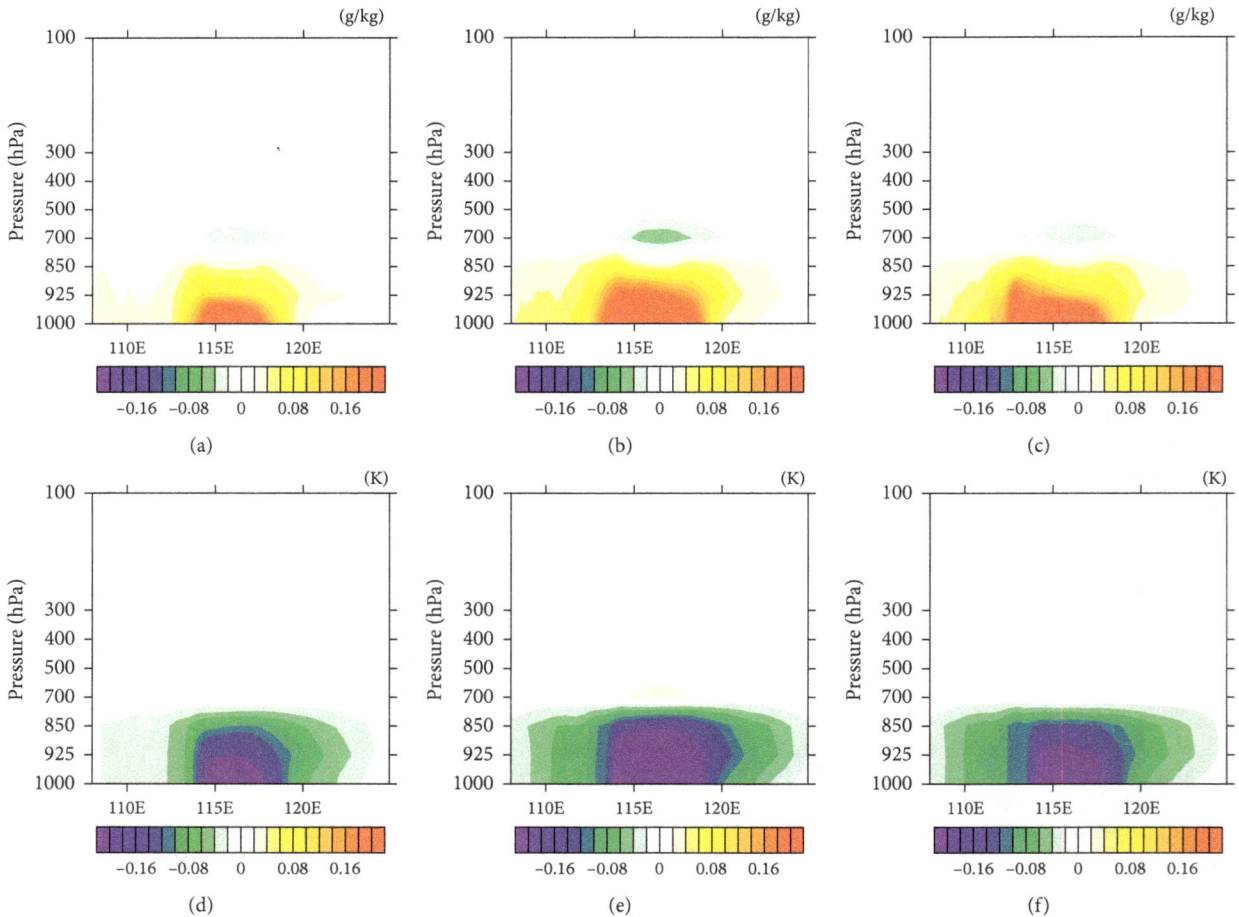

FIGURE 8: The distributions of longitude vs pressure for mean air humidity difference (a) T1−CTL, (b) T2−CTL, (c) T3−CTL; and air temperature difference (d) T1−CTL, (e) T2−CTL, (f) T3−CTL averaged between 35°N and 40°N.

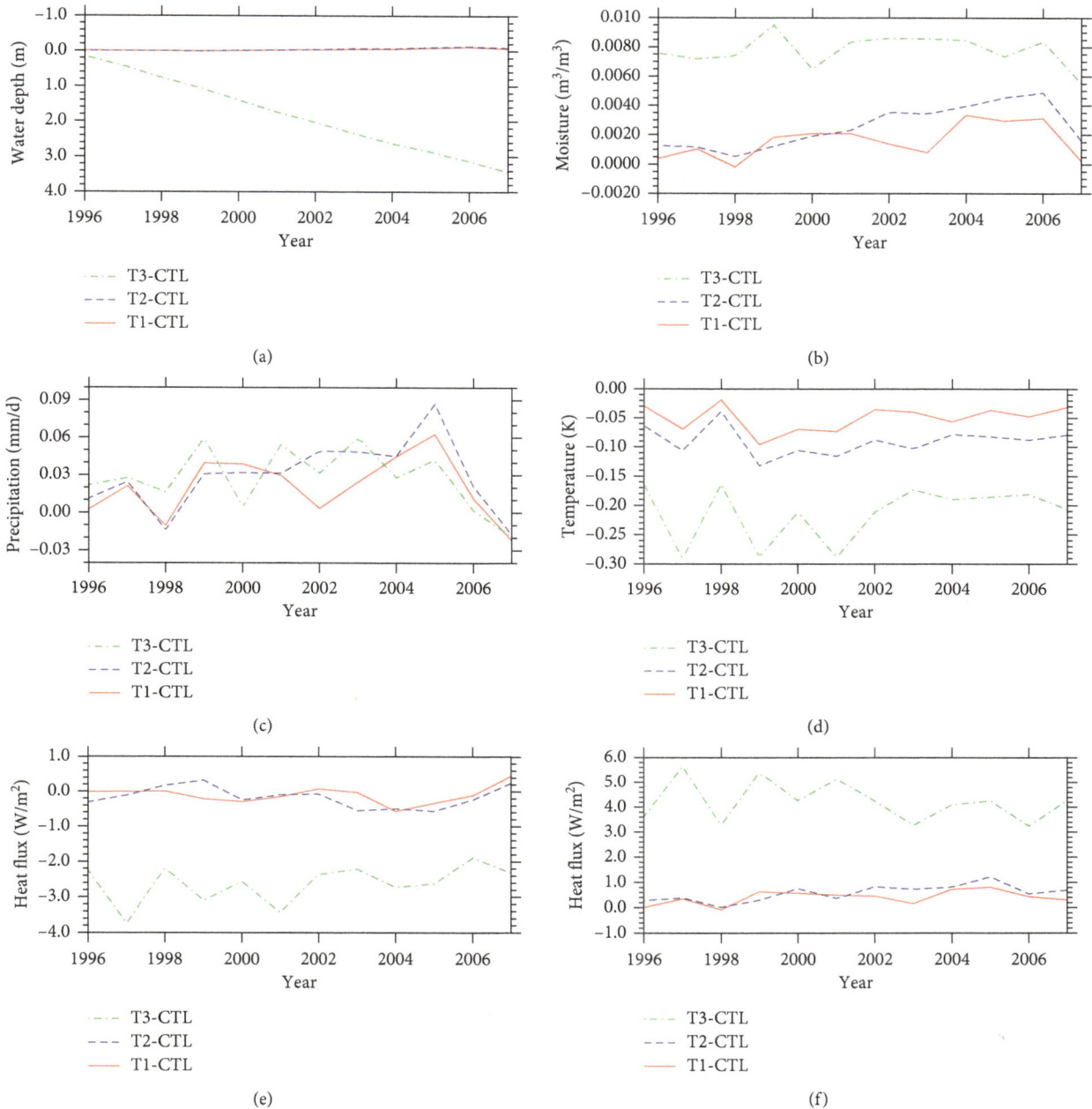

FIGURE 9: Annual series of (a) groundwater depth differences, (b) soil moisture differences at 10 cm depth, (c) precipitation differences, (d) 2 m air temperature differences, (e) sensible heat flux differences, (f) latent heat flux differences in the Taihang Mountains.

Mountains. For example, the precipitation data series of the T1 test (Figure 9(c)) was approximately 35% lower than the T3 series. The differences between the T1 and T3 tests were higher for the land surface variables, which were further affected by the cooler and wetter atmosphere. For example, the T1 series of 10 cm soil moisture (Figure 9(b)) and 2 m air temperature (Figure 9(d)) were nearly 20–25% of the T3 series in magnitude. For the heat fluxes emitted from the land surface, sensible heat flux (Figure 9(e)), and latent heat flux (Figure 9(f)) were strongly affected by changes in evaporation associated with local water consumption processes. The indirect effects from the neighboring Haihe Plain

were nearly negligible, and the series of the two variables in the T1 test were nearly 90–95% lower in magnitude than those of the T3 test.

The mean monthly differences between the three exploitation tests and the control test in the Taihang Mountains are shown in Figure 10. The monthly differences in groundwater depth are not shown because their changes in the T1 and T2 tests were tiny, and the monthly changes in the T3 test had no obvious seasonal variations due to continuous exploitation. As shown in Figure 10(a), the monthly differences of 10 cm soil moisture in the T1 and T2 tests did not vary greatly. The increases in soil moisture were slightly

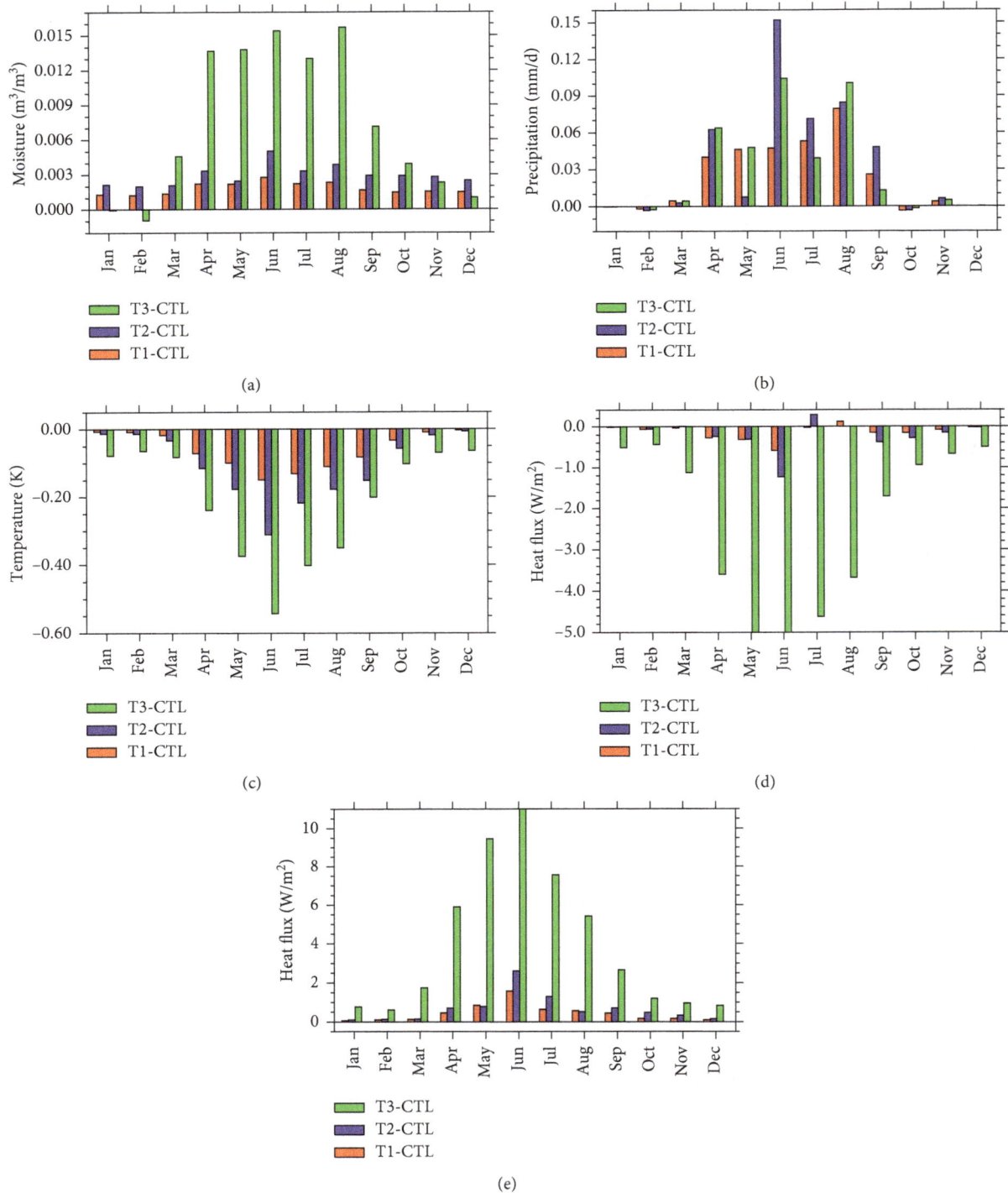

FIGURE 10: Mean monthly (a) soil moisture differences at 10 cm depth, (b) precipitation differences, (c) 2 m air temperature differences, (d) sensible heat flux differences, and (e) latent heat flux differences in the Taihang Mountains.

greater during the irrigation seasons (primarily spring and summer) due to increased precipitation, and during the nonirrigation seasons (primarily autumn and winter), the increased soil moisture reduced gradually due to the water-holding capacity of soil. When considering local water

consumption processes, 10 cm soil moisture was detected to increase clearly due to local irrigation.

For the other four variables, most differences occurred during the irrigation seasons, regardless of whether local water consumption processes were considered. During nonirrigation seasons, the wetter and cooler differences still existed in the T3 test, because increased evaporation due to domestic and industrial consumption can also lead to wetter and cooler differences to a certain level. However, the differences in the T1 and T2 tests were almost negligible during the nonirrigation seasons because the cooling and wetting differences were weak, even in the Haihe Plain, and low temperatures in nonirrigation seasons also suppress convection and moisture transfer from the Haihe Plain to the neighboring Taihang Mountains.

4. Conclusion and Discussion

In this study, the regional climate model RegCM4 was incorporated with a scheme of human-induced water exploitation and consumption to investigate the impact of water consumption activities in the Haihe Plain on the local climate in the Taihang Mountains. Four simulation tests (i.e., three exploitation tests and one control test) were conducted. One Exploitation Test (T1) considered water consumption activities in the Haihe Plain and another (T2) doubled these water demands. The third Exploitation Test (T3) considered the combined water demands of the Haihe Plain and Taihang Mountains, and a control test (CTL) simulated a case without any water exploitation or consumption. By comparing the differences between these three exploitation tests and the control test, we were able to analyze changes to climatic variables in the Taihang Mountains due to water consumption activities in the neighboring Haihe Plain.

The main conclusions are as follows. (1) The processes of water exploitation and consumption in the Haihe Plain cause rapid increase of groundwater depth and local cooling and wetting effects at the land surface. This causes a cooler and wetter atmosphere that transfers to surrounding areas, including the Taihang Mountains, leading to decreases in temperature and increases in surface moisture. The cooling and wetting changes at the land surface of the Taihang Mountains affect the local energy balance, increasing latent heat flux, and decreasing sensible heat flux emitted from the land surface. (2) These wetting and cooling effects are positively related to the volume of water consumed. However, the cooling effects do not enhance the development of lower atmospheric convection and moisture transfer. For the T2 test with double the water demands of the T1 test, the wetting and cooling changes were slightly suppressed and were less than double the changes estimated in the T1 test. (3) In the simulations, water consumption activities in the Haihe Plain-induced wetting and cooling effects in the Taihang Mountains via atmospheric transfer. However, these wetting and cooling effects were rather weak at the land surface and caused less change than that caused by local water exploitation and consumption in the Taihang Mountains. For the T1 and T2 tests, changes at the land surface in the Taihang Mountains were entirely caused by changes in atmospheric variables. For comparison, the conditions of the T3 test were the opposite—local water consumption activities first led to changes of surface variables, then atmospheric variables changed due to land-

atmosphere interactions. (4) The cooling and wetting effects of water consumption activities mainly occur during irrigation periods, because irrigation comprises most of the total water demand in the Haihe Plain and Taihang Mountains.

This study simulated the impacts of water consumption activities in the Haihe Plain on the local climate in the Taihang Mountains, located immediately to the west. However, many aspects of the tests, including data estimation, scheme design, and structural defects in RegCM4/CLM3.5, introduced uncertainties to the simulations and are discussed next.

Although the resolution of the water demand data used in this study was higher than the one used in our previous studies, there was potential underestimation of the total demand according to Table 1. The largest uncertainty remains the estimation of irrigation water use. In this study, the ratio between the irrigated and total cropland areas, α, was constant derived from the *China Statistical Yearbook 2000*. Future work should address the acquisition of more reasonable distributions of α by using remote sensing data. The *Water Resource Bulletin of China* further indicates that the per capita water use differs greatly between urban and rural regions. The lack of consideration about the rural-urban difference may lead to excessive domestic water demand estimation over rural regions; thus, a further division between rural and urban domestic demand should be addressed to differentiate the allocation proportion of domestic water consumption. The scheme of water exploitation and consumption used in this study was highly simplified, and more detailed data should be obtained to improve the spatial heterogeneity of water demand estimation and the allocation process of water consumption.

Structural defects in the RegCM4/CLM3.5 models also introduced uncertainties. For example, water in lakes and reservoirs was not included in the water cycle process of the model. Although significant improvements were made by introducing the lake depth dataset, more accurate lake parameterization, and groundwater recharge processes, the water volume in lakes and reservoirs was still maintained constant even in the newer versions of the CLM model [47–49]. Thus, the surface water supply was only withdrawn from river channels, leading to overexploited groundwater and deeper groundwater table in the simulation tests. A mechanism of reservoir regulation based on the accurate description of the reservoir water volume should be included in the river transport model component of CLM. The processes of groundwater lateral flow were also not included in the model, and the groundwater table depth was only dependent on water input/output in the vertical direction. The lack of groundwater lateral flow data input may lead to groundwater grid cells being isolated from each other and to spatial distributions of simulated groundwater depths differing from actual conditions, especially in mountainous areas. Future efforts should be made to incorporate 3-D groundwater flow into the CLM model.

Furthermore, the time frame of the study is relatively old. The year 2000 is used as the benchmark year and the simulation tests are conducted for the years 1996–2007. As there are significant demographic changes in the areas under study during the past decade, it would definitely be of interest to explore the climatic effects of water exploitation and consumption using the updated numbers of population, GDP, and water use. This could be the focus of a future study; however, here we discuss a few preliminary aspects. In fact, the water use over China in general and the Haihe Plain in particular did not increase significantly from the year 2000 onwards. The rapid increase in water use that took place in the previous decades (mainly the 1980s and 1990s) was discontinued upon the beginning of the new millennium. According to the *Water Resource Bulletin of China* over the years, the total water volume used all over China was 556.6 billion m^3 in 1997 as opposed to 604 billion m^3 in 2016. Therefore, and despite the significant increase of the Chinese GDP after the year 2000, the total water use was characterized by a relatively small raise. And it is widely known from previous studies that the climatic impacts of water consumption rely heavily on the volume of the water consumed [18, 19, 23, 25]. Moreover, our study uses multiyear mean differences of the simulation tests to discuss the mechanism and magnitude of the climatic impacts of water consumption, which are not expected to change significantly if the climate forcing data of RegCM4 is replaced with a recent dataset. Therefore, based on the aforementioned features, it would be expected that the results of our simulation tests would not change significantly if more recent data was to be used.

Conflicts of Interest

The authors declare that they have no conflicts of interest.

Acknowledgments

The authors would like to thank Prof. Theodore Karacostas and the reviewers for the comments and suggestions on this paper. This study was partially supported by the National Key Research and Development Program of China (grant 2017YFA0603702), the National Basic Research Program of China (973 Program) (grant 2015CB452701), the Natural Science Foundation of China (grants 41705046, 41571019, and 41606112;), the Key Research and Development Program of Shandong Province of China (grant 2016JMRH0538), and the Natural Science Foundation of Shandong Province of China (grant ZR2016DB32).

References

[1] X. Li, B. Wu, H. Wang, and J. Zhang, "Regional soil erosion risk assessment in Hai Basin," *Journal of Remote Sensing*, vol. 15, no. 2, pp. 372–387, 2011.

[2] X. Liu, W. Zhang, Z. Liu, F. Qu, and W. Song, "Impacts of land cover changes on soil chemical properties in Taihang Mountain, China," *Journal of Food, Agriculture & Environment*, vol. 8, no. 3, pp. 985–990, 2010.

[3] S. Piao, P. Ciais, Y. Huang et al., "The impacts of climate change on water resources and agriculture in China," *Nature*, vol. 467, no. 7311, pp. 43–51, 2010.

[4] S. Wang and Z. Zhang, "Effects of climate change on water resources in China," *Climate Research*, vol. 47, no. 1, pp. 77–82, 2011.

[5] J. Liu, "The temporal–spatial variation characteristics of precipitation in Beijing–Tianjin–Heibei during 1961–2012," *Climate Change Research Letters*, vol. 3, no. 3, pp. 146–153, 2014, in Chinese.

[6] Z. Bao, J. Zhang, G. Wang et al., "Attribution for decreasing streamflow of the Haihe River basin, northern China: climate variability or human activities," *Journal of Hydrology*, vol. 460-461, pp. 117–129, 2012.

[7] Z. Li, X. Deng, F. Yin, and C. Yang, "Analysis of climate and land use changes impacts on land degradation in the North China Plain," *Advances in Meteorology*, vol. 2015, Article ID 976370, 11 pages, 2015.

[8] J. Liu, Q. Zhang, V. Singh, and P. Shi, "Contribution of multiple climatic variables and human activities to streamflow changes across China," *Journal of Hydrology*, vol. 545, pp. 145–162, 2017.

[9] L. Ren, M. Wang, C. Li, and W. Zhang, "Impacts of human activity on river runoff in the northern area of China," *Journal of Hydrology*, vol. 261, no. 1–4, pp. 204–217, 2002.

[10] Y. Jia, X. Ding, H. Wang et al., "Attribution of water resources evolution in the highly water-stressed Hai River Basin of China," *Water Resources Research*, vol. 48, no. 2, article W02513, 2012.

[11] Department of Urban & Social Economic Survey, National Bureau of Statistics of China, *Urban Statistical Yearbook of China 2000*, China Statistics Press, Beijing, China, 2001, in Chinese.

[12] S. Foster, H. Garduno, R. Evans et al., "Quaternary aquifer of the North China Plain–assessing and achieving groundwater resource sustainability," *Hydrogeology Journal*, vol. 12, no. 1, pp. 81–93, 2004.

[13] G. Cao, C. Zheng, B. R. Scanlon et al., "Use of flow modeling to assess sustainability of groundwater resources in the North China Plain," *Water Resources Research*, vol. 49, no. 1, pp. 159–175, 2013.

[14] J. Liu, C. Zheng, and Y. Lei, "Ground water sustainability: methodology and application to the North China Plain," *Ground Water*, vol. 46, no. 6, pp. 897–909, 2008.

[15] W. Feng, M. Zhong, J. M. L. Lemoine et al., "Evaluation of groundwater depletion in North China using the Gravity Recovery and Climate Experiment (GRACE) data and ground-based measurements," *Water Resources Research*, vol. 49, no. 4, pp. 2110–2118, 2013.

[16] C. Davidsen, S. Liu, X. Mo, D. Rosbjerg, and P. Bauer-Gottwein, "The cost of ending groundwater overdraft on the North China Plain," *Hydrology and Earth System Sciences*, vol. 20, no. 2, pp. 771–785, 2016.

[17] D. Han, G. Wang, B. Xue, T. Liu, A. Yinglan, and X. Xu, "Evaluation of semiarid grassland degradation in North China from multiple perspectives," *Ecological Engineering*, vol. 112, pp. 41–50, 2018.

[18] K. J. Harding and P. K. Snyder, "Modeling the atmospheric response to irrigation in the Great Plains. Part I: general impacts on precipitation and the energy budget," *Journal of Hydrometeorology*, vol. 13, no. 6, pp. 1667–1686, 2012.

[19] E. M. Douglas, A. Beltran-Przekurat, D. Niyogi, A. Pielke Sr., and C. J. Vörösmarty, "The impact of agricultural intensification and irrigation on land-atmosphere interactions

and Indian monsoon precipitation—a mesoscale modeling perspective," *Global and Planetary Change*, vol. 67, no. 1-2, pp. 117–128, 2009.

[20] M. J. Puma and B. I. Cook, "Effects of irrigation on global climate during the 20th century," *Journal of Geophysical Research*, vol. 115, no. D16, article D16120, 2010.

[21] L. M. Kueppers and M. A. Snyder, "Influence of irrigated agricultural on diurnal surface energy and water fluxes, surface climate, and atmospheric circulation in California," *Climate Dynamics*, vol. 38, no. 5-6, pp. 1017–1029, 2012.

[22] Y. Qian, M. Huang, B. Yang, and L. K. Berg, "A modeling study of irrigation effects on surface fluxes and land-air-cloud interactions in the Southern Great Plains," *Journal of Hydrometeorology*, vol. 14, no. 3, pp. 700–712, 2013.

[23] F. Saeed, S. Hagemann, and D. Jacob, "Impact of irrigation on the South Asian summer monsoon," *Geophysical Research Letters*, vol. 36, no. 20, article L20711, 2009.

[24] A. DeAngelis, F. Dominuez, Y. Fan, A. Robock, M. Deniz Kustu, and D. Robinson, "Evidence of enhanced precipitation due to irrigation over the Great Plains of the United States," *Journal of Geophysical Research*, vol. 115, no. D15, article D15115, 2010.

[25] F. Chen and Z. Xie, "Effects of interbasin water transfer on regional climate: a case study of the Middle Route of the South-to-North Water Transfer Project in China," *Journal of Geophysical Research*, vol. 115, no. D11, article D11112, 2010.

[26] G. Leng, Q. Tang, M. Huang, and L. R. Leung, "A comparative analysis of the impacts of climate change and irrigation on land surface and subsurface hydrology in the North China Plain," *Regional Environmental Change*, vol. 15, no. 2, pp. 251–263, 2015.

[27] F. Giorgi, E. Coppola, F. Solmon et al., "RegCM4: model description and preliminary tests over multiple CORDEX domains," *Climate Research*, vol. 52, pp. 7–29, 2012.

[28] K. W. Oleson, G. Niu, Z. Yang et al., "Improvements to the Community Land Model and their impact on the hydrological cycle," *Journal of Geophysical Research*, vol. 113, no. G1, article G01021, 2008.

[29] G. Niu, Z. Yang, R. E. Dickinson, and L. E. Gulden, "A simple TOPMODEL-based runoff parameterization (SIMTOP) for use in global climate models," *Journal of Geophysical Research*, vol. 110, no. D21, article D21106, 2005.

[30] G. Y. Niu, Z. L. Yang, R. E. Dickinson, L. E. Gulden, and H. Su, "Development of a simple groundwater model for use in climate models and evaluation with Gravity Recovery and Climate Experiment data," *Journal of Geophysical Research*, vol. 112, no. D7, article D07103, 2007.

[31] X. Wang, M. Yang, and G. Pang, "Influences of two land-surface schemes on RegCM4 precipitation simulations over the Tibetan Plateau," *Advances in Meteorology*, vol. 2015, Article ID 106891, 12 pages, 2015.

[32] M. Ashfaq, D. Rastogi, R. Mei et al., "High-resolution ensemble projections of near-term regional climate over the continental United States," *Journal of Geophysical Research*, vol. 121, no. 17, pp. 9943–9963, 2016.

[33] A. J. K. Mbienda, C. Tchawoua, D. A. Vondou, P. Choumbou, C. Kenfack Sadem, and S. Dey, "Sensitivity experiments of RegCM4 simulations to different convective schemes over Central Africa," *International Journal of Climatology*, vol. 37, no. 1, pp. 328–342, 2017.

[34] J. Zou, Z. Xie, C. Zhan et al., "Effects of anthropogenic groundwater exploitation on land surface processes: a case study of the Haihe River Basin, northern China," *Journal of Hydrology*, vol. 524, pp. 625–641, 2015.

[35] J. Zou, C. Zhan, Z. Xie, P. Qin, and S. Jiang, "Climatic impacts of the middle route of the south-to-North water transfer project over the Haihe River basin in North China simulated by a regional climate model," *Journal of Geophysical Research: Atmospheres*, vol. 121, no. 15, pp. 8983–8999, 2016.

[36] Y. Mao, A. Ye, X. Liu, F. Ma, X. Deng, and Z. Zhou, "High-resolution simulation of the spatial pattern of water use in continental China," *Hydrological Sciences Journal*, vol. 61, no. 14, pp. 2626–2638, 2016.

[37] N. Hanasaki, S. Kanae, T. Oki et al., "An integrated model for the assessment of global water resources–part 1: model description and input meteorological forcing," *Hydrology and Earth System Sciences*, vol. 12, no. 4, pp. 1007–1025, 2008.

[38] I. A. Shiklomanov, "Appraisal and assessment of world water resources," *Water International*, vol. 25, no. 1, pp. 11–32, 2000.

[39] F. Zhang, D. Wang, and B. Qiu, *China Agricultural Phenology Atlas*, Science Press, Beijing, China, 1987, in Chinese.

[40] D. Xiao, F. Tao, Y. Liu et al., "Observed changes in winter wheat phenology in the North China Plain for 1981–2009," *International Journal of Biometeorology*, vol. 57, no. 2, pp. 275–285, 2013.

[41] F. Tao, S. Zhang, Z. Zhang, and R. Rotter, "Maize growing duration was prolonged across China in the past three decades under the combined effects of temperature, agronomic management, and cultivar shift," *Global Change Biology*, vol. 20, no. 12, pp. 3686–3699, 2014.

[42] Y. Liu, R. Xie, P. Hou et al., "Phenological responses of maize to changes in environment when grown at different latitudes in China," *Field Crops Research*, vol. 144, pp. 192–199, 2013.

[43] Y. Ran, X. Li, L. Lu, and Z. Li, "Large-scale land cover mapping with the integration of multi-source information based on the Dempster-Shafer theory," *International Journal of Geographical Information Science*, vol. 26, no. 1, pp. 169–191, 2010.

[44] National Bureau of Statistics of China, *China Statistical Yearbook 2000*, China Statistics Press, Beijing, China, 2001, in Chinese.

[45] Z. Xie, F. Yuan, Q. Duan, J. Zheng, M. Liang, and F. Chen, "Regional parameter estimation of the VIC land surface model: methodology and application to river basins in China," *Journal of Hydrometeorology*, vol. 8, no. 3, pp. 447–468, 2007.

[46] J. Xie, Y. Zeng, M. Zhang, and Z. Xie, "Detection and attribution of the influence of climate change and human activity on hydrological cycle in China's eastern monsoon area," *Climatic and Environmental Research*, vol. 21, no. 1, pp. 87–98, 2016, in Chinese.

[47] E. Kourzeneva, H. Asensio, E. Martin, and S. Faroux, "Global gridded dataset of lake coverage and lake depth for use in numerical weather prediction and climate modeling," *Tellus A: Dynamic Meteorology and Oceanography*, vol. 64, no. 1, p. 15640, 2012.

[48] Z. M. Subin, W. J. Riley, and D. Mironov, "An improved lake model for climate simulations: model structure, evaluation, and sensitivity analyses in CESM1," *Journal of Advances in Modeling Earth Systems*, vol. 4, article M02001, 2012.

[49] M. Zampieri, E. Serpetzoglou, E. N. Anagnostou et al., "Improving the representation of river-groundwater interactions in land surface modeling at the regional scale: observational evidence and parameterization applied in the Community Land Model," *Journal of Hydrology*, vol. 420-421, pp. 72–86, 2012.

6

Climatology and Teleconnections of Mesoscale Convective Systems in an Andean Basin in Southern Ecuador: The Case of the Paute Basin

Lenin Campozano [iD],[1,2,3] Katja Trachte,[4] Rolando Célleri,[1,2] Esteban Samaniego,[1,2] Joerg Bendix,[4] Cristóbal Albuja [iD],[1,2] and John F. Mejia[5]

[1]*Departamento de Recursos Hídricos y Ciencias Ambientales, Universidad de Cuenca, Cuenca, Ecuador*
[2]*Facultad de Ingeniería, Universidad de Cuenca, Cuenca, Ecuador*
[3]*Depto. de Ingeniería Civil y Ambiental, Escuela Politecnica Nacional, Quito, Ecuador*
[4]*Laboratory for Climatology and Remote Sensing (LCRS), Faculty of Geography, Philipps-University Marburg, Deutschhausstraße 10, 35032 Marburg, Germany*
[5]*Department of Atmospheric Sciences, Desert Research Institute, Reno, NV, USA*

Correspondence should be addressed to Lenin Campozano; lenin.campozano@epn.edu.ec

Academic Editor: Efthymios I. Nikolopoulos

Mesoscale convective systems (MCSs) climatology, the thermodynamic and dynamical variables, and teleconnections influencing MCSs development are assessed for the Paute basin (PB) in the Ecuadorian Andes from 2000 to 2009. The seasonality of MCSs occurrence shows a bimodal pattern, with higher occurrence during March-April (MA) and October-November (ON), analogous to the regional rainfall seasonality. The diurnal cycle of MCSs shows a clear nocturnal occurrence, especially during the MA and ON periods. Interestingly, despite the higher occurrence of MCSs during the rainy seasons, the monthly size relative frequency remains fairly constant throughout the year. On the east of the PB, the persistent high convective available potential and low convective inhibition values from midday to nighttime are likely related to the nocturnal development of the MCSs. A significant positive correlation between the MCSs occurrence to the west of the PB and the Trans-Niño index was found, suggesting that ENSO is an important source of interannual variability of MCSs frequency with increasing development of MCSs during warm ENSO phases. On the east of the PB, the variability of MCSs is positively correlated to the tropical Atlantic sea surface temperature anomalies south of the equator, due to the variability of the Atlantic subtropical anticyclone, showing main departures from this relation when anomalous conditions occur in the tropical Pacific due to ENSO.

1. Introduction

Precipitation in tropical South America (TSA) is modulated by the spatial oscillation of the intertropical convergence zone (ITCZ), and an important fraction of this precipitation is generated by organized convection in the form of mesoscale convective systems (MCSs) [1–3]. MCSs are defined as groups of organized convective storms lasting at least 3 hours [4], producing high rates of precipitation, which enhance the likelihood of flooding [5]. MCSs have also been linked to flash floods and rainfall-induced landslides [6], negatively impacting the infrastructure and socioeconomic

welfare. The interaction of the prevailing air flow with the eastern inter-Andean region triggers convection and the development of MCSs. In summary, MCS-related precipitation impacts the regional distribution of water, the infrastructure and management of water resources, and the use of water by humans and ecosystems.

The climate in Ecuador presents a high spatiotemporal variability. For instance, the Andes cordillera in Ecuador acts as a weather divide [7], which also modulates precipitation regimes and seasonality. In the coastal region on the west side of the Andes range, precipitation is influenced by the ITCZ, with one rainy season peaking between December and

May and the South Pacific anticyclone inhibiting precipitation during the second half of the year [7]. The Pacific Ocean sea surface temperature, especially over the Niño 1 + 2 region, and the Southeast Pacific anticyclone have been related to interannual variability of temperature and rainfall in this region mostly related to El Niño-Southern Oscillation (ENSO) [7]. Eichler et al. [8] found that, during the warm phase of the ENSO, precipitation synoptic-scale variability during the wet season December–February (DJF) was linked to the activity of northern hemisphere extratropical cyclones. On the contrary, climate in the Amazon is mostly affected by convection activity related to the ITCZ [9] and the relatively higher humidity as well as easterly winds that produce orographic precipitation on the eastern flanks of the cordillera [10]. Therefore, the Amazon region of Ecuador is rainy along the year, with two main rainy seasons from February to May and from October to November [11]. Despite the potential impacts of the MCS on societal aspects, studies on the MCS in Ecuador are limited. Previous studies have documented the existence of katabatic cold air flows descending from the high Andean mountain region. This downslope flow encounters warm, moist air from the Amazon basin, leading to the formation of unstable convective clouds [12, 13]. Over the foothills of the southeastern Andes in Ecuador, Bendix et al. [14] showed the occurrence of convective clouds late afternoon and the nocturnal enhancement of cell development from 1:00 to 4:00 LST during the austral summer. The late afternoon mesoscale convective complexes (MCCs) are related to typical thermal convection [14]. In the study by Bendix et al. [14], the frequency of the MCC weakens at 1 h and strengthens again at 4 h, which is the second phase of MCC development (less frequent than thermal-induced MCCs). The MCCs formed on the second phase, which are related to katabatic flows, are responsible for the early morning rains [14]. As climate variability is complex in Ecuador, the variability of MCSs arises in several scales. For instance, the topographic influence is evident in the diurnal cycle of MCSs, whereas the influence of the ITCZ may be revealed in intra-annual time scales, which in turn may be modulated at intra-annual time scales for phenomena such as the 3- to 11-year quasi-oscillating ENSO.

In this paper, we study the spatial-temporal distribution of MCSs over the Paute basin, which is one of the most monitored basins in Ecuador. The Paute basin is strategically important because it harbours several important hydropower generation plants, accounting for nearly 40% of nation's hydroelectrical production. Furthermore, several cities in the region depend on the PB for their water services. Nearly 40% of the basin area is covered by páramo, a neotropical alpine wetland that occurs in the upper Andean region of Venezuela, Colombia, Ecuador, and Perú [15]. The soil-vegetation system of the páramo is considered an important water-regulating ecosystem that helps to regulate downstream water needs during the low precipitation seasons, although some authors argue that seasonal rainfall variation in these regions is low (e.g., Buytaert et al. [16]). From a scientific perspective, the PB is an important zone because of its complex spatial-temporal climate variability due to the complex topography [11, 17].

To contribute to the understanding of MCSs in the inter-Andean region of Southern Ecuador, the present study estimates the climatology of MCSs and their climatological thermodynamic and dynamical related variables such as the convective available potential energy (CAPE), convective inhibition (CIN), and Omega and provides an initial exploration of the most relevant teleconnections affecting interannual variability of MCS occurrence in the Paute River basin. Based on the previous research, a relationship between MCS occurrence and the seasonality of rainfall in the Paute basin is expected. As a consequence, the study of the relationship between different climate signals and precipitation, which will be called precipitation teleconnections from now on, is a key step to unravel the climatic drivers which are most relevant for MCS occurrence.

2. Study Area

The Andes cordillera in Ecuador stretches between 2°N and 5°S latitude and includes an inter-Andean depression with an elevation around 2500 m ASL, including peaks over 6000 m ASL (e.g., Chimborazo Glacier, 6270 m ASL). The complex topography, the influence of the ITCZ, and the influence of Pacific and Atlantic Ocean SST result in complex climatic spatial gradients [18]. The easterlies enhance precipitation along the windward side of the Andes (eastern slopes), which is also the case for the Paute basin (PB, an area of 6481 km^2) [10]. In addition, in the PB, the western cordillera is higher than the eastern cordillera sheltering the PB from the coastal plain circulation. This study is part of a larger research project aiming to study the influence of MCSs on the precipitation over the PB, in which an area of 150 km radius from the Paute station (2.8°S, 78.76°W, 2194 m ASL) was designated as the study area (Figure 1). To determine the radius of 150 km, the time resolution of GridSat data (3 hours) together with the average travel speed of clouds (50 km/h) was considered. The selected region allows to detect MCSs that developed over (i) the PB and (ii) the coastal plains west of the PB and (iii) from the Amazon region east of the PB.

The elevation of the PB ranges between 900 and 4200 m ASL. Precipitation variability in the PB can be classified into three distinctive regimes [11]. Over the inter-Andean valleys, there is a bimodal (BM) regime, related to the ITCZ passage, with the rainy months in the periods March-April and October-November. Along the eastern slopes of the eastern cordillera, the precipitation regime is rather uniform with a unimodal (UM) peak in July, coinciding with the relatively drier season of the BM regime. These regions are directly exposed to strong easterlies, bringing moist air from the Amazon during boreal summer, fostering the formation of orographic clouds and precipitation. The transition zones between the UM and BM regimes have been related to a third precipitation regime (TM) [11]. Annual precipitation in the UM regions varies between 1100 and 3400 mm, between 660 and 1100 mm in the BM inter-Andean valleys, and between 1000 and 1800 mm in the high mountain BM regions [17]. The temperature in the Andes region of Ecuador is highly correlated with the sea surface temperature anomalies (SSTA) in the tropical Pacific in El Niño 3

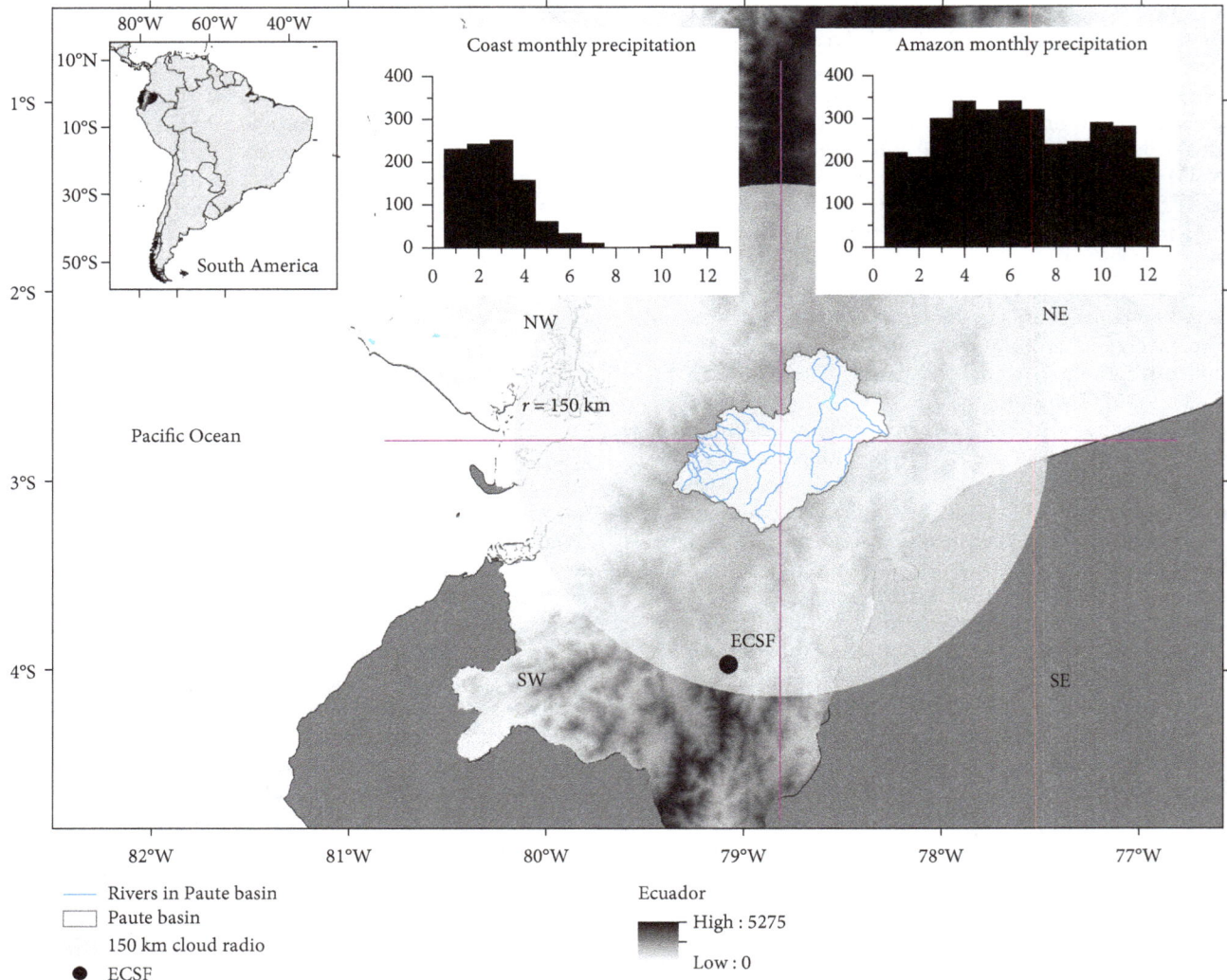

FIGURE 1: The study area includes the Paute basin, 150 km radius around the Paute station. San Francisco scientific station (ECSF).

and 3.4 regions with a one-month lag [19]. However, in the PB, due to the wide range of altitudinal gradients, the temperature is highly variable [17], with an average daily temperature of 6°C in El Labrado station, situated on the western cordillera, and 24°C in the inter-Andean valleys.

3. Data

MCS detection was based on the infrared window (IRWIN) channel (near 11 μm) brightness temperature (BRT) of the GridSat-B1 data set [20]. The GridSat-BRT data are on a latitude equal-angle grid with a 0.07-degree spatial resolution and a spatial extent of 70°N to 70°S. They are available from 01 Jan 1980 to present with a temporal resolution of 3 hours. For this study, we used the spatial subset covering 85°W to 69°W and 15°S to 5°N (Figure 2) and selected the period 01 Jan 2000 to 31 Dec 2009 which contains 29,224 images. The temporal resolution of the GridSat-B1 data certainly speaks against their use for the detection of the MCS, especially if one wants to make statements about their

life cycle, for example, the study of Rehbein et al. [21] with focus on the Amazon basin in the region between 44.96°S and 13.68°N and 82.00°W and 32.96°W. In this study, the occurrence of the MCS in the Paute basin is investigated; therefore, the time resolution of 3 hours of GridSat-B1 is sufficient for the spatial detection of the MCS.

For the analysis of the climatology of the thermodynamic (CAPE in J/kg and CIN in J/kg) and dynamical variable Omega in Pa/s on 2.5° × 2.5° resolution, data were downloaded from NOAA-CIRES 20th Century Reanalysis version 2 [22] (http://www.esrl.noaa.gov/psd/data/gridded/data.20thC_ReanV2.html). COBE-SST data in 1° × 1° resolution were provided by the NOAA/OAR/ESRL PSD (Boulder, Colorado, USA) under https://www.esrl.noaa.gov/psd/ [23].

4. Methods

Features such as the cloud area maximum extent size, cloud area maximum extent time occurrence, and the geographical location were analyzed to assess the climatology of MCSs in

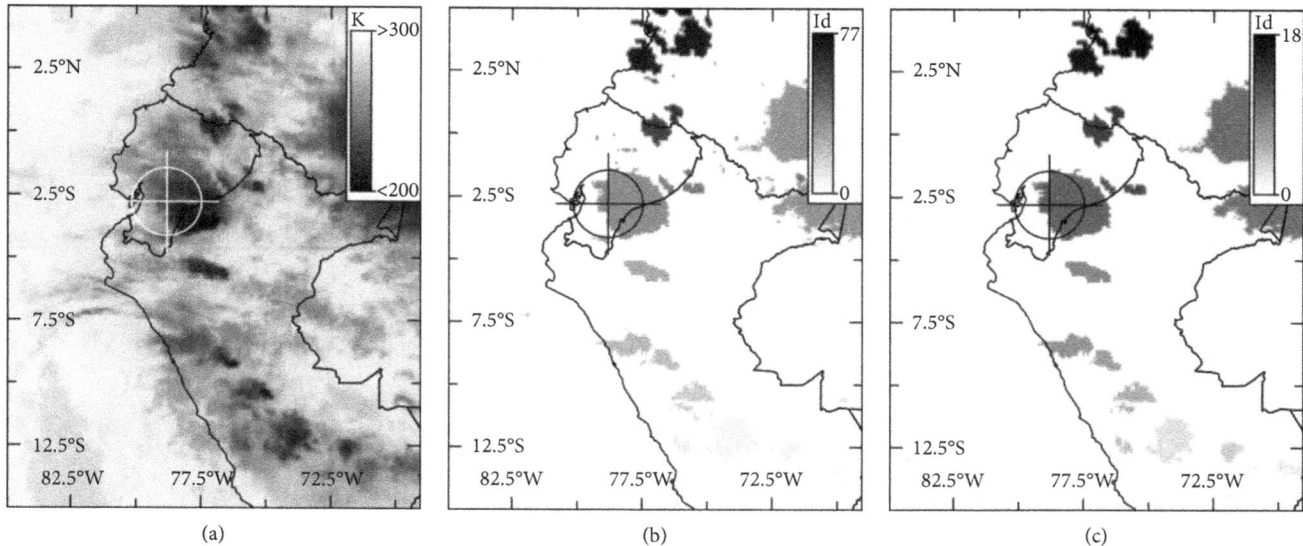

FIGURE 2: Example of image segmentation (after Bendix et al. [14]): (a) GridSat-B1 BRT at 1200 UTC 31 Mar 2009, (b) all convective entities, and (c) MCS-like entities only.

the PB. To characterize the climatology of thermodynamic and dynamical variables which may influence MCS development, the diurnal cycle of CAPE, CIN, and Omega was studied using information at 01:00, 07:00, 13:00, and 19:00 LST (LST Ecuador = UTC − 5 h). Then, it was linked to the diurnal cycle of MCSs and further related to their variability during the year. To study the MCS teleconnections, the previously studied precipitation teleconnections such as SST in the Pacific and the Atlantic Ocean were evaluated. Thus, for MCSs with maximum extend on the west of the PB, towards the coastal plains, the teleconnections with tropical Pacific SST were studied [24, 25]. On the contrary, for MCSs developed on the east of the PB, the teleconnections related to tropical Atlantic SST were studied [26].

A concise description of the different methods used is given in the following sections.

4.1. MCS Detection.

For the detection of the MCS, the method presented by Bendix et al. [14] was used with slight adaptations to the specifications of this study. The BRT thresholds for the cloud shield and for active convection were set to 241 K and 220 K, respectively, while the thresholds for the cloud area ($2000 \, km^2$) and eccentricity were maintained (Figure 2). Furthermore, the maximum distance from the center of the Paute basin (78.8177°W, 2.7874°S) to the center of the detected cloud, here the centroid of the cloud area is meant, was set to 150 km. For each classified MCS entity, morphological and intensity parameters are estimated, including the cloud area (km^2), the perimeter (km), the coldest BRT of the cloud (K), the distance to the reference point (km), and the position in the northeast (NE), northwest (NW), southeast (SE), and southwest (SW) to the reference point, and are returned by the algorithm.

4.2. Climatology of the MCS.

To study the climatology of MCSs influencing the PB, the seasonality of MCSs was first determined using box-and-whiskers plot for each month (Figure 3). Afterwards, the monthly mean number of MCSs occurring in NE, SE, NW, and SW quadrants was calculated in order to identify MCS main geographic sources of influence, which in turn may be linked to different synoptic conditions either convective or subsiding (Figure 4). Then, the diurnal cycle of MCS occurrence in the PB was evaluated by month. This was determined by calculating the average number of MCSs at its maximum extend occurring in each hour of the day on each month of the year (Figure 5). Finally, the MCS size absolute frequency and relative frequency were determined for each month. For the absolute frequency, the number of MCSs for each bin on each month was calculated. For the relative frequency, the number of MCSs for each bin on each month was divided by the total number of MCSs occurring on the respective month (Figure 6).

4.3. Climatology of Thermodynamic and Dynamical Variables Influencing MCS Development.

Variables used for the explanation of MCS formation are CAPE, CIN, and Omega. CAPE is the maximum buoyancy of an air parcel which is related to the potential updraft strength of a thunderstorm. CIN is a measure of the negative buoyancy that needs to be exceeded to initiate convection. Given that mesoscale convection may be constrained by large-scale conditions, the vertical velocity in pressure coordinates (omega (dp/dt)) at 600 hPa was analyzed, considering that upward (downward) air motion enhances (inhibits) convection. The monthly climatology of CAPE, CIN, and Omega (Figures 7–9, resp.) was generated for 01:00 LST, 07:00 LST, 13:00 LST, and 19:00 LST for the area between 82°W and 76°W and zonally averaged from 0° to 4°S.

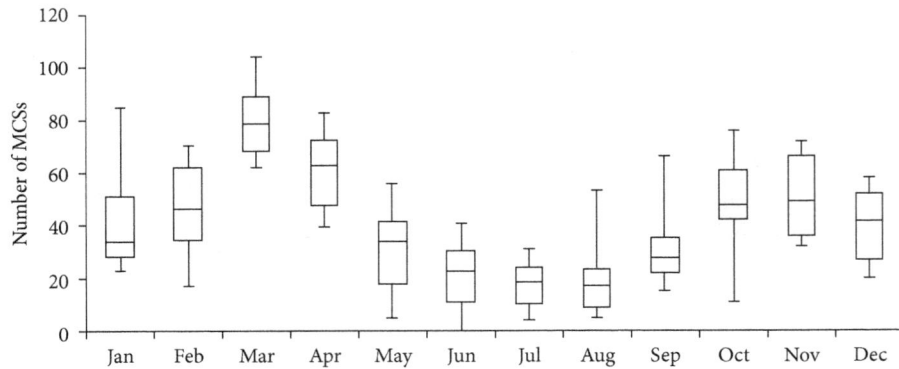

FIGURE 3: Monthly mean MCS occurrence in the Paute basin during January 2000 to December 2009.

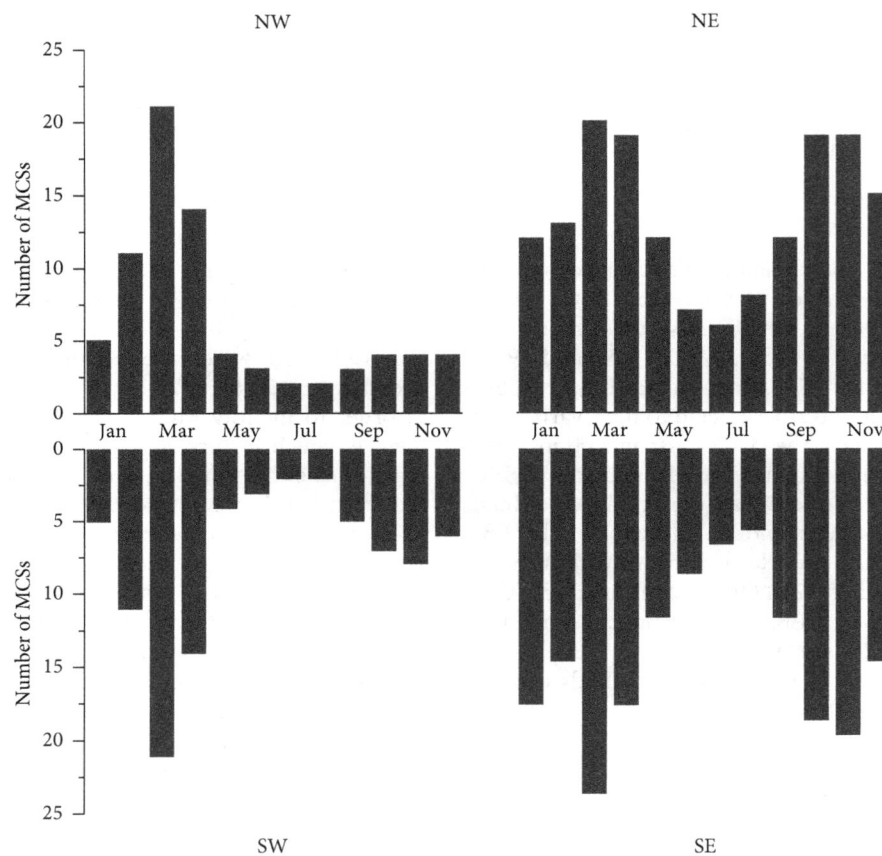

FIGURE 4: Monthly mean number of MCSs in NE, SE, NW, and SW quadrants around the Paute basin.

4.4. Teleconnection Patterns and MCS Development.
Precipitation teleconnections in Ecuador vary depending on the region [24, 26, 27]. For instance, precipitation in the coastal plains, west of the PB, is related to the sea surface temperature along the tropical Pacific. The SST in the El Niño 1 + 2 region was related to strong anomalies of precipitation in the coast mainly before the year 2000 [24, 26]. Campozano et al. [28], using the January-to-April cumulative TRMM precipitation estimates from 1999 to 2014, showed that the first principal component of precipitation in Ecuador is correlated to the Trans El Niño index, which is given by the difference between the normalized SST anomalies averaged in the Niño 1 + 2 and Niño-4 regions [29], thus measuring the gradient of anomalous SST between both regions.

The teleconnections between the cumulative area of the MCS and ENSO indices such as SST in the regions Niño 1 + 2, Niño 3.4, and TNI were evaluated on the west of the PB (MCS-W). The MCS monthly time-series anomalies were obtained by removing the trend. It was achieved by using a Fourier fitting to the time series in the first place and then removing the lower frequencies. Finally, a 3-month moving average was applied.

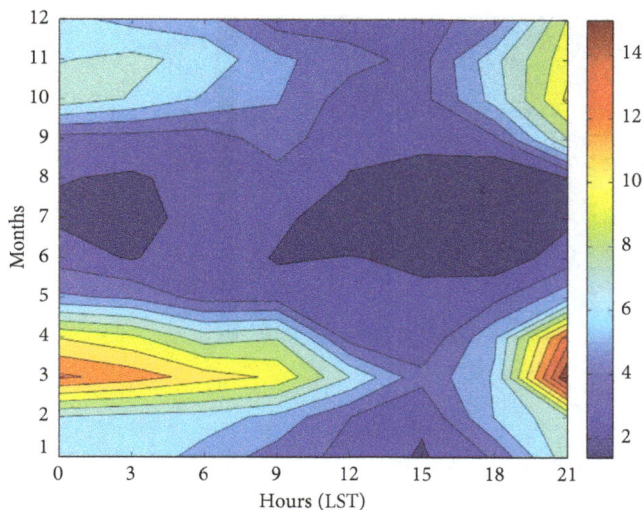

FIGURE 5: Monthly mean number of MCSs in the Paute basin.

The teleconnections were evaluated by correlating the MCS-normalized time series with SST Niño 1 + 2, SST Niño 3.4, and TNI. To study the teleconnections of MCSs east of the PB, the approach of Vuille et al. [30] was followed. Vuille et al. [30] showed that, during most of the year, the precipitation variability along the east of the Ecuadorian Andes is positively correlated to the tropical Atlantic SST anomalies south of the intertropical convergence zone and negatively correlated to the north. On the contrary, it is well known [24, 26] that SST in the tropical Pacific inversely affects the precipitation in the western Amazon. Therefore, following Vuille et al. [30], Atlantic SST data from 60°W to 10°E and 30°N to 30°S monthly anomalies were obtained by subtracting the mean monthly values for the 01-2000 to 12-2009 periods. Then, the principal components of the anomalies were calculated using the library Climate Data Operators (CDO 2018: Climate Data Operators, available at: http://www.mpimet.mpg.de/cdo). Finally, the 3-month moving average of main PC score time series was correlated with 3-month moving average normalized anomalies of MCSs on the east of the PB (MCS-E).

5. Results and Discussion

5.1. Climatology of the MCS

5.1.1. Seasonality. The multiyear annual mean of MCSs in the PB is ca. 487 MCSs/yr. At the monthly scale, the monthly mean MCS occurrence in the PB is shown in Figure 3. The bimodal intra-annual variability of MCSs in the study region follows the bimodal rainfall seasonality, which is influenced by the ITCZ displacement [11]. Bendix et al. [14] showed a similar seasonality of MCS occurrence in a study conducted in the Rio San Francisco Valley, close to the ECSF meteorological station, 3°5′22″S, 79°04′52″W, 1957 m ASL, situated in the foothills of the southern Andes in Ecuador. However, their study was based on only 3 years of satellite images. Similarly, Jaramillo et al. [3] found that the

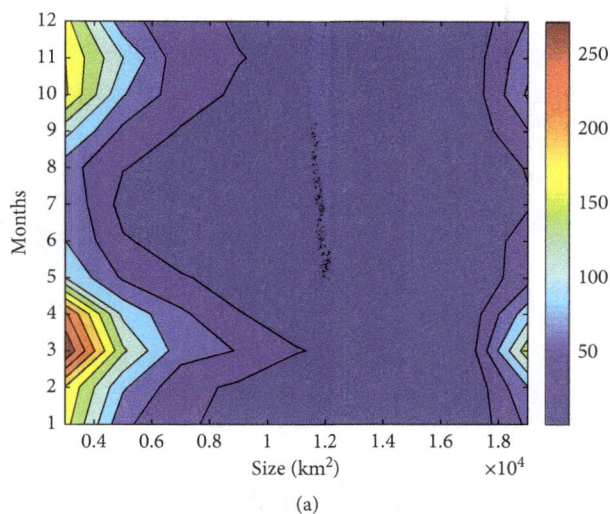

(a)

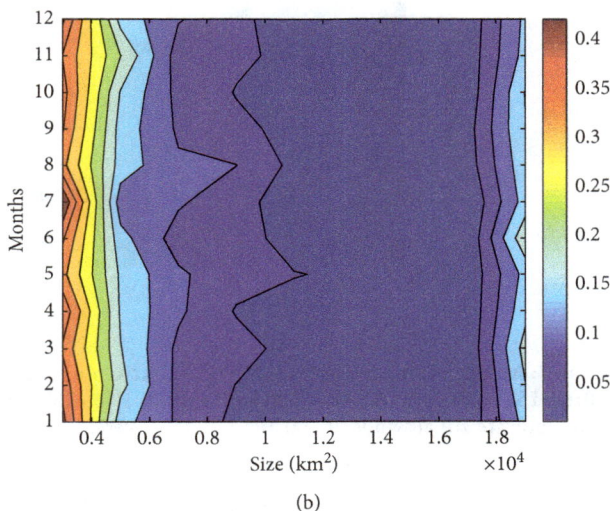

(b)

FIGURE 6: Number of MCSs by size and month (a) and monthly fraction of MCSs by size and month (b).

seasonality of MCSs is bimodal across the western Amazon using TRMM data. The results reveal that the annual course of ITCZ-related atmospheric conditions, such as atmospheric stability and changes in wind field, generally modifies the local occurrence of MCS formation.

5.1.2. MCS Geographical Location. Figure 4 shows the number of MCS occurrences per *NE, SE, NW, and SW quadrants* around the center of the PB. A striking feature is that the eastern quadrants present much more MCS occurrences, suggesting that MCSs coming from the Amazon and the eastern flanks of the eastern cordillera influence tend to dominate PB's convective systems. Only during February to April (wet season over the coast), the western quadrants show a similar occurrence of MCSs to the eastern quadrants. The predominance of MCSs throughout the year in the eastern quadrants (Figure 4) supports the hypothesis that precipitation in the PB is mainly due to the Amazon influence enhanced by the easterlies [11]. On the contrary, the

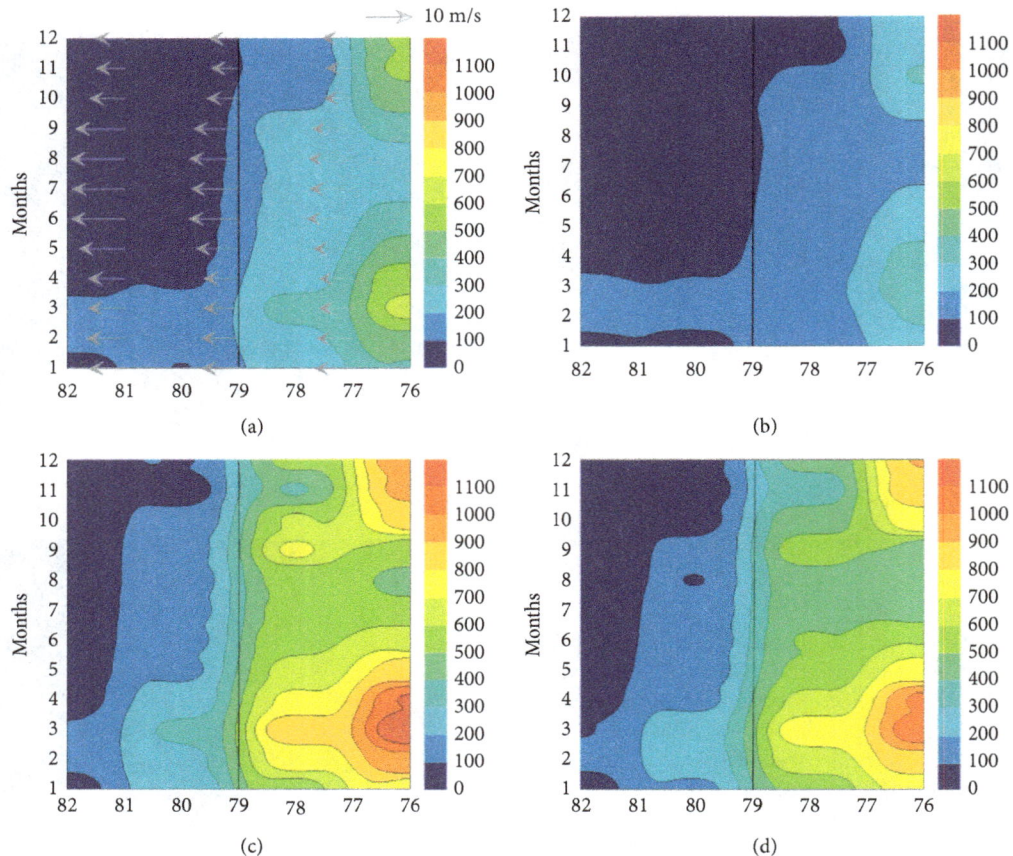

FIGURE 7: Monthly climatology of CAPE (J/kg) across a longitudinal transect from 82°W to 76°W averaged from the equator to 4°S at (a) 01:00 LST, (b) 07:00 LST, (c) 13:00 LST, and (d) 19:00 LST. The black line represents the longitudinal position of the Paute basin. In (a), arrows indicate zonal winds at 600 hPa monthly average from 2000 to 2009.

western influence of MCSs is important for the PB, from February to April, especially in March, during the rainy season of the coastal plains. The influence of the South Pacific anticyclone, which is responsible for atmospheric stable stratification during the second half of the year in the coastal plains, suppresses MCS development in the dry season in the western quadrants of the PB.

5.1.3. Monthly Diurnal Cycle.
The diurnal cycle of the MCS maximum extent averaged by month is presented in Figure 5. It is clearly depicted that nighttime occurrence is much higher, with an occurrence peak at 00:00 LST. This fact is more evident during the rainy seasons March-April and October-November. In order to examine the factors leading to MCS development during the nighttime, further analyses of thermodynamic and dynamical variables is conducted in Section 5.2. Bendix et al. [14] found a similar nighttime occurrence of MCSs in the Rio San Francisco Valley in the southern part of Ecuador. The Rio San Francisco Valley is oriented in the southwest-northeast direction, located close to the ECSF meteorological station. Bendix et al. [14] reported a secondary maximum occurrence of MCCs from 01:00 to 04:00 LST. It is important to point out that, despite the nocturnal occurrence of MCSs, the diurnal cycle of precipitation in

the inter-Andean valleys is mainly driven by the afternoon convection, with a peak around 16:00 LST [7, 11, 31].

Romatschke and Houze [32] showed that, in the eastern foothills of the Andes towards the Amazon, nocturnal convective systems are more frequent than daytime convective systems. It was hypothesized that local small-scale effects, specifically nocturnal downslope winds, a result of the divergence over the mountains, lead to convection at the foothills. Indeed, Trachte et al. [13] conducted a detailed study of the development of an MCS in the Rio San Francisco Valley located in the eastern escarpments of the Andes. They proved that the interaction of the drainage katabatic flows of the Andean drainage system with the warm-moist air of the Amazon induces flow confluence causing compressional lifting, which later leads to nocturnal cloud clusters.

5.1.4. MCS's Size Distribution.
Figure 6(a) shows the monthly absolute frequency per bin size. Of note is that February-April and October-November show larger MCSs on each bin size. However, the relative size distribution (Figure 6(b)) is more uniform throughout the year suggesting that, during the months with more MCSs, more repetitions of the same distribution occur. A more detailed analysis of the MCS monthly 25th, 50th, and 75th size

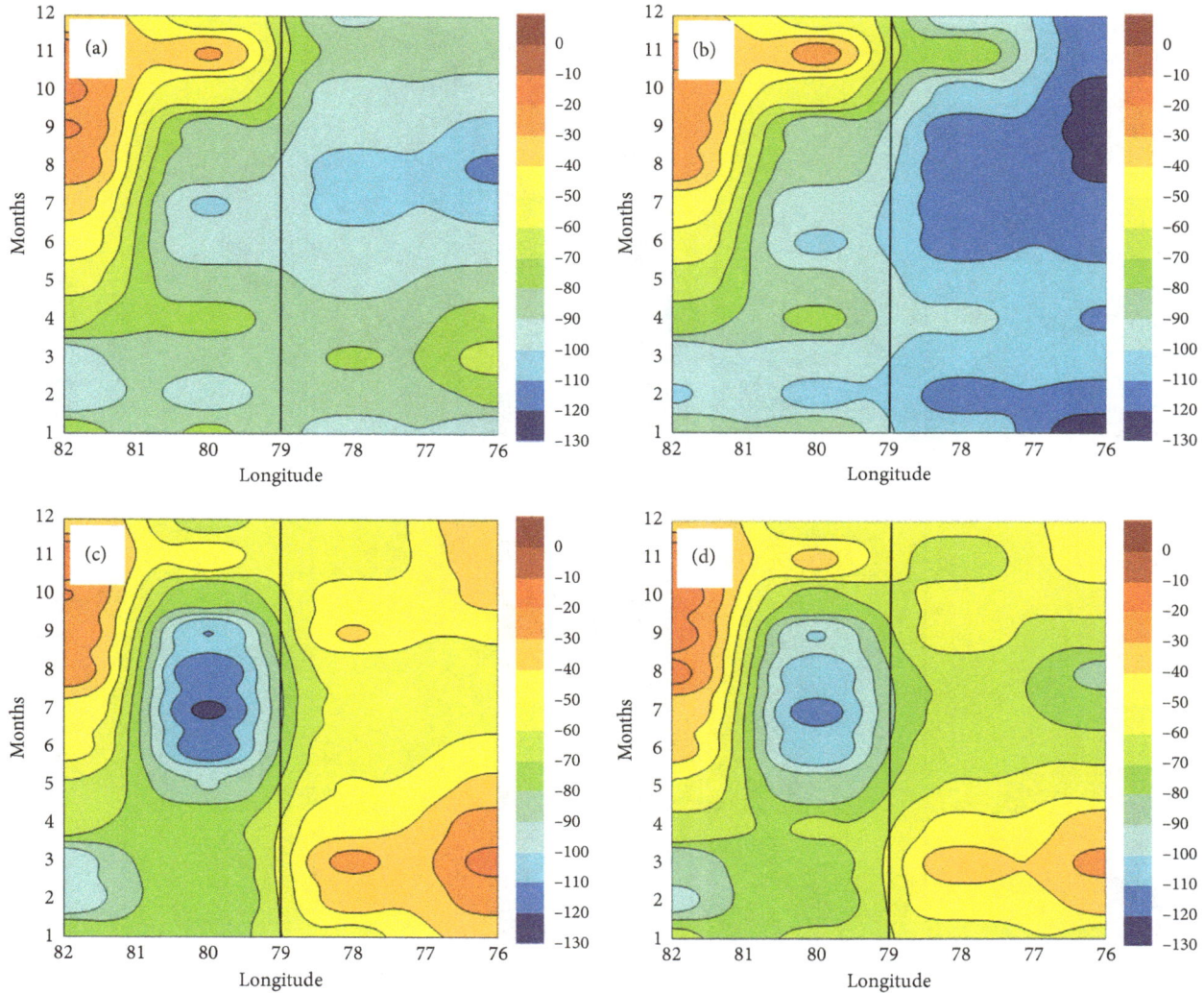

FIGURE 8: Monthly climatology of CIN (J/kg) across a longitudinal transect from 82°W to 76°W averaged from 0° to 4°S at (a) 01:00 LST, (b) 07:00 LST, (c) 13:00 LST, and (d) 19:00 LST. The black line represents the longitudinal position of the Paute basin.

percentiles is presented in Table 1. The multiannual 25th, 50th, and 75th percentiles of the MCS's area are 2989 km², 5063 km², and 11,712 km², respectively. Notwithstanding, the 25th and 50th percentiles show a small variation throughout the year. However, the 75th size percentile is 6% higher than the multiannual median in March and September, 19% higher in June (largest month), and 15% lower in December. Note that the highest 75th size percentile occurs in June. The large MCS's size is coincident with the rainy seasons February-April and October-November. The study of MCS seasonal size variation was conducted by [3] with special focus over Colombia. They found that, over the Caribbean Sea and the Pacific Ocean, MCS's size showed a small variation during the year. In contrast, the seasonal variation of MCSs over the continent was considerably large. Specifically, over the Colombian territory, Jaramillo et al. [3] showed that the largest MCSs occurred during the March-April season (average area 21,442 km²). In the Amazon region, they found that the largest MCSs tend to occur during December-February (average area 27,017 km²).

5.2. *Climatology of CAPE, CIN, and Omega.* The monthly climatology of CAPE (in J/kg), CIN (in J/kg), and Omega at 01:00 LST, 07:00 LST, 13:00 LST, and 19:00 LST was used to relate the MCS formation to the atmospheric conditions on longer terms average. The values of the CAPE are depicted in Figures 7(a)–7(d). The black line in the figure depicts the longitudinal location of the PB. In general, the CAPE pattern at 01:00, 07:00, 13:00, and 19:00 LST differs between both sides of the PB, with higher values over the eastern Amazon sectors (longitude 76W). CAPE values at 01:00 LST (Figure 7(a)) are low across the longitudinal transect, limiting the capabilities of the formation of MCSs. CAPE values at 07:00 LST (Figure 7(b)) display low values on the western and eastern sides of the PB, suggesting that a relatively low energy for convection is available early in the morning, hindering the development of MCSs. This condition is more common during June-August, when low CAPE values are observed over the Amazon. CAPE at 13:00 LST (Figure 7(c)) shows a large longitudinal contrast of CAPE with the Andes as boundary. To the east of the Andes and into the Amazon,

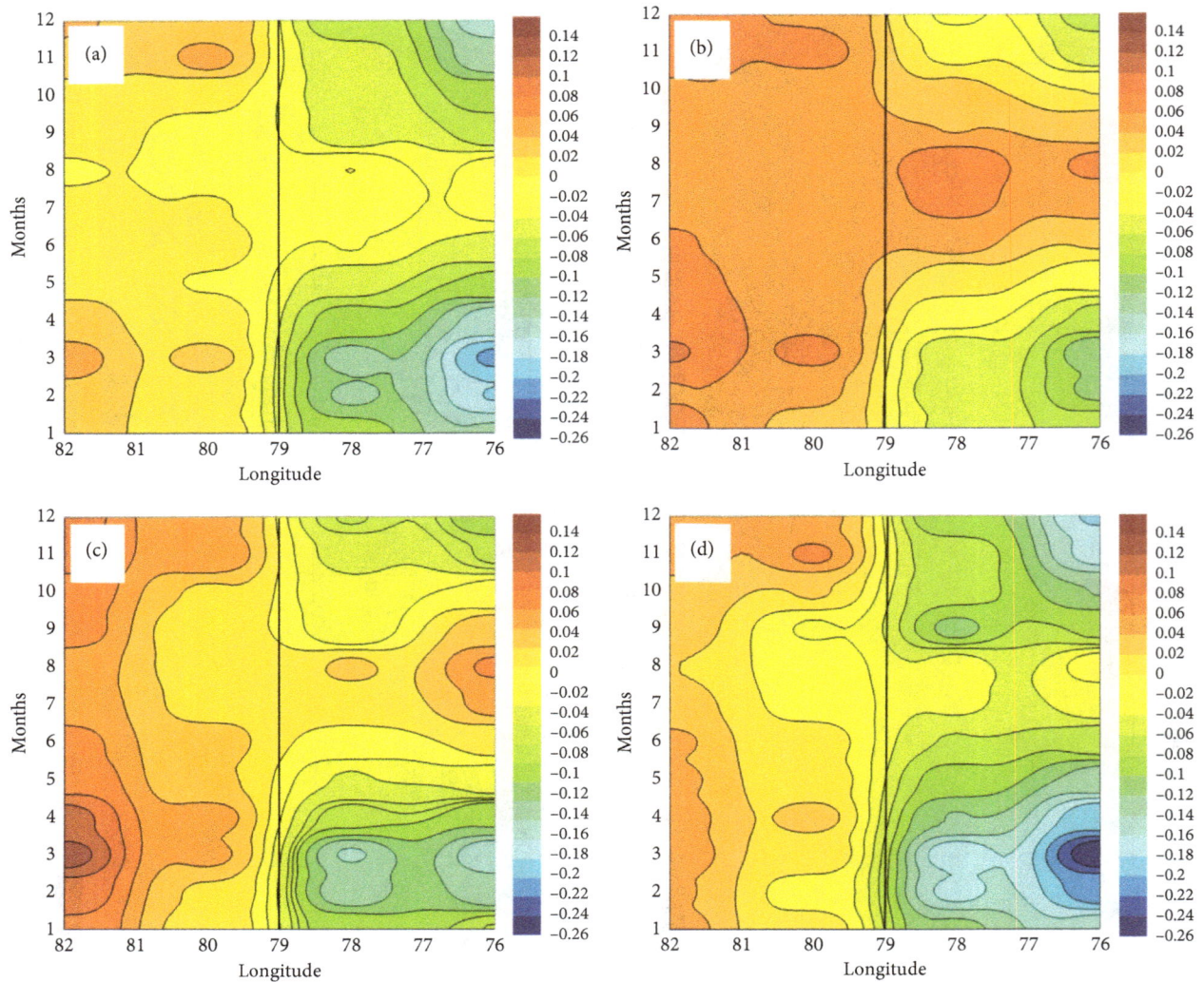

FIGURE 9: Monthly climatology of Omega (Pa/s) across a longitudinal transect from 82°W to 76°W averaged from 0° to 4°S at (a) 01:00 LST, (b) 07:00 LST, (c) 13:00 LST, and (d) 19:00 LST. The black line represents the longitudinal position of the Paute basin.

values of CAPE over 800 J/kg are displayed from February to April, with CAPE increasing eastward. These high values of CAPE are related to the increase of MCS development during the afternoon. From June to August, CAPE reduces to around 600 J/kg, with a corresponding reduction of MCS activity in the afternoon. CAPE values increase again in September and October, when there is a corresponding increase in MCS activity. From 13:00 to 19:00 LST and over the west side of the PB, lower CAPE values are observed likely due to midlevel warm air advection from the Andes. The latter likely explains the higher occurrence of MCSs during the night, especially in the periods February to April and September to December in the Amazon region.

The monthly climatology of CIN is shown in Figures 8(a)–8(d), averaged for 01:00 LST, 07:00 LST, 13:00 LST, and 19:00 LST. CIN values from 0 to −50 are considered weak, from −50 to −200 moderate, and <−200 strong. Throughout the year, the CIN tends to be moderately intense during the night, except during February-April, when the CIN decreases

between −60 and −80. Over the western side of the Andes and early in the morning, CIN is moderately intense during January-September and weak during October–December. Notwithstanding, CAPE values at this time are lowest in the Amazon side limiting the possibility of the development of MCSs. At 13:00 LST (Figure 8(c)), throughout the year, the CIN values at the west side of the PB are moderate but are weak between November and December, coinciding with the beginning of the rainy season in the coastal region. Similarly, the CIN values are weak, all year round, on the Amazonian-influenced east side of the PB, which together with the bi-modal signal of CAPE might account for the afternoon convection. The CIN values at 19:00 LST (Figure 8(d)) are similar to the values obtained at 13:00 LST, which affects sides of the PB and throughout the year.

The situations depicted by the CAPE and CIN analyses may be conditioned by Omega. Figures 9(a)–9(d) show the Omega monthly climatology at 01:00 LST, 07:00 LST, 13:00 LST, and 19:00 LST. As it can be observed from Figures 9(a)

TABLE 1: MCS's size (km^2): 25th, 50th, and 75th percentiles by month.

Percentile	Jan	Feb	Mar	Apr	May	Jun	Jul	Aug	Sep	Oct	Nov	Dec	Mean
P25	2806	2867	3050	3050	3187	3050	2928	3142	3004	2928	2928	2867	2989
P50	4728	4758	5246	5002	5612	5246	4758	5734	5368	5368	5002	4514	5063
P75	11,605	10,797	12,444	11,163	12,292	13,939	10,919	12,109	12,444	12,154	11,041	9897	11,712

and 9(d), a considerable modification of mesoscale features is the likely consequence of the ascendant air velocity in February to April at 01:00 and 19:00 LST. This fact may support the development of convective activity. On the contrary, much lower values of Omega ascending velocity are shown during the same months at 7:00 LST and 13:00 LST (Figures 9(b) and 9(c)). Interestingly, subsidence is present at 07:00 LST and 13:00 LST (Figures 9(b) and 9(c)) during June, July, and August, which is in agreement with the least active season of MCS development.

5.3. Teleconnections for MCS Development in the Paute Basin

5.3.1. Tropical Atlantic SST Influence on MCS Development.
In Table 2, the statistics of the first 10 PCs are shown. The highest correlation (Pearson correlation 0.51) between MCS-E and PC score time series is achieved with the 2nd PC, which accounts for 17.70% of the explained variance and together with the PC-1 accounts for 40.25% of the explained variance. In Figure 10, the PC2 of the tropical Atlantic SSTA region 60°W to 10°E and 30°N to 30°S is presented. The spatial loading pattern of PC-2 depicts positive Atlantic SST anomalies from 5°N to 15°S (Figure 10). This pattern corresponds to the second varimax-rotated PC of Vuille et al. [30], with the first EOF of [33] and the EOF1 of [34] for the South Atlantic domain. This mode represents an in-phase SST response over the entire tropical South Atlantic with the strengthening and weakening of the subtropical anticyclone [34].

The normalized PC-2 score time series and detrended MCS-E time series are shown in Figure 11(a). The normalized values greater/lower than 0.5/−0.5 standard deviations are shown. Values less than this threshold are set to zero. Important departures of MCS-E occurrence are especially during November 2001 and April 2008. Both events may be related to anomalous SST conditions in the tropical Pacific (Figure 11(b)). For instance, in November 2001, TNI registers 1.3 negative standard deviation anomaly, which is the second lower value behind June 2007 during the study period. For April 2008, TNI shows a positive anomalous value of 2.25 standard deviations, which belongs to the extremely wet episode in the coast of Ecuador [26]. Moreover, following Campozano et al. [11], extreme positive values of TNI imply wet conditions in Ecuador, whereas extreme positive values of Niño 1 + 2 represent wet conditions in the coast and dry conditions in the Amazon; thus, both are Pacific competing influences for rainfall in Ecuador.

5.3.2. Tropical Pacific Influence on MCS Development.
The influence of SST along the tropical Pacific on the MCS developed on the western region of the PB can be best

TABLE 2: Eigenvalues, percentage of the explained variance, and percentage of cumulative explained variables for 10 principal components, based on the principal component analysis of the tropical Atlantic region

Principal component	Eigenvalue	Explained variance (%)	Cumulative explained variance (%)
1	0.0336	22.55	22.55
2	**0.0263**	**17.70**	**40.25**
3	0.0212	14.25	54.50
4	0.0106	7.15	61.65
5	0.0093	6.26	67.91
6	0.0081	5.46	73.37
7	0.0053	3.57	76.94
8	0.0045	3.01	79.94
9	0.0040	2.68	82.63
10	0.0028	1.88	84.50

Bold values indicate the highest correlation patterns with MCS-E.

illustrated by Figure 11(b). In Figure 11(b), the time series of TNI Fourier-detrended normalized anomalies and MCS-normalized detrended anomalies is shown (Pearson correlation 0.44). The normalized values greater/lower than 0.5/−0.5 standard deviations are shown. The contrasting patterns between the MCS-E and MCS-W (Figures 11(a) and 11(b)), respectively, are due to the different climatic influences on both sides of the Andes. Given that the effect of ENSO in Ecuador is mainly confined to the coastal area [24, 26], it is expected that, during ENSO phases, mainly MCS-W be affected. A clear sign of ENSO influence on MCS-W during the 2008 event is shown in Figure 11(b). The event of 2008, characterized by a cold central Pacific sea surface temperature and hot sea surface temperature along the east of the tropical Pacific, is typical for La Niña Modoki (LNM) [35]. On the contrary, Pearson correlation of MCS-W with SST Niño 1 + 2 and Niño 3.4 is 0.119 and 0.130, respectively, showing that MCS-W presents a much higher relation to the TNI ENSO index (Pearson correlation 0.44). Therefore, the increase of MCS-W may be related to an enhancement of convective conditions as a consequence of a positive SST gradient between east and west tropical Pacific, which may change the typical pattern of subsidence on the tropical eastern Pacific of the Walker circulation [35].

On the seminal study of Velasco and Fritsch [36], they used one year (05-1982/05-1983) of MCS data during an El Niño year. In South American midlatitudes, they found that, during the El Niño period, the number of MCCs was more than double the number than in the non-El Niño year (05-1982/05-1983), showing the impact of ENSO on MCC development. Moreover, they found an increase of MCS along the Peruvian coasts, due to the anomalous increase of SST in the Niño 1 + 2 region.

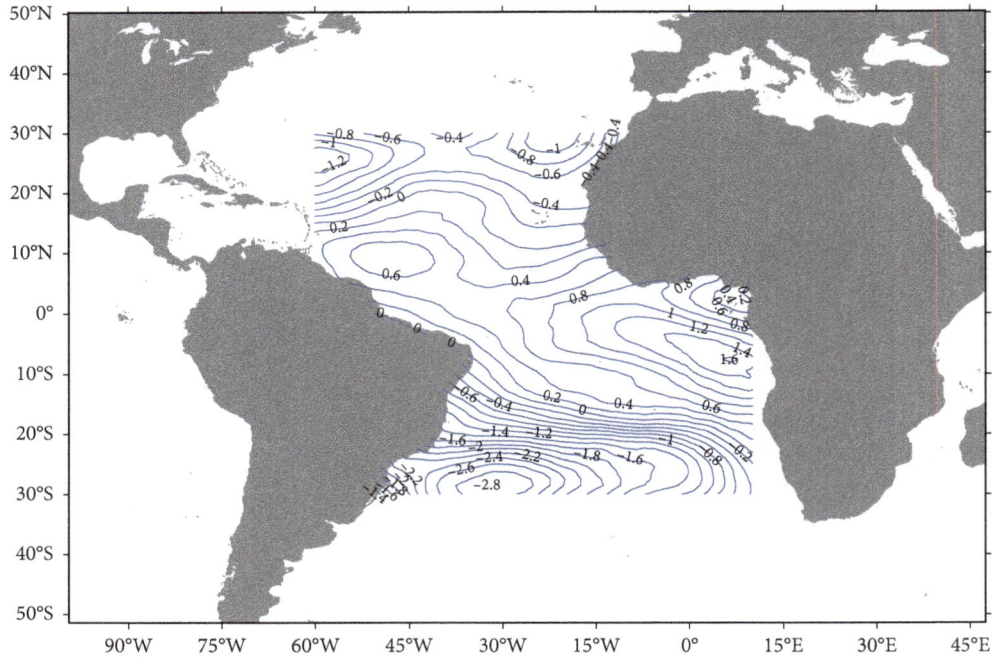

FIGURE 10: PC2 of the tropical Atlantic SSTA region 60°W to 10°E and 30°N to 30°S.

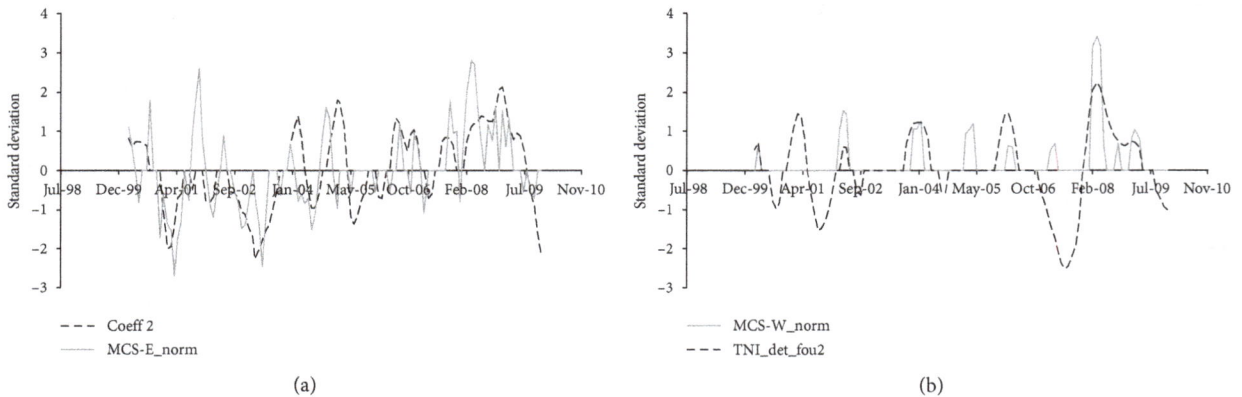

(a)

(b)

FIGURE 11: (a) 3-month moving average anomalies of MCSs monthly cumulative area east of the Paute basin (MCS-E_norm) and second principal component coefficients of tropical Atlantic SST from January 2000 to December 2009. (b) 3-month moving average anomalies of MCSs monthly cumulative area west of the Paute basin (MCS-W_norm) and 3-month moving average anomalies of Trans-Niño index detrended.

6. Conclusions

Mesoscale convective systems (MCSs) are related to large amounts of precipitation often inducing flash floods and rain-induced landslides, with further negative impacts on the infrastructure and socioeconomic welfare. This paper examined the climatology of MCSs, the climatology of thermodynamic and dynamical variables, and teleconnections of MCS development in the Paute basin, an inter-Andean river basin in southern Ecuador.

The climatologic analysis of MCS occurrence shows a clear bimodal pattern, with higher occurrences of MCSs in March-April (MA) and October-November (ON),

analogous with the bimodal character of the large-scale rainfall seasonality in the region. The easterlies are responsible for the basin preferentially receiving MCSs mainly from the east all year round, except during the El Niño periods when the number of MCSs developed in the coastal plains increases considerably. The diurnal cycle of the MCSs reveals a clear nocturnal occurrence of MCSs during both convectively active periods. Despite the fact that the MCS occurrence is bimodal, the seasonal distribution of the size of the MCSs remains almost constant throughout the year. The climatology of thermodynamic variables in the Amazon shows persistent high convective available potential energy and weak convective inhibition values from midday to nighttime, which seems to be related to the nocturnal formation of MCSs during MA and ON.

With respect to MCS teleconnections, the occurrence of MCS variability in the east of the Paute basin is positively correlated to the tropical Atlantic sea surface temperature south of the equator, linked to the strengthening and weakening of the Atlantic subtropical anticyclone. Main departures from this correlation are linked to anomalous conditions in the tropical Pacific, which are mainly related to ENSO. On the western side of the Paute basin, using the Trans-Niño index, it was shown that the MCSs developed on the west of the PB respond positively to the dynamics of ENSO, thus explaining, to a certain extent, the interannual variability of MCSs in the region of study.

Conflicts of Interest

The authors declare that there are no conflicts of interest regarding the publication of this article.

Acknowledgments

The GridSat CDR used in this study was acquired from NOAA's National Climatic Data Center (http://www.ncdc.noaa.gov). This CDR was originally developed by Ken Knapp and colleagues for NOAA's CDR Program. The authors thank the German Research Foundation (DFG) for funding our project C12 in the framework of the PAK 823–825 "Platform for Biodiversity and Ecosystem Monitoring and Research in South Ecuador." The authors acknowledge DRI for their support (Mejia) in helping develop this manuscript.

References

[1] G. Poveda and O. Mesa, "Feedbacks between hydrological processes in tropical South America and large-scale ocean–atmospheric phenomena," *Journal of Climate*, vol. 10, no. 1981, pp. 2690–2702, 1997.

[2] R. D. Garreaud, M. Vuille, R. Compagnucci, and J. Marengo, "Present-day South American climate," *Palaeogeography, Palaeoclimatology, Palaeoecology*, vol. 281, no. 3-4, pp. 180–195, 2009.

[3] L. Jaramillo, G. Poveda, and J. Mejía, "Mesoscale convective systems and other precipitation features over the tropical Americas and surrounding seas as seen by TRMM," *International Journal of Climatology*, vol. 37, pp. 380–397, 2017.

[4] R. A. Houze Jr., *Cloud Dynamics*, Academic Press, San Diego, CA, USA, 1995.

[5] R. A. Houze Jr., "Mesoscale convective systems," *Reviews of Geophysics*, vol. 42, no. 4, pp. 1–43, 2004.

[6] H. W. Kim and D. K. Lee, "An observational study of mesoscale convective systems with heavy rainfall over the Korean Peninsula," *Weather and Forecasting*, vol. 21, no. 2, pp. 125–148, 2006.

[7] J. Bendix and W. Lauer, "Die Niederschlagsjahreszeiten in Ecuador und ihreklimadynamische Interpretation," *Erdkunde*, vol. 46, pp. 118–134, 1992.

[8] T. Eichler, W. Higgins, T. Eichler, and W. Higgins, "Climatology and ENSO-related variability of North American

[9] R. V. Andreoli, R. A. Ferreira de Souza, M. T. Kayano, and L. A. Candido, "Seasonal anomalous rainfall in the central and eastern Amazon and associated anomalous oceanic and atmospheric patterns," *International Journal of Climatology*, vol. 32, no. 8, pp. 1193–1205, 2012.

[10] T. J. Killeen, M. Douglas, T. Consiglio, P. M. Jørgensen, and J. Mejia, "Dry spots and wet spots in the Andean hotspot," *Journal of Biogeography*, vol. 34, no. 8, pp. 1357–1373, 2007.

[11] L. Campozano, R. Célleri, K. Trachte, J. Bendix, and E. Samaniego, "Rainfall and cloud dynamics in the Andes: a Southern Ecuador case study," *Advances in Meteorology*, vol. 2016, Article ID 3192765, 15 pages, 2016.

[12] C. F. Angelis, G. R. McGregor, and C. Kidd, "A 3-year climatology of rainfall characteristics over tropical and subtropical South America based on tropical rainfall measuring mission precipitation radar data," *International Journal of Climatology*, vol. 24, no. 3, pp. 385–399, 2004.

[13] K. Trachte, R. Rollenbeck, and J. Bendix, "Nocturnal convective cloud formation under clear-sky conditions at the eastern Andes of south Ecuador," *Journal of Geophysical Research*, vol. 115, no. D24203, 2010.

[14] J. Bendix, K. Trachte, J. Cermak, R. Rollenbeck, and T. Nauß, "Formation of convective clouds at the foothills of the tropical eastern Andes (South Ecuador)," *Journal of Applied Meteorology and Climatology*, vol. 48, no. 8, pp. 1682–1695, 2009.

[15] W. Buytaert and B. De Bievre, "Water for cities: the impact of climate change and demographic growth in the tropical Andes," *Water Resources Research*, vol. 48, no. 8, 2012.

[16] W. Buytaert, R. Celleri, P. Willems, B. De Bièvre, and G. Wyseure, "Spatial and temporal rainfall variability in mountainous areas: a case study from the south Ecuadorian Andes," *Journal of Hydrology*, vol. 329, no. 3-4, pp. 413–421, 2006.

[17] R. Celleri, P. Willems, W. Buytaert, and J. Feyen, "Space–time rainfall variability in the Paute basin, Ecuadorian Andes," *Hydrological Processes*, vol. 21, no. 24, pp. 3316–3327, 2007.

[18] J. Espinoza, J. Ronchail, L. Guyot et al., "Spatio-temporal rainfall variability in the Amazon basin countries (Brazil, Peru, Bolivia, Colombia)," *International Journal of Climatology*, vol. 29, no. 11, pp. 1574–1594, 2009.

[19] J. Bendix, "Precipitation dynamics in Ecuador and northern Peru during the 1991/92 El Niño: a remote sensing perspective," *International Journal of Remote Sensing*, vol. 21, no. 3, pp. 533–548, 2000.

[20] K. R. Knapp and NOAA CDR Program, *NOAA Climate Data Record (CDR) o of Gridded Satellite Data from ISCCP B1 (GridSat-B1) 11 Micron Brightness Temperature, Version 2*, NOAA National Climatic Data Center, Asheville, NC, USA, 2014.

[21] A. Rehbein, T. Ambrizzi, and C. R. Mechoso, "Mesoscale convective systems over the Amazon basin. Part I: climatological aspects," *International Journal of Climatology*, vol. 38, no. 1, pp. 215–229, 2018.

[22] J. Bendix, R. Rollenbeck, and W. E. Palacios, "Cloud detection in the Tropics–a suitable tool for climate-ecological studies in the high mountains of Ecuador," *International Journal of Remote Sensing*, vol. 25, no. 21, pp. 37–41, 2004.

[23] M. Ishii, A. Shouji, S. Sugimoto, and T. Matsumoto, "Objective analyses of sea-surface temperature and marine meteorological variables for the 20th century using ICOADS and the Kobe Collection," *International Journal of Climatology*, vol. 25, no. 7, pp. 865–879, 2005.

[24] F. Rossel, R. Mejía, G. Ontaneda et al., "Régionalisation de l'influence du el Niño sur les précipitations de l'Équateur,"

Bulletin de l'Institut Français d'Études Andines, vol. 27, no. 3, pp. 643–654, 1998.

[25] G. P. Compo, J. S. Whitaker, and P. D. Sardeshmukh, "Feasibility of a 100 year reanalysis using only surface pressure data," *Bulletin of the American Meteorological Society*, vol. 87, pp. 175–190, 2006.

[26] J. Bendix, K. Trache, E. Palacios et al., "El Niño meets La Niña–anomalous rainfall patterns in the "traditional" El Niño region of southern Ecuador," *Erdkunde*, vol. 65, no. 2, pp. 151–167, 2011.

[27] D. Mora and P. Willems, "Decadal oscillations in rainfall and air temperature in the Paute River Basin—Southern Andes of Ecuador," *Theoretical and Applied Climatology*, vol. 108, pp. 267–282, 2012.

[28] L. Campozano, D. Ballari, and R. Célleri, "Imágenes TRMM para identificar patrones de precipitación e índices ENSO en Ecuador," *MASKANA*, vol. 5, pp. 185–191, 2014.

[29] K. E. Trenberth and D. P. Stepaniak, "Indices of El Niño evolution," *Journal of Climatology*, vol. 14, no. 8, pp. 1697–1701, 2001.

[30] M. Vuille, R. Bradley, and F. Keimig, "Climate variability in the Andes of Ecuador and its relation to tropical Pacific and Atlantic sea surface temperature anomalies," *Journal of Climate*, vol. 13, no. 14, pp. 2520–2535, 1999.

[31] R. S. Padrón, B. P. Wilcox, P. Crespo, and R. Célleri, "Rainfall in the Andean Páramo: new insights from high-resolution monitoring in Southern Ecuador," *Journal of Hydrometeorology*, vol. 16, no. 3, pp. 985–996, 2015.

[32] U. Romatschke and R. A. Houze Jr., "Characteristics of precipitating convective systems accounting for the summer rainfall of tropical and subtropical South America," *Journal of Hydrometeorology*, vol. 14, no. 1, pp. 25–46, 2013.

[33] D. B. Enfield and D. A. Mayer, "Tropical Atlantic SST variability and its relation to El Niño–Southern Oscillation," *Journal Geophysical Research*, vol. 102, pp. 929–945, 1997.

[34] S. A. Venegas, L. Mysak, and D. Straub, "Evidence for interannual and interdecadal climate variability in the South Atlantic," *Geophysical Research Letters*, vol. 23, no. 19, pp. 2673–2676, 1996.

[35] K. Ashok, S. K. Behera, S. A. Rao, H. Weng, and T. Yamagata, "El Niño Modoki and its possible teleconnection," *Journal of Geophysical Research: Oceans*, vol. 112, no. 11, pp. 1–27, 2007.

[36] I. Velasco and J. M. Fritsch, "Mesoscale convective complexes in the Americas," *Journal of Geophysical Research*, vol. 92, no. D8, pp. 9591–9613, 1987.

7

Continentality and Oceanity in the Mid and High Latitudes of the Northern Hemisphere and their Links to Atmospheric Circulation

Edvinas Stonevicius (iD), Gintautas Stankunavicius, and Egidijus Rimkus (iD)

Institute of Geosciences, Vilnius University, Vilnius, Lithuania

Correspondence should be addressed to Egidijus Rimkus; egidijus.rimkus@gf.vu.lt

Academic Editor: Roberto Fraile

The climate continentality or oceanity is one of the main characteristics of the local climatic conditions, which varies with global and regional climate change. This paper analyzes indexes of continentality and oceanity, as well as their variations in the middle and high latitudes of the Northern Hemisphere in the period 1950–2015. Climatology and changes in continentality and oceanity are examined using Conrad's Continentality Index (CCI) and Kerner's Oceanity Index (KOI). The impact of Northern Hemisphere teleconnection patterns on continentality/oceanity conditions was also evaluated. According to CCI, continentality is more significant in Northeast Siberia and lower along the Pacific coast of North America as well as in coastal areas in the northern part of the Atlantic Ocean. However, according to KOI, areas of high continentality do not precisely correspond with those of low oceanity, appearing to the south and west of those identified by CCI. The spatial patterns of changes in continentality thus seem to be different. According to CCI, a statistically significant increase in continentality has only been found in Northeast Siberia. In contrast, in the western part of North America and the majority of Asia, continentality has weakened. According to KOI, the climate has become increasingly continental in Northern Europe and the majority of North America and East Asia. Oceanity has increased in the Canadian Arctic Archipelago and in some parts of the Mediterranean region. Changes in continentality were primarily related to the increased temperature of the coldest month as a consequence of changes in atmospheric circulation: the positive phase of North Atlantic Oscillation (NAO) and East Atlantic (EA) patterns has dominated in winter in recent decades. Trends in oceanity may be connected with the diminishing extent of seasonal sea ice and an associated increase in sea surface temperature.

1. Introduction

Continentality and oceanity are important parameters which describe local climatic conditions. They demonstrate the extent to which local climate is influenced by sea-landmass interactions. Like most other climate indicators, these parameters are dynamic and are related to both global climate change and consequently changes in atmospheric circulation.

Continentality is primarily affected by a range of climatic variables, such as latitude, distance to sea, and atmospheric circulation. In most cases, continentality index calculations are based on the annual air temperature range and latitude. A larger annual air temperature range is associated with higher thermal contrasts and greater continentality.

Under changing climate conditions, continentality could be affected in different ways [1]. Due to global climate change, air temperature tends to increase in most parts of the world. In recent decades, the fastest warming was observed in the mid latitudes of the Northern Hemisphere [2]. Therefore, analysis of changes in different climate indices, including of continentality changes in the mid and high latitudes of the Northern Hemisphere, is of considerable importance. In areas where winter air temperatures have a more substantial positive trend than their summer counterparts, corresponding values of the continentality index decline, and vice versa. An increase in the annual cycle's amplitude in the mid latitudes of the Northern Hemisphere has been identified over the past two decades:

i.e., winter air temperatures increased slightly, while changes in the summer were more significant [2–4].

However, changes in the annual air temperature range vary considerably in different regions, and therefore, trends in climate continentality also differ. Regional investigations of climate continentality began in the first half of the twentieth century. Gorczynski [5], Brunt [6], Raunio [7], and others described the climate continentality of different localities on the basis of the annual air temperature range. Hirschi et al. [1] analysed the global continentality change using NCEP/NCAR reanalysis data in the period 1948–2005. A significant decline in continentality was noted in the Arctic and Antarctic due to a large increase in the temperature of the coldest month. However, the continentality index in Southeastern Europe also increased [1].

In recent years, regional features of continentality and oceanity have been analysed in Greece [8], Turkey [9], and Pakistan [10]. It has been determined that climate continentality has intensified in the Iberian Peninsula [11]. Negligible increases in continentality were also observed in Slovakia [12] and no significant changes were found in the Czech Republic [13], while a statistically significant increase in continentality was identified in the Middle East and North Africa [14]. Moreover the authors of [14] argue that regional circulation patterns (e.g., over the Mediterranean) do not play a critical role in determining the trends identified in continentality. Rather, they refer to changes in large-scale atmospheric circulation over the North Atlantic [14].

Moving poleward in the Northern Hemisphere, landmasses become larger, so continentality tends to be less pronounced in periods with enhanced zonal circulation. In contrast, it becomes more noteworthy with enhanced meridional circulation [15] and the greater influence of the continental Arctic air masses [16].

Changes in continentality affect both natural (such as vegetation zones) and anthropogenic (e.g., water resources and agriculture) systems, so investigations of changes in continentality are of great importance [8, 14, 17]. Furthermore, relatively few studies have analysed continentality and its changes on a global scale [1]. In addition, a lack of research exists analysing the effect of atmospheric circulation on continentality index values.

Therefore, the aim of this research is to evaluate the spatial distribution of the widely accepted Conrad's Continentality Index (CCI) and Kerner's Oceanity Index (KOI) in the middle and high latitudes of land areas of the Northern Hemisphere, as well as to evaluate the changes in these indexes since the middle of the twentieth century and their connections to atmospheric circulation.

2. Methods

The annual air temperature range and latitude were included in the continentality index formulas developed by Gorczynski [18], Johansson [19], Conrad [20], Raunio [7], Marsz and Rakusa-Suszczewskis [21], and others.

In this research, continentality was evaluated using the CCI proposed by Conrad [20]:

$$\text{CCI} = \frac{1.7\left(T_{\max} - T_{\min}\right)}{\sin\left(\varphi + 10\right)} - 14, \tag{1}$$

where $T_{\max}$ (°C) is the mean temperature of the warmest months of the year, $T_{\min}$ (°C) is the mean temperature of the coldest months of the year, and φ is the latitude.

A large annual range of air temperatures results in larger index values and consequently indicates a more continental climate. The smallest differences can be observed in the most oceanic climate conditions. The territories where index values range from −20 to 20 can be described as hyperoceanic, from 20 to 50 as oceanic, from 50 to 60 as subcontinental, from 60 to 80 as continental, and from 80 to 120 as hypercontinental [11].

In 1905, Kerner proposed an oceanity index [22]. This index represents the ratio of the mean monthly air temperature difference between October and April and the difference between mean monthly temperatures of the warmest and coldest months. Small or negative values indicate high continentality, whereas high index values indicate marine climate conditions [10]. The oceanity index (KOI) according to Kerner was evaluated as follows:

$$\text{KOI} = \frac{100\left(T_{\text{oct}} - T_{\text{apr}}\right)}{T_{\max} - T_{\min}}, \tag{2}$$

where T_{oct} and T_{apr} (°C) are the mean monthly temperature in October and April, respectively, and $T_{\max}$ and $T_{\min}$ (°C) are the same as in Equation (1). This index is based on the assumption that due to higher thermal water inertia in marine climates, springs are colder than autumns, whereas in continental climates, springs tend to demonstrate higher or similar temperatures as in autumn. The oceanity of the climate increases with index values. Small or negative values demonstrate continental climate conditions, while large values indicate a marine climate [8]. In order to visualise the spatial distribution of KOI, the following classes of index were used in this research: less or equal to −10 = hypercontinental; from −9 to 0 = continental; from 1 to 10 = subcontinental; from 11 to 20 = oceanic; and from 21 to 50 = hyperoceanic.

The CCI and especially KOI are only feasible in regions with distinct seasonal air temperature changes. We opted to analyse continentality and oceanity above a latitude of 30° in the Northern Hemisphere, where temperature seasonality is high.

The mean monthly air temperature values for the period 1950–2015 above the land were derived from the CRU TS4.00 database [23]. The grid cell size was 0.5 × 0.5°. CRU TS is a high-resolution global data set, covering all landmasses between 60°S and 80°N. The priority of the CRU TS data set is its completeness, having no missing data over the land. Particular attention is paid to data quality control [24]. However, the data set is not strictly homogeneous, and larger uncertainties can be found over regions with a sparse network of meteorological stations, especially deserts and mountains [25, 26]. Nevertheless, in spite of some limitations, the CRU TS database is widely used for climate investigations [27–29].

The long-term trends of the continentality/oceanity index during the period 1950–2015 were calculated using Sen's slope test. The statistical significance of the trend values was evaluated using the Mann–Kendall test. Changes with p values of less than 0.05 were considered statistically significant. 1981–2010 continentality/oceanity index normals were also determined.

We also analysed the impact of atmospheric circulation on seasonal temperature indicators and thus on the variability of climate continentality and oceanity. The Northern Hemisphere teleconnection patterns (NHTPs) derived from 500 hPa height field are the leading modes of low-frequency atmospheric circulation variability in the Northern Hemisphere. The data are available from the website of NOAA Center for Weather and Climate Prediction. We selected eight of the 10 available NHTPs because they alone can explain two-thirds of low-frequency atmospheric circulation variability within NH extratropics, and they are active all year and have the same retrieval procedure (Table 1).

One group of NHTPs (NAO and EA) is prominent over the North Atlantic and Europe. Others—SCA, POL, and EA/WR—span over the mid and high latitudes of Eurasia, and PNA, EP/NP, and WP represent the North Pacific and North America.

The correlations between T_{min} and the mean January–March NHTP values, T_{max} and July–September NHTP values, T_{apr} and March–May NHTP values and T_{oct} and September–November NHTP values were analysed to determine the effect of atmospheric circulation on the variation of surface air temperatures as well as on CCI and KOI. The three-month average of NHTP indexes in correlations was used in order to avoid mismatching T_{min} and T_{max} with the particular coldest winter/warmest summer month. The same procedure was subsequently applied to T_{apr} and T_{oct} in order to unify the assessment of the impact of atmospheric circulation on both CCI and KOI.

NHTP indexes are available at a monthly time scale. However, every index value represents the three-month period centred on a particular month owing to its calculation procedure.

3. Results and Discussion

3.1. Climate Norm and Determinant. In the climatological standard normal period (1981–2010), the hypercontinental climate (CCI values >80) was in Northeast Siberia, while the hyperoceanic climate (CCI values <20) was identified along the Pacific coast of North America and in coastal areas in northern parts of the Atlantic Ocean (Figure 1). The surface air temperature of the coldest month represented the most important determinant of CCI values in almost the whole study area (Figure 2). This can be explained by the fact that, in a substantial part of the analysed territory, winter temperature fluctuations were greater than their summer counterparts. Meanwhile, the temperature of the warmest month was the principal factor for CCI only in the western part of the Mediterranean basin.

The high continentality (CCI) within central-northern North America and Northeastern Eurasia (East Siberia) was

TABLE 1: Standardised Northern Hemisphere teleconnection indexes used in the study.

Abbreviation	Full name
NAO	North Atlantic Oscillation
EA	East Atlantic pattern
WP	West Pacific pattern
EP/NP	East Pacific/North Pacific pattern
PNA	Pacific/North American pattern
EA/WR	East Atlantic/West Russian pattern
SCA	Scandinavian pattern
POL	Polar/Eurasian pattern

primarily influenced by very low air temperatures in the coldest month of the year. The Siberian High (SH) and North American High (NAH) favour extreme negative surface temperatures during the winter in the larger part of Northern Asia and the northernmost parts of North America. These are seasonal high-pressure systems composed of cold and dry air; however, SH is much more persistent than NAH, and due to local topography (mountain valleys), it initiates the largest temperature inversions over the northeastern part of Siberia [30, 31]. Relatively high coldest month temperatures in the larger part of Europe, Southeastern USA, and the Pacific coast of North America seem to be responsible for the low CCI values there.

The strongest oceanity (large KOI) was observed within coastal areas of the Arctic Ocean, North Atlantic, Mediterranean, and the Far East (Figure 1). The lowest KOI was found within the inner part of Eurasia (particularly Central Asia and the Tibetan Plateau), the Canadian Prairies, and Yukon. Such spatial variation of KOI can partly be explained by sea surface temperature (SST) differences in October and April: October SST was always higher than April SST in the Arctic, North Atlantic, Mediterranean, and so forth; moreover, many coastal areas in high latitudes in April are covered by sea ice, but in October they are ice-free.

The relationship between KOI and the average temperature of the warmest (T_{max}) and coldest (T_{min}) months is weak ($R^2 < 0.15$). Both the April and October temperatures have a greater effect on variations in the KOI (Figure 3). Fluctuations of air temperature in April play a leading role (especially in the central parts of continents), whereas October temperatures are more important in coastal areas. The land surface temperature in April is usually higher than in October at the same latitude over inland areas as long as there is no snow cover.

The areas of low KOI do not exactly correspond with those of high CCI; indeed, low KOI areas are situated to the south and west of their high CCI counterparts (Figure 1). Cyclonic activity as well as mean wind speed over the Arctic, North Atlantic, and North Pacific is higher in October than in April and represents one of the main drivers of heat flux to the high latitudes, thus contributing to the high KOI in the coastal areas of these regions [32, 33].

3.2. Long-Term Trends. Annual minimum and maximum monthly air temperatures and temperatures in April and October increased across the majority of the study area in the

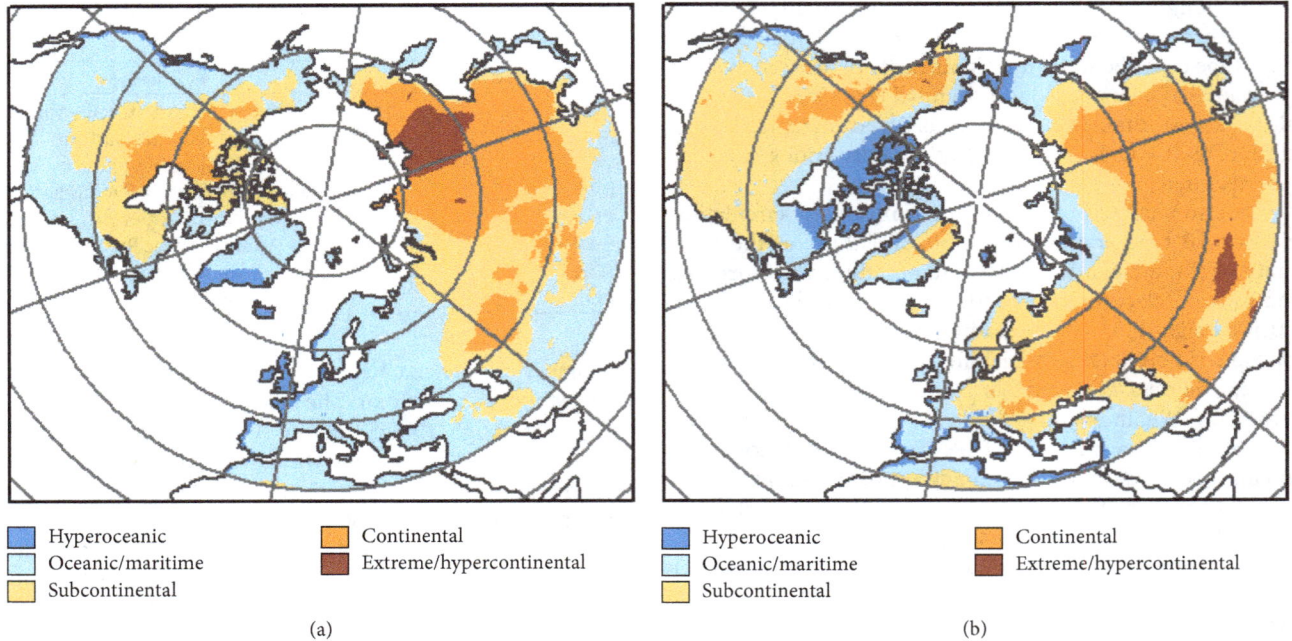

■ Hyperoceanic	■ Continental
■ Oceanic/maritime	■ Extreme/hypercontinental
■ Subcontinental	

(a)

■ Hyperoceanic	■ Continental
■ Oceanic/maritime	■ Extreme/hypercontinental
■ Subcontinental	

(b)

FIGURE 1: The climatological standard normal (1981–2010) of (a) Conrad's Continentality Index (CCI) and (b) Kerner's Oceanity Index (KOI).

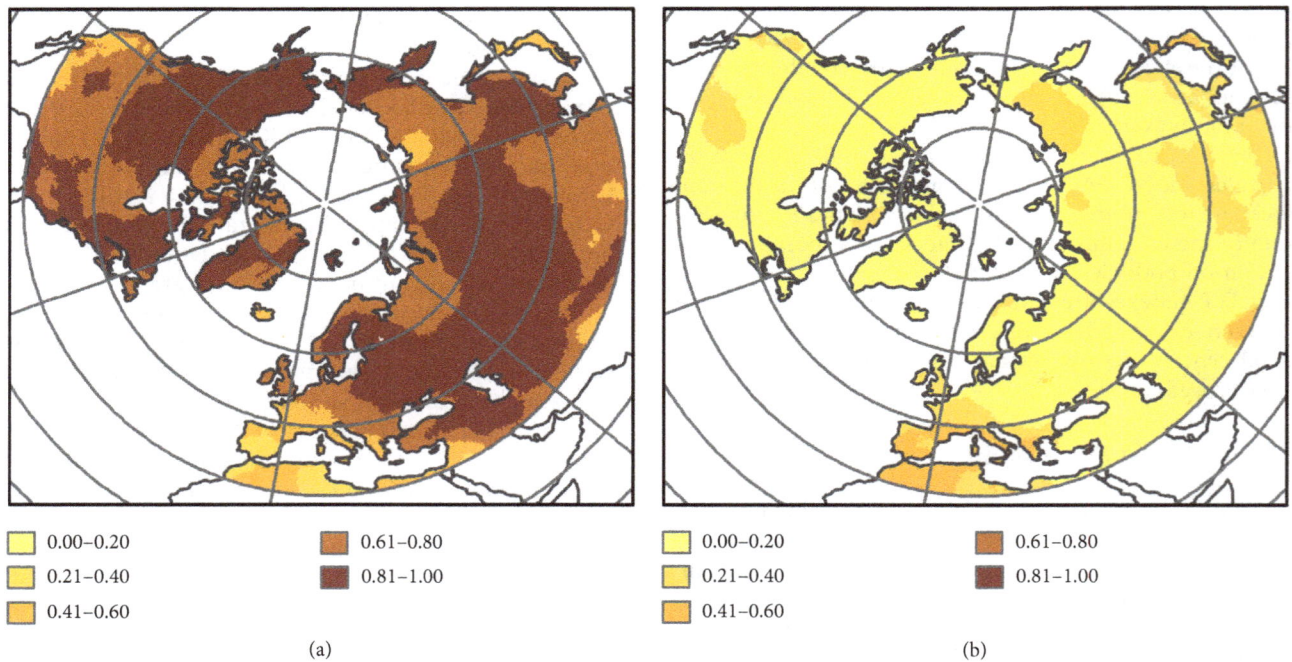

☐ 0.00–0.20	■ 0.61–0.80
☐ 0.21–0.40	■ 0.81–1.00
☐ 0.41–0.60	

(a)

☐ 0.00–0.20	■ 0.61–0.80
☐ 0.21–0.40	■ 0.81–1.00
☐ 0.41–0.60	

(b)

FIGURE 2: The determination coefficient (R^2) between Conrad's Continentality Index (CCI) and the average temperature of the (a) coldest (T_{min}) and (b) warmest (T_{max}) months in 1950–2015.

period 1950–2015 (Figure 4). The annual minimum monthly temperature increased by more than 0.5°C/10 years in Western Russia, East Siberia, and in some parts of Central Asia (Figure 4), while the largest rise of T_{min} was found in the northwestern part of North America (more than 1.0°C/10 years). The annual minimum monthly temperature declined slightly only in the northeastern part of Siberia and in the eastern part of North America.

The magnitude of the annual maximum monthly temperature (T_{max}) trend was smaller than that of T_{min} in 1950–2015. Trends above 0.25°C/10 years were observed in the northeastern part of Siberia, in large parts of Central Asia

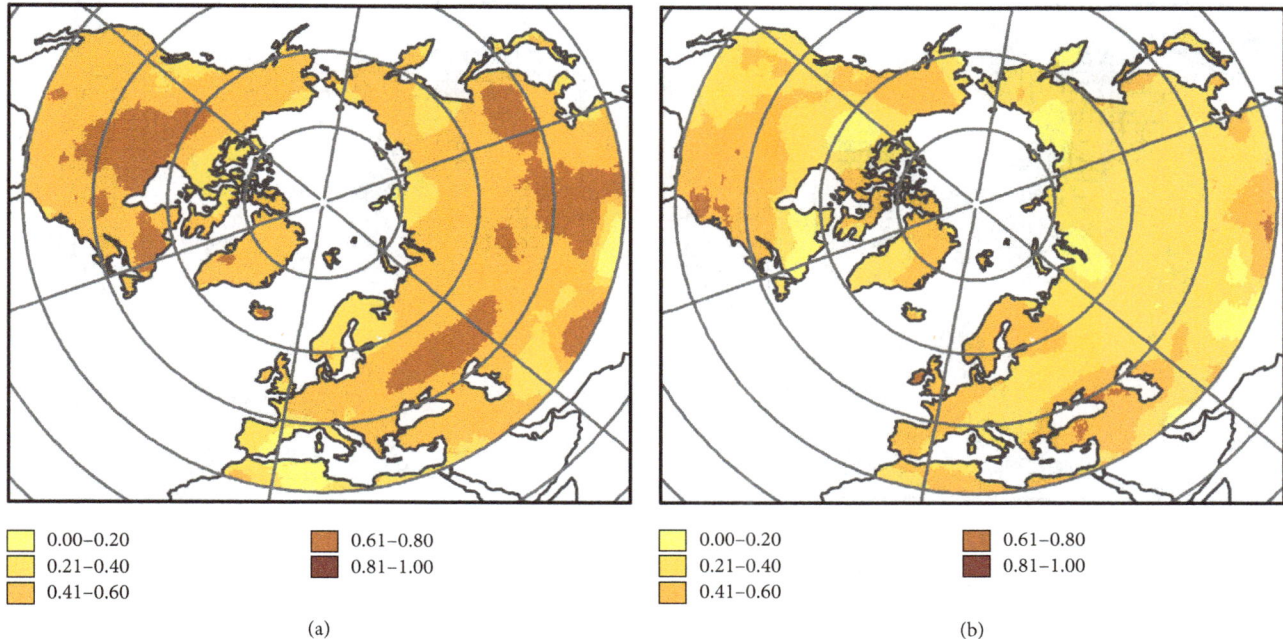

0.00–0.20		0.61–0.80	
0.21–0.40		0.81–1.00	
0.41–0.60			

(a)

0.00–0.20		0.61–0.80	
0.21–0.40		0.81–1.00	
0.41–0.60			

(b)

FIGURE 3: The determination coefficient (R^2) between Kerner's Oceanity Index (KOI) and the average temperature in (a) April (T_{apr}) and (b) October (T_{oct}) in 1950–2015.

and Europe, and northern North America. A larger T_{min} increase rate relative to T_{max} reduced the annual temperature amplitude and CCI over the majority of North America, Asia, and Eastern Europe (Figure 5). In Southwestern Europe, the CCI increased in the areas where T_{max} grew more than T_{min}. In the northeastern part of Siberia and the eastern part of the USA, the increase of CCI was related to a decrease of T_{min} and a rise of T_{max}.

Some teleconnection patterns also demonstrated clear tendencies: the prevalence of a certain phase in the last few decades—EA (positive), EA/WR (negative), and EP/NP (negative) due to the same reasons as for NAO.

A strong reduction in oceanity in most of Europe and Mongolia in addition to an increase in the Caspian Sea-Caucasus region as well as in a large part of North America during recent decades also indicates the prevalence of certain phases of particular circulation patterns in April and October: EA, EA/WR, SCA, POL, and PNA. However, recent research has argued that heating anomalies over the subtropical Northwestern Atlantic, as well as storm-track activity over the North Atlantic, are able to produce well-organised EA/WR-like wave patterns with associated widespread anomalies from the continental USA to Central Asia, with the strongest impact on the Caspian Sea and Western European regions [34].

In 1950–2015, through April (T_{apr}) and October (T_{oct}), the increase in monthly temperatures was greatest in areas near the Arctic Ocean (>0.50°C/10 years) (Figure 4). In lower latitudes, both April and October temperatures increased, but the spatial pattern of trends was very different, especially in Asia and North America. The April temperature trend was highest in East Siberia and the eastern part of Central Asia, while October temperatures increased more significantly in

the northern and northeastern parts of Siberia and in some areas of Central Asia. Insignificant negative changes were observed across a large part of North America in October. The differences in the trends of spatial patterns resulted in statistically significant changes in the KOI over the Baltic Sea region and parts of Siberia and Mongolia (Figure 5). The climate became more oceanic in the northern part of Canada, outermost parts of the Far East and Africa, and large parts of the Mediterranean region.

3.3. Atmospheric Circulation. Atmospheric circulation is an important driver of spatial distribution and temporal variation of selected temperature parameters: T_{min}, T_{max}, T_{apr}, and T_{oct}. The correlation between Northern Hemisphere teleconnection patterns and the analysed temperature parameters permits identification of the areas where atmospheric circulation has a significant effect on the temporal variation of seasonal temperature differences and thus CCI and KOI (Figure 6). The used teleconnection patterns are identified using rotated principal component analysis, and in theory, there should not be multicollinearity between different patterns and their effects on temperature indicators.

Atmospheric circulation had the largest effect on T_{min} variation in latitudes between 40° and 60° (Figure 6). NAO had a statistically significant positive correlation with T_{min} in the larger part of Eurasian mid and high latitudes. Therefore, CCI tends to decrease in Northern Eurasia during winters with a prevailing positive NAO phase and vice versa during a negative NAO phase (Figure 6). NAO, or its hemispheric counterpart Arctic oscillation (AO), has a significant impact on the shape and strength of the Siberian High (SH) and thus

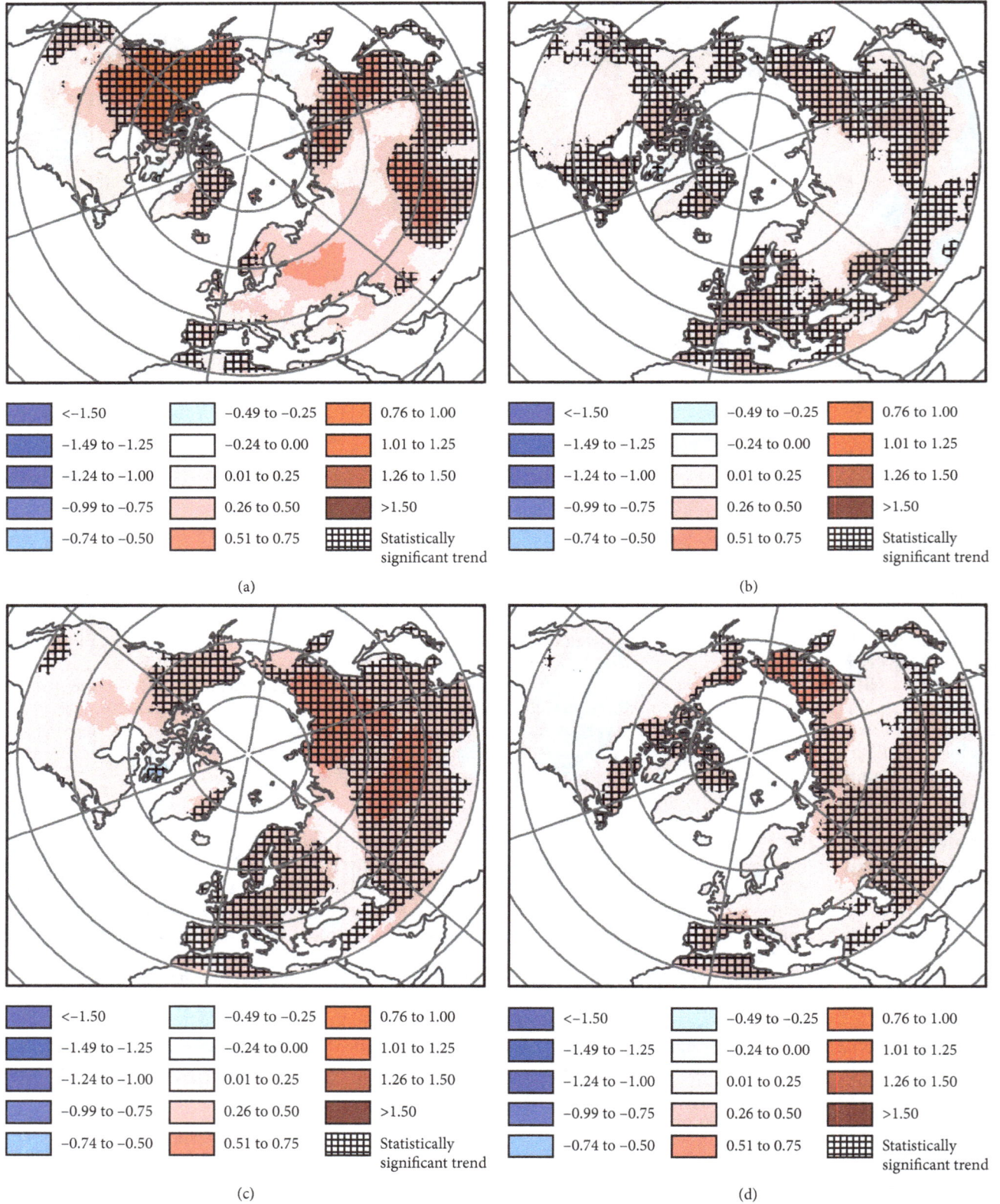

FIGURE 4: Sen's slope trends of annual (a) minimum (T_{min}) and (b) maximum (T_{max}) monthly temperatures and temperature in (c) April (T_{apr}) and (d) October (T_{oct}) in the Northern Hemisphere above 30°N latitude in 1950–2015. The magnitude of trends is expressed as a change of temperature in °C over 10 years.

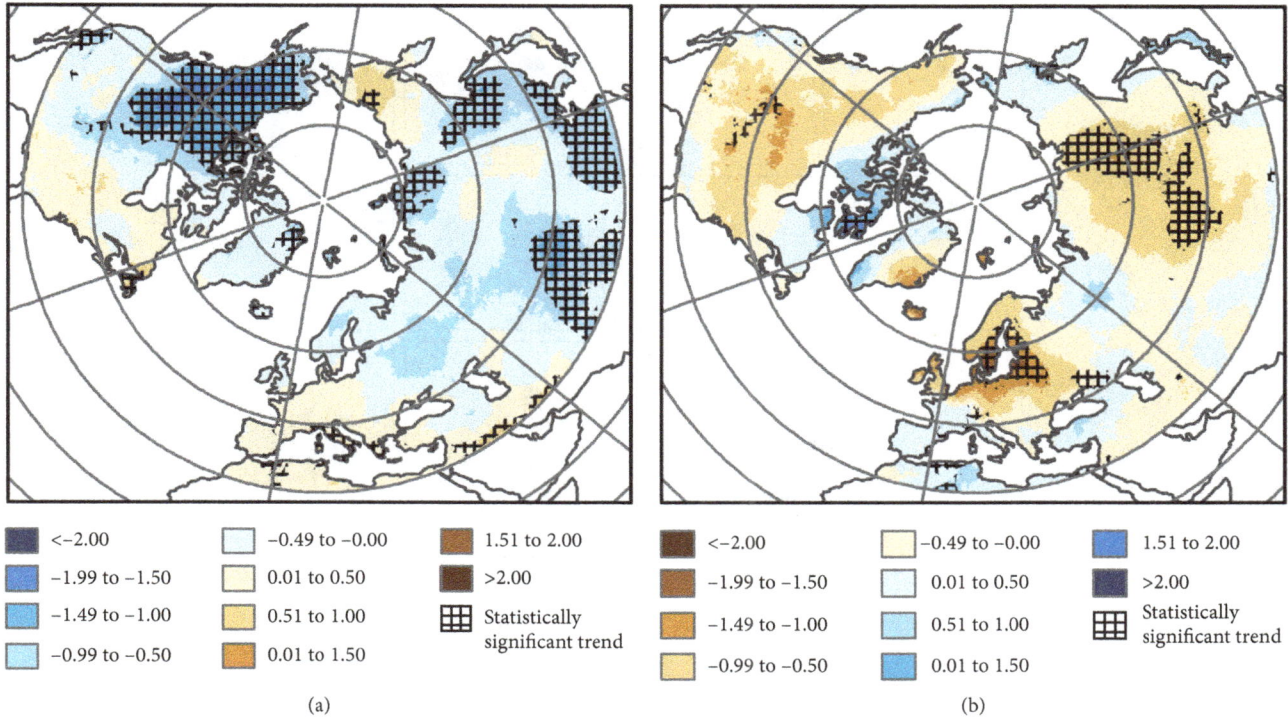

■ <−2.00	□ −0.49 to −0.00	■ 1.51 to 2.00
■ −1.99 to −1.50	□ 0.01 to 0.50	■ >2.00
■ −1.49 to −1.00	■ 0.51 to 1.00	▦ Statistically significant trend
□ −0.99 to −0.50	■ 0.01 to 1.50	

(a)

■ <−2.00	□ −0.49 to −0.00	■ 1.51 to 2.00
■ −1.99 to −1.50	□ 0.01 to 0.50	■ >2.00
■ −1.49 to −1.00	■ 0.51 to 1.00	▦ Statistically significant trend
□ −0.99 to −0.50	■ 0.01 to 1.50	

(b)

FIGURE 5: Sen's slope trends of (a) continentality (CCI) and (b) oceanity (KOI) indexes in the Northern Hemisphere above 30°N latitude in 1950–2015. The magnitude of trends is expressed as a change of index over 10 years. Blue colours indicate a change towards oceanity and brown towards continentality.

on land surface winter temperatures [35]. PNA and WP patterns have a similar effect on T_{min} in the northern part of North America. Other NHTPs seem to only have a regional effect on T_{min}: EA in Europe, SCA in the western part of Eurasia, POL in parts of Siberia, and EP/NP in the Eastern Arctic and Siberia. Positive phases of NAO (AO) and to an extent EA imply larger equator-to-pole temperature gradients during the winter season, which is linked with stronger zonal winds bringing maritime air masses far into the interior parts of continents [36, 37]. In summer, NAO also seems to play a significant role in determining the distribution of surface temperature anomalies across Northern Hemisphere continents, especially over Eurasia and the North Atlantic [38].

The correlations between NHTPs and T_{max} contribute less to CCI than does T_{min} (Figure 2). Furthermore, almost all of the selected NHTPs have a merely regional effect on T_{max}. The most important ones are EA for Europe and the Far East, POL for Europe and Southern Siberia, EA/WR for Eastern Europe and the Ural region, and EP/NP primarily for North America and some parts of Eurasia [39]. The most important circulation modes in winter and NAO and PNA in summer appear to have significant correlations only in very discrete and local land surface areas in the Northern Hemisphere (Figure 6).

For KOI, according to correlations between NHTPs and T_{oct} and T_{apr}, the most important patterns appear to be SCA, EA/WR, POL, and EA for Eurasian regions, PNA for North American regions, NAO for Greenland and Northeastern

Canada, and EP/NP and WP for both Eurasia and North America (Figure 6). The most crucial factor influencing KOI, with reference to its formula, may be patterns that have an opposite effect on temperature in October and April in the same areas during their different phases. This is particularly important within inland regions of Eurasia and North America (low KOI regions) as well as in coastal areas in high and mid latitudes (high KOI regions). Such NHTP patterns are NAO and SCA for Siberia and the Ural region, EA/WR for Eastern Europe, Caucasus, and Turkey, EA for Central Europe and Eastern China, EP/NP for eastern North America, and WP for Northeast Siberia and the Great Plains (Figure 6).

The sum of eight determination coefficients describing the relationship between T_{min}, T_{max}, T_{apr}, and T_{oct} and teleconnection indexes was used as the measure of cumulative effect of selected Northern Hemisphere teleconnection patterns on T_{min}, T_{max}, T_{oct}, and T_{apr} and hence CCI and KOI values (Figure 7). The impact on T_{min} has a latitudinal extension between 40°N and 60°N in Eurasia and between 50°N and 70°N in North America (Figure 7). Such a spatial effect coincides with the extension of the Siberian High to the west in Eurasia and the winter Arctic anticyclone in North America. These areas appear to be sensitive to signs of an NAO phase as well as to a POL, EA, and PNA phase. The cumulative effect on T_{max} is rather discrete and consequently depends on NHTP patterns representing the Rossby wavetrain: EA, EA/WR, EP/NP, WP, and POL. Given that the CCI is largely contingent on T_{min}, NAO seems to be the

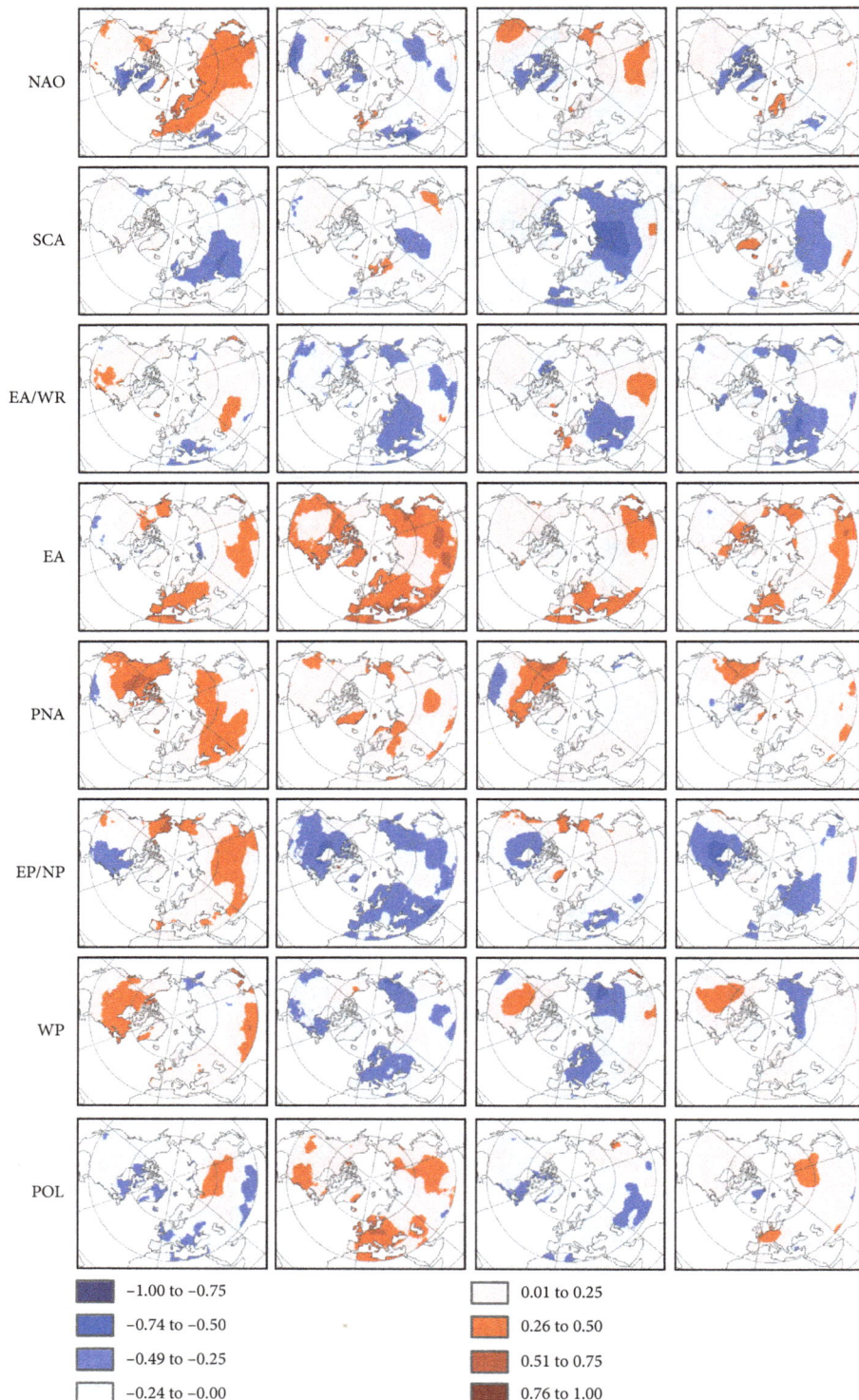

FIGURE 6: Spatial variation of correlation coefficients between indexes of the Northern Hemisphere teleconnection patterns and T_{min}, T_{max}, T_{apr}, and T_{oct} in 1950–2015. Correlation coefficients higher than 0.25 and lower than −0.25 are statistically significant ($p < 0.05$).

main contributor to its temporal variation in Eurasia and NAO and PNA in North America. For KOI, the cumulative effect of NHTPs on temperature seems to be strong in Eastern Canada and the northeastern part of Siberia (T_{apr} and T_{oct}) and a little weaker in Western Siberia, the Ural region, and Northern Kazakhstan (only T_{apr}). Therefore, the main contributors to the cumulative effect here are EP/NP and WP, while other teleconnection patterns contribute to the cumulative effect only in a particular season, e.g., EA/WR in October and SCA in April (Figure 7).

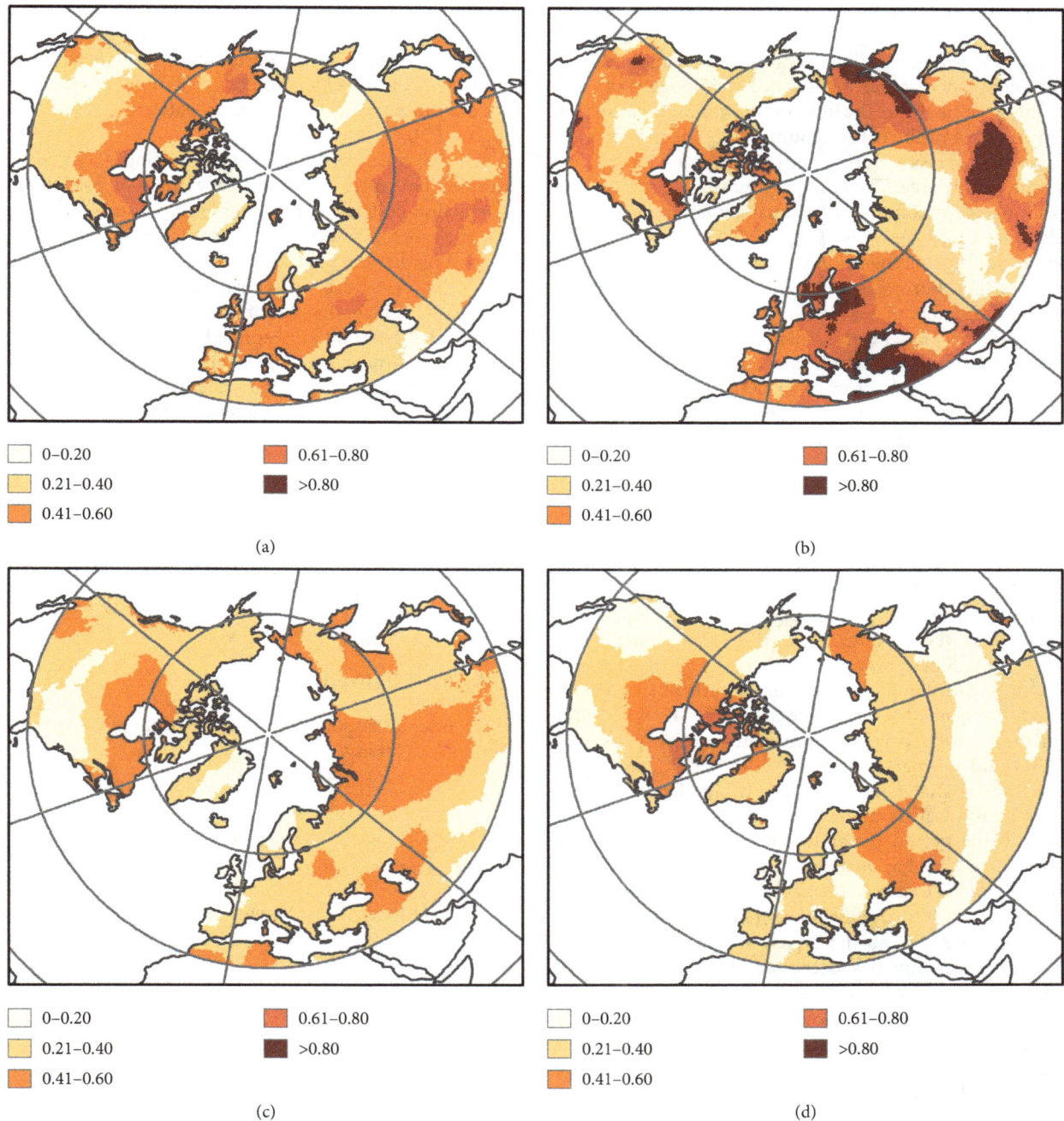

FIGURE 7: The spatial distribution of the sum of eight determination coefficients describing the relationship between (a) T_{min}, (b) T_{max}, (c) T_{apr}, and (d) T_{oct} and selected Northern Hemisphere teleconnection indexes.

Large-scale atmospheric circulation patterns and oscillations affect continentality (CCI) and oceanity (KOI) of many regions in the Northern Hemisphere. Their asymmetry in phases during certain time periods can affect the trends of CCI and KOI. For instance, the NAO exhibited the predominance of its positive phase in the final three decades of the twentieth century, with a peak in the early-1990s [40–42]. This coincided with the change in magnitude and shift in position of the centres of action, especially in the cold season of the year: the Azores High, Siberian High, North American High, Icelandic Low, and Aleutian Low. This is also confirmed by trends of CCI and KOI (Figure 5).

4. Conclusions

This paper has investigated variations in continentality and oceanity in the middle and high Northern Hemisphere latitudes in the period 1950–2015. The Conrad Continentality Index (CCI) and Kerner Oceanity Index (KOI) were employed for this purpose. The impacts of atmospheric circulation on the variability of these indexes were also analysed.

The spatial pattern of climate continentality and oceanity is contingent on distance to the ocean, topography, and atmospheric circulation. The interannual variation of surface air temperatures of the coldest month (T_{min}) is greater than temperature variations during summer months in the majority of the study area. Therefore, the most important factor determining the magnitude of annual air temperature amplitude and CCI values is T_{min}. The warmest month's temperature (T_{max}) is the leading factor for CCI only in the western part of Europe and North Africa. The temporal variation of differences between April (T_{apr}) and October (T_{oct}) surface air temperatures is larger than the variation of annual air temperature amplitude, and it has a greater impact on the temporal dynamics of KOI. The KOI in central parts of continents better correlates with air temperatures in April, while in coastal areas KOI is closely connected to October temperature fluctuations.

Since 1950 in many regions of the Northern Hemisphere, positive and statistically significant trends of T_{min}, T_{max}, T_{apr}, and T_{oct} were recorded. The direction and magnitude of the CCI trend were determined by the ratio of T_{min} and T_{max} trends. Climate continentality has declined in areas where the difference between T_{min} and T_{max} trend values is positive, and vice versa. For example, the dramatic reduction in strength of the Siberian High, which is responsible for extreme continental conditions over larger parts of Siberia and East Asia, has been observed since the 1980s [43, 44], and it was primarily influenced by dominant positive phases of NAO/AO and EA patterns. Meanwhile, the T_{apr} and T_{oct} positive trend spatial patterns resulted in significant changes in KOI over most of Eurasia and the northern part of North America in the period researched. According to KOI, climate continentality has increased in the Baltic Sea region and in parts of East Siberia, Mongolia, and the Great Plains, while the statistically significant increase in oceanity was mostly found over the northern part of Canada. According to CCI, continentality has decreased in western parts of Canada and the USA as well as in parts of Central and Eastern Asia. In summary, we can highlight that statistically significant CCI trends on larger areas display reduced continentality, while statistically significant KOI trends show decreased oceanity from 1950 until 2015. This can be explained by the fact that, during the study period, larger temperature increases were noted in winter and spring. A statistically significant decrease of CCI in areas surrounding the North Atlantic and Eastern Arctic and within the Southeastern USA and an increase of CCI in Inner Mongolia and the Mediterranean seem to be the consequence of the change in position and magnitude of centres of action, both semipermanent (e.g., NAO) and seasonal (e.g., the Siberian High) [45, 46].

El Kenawy et al. [14] have asserted that changes in the spatial variability of continentality are closely coupled with the Atlantic modes of variability, especially with the Eastern Atlantic pattern (in the Mediterranean, Middle East, and northern part of Africa). Similar findings concerning the North Atlantic Oscillation were also detected at an earlier point in the large domain extending from Eastern Canada to the Central Arctic via Europe [47–49]. On the contrary, the high spatial variability of continentality as well as changes in its gradients in specific areas (e.g., Greenland) could be due not only to large-scale circulation patterns but also to local effects [50].

Not all areas of decreasing (increasing) CCI (KOI) can however be explained by the prevalence of particular teleconnection patterns. The higher latitudes of North America and the most northeastern parts of Siberia were most likely affected by the retreat of seasonal sea ice (later freezing time) driven by an increase in sea surface temperatures [51].

It is very likely that continentality will alter in the future and its changes may be amplified in the following decades. Therefore, climate projections are important to assess potential continentality/oceanity changes and to evaluate the associated impact on natural and anthropogenic systems.

Conflicts of Interest

The authors declare that there are no conflicts of interest regarding the publication of this paper.

Acknowledgments

This work was supported by the Institute of Geosciences of Vilnius University.

References

[1] J. J. M. Hirschi, B. Sinha, and S. A. Josey, "Global warming and changes of continentality since 1948," *Weather*, vol. 62, no. 8, pp. 215–221, 2007.

[2] B. D. Santer, P.-C. Stephen, M. D. Zelinka et al., "Human influence on the seasonal cycle of tropospheric temperature," *Science*, vol. 361, no. 6399, article eaas8806, 2018.

[3] J. L. Cohen, J. C. Furtado, M. Barlow, V. A. Alexeev, and J. E. Cherry, "Asymmetric seasonal temperature trends," *Geophysical Research Letters*, vol. 39, no. 4, 2012.

[4] F. Ji, Z. Wu, J. Huang, and E. P. Chassignet, "Evolution of land surface air temperature trend," *Nature Climate Change*, vol. 4, pp. 462–466, 2014.

[5] W. Gorczyński, "O wyznaczeniu stopnia kontynentalizmu według amplitud temperatury," Sprawozdanie z posiedzeń Towarzystwa Naukowego Warszwskiego, Warsaw Scientific Society, Warsaw, Poland, 1918.

[6] D. Brunt, "Climatic continentality and oceanity," *The Geographical Journal*, vol. 64, no. 1, pp. 44–43, 1924.

[7] N. Raunio, "The effect of local factors on the meteorological observations at Torshavn," *Geophysica*, vol. 3, pp. 173–179, 1948.

[8] E. Baltas, "Spatial distribution of climatic indices in northern Greece," *Meteorological Applications*, vol. 14, no. 1, pp. 69–78, 2007.

[9] A. Deniz, H. Toros, and S. Incecik, "Spatial variations of

climate indices in Turkey," *International Journal of Climatology*, vol. 31, no. 3, pp. 394–403, 2011.

[10] S. M. Gadiwala, F. Burke, T. M. Alam, S. Nawaz-ul-Huda, and M. Azam, "Oceanity and continentality climate indices in Pakistan," *Malaysian Journal of Society and Space*, vol. 9, no. 4, pp. 57–66, 2013.

[11] C. Andrade and J. Corte-Real, "Assessment of the spatial distribution of continental-oceanic climate indices in the Iberian Peninsula," *International Journal of Climatology*, vol. 37, no. 1, pp. 36–45, 2017.

[12] J. Vilček, J. Škvarenina, J. Vido, P. Nalevanková, R. Kandrík, and J. Škvareninová, "Minimal change of thermal continentality in Slovakia within the period 1961–2013," *Earth System Dynamics*, vol. 7, no. 3, pp. 735–744, 2016.

[13] R. Brázdil, K. Chromá, P. Dobrovolný, and R. Tolasz, "Climate fluctuations in the Czech Republic during the period 1961–2005," *International Journal of Climatology*, vol. 29, no. 2, pp. 223–242, 2009.

[14] A. M. El Kenawy, M. F. McCabe, S. M. Vicente-Serrano, S. M. Robaa, and J. I. Lopez-Moreno, "Recent changes in continentality and aridity conditions over the Middle East and North Africa region, and their association with circulation patterns," *Climate Research*, vol. 69, no. 1, pp. 25–43, 2016.

[15] B. V. Poltaraus and D. B. Staviskiy, "The changing continentality of climate in Central Russia," *Soviet Geography*, vol. 27, no. 1, pp. 51–58, 1986.

[16] S. A. Harris, "Continentality index: its uses and limitations applied to permafrost in the Canadian Cordillera," *Physical Geography*, vol. 10, no. 3, pp. 270–284, 1989.

[17] D. I. Nazimova, V. G. Tsaregorodtsev, and N. M. Andreyeva, "Forest vegetation zones of Southern Siberia and current climate change," *Geography and Natural Resources*, vol. 31, no. 2, pp. 124–131, 2010.

[18] L. Gorczynski, "The calculation of the degree of continentality," *Monthly Weather Review*, vol. 50, no. 7, pp. 369-370, 1922.

[19] O. V. Johansson, "Uber die Asymmetrie der meteorologischen Schwankungen," *Societas Scientiarum Fennica Commentationes Physico-Mathematicae*, vol. 3, 1926.

[20] V. Conrad, "Usual formulas of continentality and their limits of validity," *Transactions of the American Geophysical Union*, vol. 27, no. 5, pp. 663-664, 1946.

[21] A. Marsz and S. Rakusa-Suszczewskis, "Charakterystyka ekologiczna rejonu Zatoki Admiralicji (King George Island, South Shetland Islands). 1. Klimat i obszary wolne od lodu," *Kosmos*, vol. 36, pp. 103–127, 1987.

[22] F. Kerner, "Thermoisodromen, Versuch einer kartographischen Darstellung des jahrlichen Ganges der Lufttemperatur (Wien)," *K. K. Geographische Gesellschaft*, vol. 6, no. 3, p. 30, 1905.

[23] I. C. Harris and P. D. Jones, "CRU TS4.00: Climatic Research Unit (CRU) Time-Series (TS) version 4.00 of high-resolution gridded data of month-by-month variation in climate (Jan. 1901–Dec. 2015)," *Centre for Environmental Data Analysis*, 2017.

[24] I. C. Harris, P. D. Jones, T. J. Osborn, and D. H. Lister, "Updated high-resolution grids of monthly climatic observations—the CRU TS3.10 Dataset," *International Journal of Climatology*, vol. 34, no. 3, pp. 623–642, 2014.

[25] M. New, M. Hulme, and P. Jones, "Representing twentieth-century space-time climate variability. Part I: development of a 1961–90 mean monthly terrestrial climatology," *Journal of Climate*, vol. 12, no. 3, pp. 829–856, 1999.

[26] M. Tanarhte, P. Hadjinicolaou, and J. Lelieveld, "Intercomparison of temperature and precipitation data sets based on observations in the Mediterranean and the Middle East," *Journal of Geophysical Research: Atmospheres*, vol. 117, no. D12, 2012.

[27] R. G. Lauritsen and J. C. Rogers, "US diurnal temperature range variability and regional causal mechanisms, 1901–2002," *Journal of Climate*, vol. 25, no. 20, pp. 7216-7231, 2012.

[28] M. Belda, E. Holtanova, T. Halenka, and J. Kalvova, "Climate classification revisited: from Koppen to Trewartha," *Climate Research*, vol. 59, no. 1, pp. 1–13, 2014.

[29] M. A. A. Zarch, B. Sivakumar, and A. Sharma, "Droughts in a warming climate: a global assessment of standardized precipitation index (SPI) and Reconnaissance drought index (RDI)," *Journal of Hydrology*, vol. 526, pp. 183–195, 2015.

[30] K. Takaya and H. Nakamura, "Mechanisms of intraseasonal amplification of the cold Siberian high," *Journal of the Atmospheric Sciences*, vol. 62, no. 12, pp. 4423–4440, 2005.

[31] J. E. Jones and J. Cohen, "A diagnostic comparison of Alaskan and Siberia strong anticyclones," *Journal of Climate*, vol. 24, no. 10, pp. 2599–2611, 2011.

[32] M. D. S. Mesquita, D. E. Atkinson, and K. I. Hodges, "Characteristics and variability of storm tracks in the north Pacific, Bering Sea, and Alaska," *Journal of Climate*, vol. 23, no. 2, pp. 294–311, 2010.

[33] J. C. Stroeve, M. C. Serreze, A. Barrett, and D. N. Kindig, "Attribution of recent changes in autumn cyclone associated precipitation in the Arctic," *Tellus*, vol. 63, no. 4, pp. 653–663, 2011.

[34] Y. K. Lim, "The East Atlantic/West Russia (EA/WR) teleconnection in the North Atlantic: climate impact and relation to Rossby wave propagation," *Climate Dynamics*, vol. 44, no. 11-12, pp. 3211–3222, 2015.

[35] J. Cohen, K. Saito, and D. Entekhabi, "The role of the Siberian high in Northern Hemisphere climate variability," *Geophysical Research Letters*, vol. 28, no. 2, pp. 299–302, 2001.

[36] D. W. J. Thompson and J. M. Wallace, "The Arctic Oscillation signature in the wintertime geopotential height and temperature fields," *Geophysical Research Letters*, vol. 25, no. 9, pp. 1297–1300, 1998.

[37] S. He, Y. Gao, F. Li, H. Wang, and Y. He, "Impact of Arctic Oscillation on the East Asian climate: a review," *Earth-Science Reviews*, vol. 64, pp. 48–62, 2017.

[38] C. K. Folland, J. Knight, H. W. Linderholm, D. Fereday, S. Ineson, and J. W. Hurrell, "The summer North Atlantic Oscillation: past, present, and future," *Journal of Climate*, vol. 22, no. 5, pp. 1082–1103, 2009.

[39] B. K. Tan, J. C. Yuan, Y. Dai, and S. B. Feldstein, "The linkage between the eastern Pacific teleconnection pattern and convective heating over the tropical western Pacific," *Journal of Climate*, vol. 28, no. 14, pp. 5783–5794, 2015.

[40] J. W. Hurrell and C. Deser, "North Atlantic climate variability: the role of the North Atlantic Oscillation," *Journal of Marine Systems*, vol. 78, no. 1, pp. 28–41, 2009.

[41] T. L. Delworth, F. Zeng, G. A. Vecchi, X. Yang, L. Zhang, and R. Zhang, "The North Atlantic Oscillation as a driver of rapid climate change in the Northern Hemisphere," *Nature Geoscience*, vol. 9, no. 7, pp. 509–512, 2016.

[42] A. Hannachi, D. M. Straus, C. L. Franzke, S. Corti, and T. Woollings, "Low-frequency nonlinearity and regime

behavior in the Northern Hemisphere extratropical atmosphere," *Reviews of Geophysics*, vol. 55, no. 1, pp. 199–234,
2017.

[43] F. Panagiotopoulos, M. Shahgedanova, A. Hannachi, and
D. B. Stephenson, "Observed trends and teleconnections of
the Siberian high: a recently declining center of action,"
Journal of Climate, vol. 18, no. 9, pp. 1411–1422, 2005.

[44] J. Yun, K.-J. Ha, and Y.-H. Jo, "Interdecadal changes in winter
surface air temperature over East Asia and their possible
causes," *Climate Dynamics*, vol. 51, no. 4, pp. 1375–1390, 2017.

[45] N. C. Johnson, "The continuum of Northern Hemisphere
teleconnection patterns and a description of the NAO shift
with the use of self-organizing maps," *Journal of Climate*,
vol. 21, no. 23, pp. 6354–6371, 2008.

[46] G. W. K. Moore, I. A. Renfrew, and R. S. Pickart, "Multidecadal mobility of the North Atlantic Oscillation," *Journal of
Climate*, vol. 26, no. 8, pp. 2453–2466, 2013.

[47] D. Chen and C. Hellstrom, "The influence of the North Atlantic Oscillation on the regional temperature variability in
Sweden: spatial and temporal variations," *Tellus A: Dynamic
Meteorology and Oceanography*, vol. 51, no. 4, pp. 505–516,
1999.

[48] R. M. M. Crawford, C. E. Jeffree, and W. G. Rees, "Pal-
udification and forest retreat in northern oceanic environments," *Annals of Botany*, vol. 91, no. 2, pp. 213–226, 2003.

[49] A. Vladuț, N. Nikolova, and M. Licurici, "Evaluation of
thermal continentality within southern Romania and
northern Bulgaria (1961–2015)," *Geofizika*, vol. 35, no. 1,
pp. 1–18, 2018.

[50] J. Abermann, B. Hansen, M. Lund, S. Wacker, M. Karami, and
J. Cappelen, "Hotspots and key periods of Greenland climate
change during the past six decades," *Ambio*, vol. 46, no. 1,
pp. 3–11, 2017.

[51] C. M. Serreze, A. Crawford, J. Stroeve, A. Barrett, and
R. A. Woodgate, "Variability, trends, and predictability of
seasonal sea ice retreat and advance in the Chukchi Sea,"
Journal of Geophysical Research: Oceans, vol. 121, no. 10,
pp. 7308–7325, 2016.

8

Trends in Extreme Climate Events over Three Agroecological Zones of Southern Ethiopia

Befikadu Esayas[iD],[1] Belay Simane,[1] Ermias Teferi,[1] Victor Ongoma[iD],[2] and Nigussie Tefera[3]

[1]Center for Environment and Development Studies, Addis Ababa University, P.O. Box 1176, Addis Ababa, Ethiopia
[2]Department of Meteorology, South Eastern Kenya University, P.O. Box 170-90200, Kitui, Kenya
[3]The United Nations, World Food Programme (WFP), Addis Ababa, Ethiopia

Correspondence should be addressed to Befikadu Esayas; befikadu.esayas@aau.edu.et

Academic Editor: Marina Baldi

The study aims to assess trends in extremes of surface temperature and precipitation through the application of the World Meteorological Organization's (WMO) Expert Team on Climate Change Detection and Indices (ETCCDI) on datasets representing three agroecological zones in Southern Ethiopia. The indices are applied to daily temperature and precipitation data. Nonparametric Sen's slope estimator and Mann–Kendall's trend tests are used to detect the magnitude and statistical significance of changes in extreme climate, respectively. All agroecological zones (AEZs) have experienced both positive and negative trends of change in temperature extremes. Over three decades, warmest days, warmest nights, and coldest nights have shown significantly increasing trends except in the midland AEZ where warmest days decreased by $0.017°C/year$ ($p < 0.05$). Temperature extreme's magnitude of change is higher in the highland AEZ and lower in the midland AEZ. The trend in the daily temperature range shows statistically significant decrease across AEZs ($p < 0.05$). A decreasing trend in the cold spell duration indicator was observed in all AEZs, and the magnitude of change is 0.667 days/year in lowland ($p < 0.001$), 2.259 days/year in midland, and 1 day/year in highland ($p < 0.05$). On the contrary, the number of very wet days revealed a positive trend both in the midland and highland AEZs ($p < 0.05$). Overall, it is observed that warm extremes are increasing while cold extremes are decreasing, suggesting considerable changes in the AEZs.

1. Introduction

The Intergovernmental Panel on Climate Change (IPCC) report shows that climate change is evident by high frequency in climate extreme events including flooding, drought, sea level rise, and heat waves [1]. Various studies have shown changes in the occurrence and severity of climate extreme events, along with the variability of weather patterns, causing substantial impacts on human and natural systems [2–4]. Climate change impacts, however, are differently experienced in different parts of the world owing to various geographic settings and socioeconomic factors [3, 5]. With a projected 3 to 4°C temperature increase, climate change impact in the future will result in more hostile environments, associated with increases in the frequency

and severity of floods and droughts [3, 6]. Furthermore, the IPCC reports illustrate that the intensity and occurrence of extreme events are expected to increase in different parts of Africa [2, 3]. Climate models have shown that climate impacts will be severe in many areas of Africa, including East Africa [7, 8], primarily associated with changes in atmospheric forcing due to anthropogenic causes [9].

Due to the adverse effects of climate variability and change, Ethiopia is considered to be one of the most vulnerable countries [10, 11]. The country largely suffers from hazards linked to high rainfall variability [12, 13] and climate extreme events [14]. It has experienced droughts and floods from the 1980s onwards [10], and since 1990, the country has recorded 47 major floods that killed about 2,000 people and affected close to 2.2 million people [15]. Ethiopia also experienced 12 major droughts between 1900 and 2010 that claimed the lives of over 400,000 people, and the number of those affected was over 54 million [15]. Very recently, the 2015 El Niño-induced drought has caused food insecurity among 10.2 million people, one of the highest on the record [16]. Moreover, it is projected that Ethiopia will face serious and damaging impacts resulting from changing climate patterns in the future [17].

Owing to the high probability of changes in climate extremes and the negative economic, social, and environmental impacts [2, 18, 19], due consideration has been given to the analysis of climate extreme events in recent years [20, 21]. This is because climate extremes respond more sensitively to climate change than changes in the average climates [21, 22]. Furthermore, extreme events affect the ecosystems much more than changes in the mean climate [23, 24]. Following the IPCC [6] definition, in this study, an "extreme event" is used to illustrate the occurrence of a value of a weather or climate variable above or below a threshold value, generally occurring at the tails of the probability density function (PDF) of the range of observed values of the variable within a defined climate reference period.

Temperature and rainfall are the two most significant and sensitive climatic elements in tropical regions. Data regarding extreme climates and their characteristics are essential to identify, plan, implement, monitor, and evaluate different socioeconomic activities in developing economies such as Ethiopia. In Ethiopia, responding to the negative impacts of extreme climates on the smallholder lives and livelihoods requires detailed studies that document the extent and trends of changes in the extreme climate events. Thus, this study considers the local level analysis of extreme climate events as a case study, which may help to react timely to the associated shocks.

Some empirical studies have attempted to analyze the extreme climatic events in Ethiopia but found mixed results owing to contextual differences. Despite varied results in the magnitude of change observed, a growing body of literature now points to significant trends in precipitation and temperature extremes. For example, a negative trend was observed in seasonal extreme rainfall [25], while mixed trends of changes were reported in rainfall extremes [26–29]. Several empirical studies suggest positive trends in air temperature and negative trends in rainfall [30–33]. A study by Viste et al. [34] reported that even through the degree of drying varies spatially, all the studied areas experienced drought at annual scales in Ethiopia. Zeleke et al. [35] documented that south and southwestern regions of Ethiopia have experienced drying, while the central mountainous, north, and northwestern regions had no observed long-term trends. Recently, Worku et al. [36] reported a warming trend in extreme temperature indices while an increase in rainfall extreme events in Jemma Subbasin, Upper Blue Nile Basin. However, Kebede et al. [37] reported neither a clear monotonic trend in dry spells nor a significant variation in the rainfall duration, onset, and cessation. In Ethiopia, different studies reported that there has been inconsistency in patterns in the precipitation extremes [25–28, 31, 33, 36, 38]. Regardless of the inconsistency in the rainfall extremes in Ethiopia, recent evidence shows that the frequency of the occurrences of extreme events and their variability has increased over the last 20 years [3, 36, 39, 40]. In the context of the study area, a recent study by Degefu and Bewket [38] revealed that the geographic distribution on the occurrence and magnitude of observed drought events was complex.

Although the abovementioned studies have documented trends in climate extremes at national, regional, and local levels, the studies have been reporting different patterns in the climate extremes mostly focusing on the national level analysis with a lot of emphasis on drought. A few studies have examined variations in rainfall extremes using selected indices at national and subnational levels, which may not fully explain the situation at the local level. Though others have assessed trends of climate extremes, most of the studies are spatially confined to the northern part of Ethiopia with the exception of the recent study by Degefu and Bewket [38] that analyzed trends of climate extremes in Omo-Ghibe River basin. Therefore, the existing information on climate extremes is limited in scope, is fragmented in coverage, and does not provide complete perspectives on the complex topography, relief, and agroecological settings in Ethiopia [26, 41].

Unlike earlier studies, this study focuses on aggregation and disaggregation by agroecological zones (AEZs) because of the increasing importance of agroecological-based analysis in the face of changing climate extremes. Case studies based on such a perspective (e.g., *climate resilience to farm productivity*) in the context of climate variability and change are very important [42]. This reveals on the importance of agroecological-based approach as one of the scientific disciplines, sustainable farming approach, and social movement [42]. Moreover, with the changing climate extremes and their adverse effects on peoples' livelihoods, some attempts have been made to promote "agroecology as the sustainable alternative to climate change crisis" [43]. In the Ethiopian context, existing evidence suggest that different agroclimatic zones are found. Traditionally, the agroecologies have been classified into five categories based on altitude, rainfall, and temperature (Table 1) [44, 45].

More importantly, the design and development of local level climate adaptation options and enhancing early warning systems require us to understand the characteristics and trends in climate extreme events at different geographical scales. This study thus contributes to the rapidly growing climate extreme literature by providing first-hand evidence both on the temperature- and precipitation-based indices at AEZs in Southern Ethiopia. It also gives new insights into the application of gridded data at a microlevel where finding complete station data is a serious challenge. The study, by employing climate extreme indices—as one way to investigate climate change and variability in the study area, can serve as a reference for similar studies in the future.

In summary, this study intends to better the understanding of changes and trends in extreme climate events and their frequency, and duration, and variability over three AEZs of Wolaita Zone, Southern Ethiopia. It aims at (a) analyzing the magnitude and frequency of occurrence of extreme temperature events and (b) analyzing the magnitude and frequency of occurrence of extreme precipitation events in AEZs. The rest of

TABLE 1: Traditional agroecological zones and their physical features.

AEZs	Altitude (meters)	Rainfall (mm/year)	Growing period length (days)	Mean annual temperature (°C)
Upper highland (Wurch)	Above 3200	900–2200	211–365	<11.5
Highland (Dega)	2,300–3,200	900–1,200	121–210	17.5/16–11.5
Midland (Weyna Dega)	1,500–2,300	800–1,200	91–120	17.5/16–11.5
Lowlands (Kola)	500–1,500	200–800	46–90	27.5–20
Desert (Berha)	Below 500	Below 200	0–45	>27.5

Adapted from MoA [45].

the paper is organized into three sections: Section 2 provides a brief description of the data, methodology, and area of study; Section 3 presents and discusses the study findings; and the conclusions of the study are given in Section 4.

2. Study Area, Data, and Methodology

2.1. Study Area. Wolaita Zone is located in Southern Nation Nationalities and People (SNNP) of Ethiopia. It lies between 6.4°–7.1° N and 37.4°–38.2° E (Figure 1). It is grouped into three traditional agroecological zones: 56% of the area is midland (*Weyna Dega*); 35 % of the area is lowland (*Kola*); and the rest 9% of the area is covered by highland (*Dega*) (Table 1) [46]. The relief of the area is generally a highland that covers most parts of the midland while the peripheries are lowland areas (Figure 1). The altitude ranges from 501 meters in the lowlands at BilateTena to 3000 meters above the sea level in the highlands at the Damota mountain area. The amount, period, and frequency of rainfall vary considerably from one AEZ to the other. The mean annual rainfall ranges from 800 mm in BilateTena to the highest 1,200 mm in Wolaita Sodo. Rainfall is erratic by nature and variable, occurring in two dissimilar seasons. The pattern of rainfall distribution is bimodal. The main rainy season (*Kirmet*) starts in mid-June and extends to the end of September. The second one is the *Belg* season that spans from the end of February to March and/or early April [46]. The mean annual minimum temperature ranged from 15.1 to 25.1°C in 2015/2016. However, temperatures are usually high with little variations among seasons. In the same year, the mean annual maximum temperature was reported between 17.1°C and 29.7°C. Following Hurni [47], this study adopted the traditional AEZ grouping approach, where three meteorological stations' daily minimum and maximum temperature and daily precipitation data were extracted using the latitude and longitude to represent highland, midland, and lowland AEZs, respectively (Figure 1).

2.2. Data. The study is based on gridded dataset (4 km by 4 km spatial resolution) of daily maximum and minimum temperatures and daily total rainfall from 1983 to 2014. The gridded dataset combines two datasets: the first is station data (rainfall and temperature) from the national network managed by the Ethiopian National Meteorological Services Agency (NMA) and the second dataset is satellite rainfall and temperature estimations from European Organization for the Exploitation of Meteorological Satellites (EUMETSAT) and the US National Aeronautics and Space Administration (NASA).

In other words, the gridded dataset integrated quality-controlled station data from the national observation network with locally calibrated satellite-derived data that were used to fill spatial and temporal gaps in the Ethiopian national observations. Data reconstruction was undertaken by the NMA in partnership with International Research Institute for Climate and Society at Columbia University, USA, whereas data calibration and validation were carried out by Reading University, UK.

As mentioned by Mengistu et al. [48] in the Upper Nile Basin, Ethiopia, station-based data in the Wolaita Zone had also many missing values and measurement errors, was poor in quality, and lacks continuous data both for temperature and precipitation. Due to these reasons, the preference was given to the use of the gridded dataset, which had better data quality and daily minimum and maximum temperature, and daily total precipitation data were available between 1983 and 2014 for the studied AEZs as opposed to using the available station-based dataset. The data used for this study can be found at Ethiopian National Meteorological Services Agency (http://www.ethiomet.gov.et/) for the climatic stations located over three agroecological zones.

On the contrary, this study considered three existing stations, which are located over the AEZs using the gridded dataset for the purpose of comparison by AEZ, which in turn is assumed to represent each AEZ with the available climate data over the study period (1983–2014) (Figure 1). The stations include Bilate (lowland), Wolaita (midland), and Boditi School (highland) (Table 2). The stations were selected purposively as they have long years (over 30 years) of observed temperature and rainfall data. The analysis period, 1983–2014, was chosen due to data availability within the selected periods and to explore the recent change in extreme temperature and rainfall across the study AEZs. Data quality control was carried out using ClimPACT2 Software in R [49]. Data quality was tested in order to label potentially wrong values and to remove them from the analysis. Unrealistic values, such as daily maximum temperature less than or equal to daily minimum temperature, were identified and set to missing values. The reference period of 1983–2000 was chosen out of the full-time range (1983–2014) mainly for the calculation of the percentile-based indices.

2.3. Methodology. In an effort to investigate the existence of trends in time series of both temperature and rainfall indices obtained from daily data, we used the nonparametric Mann–Kendall (MK) test statistic [50, 51] and Sen's estimator test [52] at the 5% significant level. Detailed description on Mann–Kendall and Sen's slope estimation can be found from the related studies [53–56], as described in

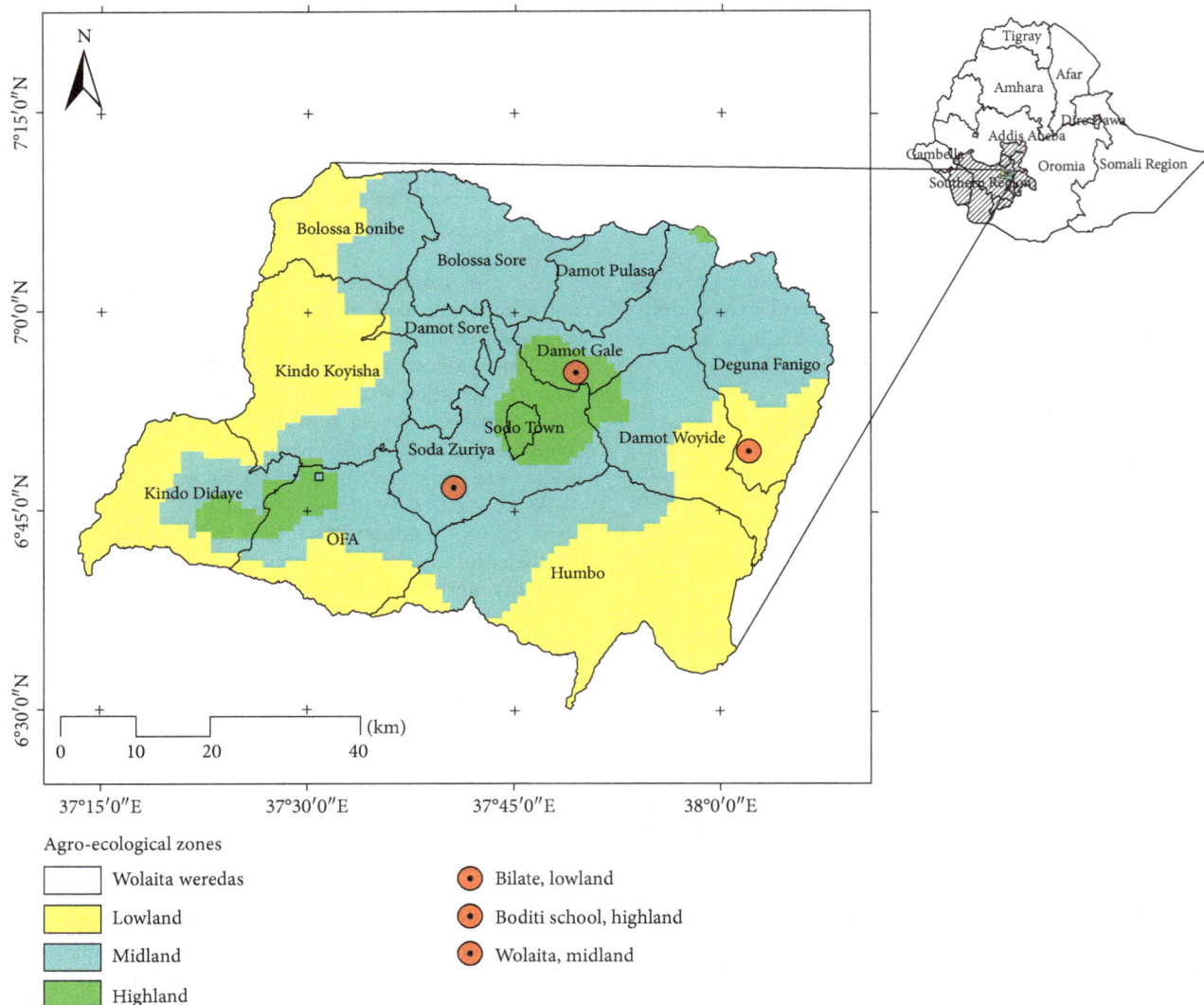

FIGURE 1: Location of agroecological zones.

TABLE 2: Selected meteorological station in Wolaita Zone.

No.	Station name	Longitude	Latitude	Altitude (m)	AEZs	Duration
1	Boditi School	37.96	6.95	2043	Highland	1983–2014
2	Wolaita	37.58	6.81	1854	Midland	1983–2014
3	Bilate	38.08	6.81	1361	Lowland	1981–2014

Supplementary Materials (available here). The stations are not only representative of elevation but also other criteria as suggested by MoA [45], which grouped AEZs into five categories based on altitude (meters), rainfall (mm/year), growing period length (days), and mean annual temperature (°C) (Table 1).

3. Results and Discussion

3.1. Trends in Temperature Extremes

3.1.1. Warm Days (TX90p) and Warm Nights (TN90p).

Trends for the temperature indices for the three AEZs are shown in Table 3. From the percentile-based temperature indices, significantly increasing trend in the frequency of warm days (TX90p) was observed both in the lowland (Figure 2(a)) and highland AEZs (Figure 2(e)) ($p < 0.05$ and $p < 0.001$), respectively. The TX90p shows an insignificant decreasing trend in the midland AEZ (Figure 2(c)). Similar to the frequency of TX90p reported in lowland and in highland AEZs, the occurrence of warm nights (TN90p) has shown very significant increasing trends ($p < 0.001$) (Figure 2(b)) for the lowland AEZ and (Figure 2(f)) for highland AEZ, respectively. The increasing trend observed in the two

TABLE 3: Trends in temperature extreme indices per year by AEZ.

Index	Units	Lowland AEZ		Midland AEZ		Highland AEZ	
		MK test (Z-test)	Sen's slope	MK test (Z-test)	Sen's slope	MK test (Z-test)	Sen's slope
TX90p	%	0.315^*	0.326^*	-0.177^{ns}	-1.501^{ns}	0.480^{***}	0.565^{***}
TX10p	%	-0.234^{ns}	-0.158^{ns}	-0.089^{ns}	-0.05^{ns}	-0.363^*	-0.284^*
TN90p	%	0.552^{***}	0.547^{***}	0.198^{ns}	0.234^{ns}	0.572^{***}	0.61^{***}
TN10p	%	-0.659^{***}	-0.349^{***}	-0.612^{***}	-0.605^{***}	-0.686^{***}	-0.264^{***}
TXx	°C	0.258^*	0.025^*	-0.252^*	-0.017^*	0.523^{***}	0.042^{***}
TXn	°C	0.137^{ns}	0.01^{ns}	0.141^{ns}	0.012^{ns}	0.290^*	0.024^*
TNx	°C	0.462^{***}	0.055^{***}	0.407^{***}	0.037^{***}	0.597^{***}	0.063^{***}
TNn	°C	0.653^{***}	0.078^{***}	0.581^{***}	0.075^{***}	0.690^{***}	0.084^{***}
WSDI	Days	0.156^{ns}	2.00^{ns}	-0.322^{ns}	-1.000^{ns}	0.270^{ns}	2.667^{ns}
CSDI	Days	-0.652^{***}	-0.667^{***}	-0.566^*	-2.259^*	-0.591^*	-1.00^*
DTR	°C	-0.367^*	-0.052^*	-0.423^{***}	-0.053^{***}	-0.282^*	-0.043^*

ns: nonsignificant, $p > 0.05$; $^*p \leq 0.05$; $^{**}p \leq 0.01$; $^{***}p \leq 0.001$.

AEZs is in agreement with results from the three ecoenvironments in Ethiopia [31] and Worku et al. [36]. On the contrary, the midland AEZ experienced insignificant increasing trend in the TN90p (Figure 2(c)). The annual number of TN90p and TX90p shows significant warming anomalies for the period 1983–2014 both in the lowland and highland AEZs. The TX90p reaches its peak in 2009 for the lowland AEZ (Figure 2(a)) and in 2012 for the highland AEZ (Figure 2(e)). On the contrary, TN90p peaked in 2010 in the lowland AEZ (Figure 2(b)) and in 2014 for the highland AEZ (Figure 2(f)). In general, from 2008 onwards, warming anomalies are consistently increasing in all AEZs.

3.1.2. Cool Days (TX10p) and Cool Nights (TN10p). Insignificant decreasing trend in the frequency of cool days (TX10p) was observed in the lowland AEZ (Figure 3(a)) and midland AEZ (Figure 3(c), respectively. However, TX10p shows a decreasing trend in the highland AEZ ($p < 0.05$) (Figure 3(e)). Concerning the frequency of cool nights (TN10p), a very significant decreasing trend was observed in all AEZs ($p < 0.001$), and the result is in line with the recent work by Worku et al. [36]. The negative anomalies are consistently declining since the late 1980s both in the lowland (Figure 3(b)) and highland AEZs (Figure 3(f)), respectively. On the contrary, except for the years 1997 to 1999, the midland AEZ also experienced negative anomalies for the TN10p between late 1980s and 2014 (Figure 3(d)). All the AEZs have experienced a very significant decreasing trend in the TN10p over the period between 1983 and 2014. The significantly increasing trends in the occurrences of TX90p and TN90p while decreasing trends in TX10p and TN10p are in agreement with results from other studies that have analyzed these trends in different parts of the world [57–60].

In summary, two of the AEZs have experienced significant increases in TX90p and TN90p over the period between 1983 and 2014 while no significant trends were observed for all indices in the midland AEZ except for TN10p. The figures are indicative of the increasing trends in the warm extremes and decreasing trends in the cold extremes. These figures clearly show significant warming. In view of this, Worku et al. [36] reported similar pattern of trends both in cold and warm extremes in Upper Blue Nile Basin. Moreover, empirical studies in East Africa, including Ethiopia, suggest that the frequencies of warm days and nights compared to the initial time showed a large increase vis-à-vis the number of cold nights per year beyond the 90th percentile threshold [61].

3.1.3. Warmest Day (TXx) and Coldest Day (TXn). The trend in the warmest day (TXx) is statistically significant with the magnitude of change being 0.025°C/year ($p < 0.05$) for the lowland AEZ, while the magnitude of change in the highland AEZ was observed to be 0.042°C/year ($p < 0.001$). The TXx in the midland AEZ showed a statistically significant decreasing trend (0.017°C/year) ($p < 0.05$) (Table 3). The TXx reached its peak in 2009 for the lowland AEZ (Figure 4(a)), in 2012 for the midland AEZ (Figure 4(c)), and in 2011 for the highland AEZ (Figure 4(e)). The temperature of the coldest day (TXn) is significantly increasing in the highland AEZ with the magnitude of change being 0.024°C/year ($p < 0.05$), while this index is insignificant in the lowland and midland AEZs (Figures 4(b) and 4(d)). A study by Mekasha et al. [31] reported a similar trend in Ethiopia where TXn values were not significantly shifting over the study periods in all the sampled stations.

3.1.4. Warmest Night (TNx) and Coldest Night (TNn). The trend in the warmest night (TNx) is significantly increasing in all AEZs ($p < 0.001$). In line with this, significant increase in warm nights was also reported in previous studies [36, 62]. The magnitude of change was reported to be 0.055°C/year for the lowland AEZ, 0.037°C/year for the midland AEZ, and 0.063°C/year for the highland AEZ. In relative terms, the highland AEZ has experienced higher magnitude of change in the TNx than both the midland and lowland AEZs (Table 3). The temperature of the coldest night (TNn) during the observation periods has increased significantly in all AEZs ($p < 0.001$). The magnitude of change in the TNn was reported to be 0.078°C/year in the lowland AEZ, 0.075°C/year in the midland AEZ, and 0.084°C/year in the highland AEZ ($p < 0.001$), respectively. Similar to the higher magnitude of change observed in the TNx in the

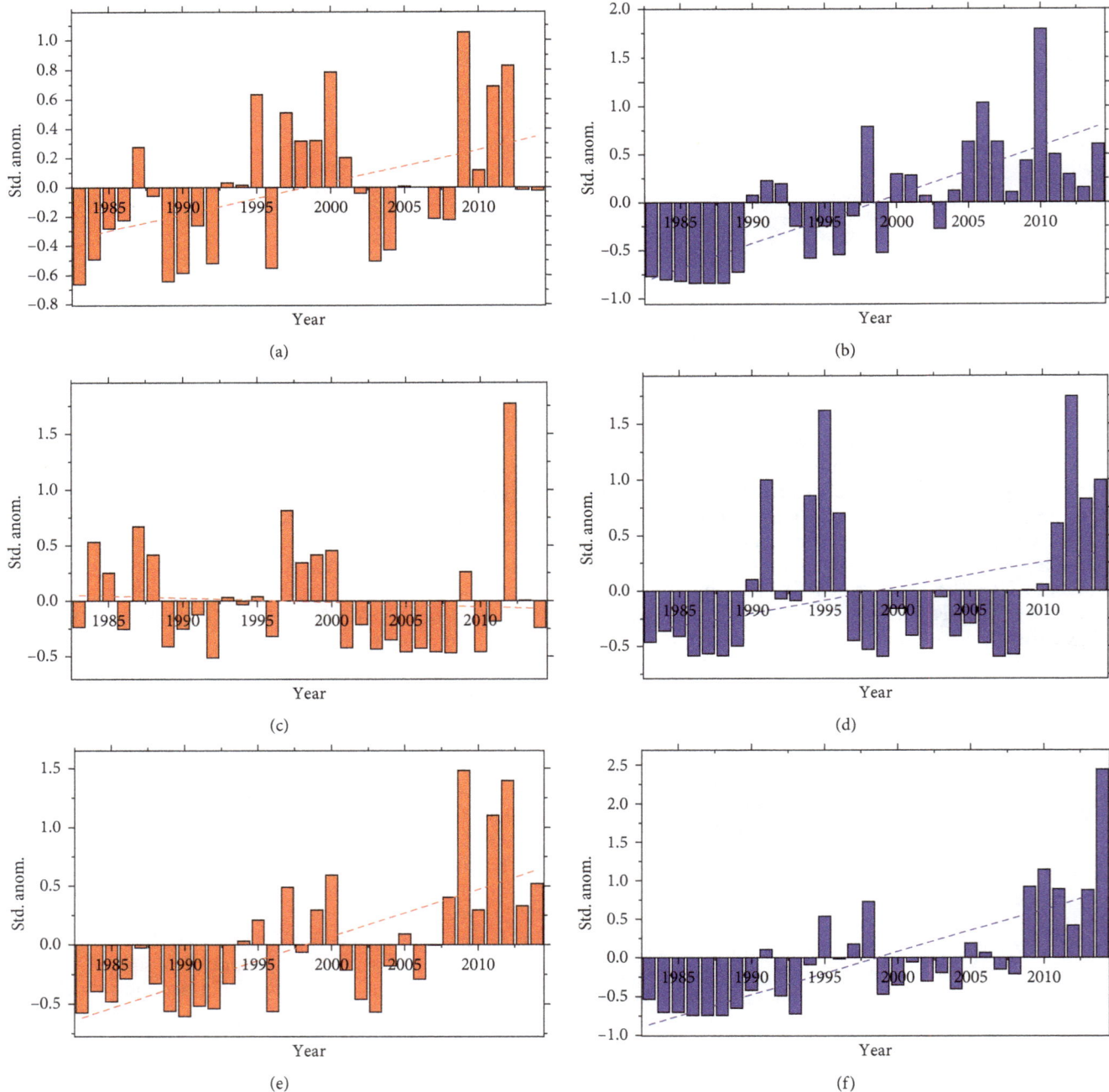

FIGURE 2: Standardized anomalies of warm days (TX90p) over three agroecological zones, (a) lowland, (c) midland, and (e) highland, and for warm nights (TN90p), (b) lowland, (d) midland, and (f) highland, for the period 1983–2014.

highland AEZ, the magnitude of change in TNn is slightly higher than the two AEZs. The anomalies of the TNx and TNn are shown in Figure 5. The TNx's highest positive value was observed in 2006 (Figure 5(a)), in 2012 (Figure 5(c)), and in 2014 (Figure 5(e)) for the lowland, midland, and highland AEZs, respectively. On the contrary, the negative anomalies of the TNn were commonly observed in the 1980s across all AEZs, while the positive anomalies have been increasing from the 1990s onwards in all AEZs (Figures 5(b), 5(d), and 5(f)).

In summary, the TXx in the midland AEZ showed a statistically significant decreasing trend of 0.017°C/year ($p < 0.05$), which is the only AEZ that experienced a decline in the trends of temperature indices (TXx, TXn, TNx, and TNn) computed for all AEZs. The higher magnitude of change in temperature extremes in the highland AEZ compared to the two other AEZs may have adverse impacts on the livelihoods of the highlanders. For example, there is evidence that continued incidence of meteorological drought episodes and famines have resulted in human and

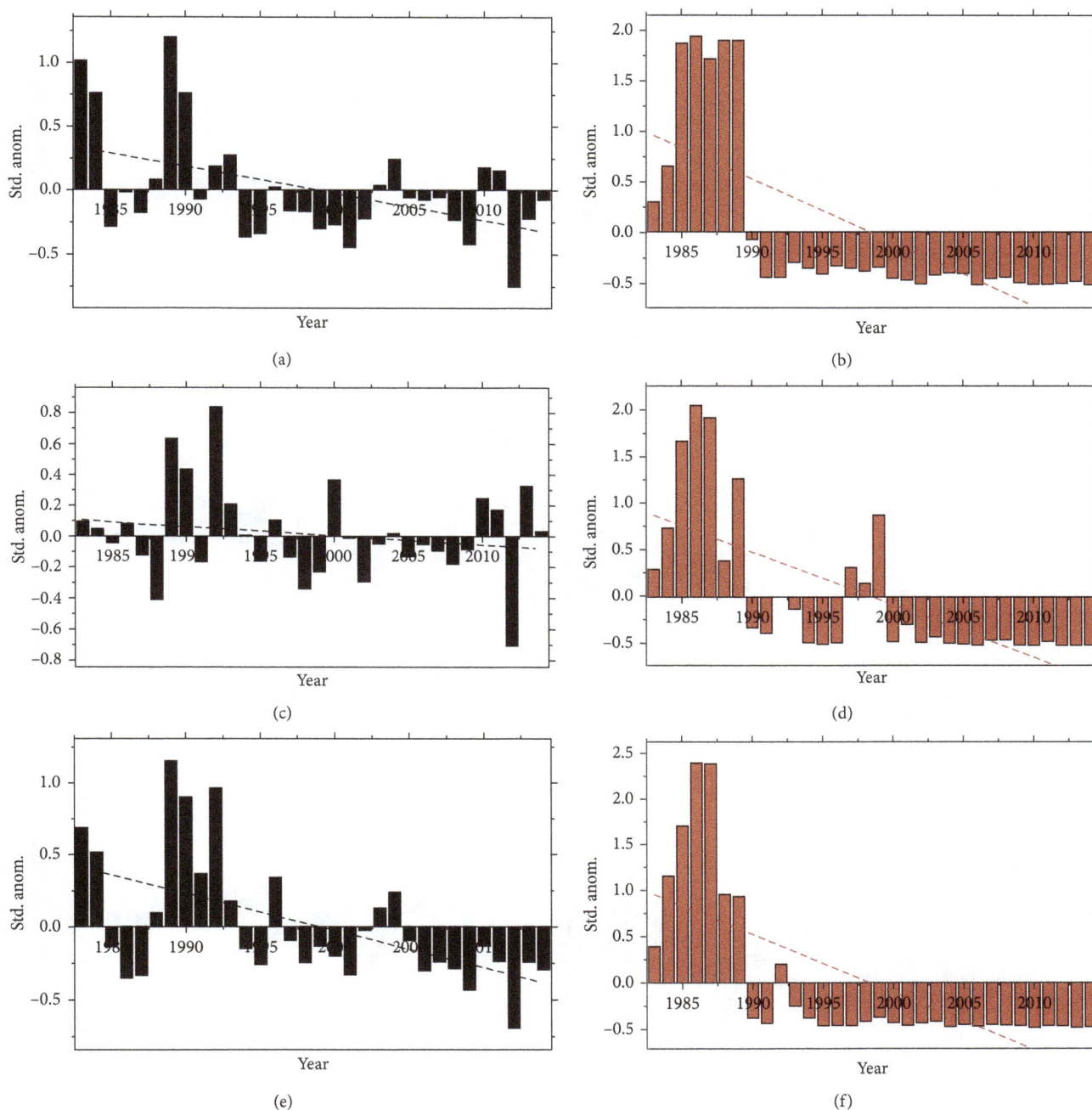

FIGURE 3: Standardized anomalies of cool days (TX10p) over three agroecological zones, (a) lowland, (c) midland, and (e) highland, and for cool nights (TN10p), (b) lowland, (d) midland, and (f) highland, for the period 1983–2014.

crop diseases, particularly in the northern highland regions of Ethiopia [63–66]. Generally, the recent IPCC projection noted that changing climatic variability will possibly result in more extreme events including flooding and drought [3], signifying that the negative impacts of climate trends have been more common than positive ones [1].

3.1.5. Warm Spell Duration Indicator (WSDI) and Cold Spell Duration Indicator (CSDI). The warm spell duration indicator (WSDI) index which represents the number of days

contributing to a warm period was statistically insignificant in all AEZs. However, a significantly decreasing trend was observed in the cold spell duration indicator (CSDI) in all AEZs. The CSDI values show a decrease of 0.667 days/year in the lowland AEZ, 2.259 days/year in the midland AEZ, and 1 day/year in highland AEZ (Table 3). In relative terms, the CSDI is higher in the midland AEZ than in the lowland and highland AEZs. Similar trends were reported by Mekasha et al. [31] for WSDI and CSDI. The CSDI revealed significantly decreasing trend in all AEZs, suggesting that the AEZs are getting warmer although WSDI is not statistically

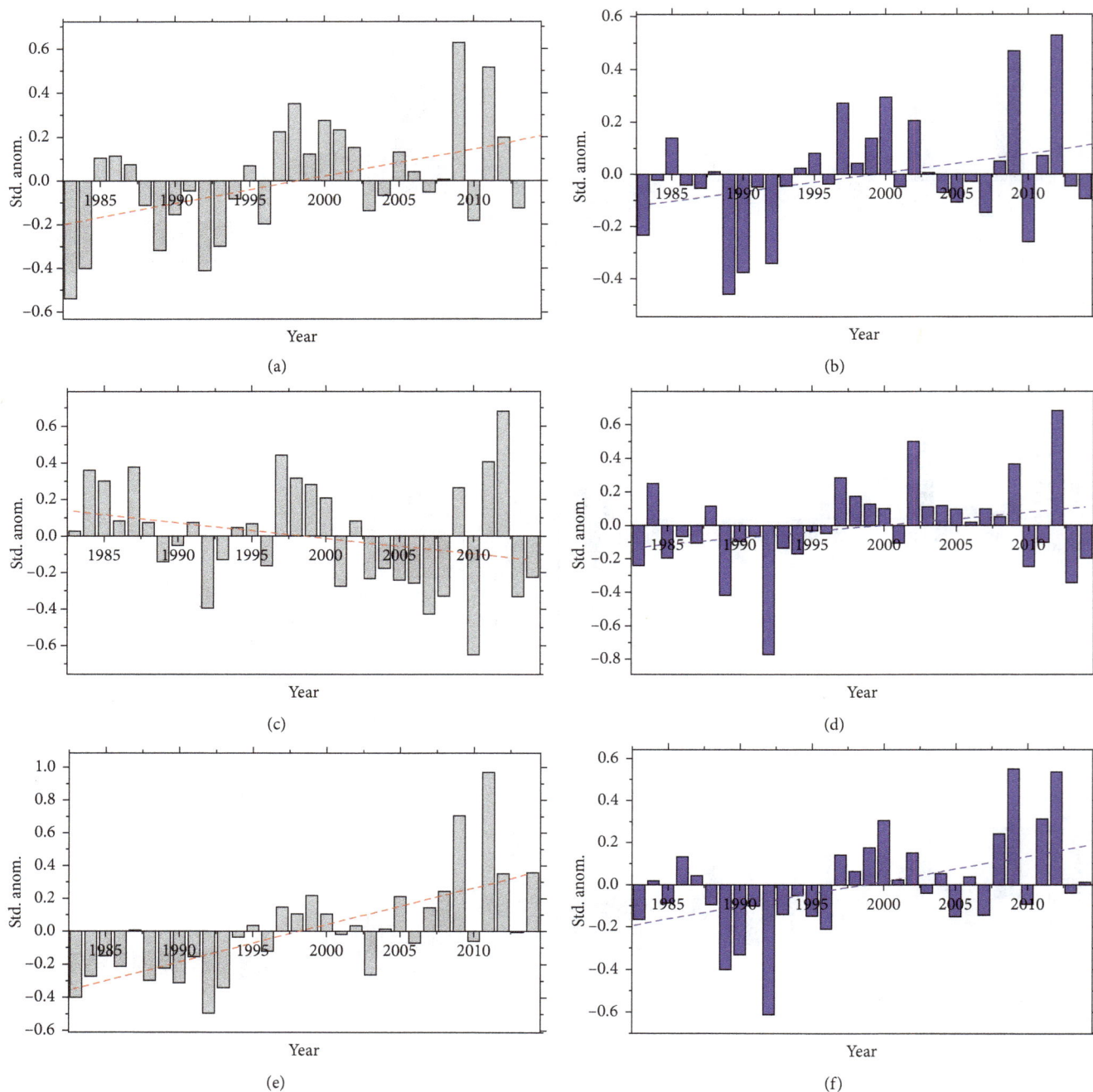

FIGURE 4: Standardized anomalies of the warmest day (TXx) over three agroecological zones, (a) lowland, (c) midland, and (e) highland, and for the coldest day (TXn), (b) lowland, (d) midland, and (f) highland, for the period 1983–2014.

significant in the same AEZs. The decrease in CSDI is in agreement with the observed warming in other studies such as the study by Donat et al. [60] carried out across the globe. However, the recent trends for CSDI and WSDI were mixed in the Upper Blue Nile Basin [36], which also suggests contextual differences between the current and previous studies.

3.1.6. Diurnal Temperature Range (DTR). With regards to the trend in the DTR, all AEZs have experienced

significantly decreasing trend (Table 3). The magnitude of change of DTR was −0.052°C/year and −0.043°C/year ($p < 0.05$) in the lowland and highland AEZs, respectively. A very significant decreasing trend in the DTR of the midland AEZ was observed (0.053°C/year) ($p < 0.001$) (Figure 6(b)). Similarly, Mekasha et al. [31] reported a significantly decreasing trend in DTR only in Negele Borena station in Ethiopia. Moreover, Zhou et al. [67] documented that there has been a decreasing trend in DTR in some other parts of the world, mainly in arid

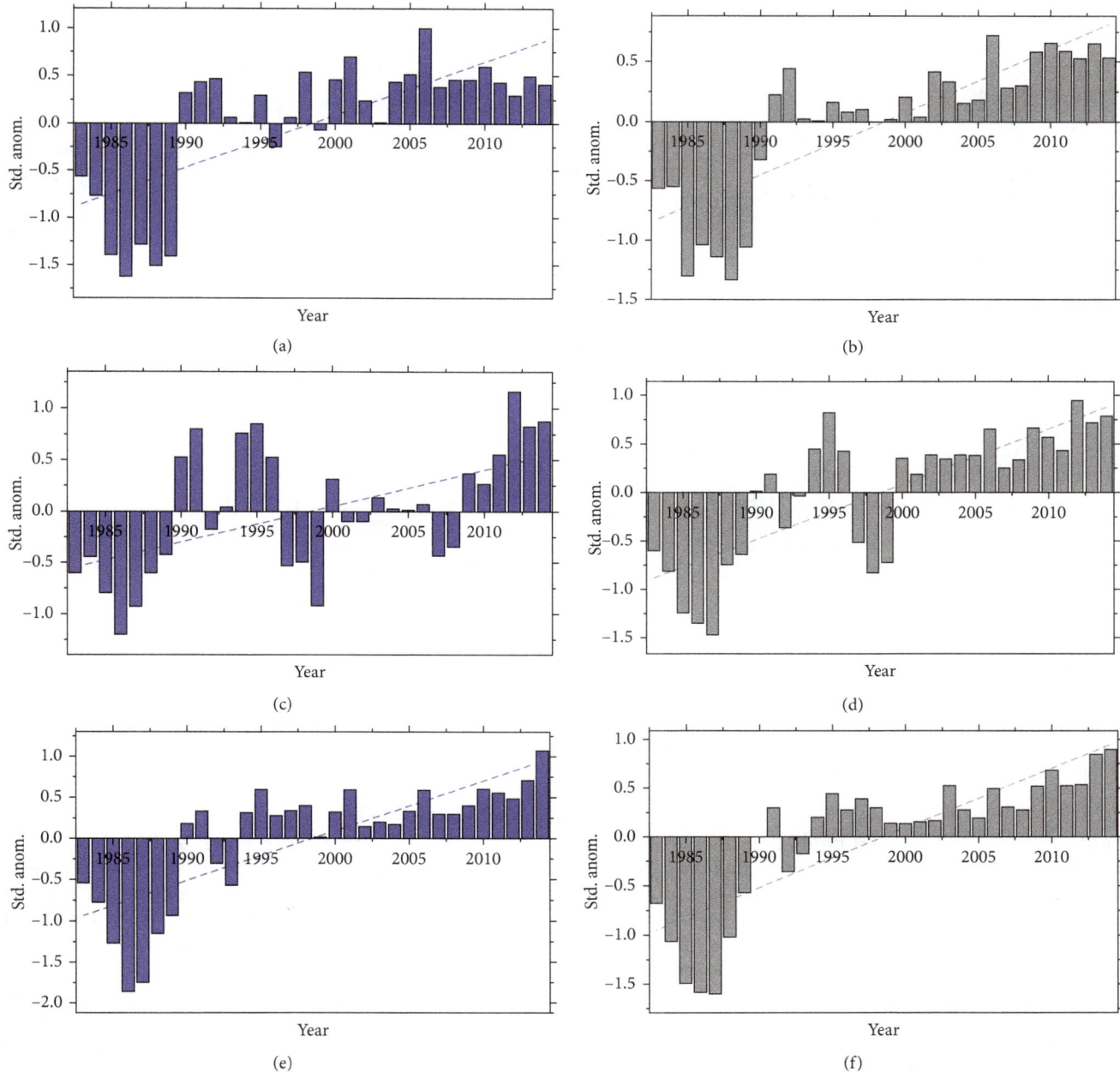

FIGURE 5: Standardized anomalies of the warmest night (TNx) over three agroecological zones, (a) lowland, (c) midland, and (e) highland, and for the coldest night (TNn), (b) lowland, (d) midland, and (f) highland, for the period 1983–2014.

and semiarid regions. Concurring to this, the mean annual DTR exhibited a reduction by 0.5 to 1°C in Sudan and Ethiopia between the 1950s and 2000 [68].

A decrease in DTR suggests that the trend in the daily maximum temperature is smaller than the trend in the daily minimum temperature. Earlier studies reported on the relevance of DTR as one of the proxy indicators for climate change. For example, a study by Makowski et al. [69] considered DTR as the important indicator of climate change.

3.2. Trends in Precipitation Extremes

3.2.1. Simple Daily Intensity Index (SDII), Consecutive Dry Days (CDD), and Consecutive Wet Days (CWD). The SDII, which monitors precipitation intensity on wet days, did not show any significant trend in all AEZs, a finding that is in line with results from [25, 31, 36]. The highest SDII was recorded in 2013 in the lowland (Figure 7(a)), in 2012 in the midland (Figure 7(b)), and in 2005 in the highland AEZ (Figure 7(c)), respectively. By the same token, the trend analysis of both consecutive dry days (CDD) and

(a)

(b)

(c)

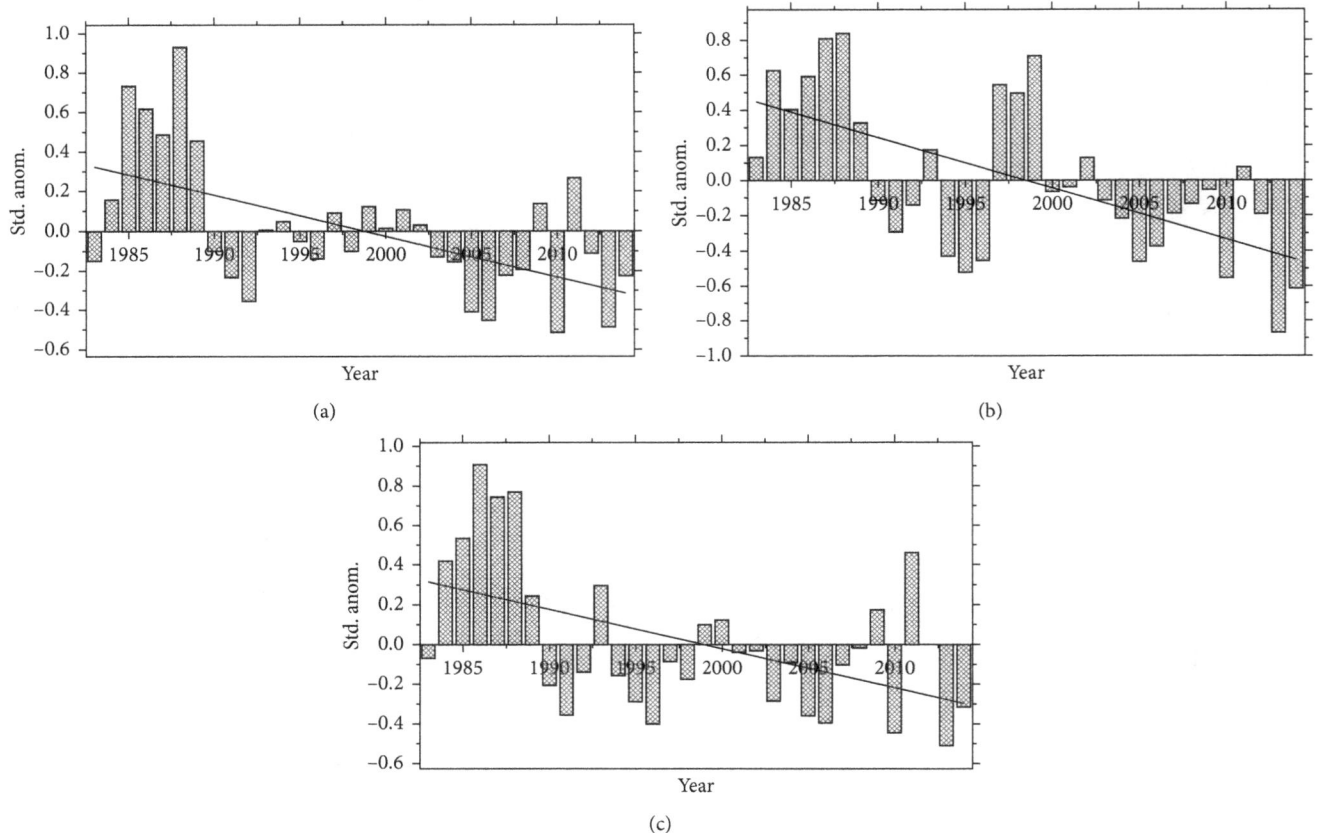

FIGURE 6: Standardized anomalies of the diurnal temperature range (DTR) over three agroecological zones, (a) lowland, (b) midland, and (c) highland, for the period 1983–2014.

consecutive wet days (CWD) imply that the results were not significant in all AEZs (Table 4). The highest CDD was observed in 2012 in the lowland (Figure 8(a)), in 2008 in the midland (Figure 8(c)), and in 2000 in highland AEZs (Figure 8(e)),respectively. On the contrary, the highest CWD was observed in the years between 1988 and 1997 in the lowland (Figure 8(b)), in 1992 in the midland (Figure 8(d)), and in 1987 in the highland AEZs (Figure 8(f)), respectively. Although insignificant trends were observed in the CDD and CWD, the early 1990s were wet years compared to the 2000s, which signify warming over the studied AEZs. The same trend was documented by previous studies such as [25, 31, 36]. However, Mekasha et al. [31] observed a decreasing trend for CDD in one station. A study by Mengistu et al. [48] reported that agroecological zones in the Upper Blue Nile River Basin experienced relatively cold years in the 1980s and warm years from the early 1990s to the 2000s.

3.2.2. Number of Heavy (R10mm) and Very Heavy (R20mm) Precipitation Days. The number of very heavy precipitation days (R20mm) was increasing with a magnitude of 0.325 days/year ($p < 0.05$) in the midland AEZ but was insignificant in the other AEZs. The highest R20mm was recorded in the year 2013 in the lowland (Figure 9(b)), in

2012 in the midland (Figure 9(d)), and in 1988 in the highland AEZs (Figure 9(f)), respectively. The increasing trend in the rainfall amount in the midland AEZ was in line with recent findings from Degefu and Bewket [38] in which Wolaita Sodo, as one of the meteorological stations, was found to experience an increasing trend in the average annual total rainfall ($p < 0.01$). Similarly, a recent study by Weldegerima et al. [70] in Northern Ethiopia has documented an increase in annual rainfall with the magnitude of change being 2.20, 3.42, 6.58, and 2.88 mm/year, in Bahir Dar, Dangila, Debre Tabor, and Gondar, respectively. On the contrary, insignificant decreasing trend was reported both in the lowland and in the highland AEZs in R10mm while it was insignificant increasing trend in the midland AEZ. The highest R10mm was recorded in the year 1987 in the lowland (Figure 9(a)), in 2006 in the midland (Figure 9(c)), and in 2007 in the highland AEZs (Figure 9(e)), respectively. In general, the midland AEZ shows both significant and insignificant increasing trends in the R20mm and R10mm between 1983 and 2014. The variations in the years of highest R10mm and R20mm suggest the trend differences between and among the AEZs.

3.2.3. Maximum 1-Day (RX1day) and 5-Day (RX5day) Precipitations. The trend of maximum 1-day precipitation

(a)

(b)

(c)

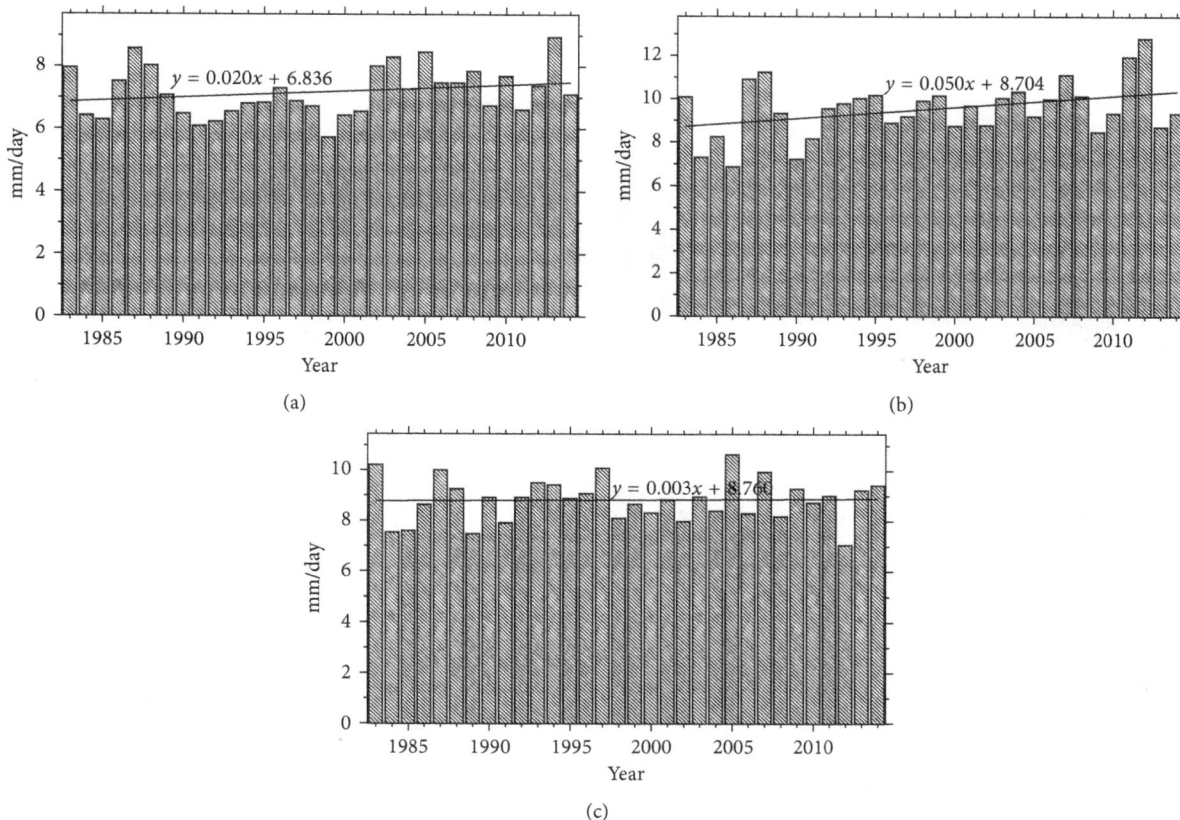

FIGURE 7: Trend of the simple daily intensity index (SDII) over three agroecological zones, (a) lowland, (b) midland, and (c) highland, for the period 1983–2014.

TABLE 4: Trends in precipitation extreme indices per year by AEZ.

Index	Units	Lowland AEZ		Midland AEZ		Highland AEZ	
		MK test (Z-test)	Sen's slope	MK test (Z-test)	Sen's slope	MK test (Z-test)	Sen's slope
SDII	mm/day	0.190^{ns}	0.026^{ns}	0.185^{ns}	0.044^{ns}	0.044^{ns}	0.006^{ns}
CDD	Days	-0.085^{ns}	-0.396^{ns}	-0.028^{ns}	-0.089^{ns}	0.020^{ns}	0.118^{ns}
CWD	Days	-0.124^{ns}	-0.049^{ns}	-0.103^{ns}	-0.04^{ns}	-0.159^{ns}	-0.017^{ns}
R10mm	Days	-0.045^{ns}	-0.057^{ns}	0.053^{ns}	0.137^{ns}	-0.060^{ns}	-0.041^{ns}
R20mm	Days	0.198^{ns}	0.078^{ns}	0.331^{*}	0.325^{*}	0.010^{ns}	0.00^{ns}
RX1day	mm	0.171^{ns}	0.143^{ns}	0.059^{ns}	0.092^{ns}	0.295^{*}	0.297^{*}
RX5day	mm	0.067^{ns}	0.111^{ns}	0.175^{ns}	0.728^{ns}	-0.067^{ns}	-0.143^{ns}
R95p	mm	0.188^{ns}	2.418^{ns}	0.275^{*}	6.048^{*}	0.244^{ns}	2.757^{*}
R99p	mm	0.188	0.00^{ns}	0.052^{ns}	0.00^{ns}	0.201^{ns}	0.125^{ns}

ns: nonsignificant, $p > 0.05$; $^{*}p \leq 0.05$; $^{**}p \leq 0.01$; $^{***}p \leq 0.001$.

amount (RX1day) was observed to increase in the highland AEZ with a magnitude of 0.297 mm/year ($p < 0.05$). On the contrary, both the lowland and midland AEZs have experienced insignificant trend in the RX1day. The maximum RX1day was recorded in the year 2005 in the lowland (Figure 10(a)), in 1988 in the midland (Figure 10(c)), and in 2014 in the highland AEZs (Figure 10(e)), respectively. An insignificant positive trend in the RX5 day was reported both in the lowland and midland AEZs, while it was a negative trend in the highland AEZ (Table 4). In relation to the maximum RX5 day, it was recorded in 2008 in the lowland (Figure 10(b)), in the years of 1987 and 2007 in the midland

(Figure 10(d)), and in 2011 in the highland AEZs (Figure 10(f)), respectively. In summary, with the exception of the highland AEZ for RX1day, all AEZs have experienced insignificant trend both in the RX1day and RX5 day. On the contrary, Worku et al. [36] found out that significant increasing trend was observed in the Rx5 day in Fichie and Mendida stations, while significant decreasing trend was observed in the Rx1day in Alem ketema station in the Upper Blue Nile Basin.

3.2.4. Very Wet Days (R95p) Extremely Wet Days (R99p). The trend of R95p was statistically significant both in the midland (6.048 mm/year) and highland AEZs (2.757 mm/

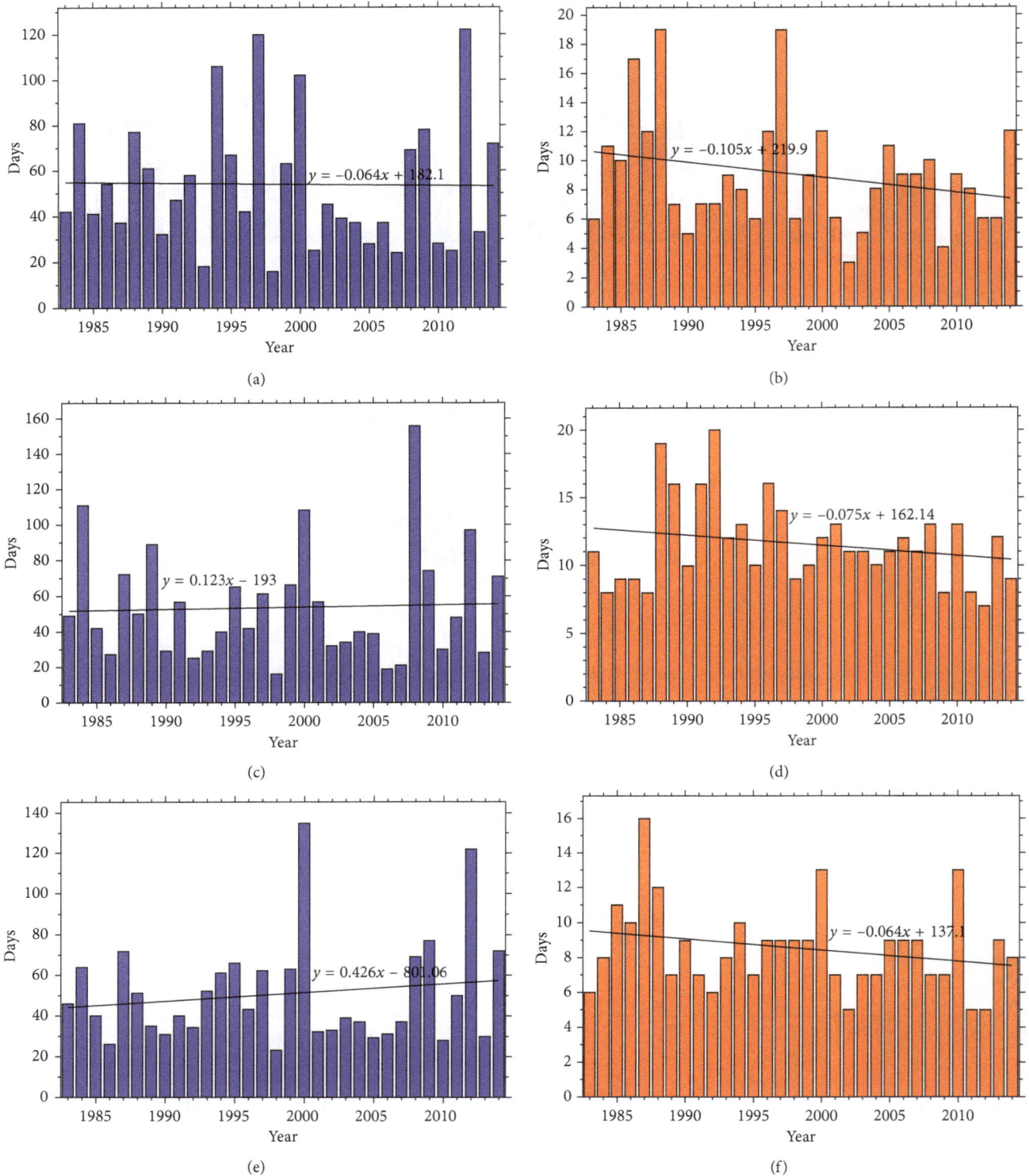

FIGURE 8: Trend of consecutive dry days (CDD) over three agroecological zones, (a) lowland, (c) midland, and (e) highland, and for consecutive wet days (CWD), (b) lowland, (d) midland, and (f) highland, for the period 1983–2014.

year) ($p < 0.05$). Insignificant increasing trend was observed in the lowland AEZ, suggesting a positive trend in very wet days in two of the AEZs. Similarly, insignificant increasing trend was observed for R99p across all AEZs. Nevertheless, with the

continuing climate change and variability, precipitation in Ethiopia is expected to decline from a mean annual value of 2.04 mm/day (1961–1990) to 1.97 mm/day (2070–2099) with a total decline in rainfall by 25.5 mm/year [71]. On this future

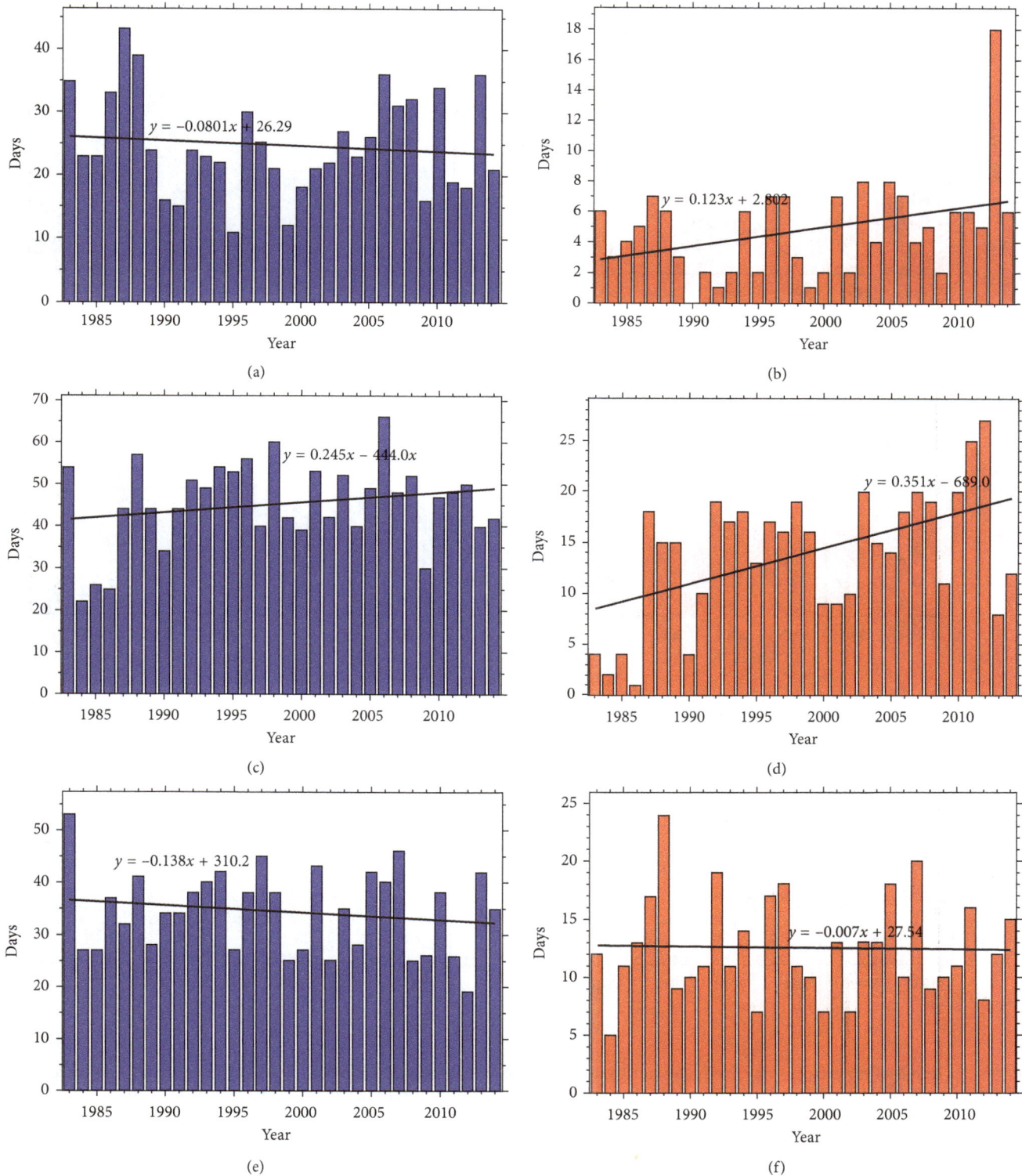

FIGURE 9: Trends of number of heavy precipitation days (R10mm) over three agroecological zones, (a) lowland, (c) midland, and (e) highland, and for number of very heavy precipitation days (R20mm), (b) lowland, (d) midland, and (f) highland, for the period 1983–2014.

change, several climate change estimation scenarios and models confirm that many parts of Ethiopia are likely to experience a reduction in the length of the growing period with higher reductions in some parts of the country [12, 66]. The observed trends in spatial and temporal variability in the climate extremes in the studied AEZs could be associated with the multiple topographic and relief structures of the country [26, 41].

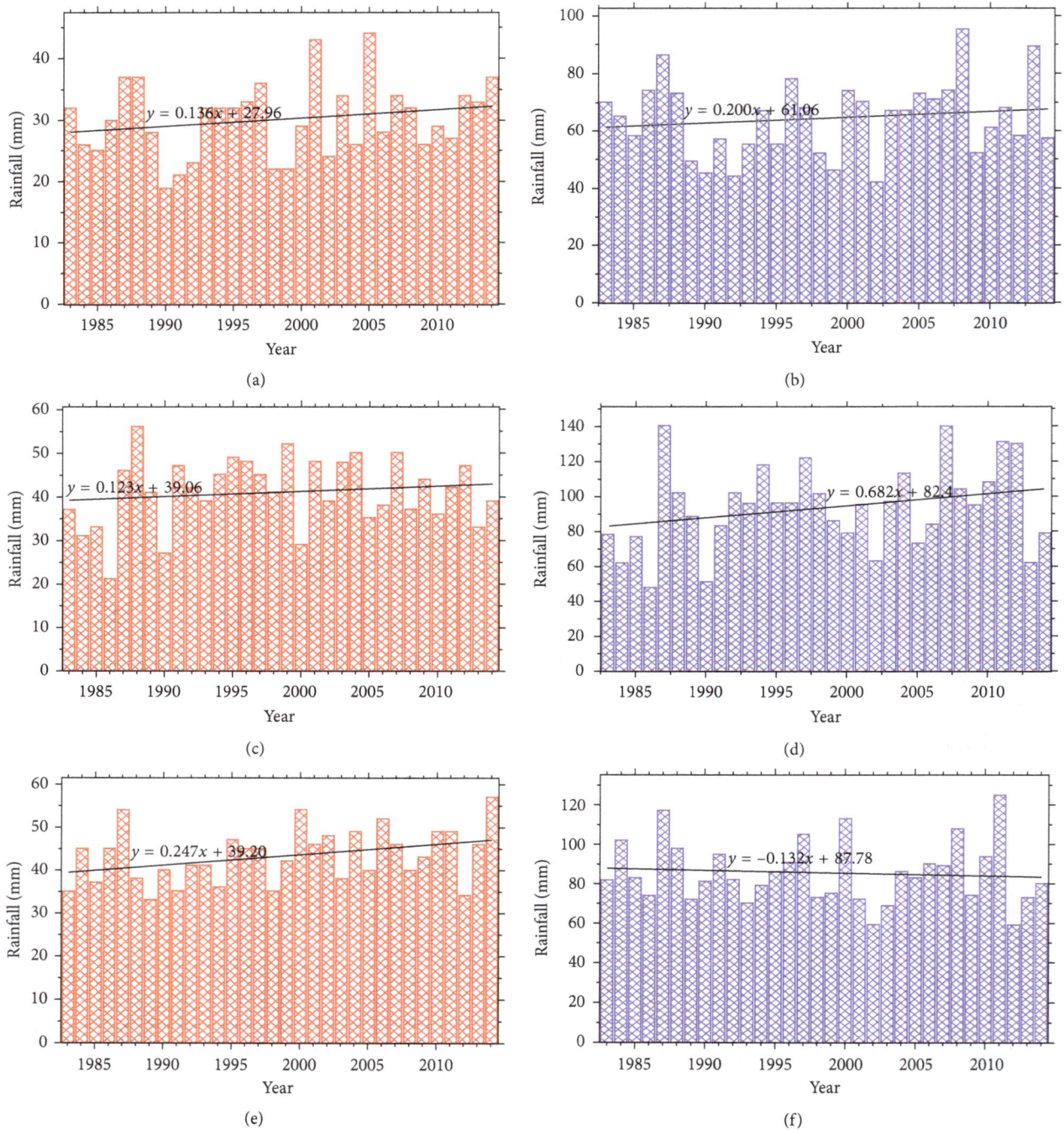

FIGURE 10: Trends of maximum 1-day precipitation (RX1day) over three agroecological zones, (a) lowland, (c) midland, and (e) highland, and for maximum 5-day precipitation (RX5 day), (b) lowland, (d) midland, and (f) highland, for the period 1983–2014.

In general, the extreme trend analysis confirmed that there is an overall propensity of increasing warm extremes and a decreasing tendency of cold extremes in the selected AEZs. The results are in line with similar findings reported in tropical environments where warm extremes showed increasing trends while the cold extremes showed decreasing trends [36, 41, 72].

4. Conclusions

This study has analyzed changes in the indices of extreme temperature and rainfall based on changes in duration, intensity, and frequency of climatic extremes in the lowland, midland, and highland AEZs in Wolaita Zone, Southern Ethiopia, over the period of 1983–2014. The trend analysis revealed that AEZs have experienced both positive and negative trends in temperature extremes. Overtime, the annual maximum value of daily maximum temperature (TXx), annual maximum value of daily minimum temperature (TNx), and annual minimum value of daily minimum temperature (TNn) have shown significant positive trends between 1983 and 2014, except in the midland where TXx deceased by 0.017°C/year. In terms of AEZ, the magnitude of change in temperature extremes is higher in the highland AEZ but lower in the midland AEZ, implying that the highland AEZ is experiencing a higher magnitude of change in the occurrence of climate extremes. This trend is likely to have adverse effects on the livelihoods of people in the highland agroecological zone. The annual number of occurrence of warm nights (TN90p) and the occurrence of warm days (TX90p) show significant warming anomalies for the period between 1983 and 2014 both in the lowland and highland AEZs. The occurrence of warm nights (TN90p) has shown a significant increasing trend ($p < 0.001$) in all AEZs except the midlands, which show insignificant increase in TN90p, while cool nights (TN10p) were consistently decreasing in all AEZs ($p < 0.001$). Insignificant decreasing trend in the frequency of cool days (TX10p) was observed both in the lowland and midland AEZs. TX10p shows a decreasing trend in the highland AEZ ($p < 0.05$). The warmest night (TNx) shows a significant increasing trend in all AEZs ($p < 0.001$). The magnitude of change in TNx is 0.055°C, 0.037°C, and 0.063°C/year for the lowland, midland, and highland AEZs, respectively. Comparatively, the highland AEZ experienced a higher magnitude of change in the TNx than the midland and lowland AEZs. DTR shows a significant decreasing trend in all AEZs over the study period and indicating warming conditions.

Overall, the trends of warm extremes are increasing, and cold extremes are decreasing, implying a significant warming in the AEZs, which is in agreement with results from other studies in different geographic scales. A negative trend in the cold spell duration indicator (CSDI) was observed in all AEZs with magnitudes of change in CSDI being 0.667 days, 2.259 days, and 1 days/year for the lowland, midland, and highland AEZs, respectively. The trend was significant ($p < 0.05$) for both midland and highland AEZs and very significant for the lowland AEZ ($p < 0.001$). Even though the trends for consecutive dry days and consecutive wet days were insignificant in all AEZs, the number of very wet days has revealed a positive trend both in the midlands (6.048 days/year) and highlands (2.757 days/year) ($p < 0.05$). R20mm (0.325 days/year) and RX1day (0.297 mm/year) have revealed positive trends in the midland and highland AEZs ($p < 0.05$), respectively. In short, most of the precipitation extreme indices have shown insignificant trend across AEZs, signifying no difference was observed between 1983 and 2014.

Generally, the inconsistency of observed patterns in the precipitation extremes and uniformity in temperature extremes reveal that the local level experiences can fit into the meso-macro levels of changes in climate extremes, which in turn can confirm the robustness of extreme indices at different levels of analysis. Therefore, results from extreme event analysis could be crucial information for the design of integrated early warning systems and development of planned adaption strategies at the local level linking the meso-macro level evidence in the face of the changing extreme climate events.

Conflicts of Interest

The authors declare that there are no conflicts of interest regarding the publication of this article.

Acknowledgments

The study was carried out with the financial support from both Wolaita Sodo University and Addis Ababa University as part of the first author's Ph.D. program. The authors appreciate the National Meteorological Agency for providing the gridded daily temperature and precipitation data.

References

[1] Intergovernmental Panel on Climate Change (IPCC), *Climate Change 2013: The Physical Science Basis. Contribution of Working Group I to the Fifth Assessment Report of the Intergovernmental Panel on Climate Change*, T. F. Stocker, D. Qin, G.-K. Plattner et al., Eds., Cambridge University Press, Cambridge, UK, 2013.

[2] Intergovernmental Panel on Climate Change (IPCC), *Climate Change 2007: Impacts, Adaptation and Vulnerability. Contribution of Working Group II to the Fourth Assessment Report of the Intergovernmental Panel on Climate Change*, M. L. Parry, O. F. Canziani, J. P. Palutikof, P. J. van der Linden, and C. E. Hanson, Eds., Cambridge University Press, Cambridge, UK, 2007.

[3] Intergovernmental Panel on Climate Change (IPCC), *Climate Change 2014: Synthesis Report. Contribution of Working Groups I, II and III to the Fifth Assessment Report of the Intergovernmental Panel on Climate Change*, R. K. Pachauri and L. A. Meyer, Eds., IPCC, Geneva, Switzerland, 2014.

[4] P. K. Thornton, P. J. Ericksen, M. Herrero, and A. J. Challinor, "Climate variability and vulnerability to climate change: a review," *Global Change Biology*, vol. 20, no. 11, pp. 3313–3328, 2014.

[5] W. N. Adger, K. Brown, D. R. Nelson et al., "Resilience implications of policy responses to climate change," *Wiley Interdisciplinary Reviews: Climate Change*, vol. 2, no. 5, pp. 757–766, 2011.

[6] Intergovernmental Panel on Climate Change (IPCC), *Managing the Risks of Extreme Events and Disasters to Advance Climate Change Adaptation. A Special Report of Working Groups I and II of the Intergovernmental Panel on Climate Change*, C. B. Field, V. Barros, T. F. Stocker et al., Eds., Cambridge University Press, Cambridge, UK, 2012.

[7] M. B. Sylla, N. Elguindi, F. Giorgi, and D. Wisser, "Projected robust shift of climate zones over West Africa in response to

anthropogenic climate change for the late 21st century," *Climatic Change*, vol. 134, no. 1-2, pp. 241–253, 2016.

[8] V. Ongoma, H. Chen, and G. W. Omony, "Variability of extreme weather events over the equatorial East Africa, a case study of rainfall in Kenya and Uganda," *Theoretical and Applied Climatology*, vol. 131, no. 1-2, pp. 295–308, 2018.

[9] T. C. Peterson, M. P. Hoerling, P. A. Stott, and S. C. Herring, "Explaining extreme events of 2012 from a climate perspective," *Bulletin of the American Meteorological Society*, vol. 94, no. 9, pp. S1–S74, 2013.

[10] World Bank, *Ethiopia: Economics of Adaptation to Climate Change*, The World Bank Group, Washington, DC, USA, 2010.

[11] D. Conway and E. L. F. Schipper, "Adaptation to climate change in Africa: challenges and opportunities identified from Ethiopia," *Global Environmental Change*, vol. 21, no. 1, pp. 227–237, 2011.

[12] Ethiopian Panel of Climate Change (EPCC), *First Assessment Report, Working Group II Agriculture and Food Security*, Ethiopian Academy of Sciences, Addis Ababa, Ethiopia, 2015.

[13] M. A. Degefu, D. P. Rowell, and W. Bewket, "Teleconnections between Ethiopian rainfall variability and global SSTs: observations and methods for model evaluation," *Meteorology and Atmospheric Physics*, vol. 129, no. 2, pp. 173–186, 2017.

[14] W. Berhanu and F. Beyene, "Climate variability and household adaptation strategies in Southern Ethiopia," *Sustainability*, vol. 7, no. 6, pp. 6353–6375, 2015.

[15] G. J. Y. You and C. Ringler, *Hydro-Economic Modeling of Climate Change Impacts in Ethiopia (No. 960)*, International Food Policy Research Institute (IFPRI), 2010.

[16] Food and Agricultural organization (FAO), *FAO in Ethiopia El Niño Response Plan 2016*, FAO, Rome, Italy, 2016, https://reliefweb.int/report/ethiopia/fao-ethiopia-el-ni-o-response-plan-2016.

[17] M. Savage, A. Mujica, F. Chiappe, and I. Ross, *Climate Finance and Water Security: Bangladesh Case Study*, Oxford Policy Management Limited, Oxford, UK, 2015.

[18] D. Frank, M. Reichstein, M. Bahn et al., "Effects of climate extremes on the terrestrial carbon cycle: concepts, processes and potential future impacts," *Global Change Biology*, vol. 21, no. 8, pp. 2861–2880, 2015.

[19] K. E. Trenberth, J. T. Fasullo, and T. G. Shepherd, "Attribution of climate extreme events," *Nature Climate Change*, vol. 5, no. 8, p. 725, 2015.

[20] K. E. Trenberth, "Attribution of climate variations and trends to human influences and natural variability," *Wiley Interdisciplinary Reviews: Climate Change*, vol. 2, no. 6, pp. 925–930, 2011.

[21] National Academies of Sciences, Engineering, and Medicine, *Attribution of Extreme Weather Events in the Context of Climate Change*, National Academies Press, Washington, DC, USA, 2016.

[22] L. V. Alexander, P. Hope, D. Collins, B. Trewin, A. Lynch, and N. Nicholls, "Trends in Australia's climate means and extremes: a global context," *Australian Meteorological Magazine*, vol. 56, no. 1, pp. 1–18, 2007.

[23] T. C. Peterson and M. J. Manton, "Monitoring changes in climate extremes: a tale of international collaboration," *Bulletin of the American Meteorological Society*, vol. 89, no. 9, pp. 1266–1271, 2008.

[24] J. E. Tierney, J. E. Smerdon, K. J. Anchukaitis, and R. Seager, "Multidecadal variability in East African hydroclimate controlled by the Indian Ocean," *Nature*, vol. 493, no. 7432, pp. 389–392, 2013.

[25] Y. Seleshi and P. Camberlin, "Recent changes in dry spell and extreme rainfall events in Ethiopia," *Theoretical and Applied Climatology*, vol. 83, no. 1–4, pp. 181–191, 2006.

[26] W. Bewket and D. Conway, "A note on the temporal and spatial variability of rainfall in the drought-prone Amhara region of Ethiopia," *International Journal of Climatology*, vol. 27, no. 11, pp. 1467–1477, 2007.

[27] G. Kebede and W. Bewket, "Variations in rainfall and extreme event indices in the wettest part of Ethiopia," *SINET: Ethiopian Journal of Science*, vol. 32, no. 2, pp. 129–140, 2009.

[28] H. Shang, J. Yan, M. Gebremichael, and S. M. Ayalew, "Trend analysis of extreme precipitation in the Northwestern Highlands of Ethiopia with a case study of Debre Markos," *Hydrology and Earth System Sciences*, vol. 15, no. 6, pp. 1937–1944, 2011.

[29] G. Kiros, A. Shetty, and L. Nandagiri, "Extreme rainfall signatures under changing climate in semi-arid northern highlands of Ethiopia," *Cogent Geoscience*, vol. 3, no. 1, article 1353719, 2017.

[30] M. R. Jury and C. Funk, "Climatic trends over Ethiopia: regional signals and drivers," *International Journal of Climatology*, vol. 33, no. 8, pp. 1924–1935, 2013.

[31] A. Mekasha, K. Tesfaye, and A. J. Duncan, "Trends in daily observed temperature and precipitation extremes over three Ethiopian eco-environments," *International Journal of Climatology*, vol. 34, no. 6, pp. 1990–1999, 2014.

[32] K. V. Suryabhagavan, "GIS-based climate variability and drought characterization in Ethiopia over three decades," *Weather and climate extremes*, vol. 15, pp. 11–23, 2017.

[33] S. Gummadi, K. P. C. Rao, J. Seid et al., "Spatio-temporal variability and trends of precipitation and extreme rainfall events in Ethiopia in 1980–2010," *Theoretical and Applied Climatology*, 2017.

[34] E. Viste, D. Korecha, and A. Sorteberg, "Recent drought and precipitation tendencies in Ethiopia," *Theoretical and Applied Climatology*, vol. 112, no. 3-4, pp. 535–551, 2013.

[35] T. T. Zeleke, F. Giorgi, G. T. Diro, and B. F. Zaitchik, "Trend and periodicity of drought over Ethiopia," *International Journal of Climatology*, vol. 37, no. 13, pp. 4733–4748, 2017.

[36] G. Worku, E. Teferi, A. Bantider, and Y. T. Dile, "Observed changes in extremes of daily rainfall and temperature in Jemma Sub-Basin, Upper Blue Nile Basin, Ethiopia," *Theoretical and Applied Climatology*, pp. 1–16, 2018.

[37] A. Kebede, B. Diekkrüger, and D. C. Edossa, "Dry spell, onset and cessation of the wet season rainfall in the Upper Baro-Akobo Basin, Ethiopia," *Theoretical and Applied Climatology*, vol. 129, no. 3-4, pp. 849–858, 2017.

[38] M. A. Degefu and W. Bewket, "Variability and trends in rainfall amount and extreme event indices in the Omo-Ghibe River Basin, Ethiopia," *Regional environmental change*, vol. 14, no. 2, pp. 799–810, 2014.

[39] E. M. Fischer and R. Knutti, "Anthropogenic contribution to global occurrence of heavy-precipitation and high-temperature extremes," *Nature Climate Change*, vol. 5, no. 6, pp. 560–564, 2015.

[40] S. C. Herring, M. P. Hoerling, J. P. Kossin, T. C. Peterson, and P. A. Stott, "Explaining extreme events of 2014 from a climate perspective," *Bulletin of the American Meteorological Society*, vol. 96, no. 12, pp. S1–S172, 2015.

[41] C. McSweeney, M. New, and G. Lizcano, *UNDP Climate Change Country Profiles Ethiopia*, 2008, http://countryprofiles.geog.ox.ac.uk.

[42] L. Silici, *Agroecology-What it is and what it has to offer. Issue Paper 14629IIED*, International Institute for Environment and Development, London, UK, 2014.

[43] H. R. Ojha, V. R. Sulaiman, P. Sultana et al., "Is South Asian agriculture adapting to climate change? Evidence from the Indo-Gangetic Plains," *Agroecology and Sustainable Food Systems*, vol. 38, no. 5, pp. 505–531, 2014.

[44] W. Zerihun, *Vegetation Map of Ethiopia*, Addis Ababa University, Addis Ababa Google Scholar, Addis Ababa, Ethiopia, 1999.

[45] Ministry of Agriculture (MoA), *Agroecological Zonations of Ethiopia*, Ministry of Agriculture (MoA), Addis Ababa, Ethiopia, 2000.

[46] Y. Gecho, G. Ayele, T. Lemma, and D. Alemu, "Rural household livelihood strategies: options and determinants in the case of Wolaita Zone, Southern Ethiopia," *Social Sciences*, vol. 3, no. 3, pp. 92–104, 2014.

[47] H. Hurni, *Agroecological Belts of Ethiopia: Explanatory Notes on Three Maps at a Scale of 1: 1,000,000. Soil Conservation Research Program of Ethiopia, Addis Ababa, Ethiopia*, Wittwer Druck AG, Bern, Switzerland, 1998.

[48] D. Mengistu, W. Bewket, and R. Lal, "Recent spatiotemporal temperature and rainfall variability and trends over the Upper Blue Nile River Basin, Ethiopia," *International Journal of Climatology*, vol. 34, no. 7, pp. 2278–2292, 2014.

[49] L. Alexander, H. Yang, and S. Perkins, *ClimPACT—Indices and Software*, 2013, http://www.wmo.int/pages/prog/wcp/ccl/opace/opace4/meetings/documents/ETCRSCI_software_documentation_v2a.doc.

[50] H. B. Mann, "Nonparametric tests against trend," *Econometrica*, vol. 13, no. 3, pp. 245–259, 1945.

[51] M. G. Kendall, "A new measure of rank correlation," *Biometrika*, vol. 30, no. 1-2, pp. 81–93, 1938.

[52] P. K. Sen, "Estimates of the regression coefficient based on Kendall's tau," *Journal of the American Statistical Association*, vol. 63, no. 324, pp. 1379–1389, 1968.

[53] E. Teferi, S. Uhlenbrook, and W. Bewket, "Inter-annual and seasonal trends of vegetation condition in the Upper Blue Nile (Abay) Basin: dual-scale time series analysis," *Earth System Dynamics*, vol. 6, no. 2, pp. 617–636, 2015.

[54] V. Ongoma, H. Chen, C. Gao, A. M. Nyongesa, and F. Polong, "Future changes in climate extremes over Equatorial East Africa based on CMIP5 multimodel ensemble," *Natural Hazards*, vol. 90, no. 2, pp. 901–920, 2018.

[55] H. Theil, "A rank-invariant method of linear and polynominal regression analysis (parts 1-3),"Proceedings of the Koninklijke Nederlandse Akademie van Wetenschappen Series A vol. 53, pp. 1397–1412, 1950.

[56] S. K. Jain and V. Kumar, "Trend analysis of rainfall and temperature data for India," *Current Science*, vol. 102, no. 1, pp. 37–49, 2012.

[57] G. Choi, D. Collins, G. Ren et al., "Changes in means and extreme events of temperature and precipitation in the Asia-Pacific Network region, 1955–2007," *International Journal of Climatology*, vol. 29, no. 13, pp. 1906–1925, 2009.

[58] M. Lin, L. W. Horowitz, R. Payton, A. M. Fiore, and G. Tonnesen, "US surface ozone trends and extremes from 1980 to 2014: quantifying the roles of rising Asian emissions, domestic controls, wildfires, and climate," *Atmospheric Chemistry and Physics*, vol. 17, no. 4, pp. 2943–2970, 2017.

[59] V. Ongoma and H. Chen, "Temporal and spatial variability of temperature and precipitation over East Africa from 1951 to 2010," *Meteorology and Atmospheric Physics*, vol. 129, no. 2, pp. 131–144, 2017.

[60] M. G. Donat, L. V. Alexander, H. Yang et al., "Updated analyses of temperature and precipitation extreme indices since the beginning of the twentieth century: the HadEX2 dataset," *Journal of Geophysical Research: Atmospheres*, vol. 118, no. 5, pp. 2098–2118, 2013.

[61] P. A. O. Omondi, J. L. Awange, E. Forootan et al., "Changes in temperature and precipitation extremes over the Greater Horn of Africa region from 1961 to 2010," *International Journal of Climatology*, vol. 34, no. 4, pp. 1262–1277, 2014.

[62] X. Zhang, L. Alexander, G. C. Hegerl et al., "Indices for monitoring changes in extremes based on daily temperature and precipitation data," *Wiley Interdisciplinary Reviews: Climate Change*, vol. 2, no. 6, pp. 851–870, 2011.

[63] A. Adem and A. Amsalu, "Climate change in the southern lowlands of Ethiopia: Local level evidences, impacts and adaptation responses," *Ethiopian Journal of Development Research*, vol. 34, no. 1, pp. 1–36, 2012.

[64] T. Gebrehiwot and A. van der Veen, "Farm level adaptation to climate change: the case of farmer's in the Ethiopian Highlands," *Environmental Management*, vol. 52, no. 1, pp. 29–44, 2013.

[65] M. A. Zaroug, E. A. Eltahir, and F. Giorgi, "Droughts and floods over the upper catchment of the Blue Nile and their connections to the timing of El Niño and La Niña events," *Hydrology and Earth System Sciences*, vol. 18, no. 3, pp. 1239-1249, 2014.

[66] W. Bewket, M. Radeny, and C. Mungai, *Agricultural Adaptation and Institutional Responses to Climate Change Vulnerability in Ethiopia. CCAFS Working Paper No. 106*, CGIAR Research Program on Climate Change, Agriculture and Food Security (CCAFS), Copenhagen, Denmark, 2015.

[67] L. Zhou, A. Dai, Y. Dai et al., "Spatial dependence of diurnal temperature range trends on precipitation from 1950 to 2004," *Climate Dynamics*, vol. 32, no. 2-3, pp. 429–440, 2009.

[68] M. Hulme, R. Doherty, T. Ngara, M. New, and D. Lister, "African climate change: 1900–2100," *Climate Research*, vol. 17, no. 2, pp. 145–168, 2001.

[69] K. Makowski, M. Wild, and A. Ohmura, "Diurnal temperature range over Europe between 1950 and 2005," *Atmospheric Chemistry and Physics*, vol. 8, no. 21, pp. 6483–6498, 2008.

[70] T. M. Weldegerima, T. T. Zeleke, B. S. Birhanu, B. F. Zaitchik, and Z. A. Fetene, "Analysis of rainfall trends and its relationship with SST signals in the Lake Tana Basin, Ethiopia," *Advances in Meteorology*, vol. 2018, Article ID 5869010, 10 pages, 2018.

[71] A. Kidanu, K. Rovin, and K. Hardee-Cleaveland, *Linking Population, Fertility and Family Planning with Adaptation to Climate Change: Views from Ethiopia*, Population Action International, Washington, DC, USA, 2009.

[72] A. C. Kruger and S. S. Sekele, "Trends in extreme temperature indices in South Africa: 1962–2009," *International Journal of Climatology*, vol. 33, no. 3, pp. 661–676, 2013.

9

Evaluation of CMIP5 Global Climate Models over the Volta Basin: Precipitation

Jacob Agyekum ⓘ,[1] Thompson Annor ⓘ,[1] Benjamin Lamptey,[2] Emmannuel Quansah ⓘ,[1] and Richard Yao Kuma Agyeman[3]

[1]Department of Physics, Kwame Nkrumah University of Science and Technology (KNUST), Kumasi, Ghana
[2]African Centre of Meteorological Applications for Development (ACMAD), Niamey, Niger
[3]Numerical Weather Prediction Unit, Ghana Meteorological Agency (GMet), Accra, Ghana

Correspondence should be addressed to Jacob Agyekum; jacoblyff@gmail.com

Academic Editor: Olivier P. Prat

A selected number of global climate models (GCMs) from the fifth Coupled Model Intercomparison Project (CMIP5) were evaluated over the Volta Basin for precipitation. Biases in models were computed by taking the differences between the averages over the period (1950–2004) of the models and the observation, normalized by the average of the observed for the annual and seasonal timescales. The Community Earth System Model, version 1-Biogeochemistry (CESM1-BGC), the Community Climate System Model Version 4 (CCSM4), the Max Planck Institute Earth System Model, Medium Range (MPI-ESM-MR), the Norwegian Earth System Model (NorESM1-M), and the multimodel ensemble mean were able to simulate the observed climatological mean of the annual total precipitation well (average biases of 1.9% to 7.5%) and hence were selected for the seasonal and monthly timescales. Overall, all the models (CESM1-BGC, CCSM4, MPI-ESM-MR, and NorESM1-M) scored relatively low for correlation (<0.5) but simulated the observed temporal variability differently ranging from 1.0 to 3.0 for the seasonal total. For the annual cycle of the monthly total, the CESM1-BGC, the MPI-ESM-MR, and the NorESM1-M were able to simulate the peak of the observed rainy season well in the Soudano-Sahel, the Sahel, and the entire basin, respectively, while all the models had difficulty in simulating the bimodal pattern of the Guinea Coast. The ensemble mean shows high performance compared to the individual models in various timescales.

1. Introduction

Rainfall is an important component of the hydrological cycle and plays an essential role in determining the amount of water available at the surface. Most of the countries in West Africa depend mainly on rainfed agriculture [1], and therefore, the amount of rainfall affects the crop yield [2]. A decrease in the amount of precipitation may lead to drought [3, 4], while an increase can cause flooding [5]. In the past years, incidents of floods in West Africa have caused a devastating impact on people's health and destruction to properties and livelihood [6]. The Volta Basin is a major source of water to a number of countries [7] in the West African region. It serves as a major driving force of the

economic progress of many countries in West Africa including Ghana, Burkina Faso, Cote d'Ivoire Mali, and Togo [8, 9]. In Ghana, the major source of energy (hydroelectric power) is generated mainly from the Akosombo Dam, Kpong Dam, and Bui Dam [9]. These dams have been constructed at different locations within the basin and contribute more than 60% [9, 10] of hydroelectric power, to the total energy needs of the country. The Volta Basin can be divided into the Guinea Coast, the Soudano-Sahel, and the Sahel [11, 12] agroecological zones (Figure 1), based on the annual amount of precipitation received by these subregions. Therefore, changes in the amount of precipitation and hence changes in the amount of available water affect the lives of the inhabitants [13]. Changes in the future climates

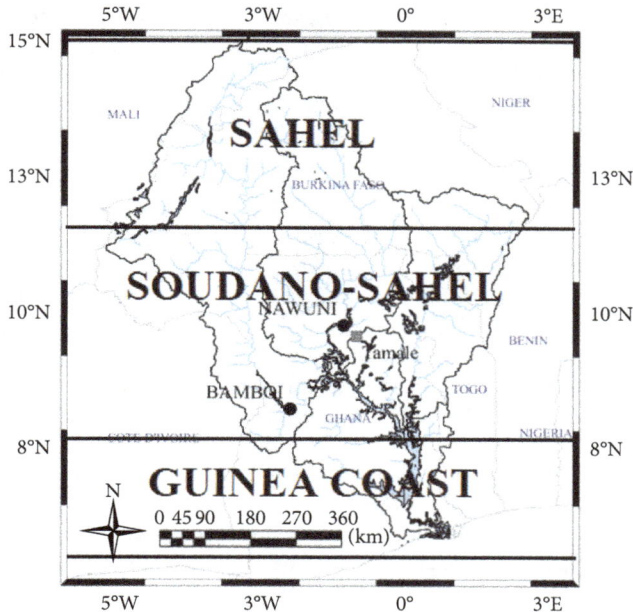

FIGURE 1: The three ecological zones in the basin, from the humid Guinea Coast to the semiarid Sahel (modified from Fujihara et al. [45]).

of the globe and the West African region imply likely changes in the hydrology and water resources of the Volta Basin which raise serious concerns that demand urgent attention for the region. Various climate studies (that mainly apply GCMs) on the future of the earth's climate system suggest changes in the climate from global (e.g., [14–16]) to the regional scales (e.g., [17, 18]).

Also, a regional climate model (RCM) takes its input from a GCM, and therefore, the performance of the driving GCM is a crucial issue that should not be ignored. In essence of that, forcing data from a poorly performed GCM can significantly affect the performances of the RCM. It is therefore important to assess the performances of GCMs over an economically important basin such as the Volta.

Mehran et al. [19] evaluated 34 CMIP5 GCMs in reproducing observed precipitation over the globe using the volumetric hit index (VHI) analysis. In their study, the GCMs showed good agreement with the observed, but reproducing the observed precipitation over arid regions and certain subcontinental regions was problematic, for the total monthly precipitation. This confirms the assertion that GCMs have limitations when it comes to the representation of observed precipitation on both short temporal and small spatial scales (e.g., [18]). They also indicated the superior performance of the multimodel ensemble mean to the individual models. Kumar et al. [20] have also evaluated temperature and precipitation trends in 19 CMIP5 GCMs, focusing on continental areas (60°S–60°N) for the 1930–2004 period. They showed that there are large uncertainties in the models in simulating local-scale temperature and precipitation trends. They also indicated the high performance of the multimodel ensemble average compared to the individual models. Other studies have evaluated the performance of CMIP5 models over other regions including North

Pacific [21], Europe [22–24], and Asia (e.g., [25–28]) and reported high variations in the performances of the GCMs over the different regions.

Over Africa, Nikulin et al. [29] evaluated the performance of CORDEX-RCMs in simulating precipitation. They evaluated the simulated precipitation at seasonal, annual, and diurnal timescales and indicated that all models simulated the seasonal and annual precipitation quite well. They also indicated the superior performance of the ensemble average of the RCMs to the individual models. Nikiema et al. [30] reported better multimodel CMIP5 and CORDEX simulations of historical summer temperature and precipitation variabilities over West Africa. They evaluated and intercompared the multimodel ensembles of the CMIP5 and the CORDEX and found that while CORDEX failed to outperform the simulated mean climatology of temperature by the CMIP5 ensembles, it substantially improved the simulation of precipitation and provided a more realistic fine-scale features tied to local topography and land use. Over the Volta Basin, Annor et al. [31] evaluated the performance of the Weather Research and Forecast (WRF) model forced by the MPI-ESM-MR in reproducing the present day (1980–2005) temperature and precipitation. They reported the transfer of bias from the GCM to the RCM and indicated that, at certain instances, the RCM minimized the bias and at other instances increased the GCM bias in both temperature and precipitation. Agyeman et al. [32] assessed the best physics parameterization scheme combination for seasonal simulation over Ghana and showed that scheme combinations are sensitive to the agroclimatic belts within the country. Aziz and Obuobie [33] looked at trend analysis in observed (1981–2010) and projected (2051–2075 and 2076–2100 under the IPCC Representative Concentration Pathways RCP4.5 and RCP8.5) precipitation and mean temperature over the Black Volta Basin using RCMs. They showed a statistically significant (at the 5% significant level) increase of 111 mm in the annual rainfall, whereas a significant increase of 0.9°C in temperature for the observed period. For the future projection, there is high uncertainty in the trend of rainfall (which is statistically nonsignificant), as some ensemble members project positive trends, while others gave negative trends. With regard to the temperature, average annual projection showed increases over the basin, with the warming being higher under the RCP8.5 scenario than under the RCP4.5 scenario.

The above studies have contributed immensely to the assessment of the performance of GCMs and RCMs in reproducing the observed climatology of various regions including Africa, but very few focus on the Volta Basin and the very few that targeted the basin in most cases assessed single GCMs. Therefore, assessing the performances of several GCMs from the CMIP5 over the Volta Basin will contribute significantly to the efforts in climate modeling over the region. Giorgi and Gutowski [34] indicated the importance of analyzing the performance of GCMs before they are used to drive RCMs. This study therefore focuses on the performance of 18 GCMs in simulating precipitation over the Volta Basin and the three belts (Guinea Coast, Soudano-Sahel, and Sahel).

In this paper, the performance of GCMs in simulating observed precipitation is presented. Section 2 presents the observational data and the models used for the study including the methods used for the evaluation of models. Section 3 discusses the results. The summary and conclusions are found in Section 4.

2. Materials and Methods

2.1. Data

2.1.1. Observational Data. In this study, the Global Precipitation Climatology Centre (GPCC) version seven (v7) precipitation data [35] on a $0.5° \times 0.5°$ grid resolution is used as reference precipitation data that depict spatial variability over the basin. This data set comprises observations based on three-dimensional variational assimilation [36]. Due to the highly varying spatial distribution of precipitation, spatial density of observed precipitation data is crucial over West Africa. Studies (e.g., [37, 38]) have indicated that the availability of high-quality data set especially in the case of precipitation over West Africa is problematic. Consequently, gridded observational open-source data sets become the practical option for model validation purposes within the region. The selection of the GPCC data as reference data for this evaluation study is based on the fact that they have been applied in similar model evaluation studies (e.g., [31, 32]) over the West African region. Other studies including Gruber et al. [39], Nicholson et al. [37], Paeth et al. [38], and Nikulin et al. [29] have demonstrated the robustness of GPCC data in reproducing observed precipitation in the West African region. These studies expressed confidence in this particular data set. Moreover, there are a significant number of gauge stations over West Africa that have been incorporated in the development of the GPCC data set. However, the GPCC data have low gauge station density over the Sahara region [29].

2.1.2. Model Data. All the model outputs evaluated in this study are from the CMIP5, used in the preparation of the Fifth Assessment Report (AR5) of the Intergovernmental Panel on Climate Change (IPCC). This is the latest phase of a coordinated effort by modeling groups across the globe to systematically perform so many prescribed climate model experiments [40]. The study uses 18 GCM simulations provided by 16 modeling centres and groups. The model names, acronyms, and their horizontal and vertical resolutions are shown in Table 1. The selection of these GCMs are based on the fact that most of the GCMs have been applied extensively over the basin (e.g., [18, 31]), and also our selection was informed by the availability of model data with similar ensemble members. Although these are not the only models with similar ensemble members, all the ones we used are of similar ensemble members for the present. All the models are long-term historical runs from the CMIP5 experiments which include a preindustrial control (piControl) experiments. They all include the same ensemble member (r1i1p1); that is, all the models are initialized from the same initial observed conditions (initialization 1 (i1)), the same

methods (realization 1 (r1)), and the same "perturbed physics" 1 (p1) [41]. For each model, total monthly precipitation data of 55 years (1950–2004) were used for the study. The Program for Climate Model Diagnosis and Intercomparison (PCMDI) collected the model data as part of the process leading to the CMIP5 [42].

2.2. Methods. The models' performances in reproducing the observed climatology are analyzed relative to the observational data. First, each of the model grids was bilinearly interpolated [43, 44] to the GPCC grid of $0.5° \times 0.5°$ resolution to facilitate direct comparison of the models with the observational data. The comparison was done over the Volta Basin (4.9°N–16°N; 6°W–3°E) and the three subregions including the Guinea Coast (4.9°N–8°N), the Soudano-Sahel (8°N–12°N), and the Sahel (12°N–16°N) [12], for the annual, seasonal, and monthly timescales. The Guinea Coast (GC) is characterized by a bimodal pattern of precipitation (a maximum peak in June and a second one in September), while the Soudano-Sahel (SD) and the Sahel (SA) regions are characterized by a unimodal pattern (the maximum peak in August in both cases). These modes are influenced by the meridional movements of the ITCZ within the year. Figure 1 shows a map of the Volta Basin including the three subregions.

2.2.1. Annual Scale Analysis. On the annual scale, biases in models were computed by taking the differences between the averages over the period (1950–2004) of the models and the observation, normalized by the average of the observed as shown in (1).

This is done to assess the spatial biases in the models over the entire basin for the study period (1950–2004):

$$\text{Relative bias} = \frac{\text{model} - \text{observed}}{\text{observed}} \times 100\%. \quad (1)$$

2.2.2. Seasonal Scale Analysis. In the seasonal total, biases in the models were computed in the same manner done for the annual totals using (1). This is done for the dry (November, December, January, February, and March (NDJFM)) and rainy (April, May, June, July, August, September, and October (AMJJASO)) seasons of the Volta Basin. Secondly, the three statistics, the standardized deviation (σ) (2), the correlation coefficient (r) (3), and the root-mean-square error (RMSE) (4) were computed, representing the temporal variability, the temporal pattern, and the temporal errors in the models, respectively. This is done for the DJF, MAM, JJA, and SON standard seasons for the entire basin and the three belts. Lastly, trend analysis was done over the Volta Basin and the three belts for the various seasons that are found over these regions. The standard deviation (σ) is given by

$$\sigma = \sqrt{\frac{1}{N} \sum_{i=1}^{N} (x_i - \overline{x})^2}, \quad (2)$$

where N is the total number of data points and $\overline{x}$ is the mean of the individual data points, x_i.

TABLE 1: Details of models used in the study (the descriptions are from Taylor et al. [42]), showing models' spatial resolution for the Atmospheric Global Climate Model (AGCM) and Oceanic Global Climate Model (OGCM) and the various institutions that produced the models.

Model name	Resolution, number of grids (lon. × lat.; levels) of AGCM (OGCM)	Institution
BCC-CSM1.1	128 × 64; 26 (360 × 232; 40)	Beijing Climate Center, China Meteorological Administration, China
BNU-ESM	128 × 64; 26 (360 × 200; 50)	College of Global Change and Earth System Science, Beijing Normal University
CanESM2	128 × 64; 26 (360 × 192; 40)	Canadian Centre for Climate Modelling and Analysis, Canada
CCSM4	228 × 192; 26 (320 × 384; 60)	National Center for Atmospheric Research, USA
CESM1-BGC	288 × 192; 26	National Center for Atmospheric Research, USA
CMCC-CM	480 × 240; 31 (182 × 149; 31)	Centro Euro-Mediterraneo per I Cambiamenti Climatici, Italy
CNRM-CM5	256 × 128; 31 (362 × 292; 42)	Centre National de Rescherches Meteorologiques/Centre Europeen de Recherche et Formation Avances en Calcu Scientifique, France
CSIRO-Mk3.6.0	192 × 96; 18 (192 × 189; 31)	Commonwealth Scientific and Industrial Research Organisation in collaboration with the Queensland Climate Change Centre of Excellence, Australia
EC-EARTH	320 × 160; 21 (182 × 149; 31)	EC-Earth (European Earth System Model)
HadGEM2-AO	192 × 145; 60 (360 × 216; 40)	National Institute of Meteorological Research, Seoul, South Korea
HadGEM2-CC	192 × 144; 60 (360 × 216; 40)	Met Office Hadley Centre, UK
INMCM4	180 × 120; 21 (360 × 340; 40)	Institute of Numerical Mathematics, Russia
IPSL-CM5A-MR	96 × 96; 39 (182 × 149; 31)	Institut Pierre Simon Laplace, France
MIROC5	256 × 128; 40 (256 × 224; 50)	Atmosphere and Ocean Research Institute (The University of Tokyo), National Institute for Environmental Studies, and Japan Agency for Marine-Earth Science and Technology, Japan
MIROC-ESM	128 × 64; 80 (256 × 192; 44)	Japan Agency for Marine-Earth Science and Technology, Atmosphere and Ocean Research Institute (The University of Tokyo), and National Institute for Environmental Studies, Japan
MPI-ESM-MR	192 × 96; 47 (256 × 220; 40)	Max Planck Institute for Meteorology (MPI-M), Germany
MRI-CGCM3	320 × 160; 48 (360 × 368; 51)	Meteorological Research Institute, Japan
NorESM1-M	144 × 96; 26 (320 × 384; 70)	Norwegian Climate Centre, Norway

The correlation coefficient r between the variables x and y is defined as follows:

$$r = \frac{(1/N)\sum_{i=1}^{N}(x_n - \overline{x})(y_n - \overline{y})}{\sigma_x \sigma_y}. \tag{3}$$

The root-mean-square error (RMSE) for the fields f and r is defined as follows:

$$\text{RMSE} = \left[\frac{1}{N}\sum_{i=1}^{N}(f_i - r_i)^2\right]^{1/2}. \tag{4}$$

The trends in the seasonal precipitation over the entire basin and the three zones are also considered using the Mann–Kendall (MK) test [46, 47]. Pettitt's test [48] was used for the change-point detection. The Theil–Sen estimator [49] was applied in the case of the slope. The MK test is a nonparametric rank-based statistical test used for detecting monotonic trends in time-series data. In comparison with other nonparametric procedures, such as

Spearman's rho test [50], the power of the Mann–Kendal test is robust and similar to the extent of giving indistinguishable results in practice [51]. The MK statistic (S_{mk}) is calculated theoretically using the following equation:

$$S_{mk} = \sum_{i=1}^{N}\sum_{j=i+1}^{N}\text{sgn}(X_j - X_i), \tag{5}$$

where X_j and X_i are the data values of j and i, such that $(j > i)$, and sgn is given as follows:

$$\begin{aligned}\text{sgn}(X_j - X_i) &= 0 && \text{if } X_j - X_i = 0, \\ \text{sgn}(X_j - X_i) &= 1 && \text{if } X_j - X_i > 0, \\ \text{sgn}(X_j - X_i) &= -1 && \text{if } X_j - X_i < 0.\end{aligned} \tag{6}$$

Under the null hypothesis of no trend and independence of the series terms, the variance of the Mann–Kendall statistic is calculated as follows:

$$\text{var}\left(S_{\text{mk}}\right) = \frac{N\left(N-1\right)\left(2N+5\right) - \sum_{i=1}^{M} U_i\left(i\right)\left(i-1\right)\left(2i+5\right)}{18},$$
$$(7)$$

where M is the number of tied groups and U_i represents the size of the Mth group. The summation term in the numerator is used only if the data series contains tied values. For the sample size $n \geq 10$, the statistic S assumes normal distribution, and the standard normal test statistic Z_S is computed as follows:

$$Z_S = 0 \quad \text{for } S = 0, \tag{8}$$

$$Z_S = \frac{S-1}{\sqrt{\text{var}\left(S\right)}} \quad \text{for } S > 0, \tag{9}$$

$$Z_S = \frac{S+1}{\sqrt{\text{var}\left(S\right)}} \quad \text{for } S < 0. \tag{10}$$

And for the p values for the MK test,

$$p = 0.5 - \varphi\left(\left|Z_S\right|\right). \tag{11}$$

The trend results in this study have been evaluated at the 5% significant level (95% confidence level). The corresponding threshold (Z) value is ± 1.96. This implies that the null hypothesis is rejected when $\left|Z_S\right| \geq Z_{\alpha/2}$ in (8)–(10) at the $\alpha = 0.05$ level of significance.

For Pettitt's test, which represents a sudden change in the statistics of a record, (12)–(14) were used:

$$K_T = \max\left|U_{tT}\right|, \tag{12}$$

where

$$U_T = \sum_{i=1}^{N} \sum_{j=i+1}^{N} \text{sgn}\left(X_j - X_i\right). \tag{13}$$

The change point is located at K_T, provided the statistic ($p < 0.05$) is significant:

$$p = 2\exp\left(\frac{-6K_T^2}{T^3 + T^2}\right). \tag{14}$$

The magnitude of the trend is estimated using the Theil–Sen estimator which robustly fits a line to sample points in the plane by choosing the median of the slopes of all lines through pairs of points. The estimation is given by

$$Q_i = \frac{X_j - X_k}{j-k} \quad \text{for all } k < j \text{ and } i = 1, \dots, N, \tag{15}$$

where Q_i = slope between the data points X_j and X_k, X_j = data values at time j, and X_k = data values at time k. $N = (n(n-1))/2$, where $1 < k < j < n$ and n is the total number of observations for each period. The N values of Q_i are ranked from the least to the largest, and the median of these N values of Q_i is Sen's estimate of the slope computed as follows:

$$Q_{\text{med}} = Q\left[\frac{N+1}{2}\right], \tag{16}$$

when N is an odd number, and

$$Q_{\text{med}} = Q\left[\frac{N}{2}\right] + Q\left[\frac{N+2}{2}\right], \tag{17}$$

when N is an even number.

The direction of the trend is given by the sign of Q_{med}, while its magnitude indicates the steepness of the trend.

2.2.3. Annual Cycle of Monthly Total Precipitation Analysis. In the annual cycle analysis, the climatological monthly mean for all the models for the whole 55-year period was compared to that of the GPCC precipitation data. First, area averages over the whole Volta Basin and the three ecological zones, the Guinea Coast, the Soudano-Sahel, and the Sahel, were computed. This is done to assess the models' ability in reproducing the bimodal pattern of precipitation over the Guinea Coast and the unimodal pattern of precipitation over the Volta Basin, the Soudano-Sahel, and the Sahel.

3. Results and Discussion

The performance of the selected CMIP5 GCMs relative to the GPCC precipitation data is presented on the annual, seasonal, and monthly timescales.

3.1. Annual Total for Climatological Mean Bias. The climatological (1950–2004) mean biases in the annual total for the individual models are presented in Figures 2–5. There are overestimation and underestimation of the observed annual precipitation over the basin by some of the models. Some models simulated precipitation close to the observed precipitation over major parts of the basin.

Figure 2 shows the results of models overestimating the observed precipitation over most parts of the basin with wet bias up to 196%. The MIROC5 overestimates the amount of precipitation over the entire basin with the highest mean relative bias of 102.5%. CSIRO-Mk3.6.0 overestimates the Sahelian part of the basin with a maximum relative bias of 116.0% and underestimates the Guinea Coast with a minimum relative bias of −42.0%, whereas the MIROC-ESM overestimates mostly the southern part of the basin with a mean relative bias over the whole basin of 23.0% where the biases are ranging from −19.0% to 195.6%. The CanESM2, the CNRM-CM5, and the EC-EARTH overestimate precipitation almost over the entire basin with biases ranging from −12.3% to 112.0%. The Volta Basin's precipitation climatology is controlled mainly by the two winds [52]: the northeasterlies, which are characterized by dry cold winds from the Sahara, and the southwesterlies, characterized by moist warm winds from the Atlantic Ocean. A proper representation of these two air masses is relevant for the simulation of the precipitation. Models that overestimate precipitation could exaggerate the effects of the moist winds from the Atlantic Ocean (e.g., [52, 53]).

Figure 3 represents the results of models generally underestimating the observed precipitation over most parts of the basin with bias up to −114%. The MRI-CGCM3,

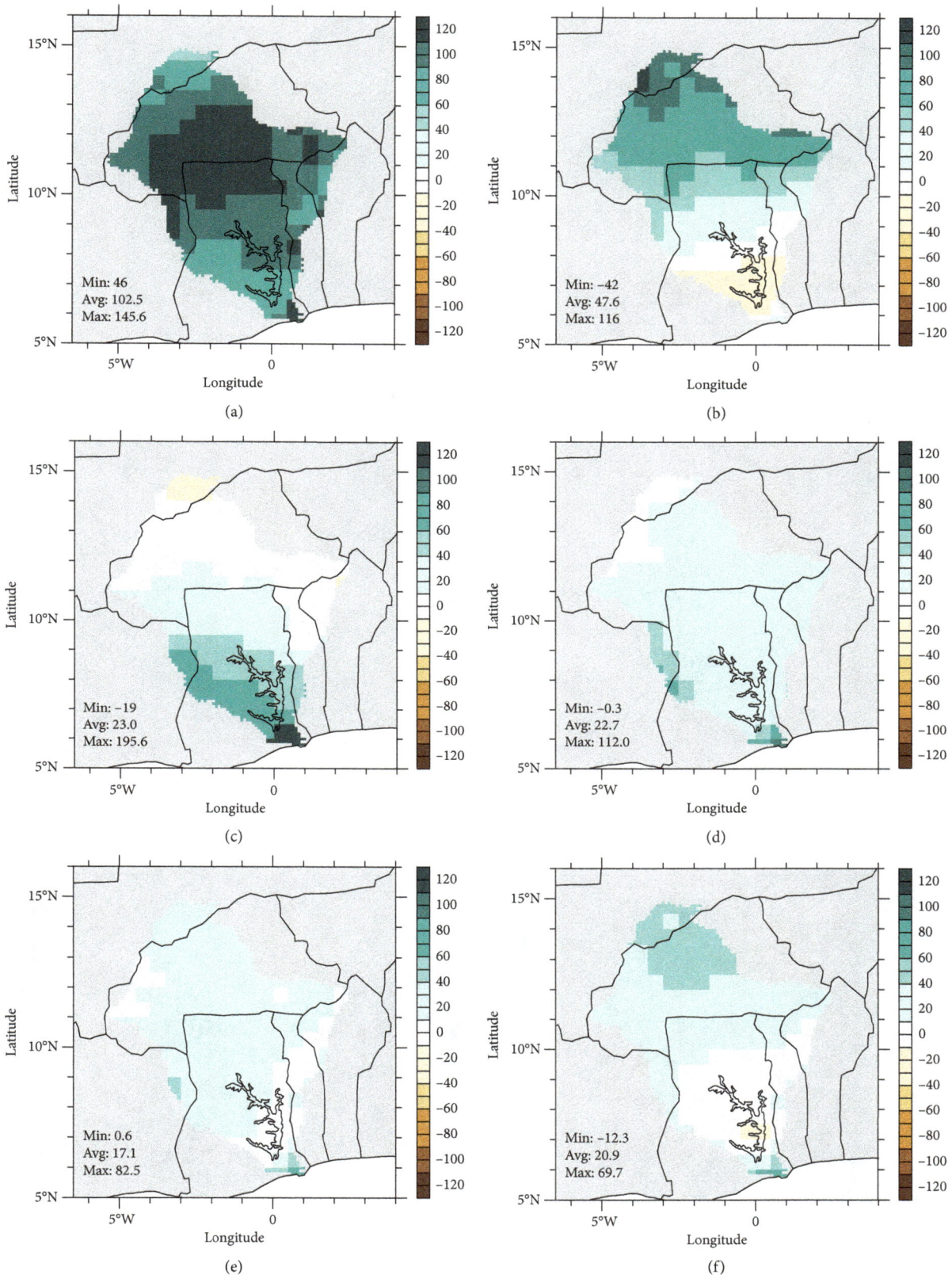

FIGURE 2: The climatological (1950–2004) mean bias in the annual total precipitation for the models: MIROC5 (a), CSIRO-Mk3.6.0 (b), MIROC-ESM (c), CNRM-CM5 (d), CanESM2 (e), and EC-EARTH (f), overestimating the observed climatological mean over major parts of the Volta Basin.

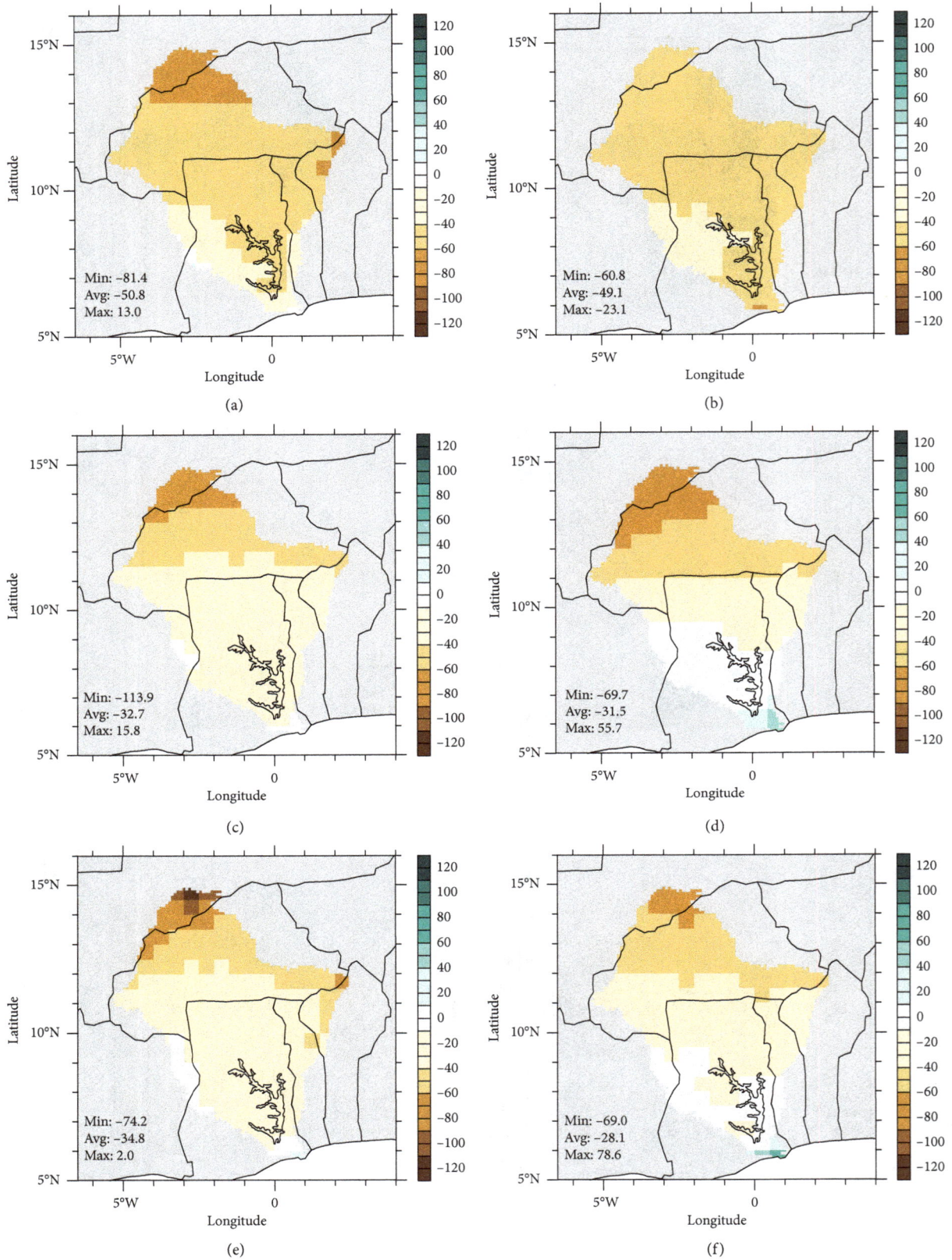

FIGURE 3: The climatological (1950–2004) mean bias in the annual total precipitation for the models: MRI-CGCM3 (a), INMCM4 (b), HadGEM2-CC (c), IPSL-CM5A-MR (d), HadGEM2-AO (e), and CMCC-CM (f), underestimating the observed climatological mean over major parts of the Volta Basin.

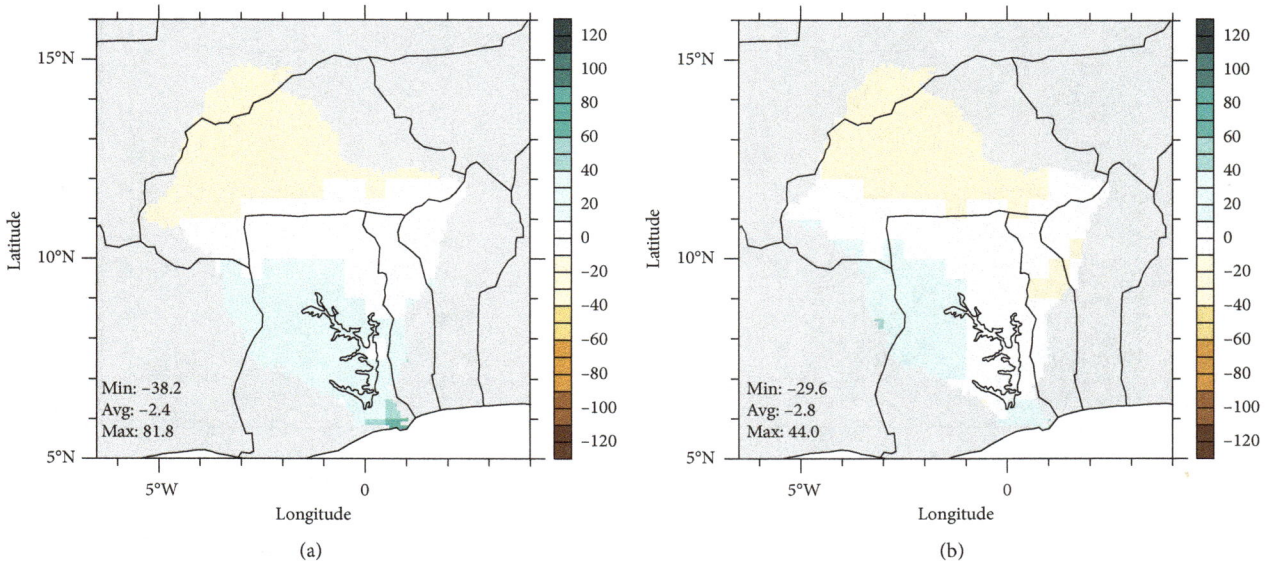

FIGURE 4: The climatological (1950–2004) mean bias in the annual total precipitation of BCC-CSM1.1 (a) and BNU-ESM (b). They averagely underestimate the observed precipitation.

INMCM4, HadGEM2-CC, IPSL-CM5A-MR, HadGEM2-AO, and CMCC-CM underestimate the observed precipitation and have mean biases of −50.8%, −49.1%, −32.7%, −31.5%, −34.8%, and −28.1%, respectively. These models could probably exaggerate the effects of the dry winds from the Sahara and also the warming of the troposphere by greenhouse gases that decrease the vertical temperature gradient inducing a stable atmosphere (e.g., [54, 55]).

Figure 4 shows the Beijing Climate Center Climate System Model version 1.1 (BCC-CSM1.1) and the Beijing Normal University Earth System Model (BNU-ESM) underestimating the observed precipitation in the Sahel while overestimating in the Guinea Coast. Averagely, both models underestimate precipitation over the entire basin.

The CESM1-BGC, CCSM4, NorESM1-M, MPI-ESM-MR, and multimodel ensemble mean are able to simulate precipitation close to the observed precipitation value, with mean biases of 4.6%, 7.5%, 2.8%, 6.1%, and 1.9%, respectively, as shown in Figure 5. These models are able to simulate the observed precipitation relatively well. Due to the ability of these models to simulate the annual climatology over the basin, they are then considered for the seasonal and monthly timescale analyses.

3.2. Seasonal Total for Climatological Mean Bias.
The climatological (1950–2004) mean biases in the seasonal totals of the four models with the least annual climatological mean biases are presented in Figures 6 and 7. The two seasons: the dry season (NDJFM) and the rainy season (AMJJASO), are considered over the entire basin.

3.2.1. The Dry Season.
This season is characterized by small amount of precipitation over the Volta Basin and coincides with the Northern Hemispheric winter season. Excluding

the MPI-ESM-MR which underestimates the observed precipitation (average bias of −47.5%) over the entire basin, all the models including the ensemble mean (EM) of all the 18 models overestimate precipitation over the Sahelian region with biases up to 164.8%, whereas the precipitation over the Guinea Coast and some parts of the Soudano-Sahel is underestimated (up to 34%), as shown in Figure 6. Averagely, the CESM1-BGC and the NorESM1-M estimate seasonal precipitation close to the observed precipitation with biases of 14.2% and 7.9%, respectively. Simulating observed climatology for shorter timescales is a major limitation of GCMs [56], as shown in the models' inability to reproduce the observed seasonal precipitation well though these four models and ability to simulate the observed precipitation well in the case of the annual total.

3.2.2. The Rainy Season.
This season follows the major dry season with a gradual increase in precipitation which peaks in August. During the rainy season, precipitation varies greatly from one latitude to the other. For this season, as shown in Figure 7, the NorESM1-M and the EM simulate minimal seasonal mean biases of 4.0% and 4.3%, respectively. All the models, including the EM, overestimate the observed seasonal precipitation over the Guinea Coast and some portions of the Soudano-Sahel. In addition to the GCMs' limitation in simulating shorter timescales for precipitation, their course grids pose a major challenge in simulating subgrid features such as orography, convective clouds, and vegetation [57] which might have impacted significantly on the simulation of the precipitation.

3.3. Regional Differences in Precipitation

3.3.1. Annual Cycles of Monthly Total Precipitation.
In the annual cycle, as shown in Figure 8, the ability of the models

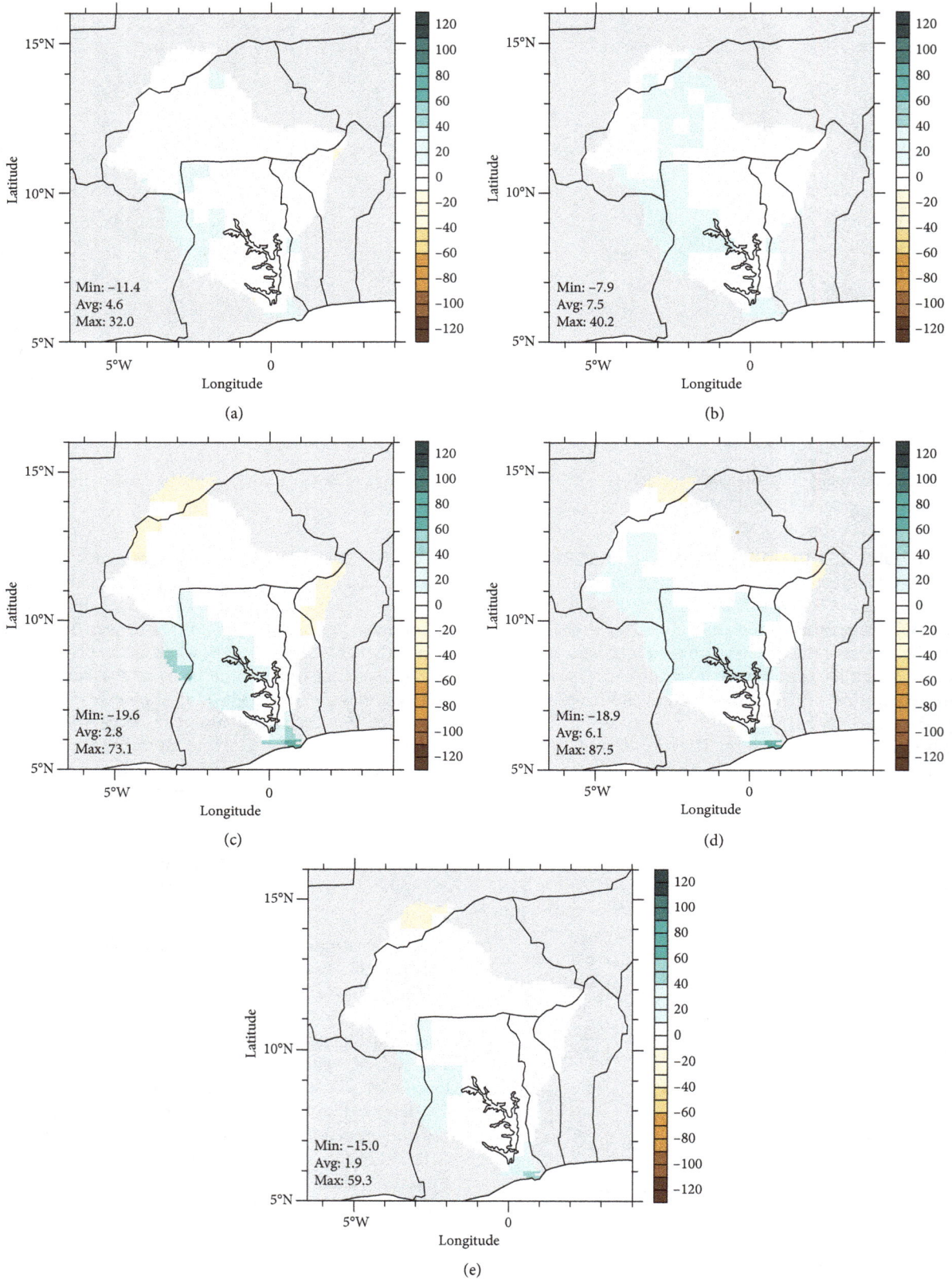

FIGURE 5: The climatological (1950–2004) mean bias in the annual total precipitation for the models: CESM1-BGC (a), CCSM4 (b), NorESM1-M (c), MPI-ESM-MR (d), and ensemble mean (e), simulating the observed climatological mean relatively well over the Volta Basin.

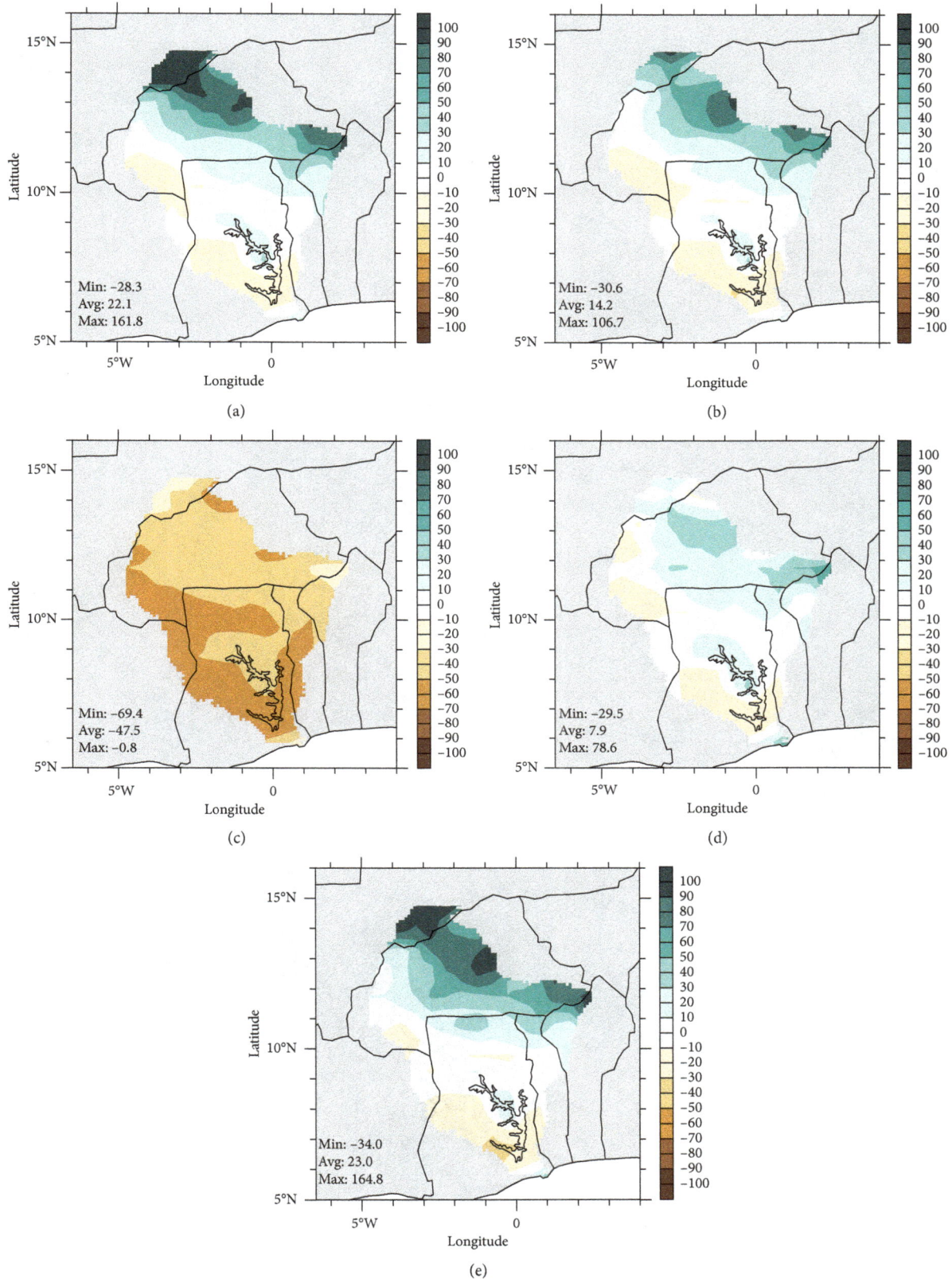

FIGURE 6: The climatological (1950–2004) mean bias in the seasonal totals of the models: CCSM4 (a), CESM1-BGC (b), MPI-ESM-MR (c), NorESM1-M (d), and ensemble mean (e), with least annual climatological mean biases for the dry season over the Volta Basin.

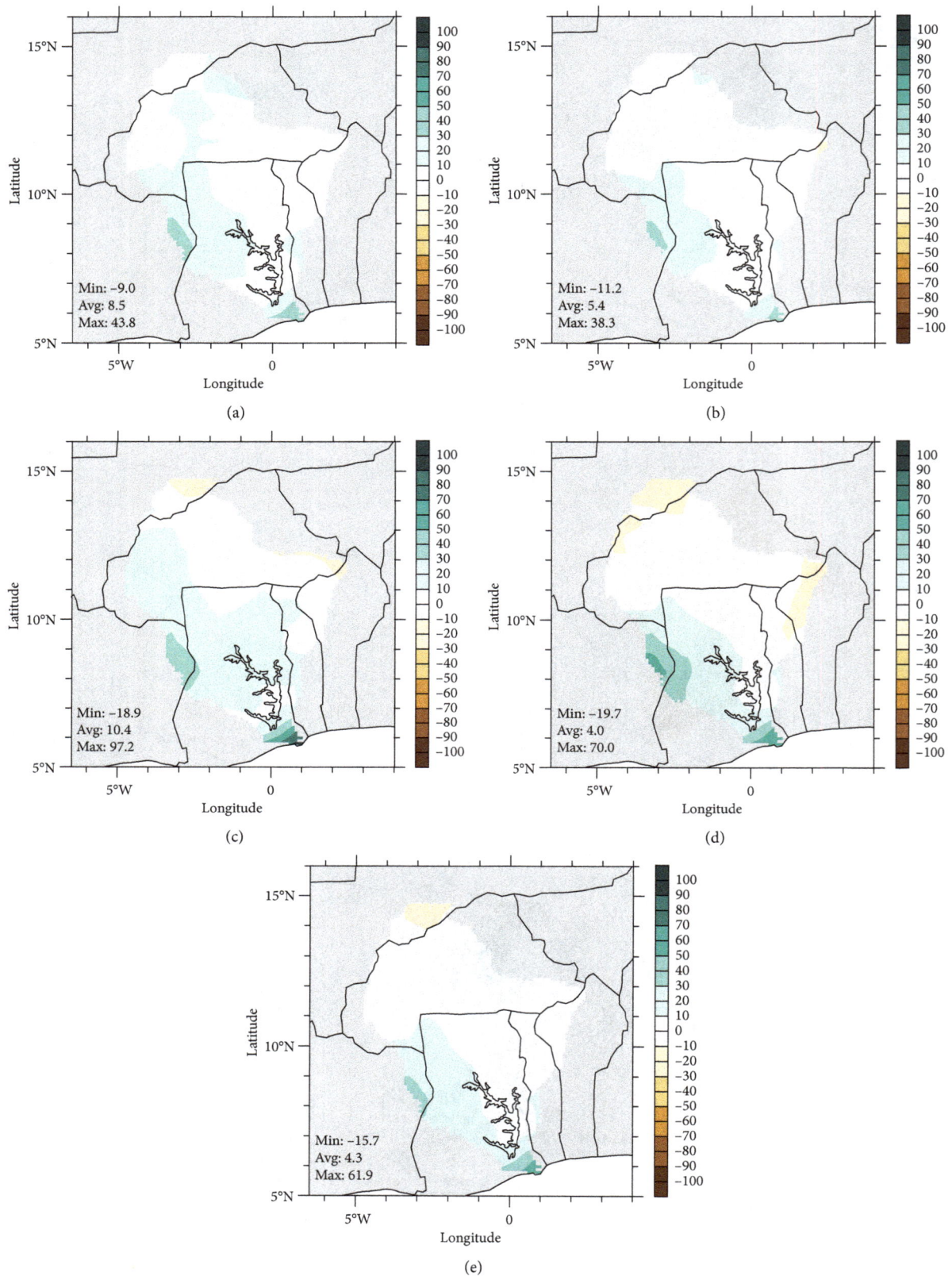

FIGURE 7: The climatological (1950–2004) mean bias in the seasonal totals of the models: CCSM4 (a), CESM1-BGC (b), MPI-ESM-MR (c), NorESM1-M (d), and ensemble mean (e), with least annual climatological mean biases for the rainy season over the Volta Basin.

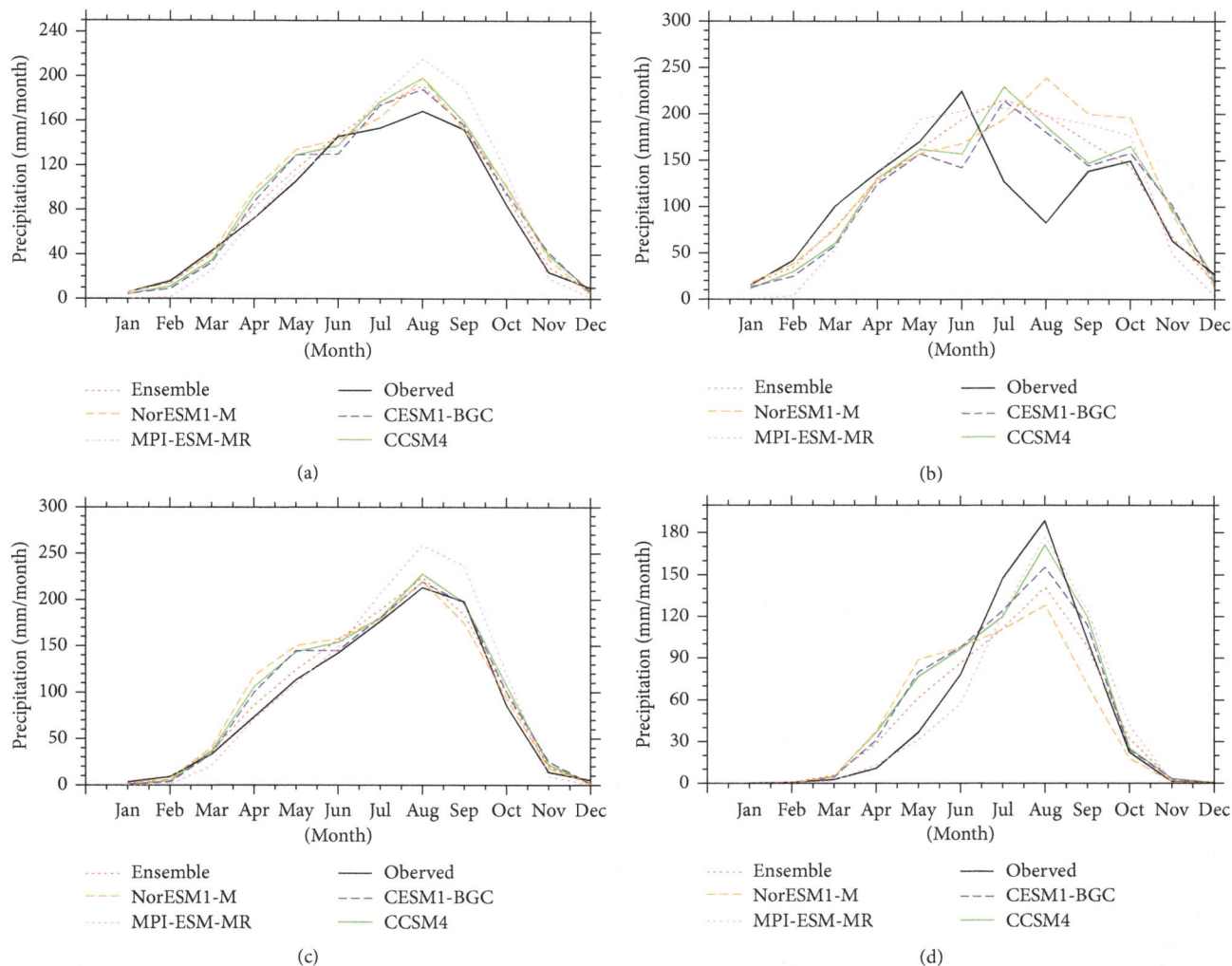

FIGURE 8: Annual cycles of monthly total precipitation for the period mean showing the bimodal (over the GC (b)) and the unimodal patterns (over the VB (a), the SD (c), and the SA (d)).

to simulate the unimodal (over the VB, the SD, and the SA) and the bimodal (over the GC) nature of the precipitation pattern [58] is examined. For the GC, precipitation increases from January and peaks in June. This is the first and major peak. Precipitation then decreases in July to a minimum in August and then increases until it peaks again in September which is the second and minor peak [58]. For the whole basin, the SD, and the SA, maximum precipitation occurs in August [59]. The observed data used (Figure 8) also confirm these cycles for the various belts.

For the entire Volta Basin, the models are able to simulate the unimodal pattern of precipitation. The MPI-ESM-MR (240.5 mm) overestimates the observed maximum precipitation (199.4 mm), which occurs in August, while the CESM1-BGC (202.8 mm), the ensemble mean (204.2 mm), and the NorESM1-M (200.8 mm) are all able to simulate the maximum observed precipitation (209.5 mm) well.

For the Guinea Coast, the models are unable to simulate the bimodal nature accurately; for example, CESM1-BGC simulates early peaks, while others, for instance, CCSM4,

simulate late peaks. Generally, most of the models overestimate the observed precipitation over the coast especially in the months of July and August. The ensemble mean of the models simulates a unimodal pattern of precipitation.

For the Soudano-Sahel, all the models are able to simulate the maximum precipitation which occurs in August. The MPI-ESM-MR although simulates the peak overestimates the observed precipitation. The NorESM1-M (214.63 mm), the CESM1-BGC (216.74 mm), the CCSM4 (224.34 mm), and the ensemble mean (220.39 mm) are all able to simulate the maximum precipitation close to the observed precipitation (216.56 mm).

For the Sahel, all the models underestimate the maximum precipitation (209.5 mm) with the exception of the MPI-ESM-MR (211.5 mm) which is able to simulate well the maximum observed precipitation.

The models' ability to simulate precipitation patterns (unimodal or bimodal) is dependent on their abilities to simulate the meridional movement of the intertropical convergent zone (ITCZ) [52, 60, 61]. The models have less

TABLE 2: Trend test for the Volta Basin rainy and dry seasons.

	Z	p value	Slope	p value (change point)	Change point year
Dry season (NDJFMA)					
Observed	−2.9	0.004	−0.85	0.003	1970
CCSM4	−1.5	0.135	−0.53	0.419	1960
CESM1-BGC	0.9	0.376	0.28	0.527	1981
MPI-ESM-MR	1.8	0.079	0.51	0.338	1995
NorESM1-M	1.3	0.182	0.30	0.527	1976
Ensemble	−0.8	0.429	−0.05	0.415	1967
Rainy season (MJJASO)					
Observed	−2.3	0.022	−1.58	0.024	1970
CCSM4	−0.1	0.919	−0.09	1.384	1965
CESM1-BGC	1.3	0.207	1.04	0.130	1983
MPI-ESM-MR	1.8	0.079	0.51	0.338	1995
NorESM1-M	−0.6	0.532	−0.56	0.603	1957
Ensemble	1.8	0.076	0.39	0.059	1985

TABLE 3: Trend test for the Guinea Coast major rainy/dry and minor rainy/dry seasons.

	Z	p value	Slope	p value (change point)	Change point year
Major dry season (DJFM)					
Observed	−4.6	$5.14E-06$	−1.83	0.0002	1979
CCSM4	−0.9	0.368	−0.27	0.513	1955
CESM1-BGC	−0.2	0.828	−0.04	0.885	1979
MPI-ESM-MR	−1.8	0.079	−0.52	0.383	1970
NorESM1-M	−1.0	0.303	−0.21	0.317	1963
Ensemble	−2.9	0.003	−0.16	0.009	1972
Major rainy season (AMJ)					
Observed	−1.2	0.212	−0.69	0.027	1970
CCSM4	0.1	0.942	0.04	1.283	1961
CESM1-BGC	0.0	0.977	0.01	1.558	1961
MPI-ESM-MR	2.0	0.050	1.31	0.269	1970
NorESM1-M	−0.8	0.400	−0.46	1.019	1967
Ensemble	0.1	0.905	0.02	1.209	1984
Minor dry season (JA)					
Observed	1.7	0.089	1.44	0.157	1962
CCSM4	−1.1	0.276	−0.88	0.603	1957
CESM1-BGC	0.1	0.885	0.12	0.701	1965
MPI-ESM-MR	0.8	0.408	0.50	0.635	1987
NorESM1-M	0.2	0.850	0.15	0.940	1984
Ensemble	−1.9	0.054	−0.3	0.020	1968
Minor rainy season (SON)					
Observed	−1.9	0.063	−1.14	0.061	1964
CCSM4	−0.7	0.486	−0.52	0.882	1975
CESM1-BGC	1.2	0.245	0.72	0.146	1983
MPI-ESM-MR	3.8	0.0001	2.43	0.002	1974
NorESM1-M	0.2	0.805	0.146	1.160	1977
Ensemble	0.4	0.687	0.06	0.620	1991

difficulty in simulating the unimodal pattern over the basin, the Soudano-Sahel, and the Sahel. Over the Guinea Coast, the bimodal pattern is poorly simulated. The models had less difficulty in simulating the dry season precipitation due to the less spatial variation in precipitation during the dry season. Compared to the dry season, precipitation varies spatially during rainy seasons due to differential occurrence of convection activities across the regions, and another factor is the coarse resolution of the GCMs which makes it difficult to simulate these mesoscale features (e.g., [57, 62, 63]).

3.3.2. Seasonal Trend Analysis. The Mann–Kendall test was used to analyze the trend in the rainy and dry seasons over the GC, SD, SA, and VB. The results are presented in Tables 2–5. The trend results in this study have been evaluated at the 5% significant level (95% confidence level), and the corresponding threshold (Z) value is ±1.96. This implies that the null hypothesis is rejected when $|Z_S| \geq Z_{\alpha/2}$ in (8)–(10) at the $\alpha = 0.05$ level of significance.

For the change-point detection (12)–(14), Pettitt's test was used to indicate the year (K_T) of change in the trend and

TABLE 4: Trend test for the Soudano-Sahel rainy and dry seasons.

	Z	p value	Slope	p value (change point)	Change point year
Dry season (NDJFMA)					
Observed	−2.3	0.024	−0.75	0.013	1970
CCSM4	−1.4	0.155	−0.55	0.513	1994
CESM1-BGC	0.7	0.459	0.26	0.651	1981
MPI-ESM-MR	1.9	0.052	0.65	0.235	1995
NorESM1-M	1.1	0.276	0.27	0.771	1986
Ensemble	−0.9	0.387	−0.05	0.556	1967
Rainy season (MJJASO)					
Observed	−1.7	0.081	−1.29	0.157	1972
CCSM4	−0.1	0.896	−0.15	1.039	1965
CESM1-BGC	1.4	0.168	1.50	0.176	1986
MPI-ESM-MR	3.6	0.0003	3.89	0.009	1984
NorESM1-M	−0.6	0.561	−0.44	0.587	1957
Ensemble	2.0	0.047	0.41	0.035	1985

TABLE 5: Trend test for the Sahel rainy and dry seasons.

	Z	p value	Slope	p value (change point)	Change point year
Dry season (NDJFMA)					
Observed	−0.2	0.873	−0.01	0.979	1969
CCSM4	−1.8	0.074	−0.40	0.042	1994
CESM1-BGC	1.1	0.264	0.17	0.110	1981
MPI-ESM-MR	2.8	0.005	0.27	0.106	1965
NorESM1-M	1.1	0.251	0.28	0.513	1974
Ensemble	−0.9	0.355	−0.04	0.76	1967
Rainy season (MJJASO)					
Observed	−4.8	$1.54E-06$	−3.84	$8.13E-06$	1970
CCSM4	0.3	0.760	0.28	1.201	1994
CESM1-BGC	1.2	0.245	0.82	0.115	1983
MPI-ESM-MR	3.3	0.001	3.40	0.106	1965
NorESM1-M	−0.6	0.561	−0.28	0.771	1959
Ensemble	2.2	0.025	0.60	0.071	1985

the significance, which is also analyzed at the 95% confidence level.

Table 2 shows results for the trends in seasonal (dry and rainy seasons) precipitation over the Volta Basin. There is a significant (99.6% confidence level) decrease in dry seasonal (November (Nov), December (Dec), January (Jan), February (Feb), March (Mar), and April (Apr)) precipitation over the basin for the study period (1950–2004) as seen in the observed data. The magnitude of the decrease is 0.85 mm per season. This denoted an increase in dryness over the basin during the dry seasons over that 55-year period. All the models are unable to simulate this decreasing trend with the exception of the CCSM4 and the EM which simulate a decreasing trend of 0.54 and 0.05 mm per season, respectively. For the rainy season (May, June (Jun), July (Jul), August (Aug), September (Sept), and October (Oct)), there is also a significant (97.8% confidence level) decreasing trend in seasonal precipitation with the magnitude of 1.58 mm per season. The CCSM4 and the NorESM1-M simulate the decreasing trend, with the NorESM1-M (0.6 mm per season) simulating the observed trend relatively well. Over the Volta Basin generally, there is a decrease in seasonal precipitation as shown in the results.

For the change-point detection (the last two columns of Table 2), there is a change in the trend which occurs in 1970 for both seasons, that is, statistically significant (about 99% and 98% confidence levels for dry and rainy seasons, resp.). All the models are unable to reproduce this change point.

Table 3 gives the results of the major dry/rainy and the minor dry/rainy seasons over the Guinea Coast. In the major dry season (Dec, Jan, Feb, and Mar), there is a significant (99.9% confidence level) decrease in seasonal precipitation of 1.83 mm per season. All the models including the EM are able to simulate this decreasing trend, with the MPI-ESM-MR (0.52 mm per season) doing relatively well as compared to the other models. A statistically significant (about 99% confidence level) change in the trend occurs in 1979, in which all the models are unable to simulate with the exception of the CESM1-BGC. In the major rainy season (Apr, May, and Jun), there is a decreasing trend in seasonal precipitation which is not statistically significant. With the exception of the NorESM1-M (decreasing at 0.46 mm per season) which is able to simulate the decreasing trend, all the models simulate an increasing trend. The change-point year as seen in the observational data occurs in 1970, and it is statistically significant (about 97% confidence level). Only the MPI-ESM-MR is able to reproduce this change point. For the minor dry season (Jul and Aug), an increasing trend, although statistically not significant, is observed. All the models simulate this increasing trend with the exception of the CCSM4 and the EM. The MPI-ESM-MR best simulates the observed trend. The change-point year occurs in 1962. Although this change point is statistically not significant, it is not simulated by the models. In the minor rainy season (Sept, Oct, and Nov), the CCSM4 simulates the decreasing trend in the observed seasonal precipitation which is statistically insignificant. All the other models simulate an increasing trend. Also, all the models are unable to simulate the change-point year of 1964 which is statistically not significant.

In the Soudano-Sahel, the trend for the dry (Nov, Dec, Jan, Feb, Mar, and Apr) and rainy (May, Jun, Jul, Aug, Sept, and Oct) seasons is shown in Table 4. In the dry season, a decreasing trend which is statistically significant (97.4% confidence level) is observed, with a magnitude of 0.75 mm per season. All the models simulate an increasing trend with the exception of the CCSM4 (decreasing trend at 0.55 mm per season) and the EM (decreasing trend at 0.1 mm per season). For the dry season, a decreasing trend of 1.3 mm per season, although statistically insignificant, is observed in the reference data. The NorESM1-M (decreasing trend at 0.44 mm per season) is able to simulate the decreasing trend well as compared to the other models.

The change-point year of the dry season occurs in 1970 which is statistically significant (about 99% confidence level), while that of the rainy season occurs in 1972 which is not significant statistically. All the models are unable to simulate these change points in both seasons.

Table 5 shows the dry (Nov, Dec, Jan, Feb, Mar, and Apr) and rainy (May, Jun, Jul, Aug, Sept, and Oct) seasonal trends over the Sahel. In the dry season, a slight decreasing trend of 0.1 mm per season which is not significant statistically is observed. The CCSM4 (0.4 mm) and the EM (0.04 mm) simulate well this trend, whereas the rest of the models simulate an increasing trend. For the rainy season, there is a strong decrease in seasonal precipitation (3.84 mm per season) which is significant (99.9% confidence level) in the observational data set. All the models simulate increasing trends with the exception of the NorESM1-M (decreasing trend at 0.28 mm per season) that simulates the decreasing trend, but the magnitude is far less than the observed trend.

The change-point year of the dry season although statistically insignificant occurs in 1969, while that of the rainy season occurs in 1970, but this is statistically significant. All the models are unable to simulate these change point years well.

Generally, there is a decrease in seasonal precipitation of the three belts and the basin. Dry seasons are becoming drier, while rainy season precipitation is decreasing. This trend, if it continues, could significantly affect the production of crops and also the shifting of the rainy season (e.g., [64–66]). The trends also show changes in the 1960s and 1970s, shown in the change-point years. This could be a result of the several droughts that occurred within the West African subregions in the 1960s–80s [67].

3.3.3. Temporal Seasonal Patterns. For the seasonal scale, the ability of the CCSM4, CESM1-BGC, NorESM1-M, MPI-ESM-MR, and ensemble mean of all the 18 models to simulate the temporal seasonal variability and the temporal precipitation patterns is presented. The first four models are selected because of their ability to simulate the observed precipitation on the annual time scale. The variability, the pattern, and the errors are represented by the normalized standard deviation, the correlation coefficient, and the root-mean-square difference, respectively, in the Taylor diagrams [68]. The results are presented for the standard seasons: winter (DJF), spring (MAM), summer (JJA), and fall (SON). In these diagrams (Figures 9–12), the correlation coefficient (r) (main arc) and the root-mean-square (RMS) difference (inner arcs) between the models and the GPCC data, along with the standard deviation (SD) (horizontal axis/vertical axis), are all indicated by points. Also, models with negative correlation are represented below the diagrams. In addition to the three statistics, the biases between the models and the observed data are also included in the diagram, with right triangles giving positive biases and left triangles giving negative biases.

(1) Over the Entire Volta Basin. For the Volta Basin (Figure 9), the performances of the models vary for the four seasons. Temporal correlation for all the seasons and for all the models is less than 0.3. This indicates the models' inability to simulate the observed pattern in seasonal precipitation over the Volta Basin. In winter (DJF), all the models underestimate the observed variability ($\sigma < 1$) with the CCSM4 ($\sigma = 0.6$) and the ensemble mean (EM) ($\sigma = 0.5$) simulating variability close to the observed. In spring (MAM), the MPI-ESM-MR ($\sigma = 2.3$), the NorESM1-M ($\sigma = 1.8$), the CESM1-BGC ($\sigma = 1.6$), and the EM ($\sigma = 1.6$) simulate relatively high variability, whereas the CCSM4 ($\sigma = 1.2$) simulates variability close to the observed variability. In summer (JJA), high variability is simulated by the EM ($\sigma = 1.5$), the CESM1-BGC ($\sigma = 1.4$), the MPI-ESM-MR ($\sigma = 1.4$), and the NorESM1-M ($\sigma = 1.3$), whereas the CCSM4 ($\sigma = 1.2$) is able to simulate variability close to the observed. In fall (SON), maximum variability is simulated by the MPI-ESM-MR ($\sigma = 1.8$) and NorESM1-M ($\sigma = 1.5$). However, the EM ($\sigma = 1.1$), the CCSM4 ($\sigma = 1.1$), and the CESM1-BGC ($\sigma = 1.2$) simulate variability close to the observed data. The CCSM4 and the EM relatively perform fairly well in simulating the observed variability over the entire basin.

(2) The Guinea Coast. Temporal correlations for all the models in all the seasons over the Guinea Coast (Figure 10) are also less than 0.3. In winter again, all the models underestimate the observed variability with the CCSM4 ($\sigma = 0.7$) and the EM ($\sigma = 0.5$) simulating variability close to the observed. In spring, the MPI-ESM-MR ($\sigma = 1.9$) and the EM ($\sigma = 1.4$) simulate fairly high variability. The NorESM1-M ($\sigma = 0.9$) and the CESM1-BGC ($\sigma = 1.0$) simulate variability close to the observed but fail totally in reproducing the observed pattern due to their negative correlations. The CCSM4 ($\sigma = 0.6$) reasonably simulates the observed variability. In summer, all the models underestimate the

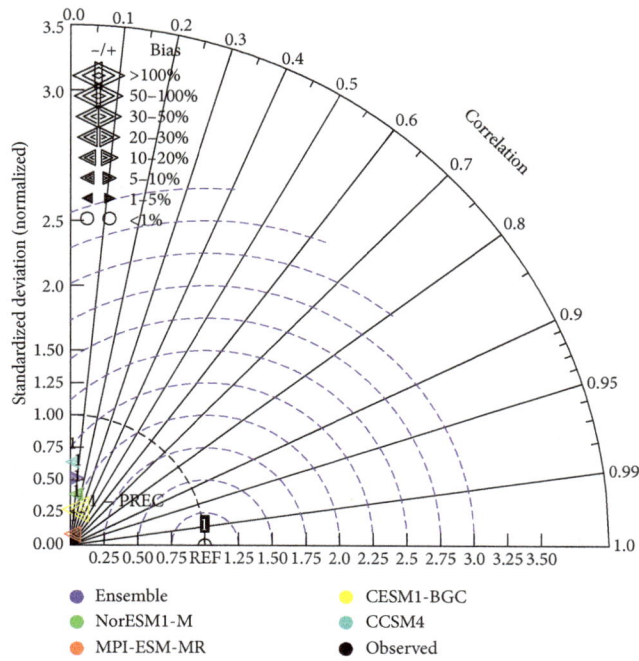

(a)

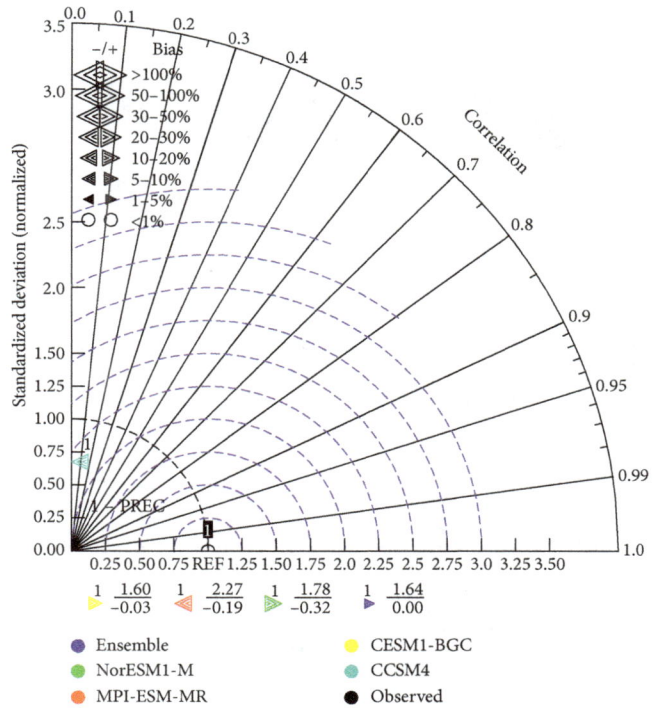

(b)

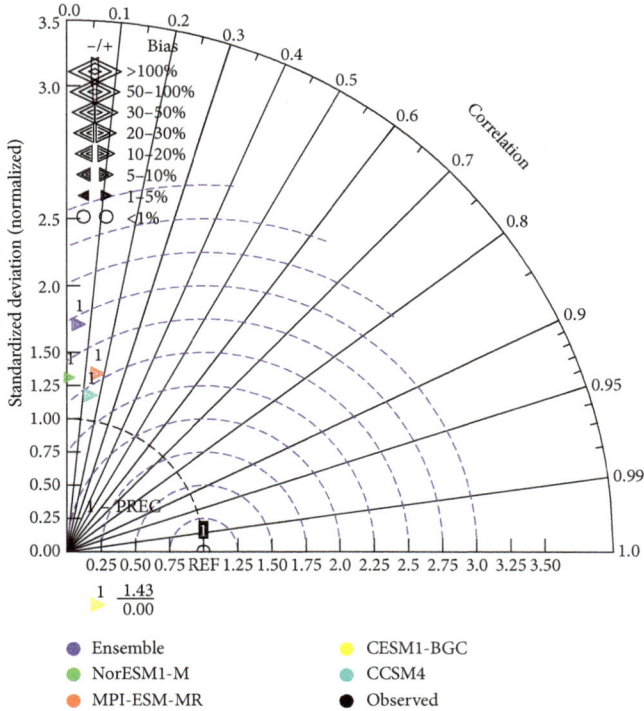

(c)

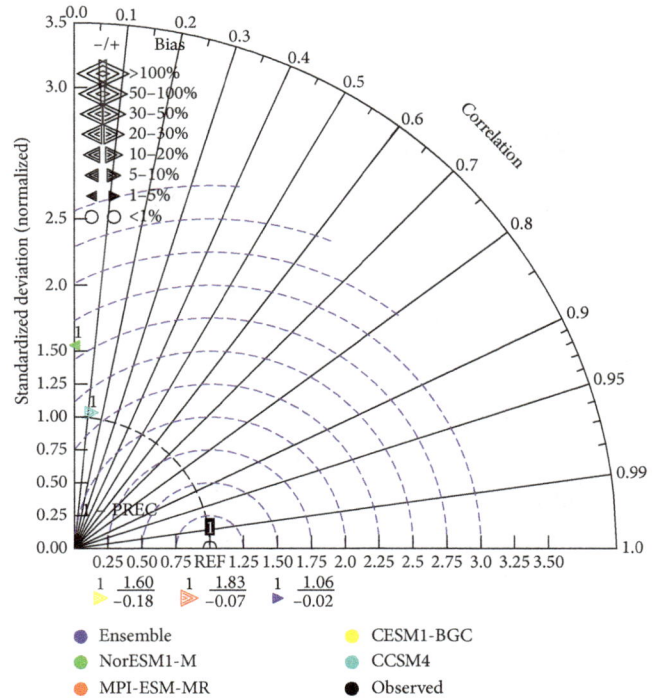

(d)

FIGURE 9: Taylor diagram for the Volta Basin showing the normalized standard deviation, the correlation, and the RMSE representing the variability, pattern, and errors, respectively, within the models and the reference data: DJF (a); MAM (b); JJA (c); SON (d).

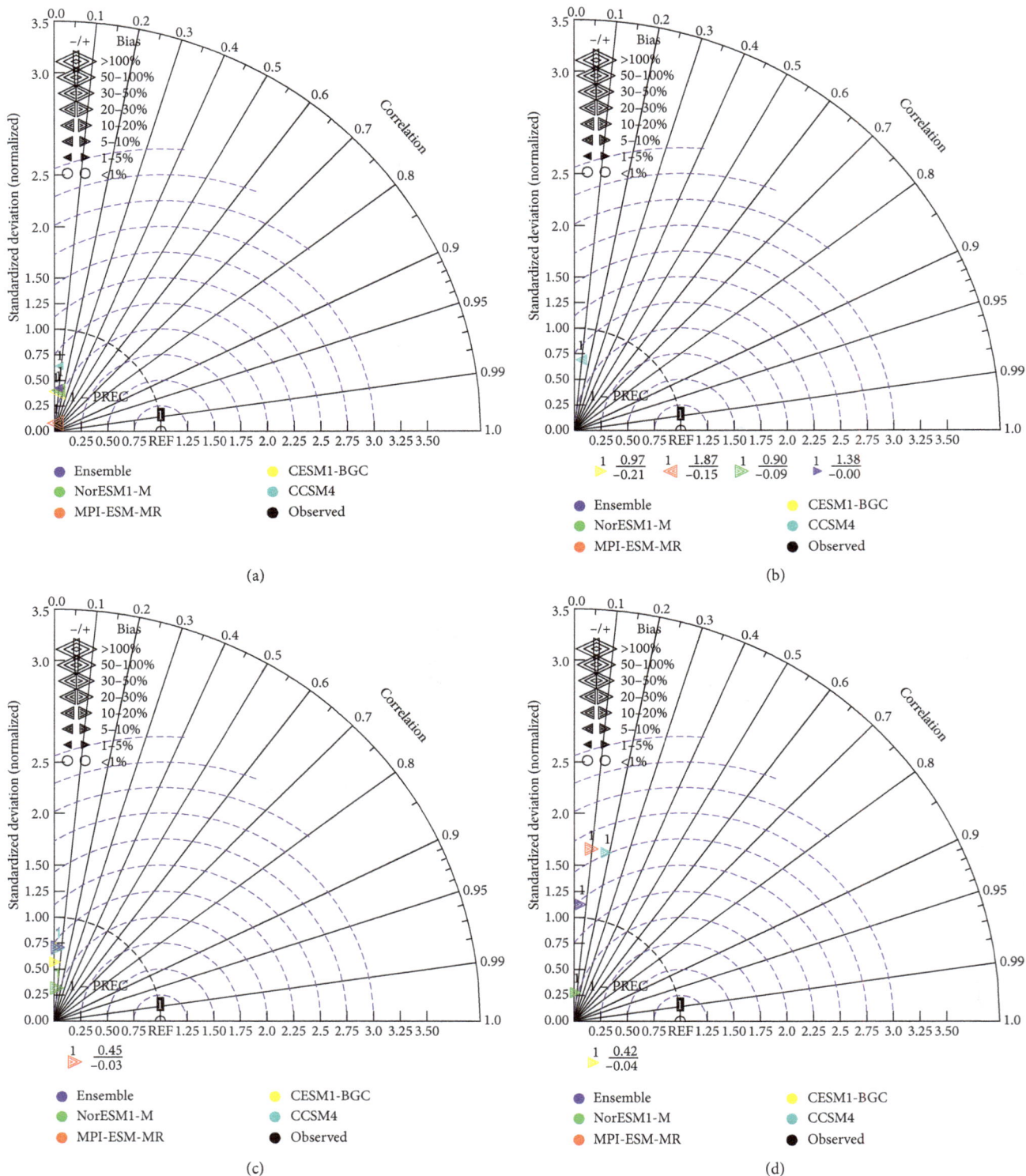

FIGURE 10: Taylor diagram for the Guinea Coast showing the normalized standard deviation, the correlation, and the RMSE representing the variability, pattern, and errors, respectively, within the models and the reference data: DJF (a); MAM (b); JJA (c); SON (d).

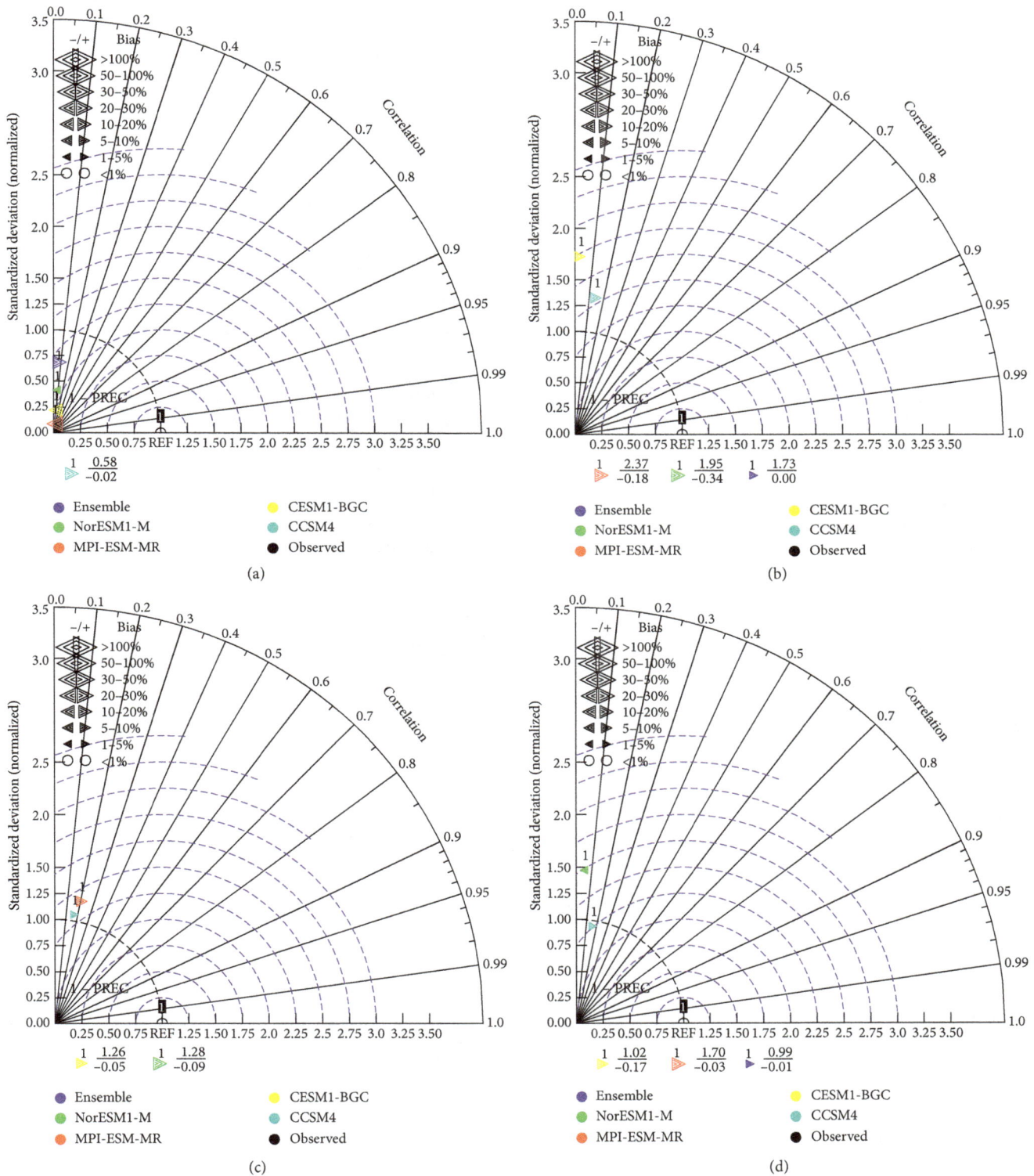

FIGURE 11: Taylor diagram for the Soudano-Sahel showing the normalized standard deviation, the correlation, and the RMSE representing the variability, pattern, and errors, respectively, within the models and the reference data: DJF (a); MAM (b); JJA (c); SON (d).

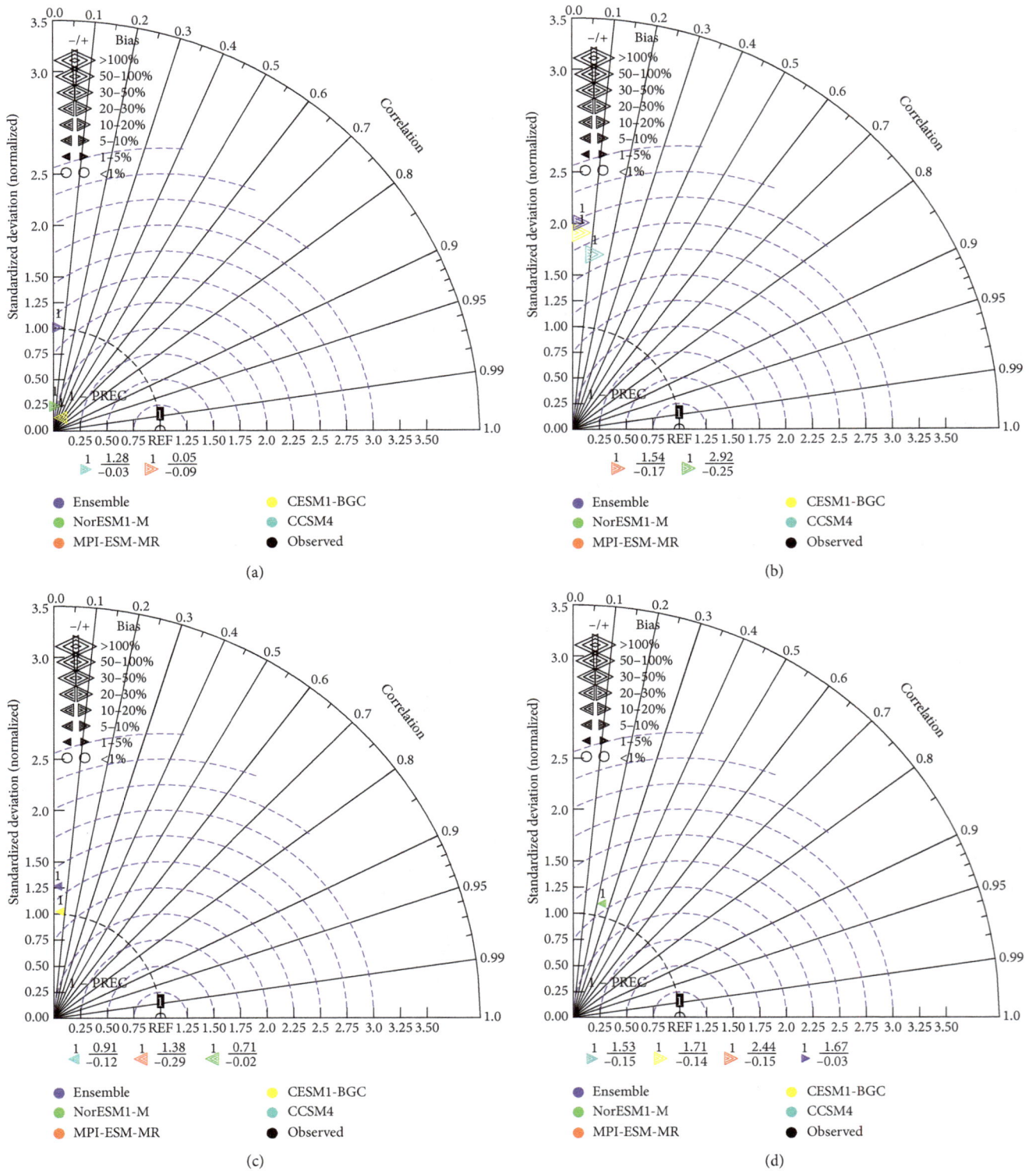

FIGURE 12: Taylor diagram for the Sahel showing the normalized standard deviation, the correlation, and the RMSE representing the variability, pattern, and errors, respectively, within the models and the reference data: DJF (a); MAM (b); JJA (c); SON (d).

observed variability with the EM ($\sigma = 0.7$) and the CCSM4 ($\sigma = 0.7$) simulating variability close to the observed. In fall, the CCSM4 ($\sigma = 1.6$) and MPI-ESM-MR ($\sigma = 1.7$) simulate high variability with the CESM1-BGC ($\sigma = 1.4$) and the EM ($\sigma = 1.1$) simulating variability close to the observed variability. Relatively, the EM and the CCSM4 simulate the observed variability over the Guinea Coast fairly well.

(3) The Soudano-Sahel. The temporal correlation for all the models and for all the seasons (Figure 11) is less than 0.4. In the winter season, all models simulate variability less than 1 with the EM ($\sigma = 0.6$) simulating variability close to the observed precipitation. The MPI-ESM-MR ($\sigma = 2.4$), NorESM1-M ($\sigma = 2.0$), CESM1-BGC ($\sigma = 1.8$), and EM ($\sigma = 1.7$) simulate relatively high variability in spring. However, the CCSM4 ($\sigma = 1.4$) simulates variability close to the observed. In summer, the CCSM4 ($\sigma = 1.1$) and the MPI-ESM-MR ($\sigma = 1.2$) simulate variability close to the observed. The CESM1-BGC ($\sigma = 1.3$) and NorESM1-M ($\sigma = 1.3$) simulate variability close to the observed with negative correlations, whereas a variability of 1.7 is recorded for the EM. The MPI-ESM-MR ($\sigma = 1.7$) and NorESM1-M ($\sigma = 1.5$) have high variability in fall with negative correlations for the CESM1-BGC ($\sigma = 1.0$) and the EM ($\sigma = 1.0$). However, the CCSM4 ($\sigma = 1.0$) has variability close to the observed. Again, the CCSM4 does relatively well in simulating the observed variability over the Soudano-Sahel.

(4) The Sahel. The correlation for all the models in the four seasons over the Sahel (Figure 12) is less than 0.4. In winter, the EM ($\sigma = 1.0$) simulates variability close to the observed, while the rest of the models simulate variability less than the observed. The CCSM4 ($\sigma = 1.3$) also simulates variability close to the observed but with a negative correlation. In spring, the NorESM1-M ($\sigma = 2.9$), CCSM4 ($\sigma = 1.7$), CESM1-BGC ($\sigma = 1.9$), and MPI-ESM-MR ($\sigma = 1.5$) including the EM ($\sigma = 2.1$) simulate high variability. In summer, the CESM1-BGC ($\sigma = 1.0$) and EM ($\sigma = 1.3$) simulate variability approximate to the observed. The CCSM4 ($\sigma = 0.9$), the MPI-ESM-MR ($\sigma = 1.4$), and the NorESM1-M ($\sigma = 0.7$) have negative correlations. In fall, the MPI-ESM-MR ($\sigma = 2.4$), CCSM4 ($\sigma = 1.5$), CESM1-BGC ($\sigma = 1.7$), and EM ($\sigma = 1.7$) simulate relatively high variability with the NorESM1-M simulating variability close to the observed variability. The variability for each season in the Sahel is simulated well by different models.

Precipitation varies highly spatially, from one region to another over the basin. The models' inability to simulate the observed pattern (low correlation) could be a result of the varying nature of spatial precipitation. The models (GCMs) are unable to capture this pattern probably because of the coarse resolutions and hence their inability to capture the subgrid features, such as orography, and its accompanied precipitation and convective clouds, that influence the varying nature of precipitation over the region (e.g., [57, 62, 63]). Generally, models evaluated in this study underestimate the observed variability in winter. Precipitation does not vary much spatially in winter over the basin due to the fact that the entire basin is under the influence of the dry northeasterly

trade winds, and the models seem to strengthen this general dryness across the basin by reproducing a smaller variability compared to the observed. Although the models are unable to simulate the observed pattern, the observed variability is simulated relatively well by the CCSM4 and the EM over the basin and the three belts. In the case of the EM, this study is consistent with previous studies (e.g., [19, 20]) that suggest better performances for EMs.

3.3.4. Interannual Variability of Precipitation. In the assessment of the interannual variability (Figure 13), the ability of the models to capture the observed deviation from the climatological mean of the period of study (1950–2004) is analyzed, for the basin and also for the three belts. The deviation shows the magnitude of how much high or low a particular year's total precipitation is from the period mean of the annual total precipitation.

Over the entire basin and the three belts, the CCSM4 does relatively well in reproducing the observed variability. The CCSM4 is able to simulate the positive or negative deviations better than the other models. All the other models generally had difficulty in simulating the observed year-to-year variability. Often, they simulate opposite deviations as compared to the observed data. This confirms the low correlations between the observed and the simulated precipitation on the seasonal scale.

4. Conclusion

This study assesses the performance of 18 GCMs in simulating present-day climatology (1950–2004) precipitation over the Volta Basin including the three belts: the Guinea Coast, the Soudano-Sahel, and the Sahel. The analyses were done on annual, seasonal, and monthly timescales.

First, the models' ability to simulate the spatial distribution of precipitation over the Volta Basin was investigated by analyzing the biases in climatological mean of the annual total precipitation. Six models (MIROC5, CSIRO-Mk3.6.0, MIROC-ESM, CNRM-CM5, CanESM2, and EC-EARTH) overestimate the observed precipitation over most parts of the basin, while six other models (MRI-CGCM3, INMCM4, HadGEM2-CC, IPSL-CM5A-MR, HadGEM2-AO, and CMCC-CM) underestimate the observed climatological mean of the annual total precipitation. Models such as the CESM1-BGC, CCSM4, NorESM1-M, MPI-ESM-MR, and EM of all the eighteen models perform relatively well in the simulation of the annual precipitation over the Volta Basin with small biases over most parts of the basin. The four models, together with the EM, were then used to assess the regional differences in interannual, seasonal, and monthly precipitation temporal patterns.

For the seasonal analysis, the three statistics: the correlation (r), the standard deviation (σ), and the root-mean-square error (RMSE), are used to evaluate the temporal pattern, variability, and error in the models. The models (CESM1-BGC, CCSM4, NorESM1-M, and MPI-ESM-MR) were assessed for the winter (DJF), spring (MAM), summer (JJA), and fall (SON) seasons for the Volta Basin and the

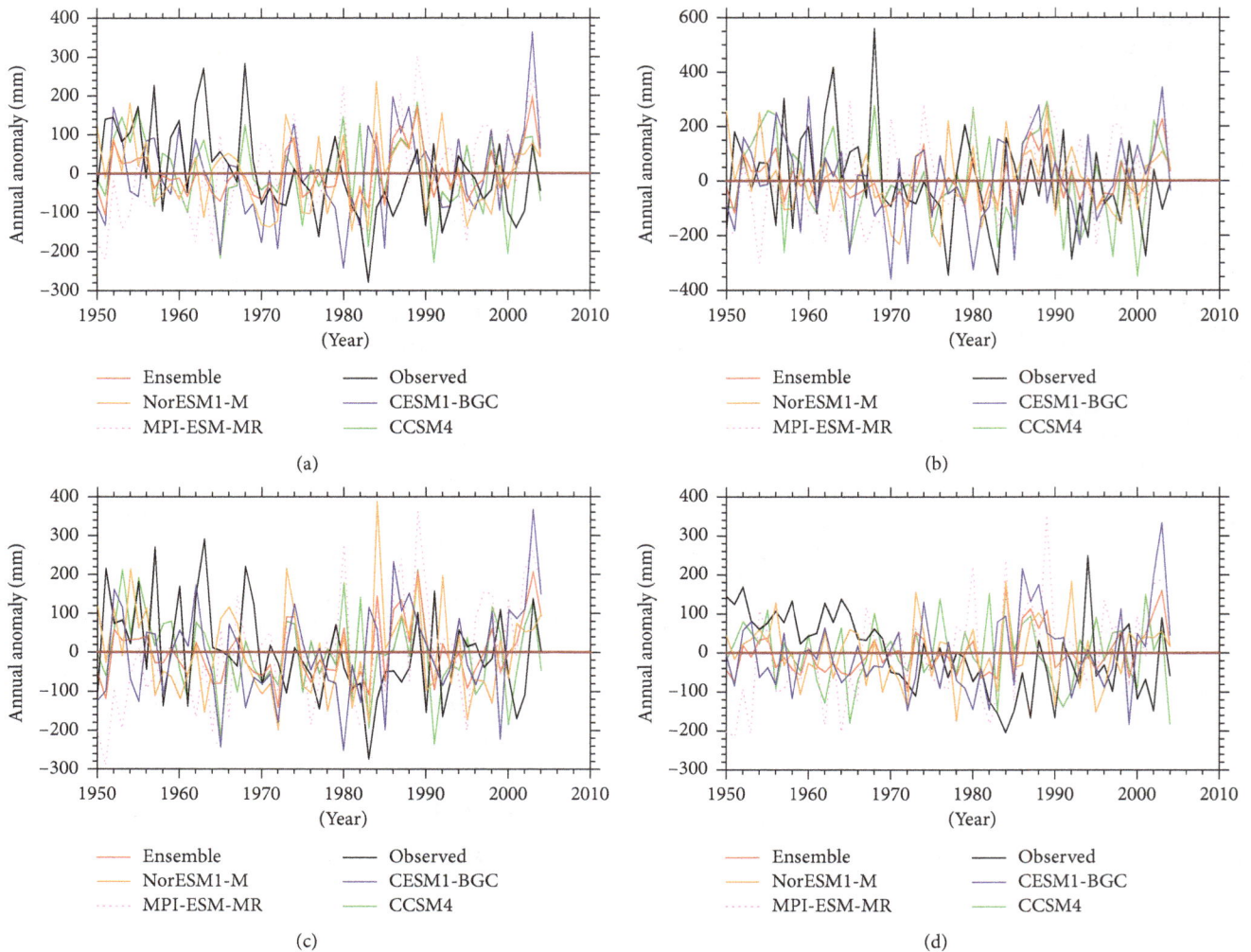

FIGURE 13: The series of interannual variability over the Volta Basin (a), the Guinea Coast (b), the Soudano-Sahel (c), and the Sahel (d). The anomaly is calculated using the 1950–2004 period mean.

three belts. Overall, all the models scored relatively low on the correlation coefficient (<0.5) but with varying standard deviations (1.0–7.3). The low correlations indicate that all the models had difficulty in simulating the observed seasonal precipitation pattern over the entire region and the three belts. In spite of all the models' inability to simulate the precipitation pattern, the CCSM4 does relatively well in simulating the observed variability over the Volta Basin, the Guinea Coast, and the Soudano-Sahel.

In the trend analysis, the Mann–Kendall test and the Pettitt test were used to analyze the seasonal total precipitation trend and the change point over the entire basin and the three belts for the dry and rainy seasons. Over the Volta Basin, the Soudano-Sahel, and the Sahel, the CCSM4 and the NorESM1-M are able to reproduce the observed trend best for the dry and rainy seasons, respectively. Over the Guinea Coast, the MPI-ESM-MR best reproduces the observed trend in the major and minor dry seasons, while the NorESM1-M and the CCSM4 reproduce the trend in the major rainy season and minor rainy season, respectively. The

performance of the EM in the seasonal precipitation trend is relatively poor.

In the case of the annual cycles, the ability of the models to simulate the bimodal precipitation pattern over the Guinea Coast and the unimodal pattern over the Soudano-Sahel, Sahel, and the whole basin is presented. The climatological mean of the monthly totals of the models was calculated and compared to the observed. In the Guinea Coast, all the four models (CESM1-BGC, CCSM4, NorESM1-M, and MPI-ESM-MR) are unable to reproduce the bimodal pattern in June for the major rainy season and September for the minor rainy season, respectively. Over the Soudano-Sahel, all the models are able to simulate the maximum precipitation in August, but the MPI-ESM-MR overestimates the observed maximum precipitation. The NorESM1-M (214.63 mm), CESM1-BGC (216.74 mm), CCSM4 (224.34 mm), and models' ensemble mean (220.39 mm) are able to simulate the maximum precipitation close to the observed (216.56 mm). In the Sahel, all the models underestimate the maximum precipitation

(209.49 mm). The MPI-ESM-MR (211.48 mm) is able to simulate well the maximum observed precipitation.

Generally, models' performances are dependent on simulation of features that influence the distribution of precipitation. Course grid size of GCMs remains the limitation in simulating some observed climatic variables. Clearly, one model could not be singled out to be the best one for all the regions and also for all timescales. Notwithstanding, the CCSM4 does relatively well in simulating the observed precipitation over the basin and the three belts for almost all the timescales used in this analysis.

Our results provide insight into CMIP5 GCMs that can be used as input data in relation to precipitation for impact studies or to drive an RCM over the Volta Basin. Future projection of precipitation for the four models over the Volta Basin needs to be considered to help in adequate planning against any future changes in precipitation over the basin.

Conflicts of Interest

The authors declare that they have no conflicts of interest.

Acknowledgments

The authors are thankful to the German Climate Computing Centre (DKRZ) for providing access to the global climate models used in this study. The authors appreciate the contributions of the Global Precipitation Climatology Centre (GPCC) for the observed data used in the research. The authors also acknowledge all individuals and institutions that made the various precipitation data sets available on the Internet. The authors also thank the reviewers for their helpful comments.

References

[1] P. Roudier, B. Sultan, P. Quirion, and A. Berg, "The impact of future climate change on West African crop yields: what does the recent literature say?," *Global Environmental Change*, vol. 21, no. 3, pp. 1073–1083, 2011.

[2] B. Sultan, C. Baron, M. Dingkuhn, B. Sarr, and S. Janicot, "Agricultural impacts of large-scale variability of the West African monsoon," *Agricultural and Forest Meteorology*, vol. 128, no. 1-2, pp. 93–110, 2005.

[3] J. Sheffield, E. F. Wood, and M. L. Roderick, "Little change in global drought over the past 60 years," *Nature*, vol. 491, no. 7424, pp. 435–438, 2012.

[4] S. E. Nicholson, C. J. Tucker, and M. Ba, "Desertification, drought, and surface vegetation: an example from the West African Sahel," *Bulletin of the American Meteorological Society*, vol. 79, no. 5, pp. 815–829, 1998.

[5] A. Tarhule, "Damaging rainfall and flooding: the other Sahel hazards," *Climatic Change*, vol. 72, no. 3, pp. 355–377, 2005.

[6] N. K. Karley, "Flooding and physical planning in urban areas in West Africa: situational analysis of Accra, Ghana," *Theoretical and Empirical Researches in Urban Management*, vol. 4, no. 13, pp. 25–41, 2009.

[7] N. van de Giesen, M. Andreini, A. van Edig, and P. Vlek, *Competition for Water Resources of the Volta Basin*, IAHS Publication, Wallingford, UK, 2001.

[8] M. Andreini, P. Vlek, and N. Van De Giesen, "Water sharing in the Volta Basin," in *FRIEND 2002—Regional Hydrology:*

Bridging the Gap between Research and Practice, no. 274, pp. 329–336, IAHS Publication, Wallingford, UK, 2002.

[9] T. Annor, *Potential impacts of climate variability and change on hydrology and water resources over the Volta Basin*, Ph.D. thesis, Department of Meteorology and Climate Science, The Federal University of Technology, Akure, Nigeria, 2015.

[10] P. K. Adom, W. Bekoe, and S. K. K. Akoena, "Modelling aggregate domestic electricity demand in Ghana: an autoregressive distributed lag bounds cointegration approach," *Energy Policy*, vol. 42, pp. 530–537, 2012.

[11] B. Barry, E. Obuobie, M. Andreini, W. Andah, and M. Pluquet, *Comprehensive Assessment of Water Management in Agriculture (Comparative Study of River Basin Development and Management)*, International Water Management Institute, Colombo, Sri Lanka, 2005.

[12] M. B. Sylla, A. T. Gaye, J. S. Pal, G. S. Jenkins, and X. Q. Bi, "High-resolution simulations of West African climate using regional climate model (RegCM$_3$) with different lateral boundary conditions," *Theoretical and Applied Climatology*, vol. 98, no. 3-4, pp. 293–314, 2009.

[13] M. Andreini, N. van de Giesen, A. van Edig, M. Fosu, and W. Andah, "Volta Basin water balance," ZEF–Discussion Papers On Development Policy No. 21, Center for Development Research, Bonn, Germany, 2000.

[14] J. Houghton, Y. Ding, D. Griggs et al., "The climate change contribution of Working Group I to the Third Assessment Report of the Intergovernmental Panel on Climate Change," in *Proceedings of the IPCC 2001: Climate Change 2001*, vol. 159, Wembley, UK, September 2001.

[15] M. L. Parry, O. F. Canziani, J. P. Palutikof, P. J. van der Linden, and C. E. Hanson, "Contribution of Working Group II to the Fourth Assessment Report of the Intergovernmental Panel on Climate Change," in *Proceedings of the IPCC 2007: Climate Change 2007: Impacts, Adaptation and Vulnerability*, Geneva, Switzerland, 2007.

[16] T. Stocker, D. Qin, G. Plattner et al., "Contribution of Working Group I to the Fifth Assessment Report of the Intergovernmental Panel on Climate Change," in *Proceedings of the IPCC 2013: Climate Change 2013: The Physical Science Basis*, Stockholm, Sweden, September 2013.

[17] R. Neumann, G. Jung, P. Laux, and H. Kunstmann, "Climate trends of temperature, precipitation and river discharge in the Volta Basin of West Africa," *International Journal of River Basin Management*, vol. 5, no. 1, pp. 17–30, 2007.

[18] H. Kunstmann and G. Jung, *Impact of Regional Climate Change on Water Availability in the Volta Basin of West Africa*, IAHS Publication, Wallingford, UK, 2005.

[19] A. Mehran, A. AghaKouchak, and T. J. Phillips, "Evaluation of CMIP5 continental precipitation simulations relative to satellite-based gauge-adjusted observations," *Journal of Geophysical Research: Atmospheres*, vol. 119, no. 4, pp. 1695–1707, 2014.

[20] S. Kumar, V. Merwade, J. L. Kinter III, and D. Niyogi, "Evaluation of temperature and precipitation trends and long-term persistence in CMIP5 twentieth-century climate simulations," *Journal of Climate*, vol. 26, no. 12, pp. 4168–4185, 2013.

[21] D. E. Rupp, J. T. Abatzoglou, K. C. Hegewisch, and P. W. Mote, "Evaluation of CMIP5 20th century climate simulations for the Pacific Northwest USA," *Journal of Geophysical Research: Atmospheres*, vol. 118, no. 19, 2013.

[22] J. Cattiaux, H. Douville, and Y. Peings, "European temperatures in CMIP5: origins of present-day biases and future

uncertainties," *Climate Dynamics*, vol. 41, no. 11-12, pp. 2889–2907, 2013.

[23] J. Perez, M. Menendez, F. J. Mendez, and I. J. Losada, "Evaluating the performance of CMIP3 and CMIP5 global climate models over the north-east Atlantic region," *Climate Dynamics*, vol. 43, no. 9-10, pp. 2663–2680, 2014.

[24] C. Miao, Q. Duan, Q. Sun et al., "Assessment of CMIP5 climate models and projected temperature changes over Northern Eurasia," *Environmental Research Letters*, vol. 9, no. 5, article 055007, 2014.

[25] J. Zhang, L. Li, T. Zhou, and X. Xin, "Evaluation of spring persistent rainfall over East Asia in CMIP3/CMIP5 AGCM simulations," *Advances in Atmospheric Sciences*, vol. 30, no. 6, pp. 1587–1600, 2013.

[26] R. Allen, J. Norris, and M. Wild, "Evaluation of multidecadal variability in CMIP5 surface solar radiation and inferred underestimation of aerosol direct effects over Europe, China, Japan, and India," *Journal of Geophysical Research: Atmospheres*, vol. 118, no. 12, pp. 6311–6336, 2013.

[27] D.-Q. Huang, J. Zhu, Y.-C. Zhang, and A.-N. Huang, "Uncertainties on the simulated summer precipitation over Eastern China from the CMIP5 models," *Journal of Geophysical Research: Atmospheres*, vol. 118, no. 16, pp. 9035–9047, 2013.

[28] J. H. Siew, F. T. Tangang, and L. Juneng, "Evaluation of CMIP5 coupled atmosphere–ocean general circulation models and projection of the Southeast Asian winter monsoon in the 21st century," *International Journal of Climatology*, vol. 34, no. 9, pp. 2872–2884, 2014.

[29] G. Nikulin, C. Jones, F. Giorgi et al., "Precipitation climatology in an ensemble of CORDEX-Africa regional climate simulations," *Journal of Climate*, vol. 25, no. 18, pp. 6057–6078, 2012.

[30] P. M. Nikiema, M. B. Sylla, K. Ogunjobi, I. Kebe, P. Gibba, and F. Giorgi, "Multi-model CMIP5 and CORDEX simulations of historical summer temperature and precipitation variabilities over West Africa," *International Journal of Climatology*, vol. 37, no. 5, pp. 2438–2450, 2017.

[31] T. Annor, B. Lamptey, S. Wagner et al., "High-resolution long-term WRF climate simulations over Volta Basin. Part 1: validation analysis for temperature and precipitation," *Theoretical and Applied Climatology*, vol. 133, no. 3-4, pp. 829–849, 2017.

[32] R. Y. K. Agyeman, T. Annor, B. Lamptey, E. Quansah, J. Agyekum, and S. A. Tieku, "Optimal physics parameterization scheme combination of the weather research and forecasting model for seasonal precipitation simulation over Ghana," *Advances in Meteorology*, vol. 2017, Article ID 7505321, 15 pages, 2017.

[33] F. Aziz and E. Obuobie, "Trend analysis in observed and projected precipitation and mean temperature over the Black Volta Basin, West Africa," *International Journal of Current Engineering and Technology*, vol. 7, no. 4, 2017.

[34] F. Giorgi and W. J. Gutowski, "Regional dynamical downscaling and the CORDEX initiative," *Annual Review of Environment and Resources*, vol. 40, no. 1, pp. 467–490, 2015.

[35] U. Schneider, T. Fuchs, A. Meyer-Christoffer, and B. Rudolf, *Global Precipitation Analysis Products of the GPCC, Vol. 112*, Global Precipitation Climatology Centre (GPCC), DWD, Internet Publikation, 2008.

[36] B. Andreas, P. Finger, A. Peter Meyer-Christoffer, B. Rudolf, and M. Ziese, "GPCC Full Data Reanalysis Version 7.0 at 0.5: monthly land-surface precipitation from rain-gauges built on GTS-based and historic data," 2011.

[37] S. E. Nicholson, B. Some, J. McCollum et al., "Validation of TRMM and other rainfall estimates with a high-density gauge dataset for West Africa. Part I: validation of GPCC rainfall product and pre-TRMM satellite and blended products," *Journal of Applied Meteorology*, vol. 42, no. 10, pp. 1337–1354, 2003.

[38] H. Paeth, A. H. Fink, S. Pohle, F. Keis, H. Mächel, and C. Samimi, "Meteorological characteristics and potential causes of the 2007 flood in sub-Saharan Africa," *International Journal of Climatology*, vol. 31, no. 13, pp. 1908–1926, 2011.

[39] A. Gruber, X. Su, M. Kanamitsu, and J. Schemm, "The comparison of two merged rain gauge–satellite precipitation datasets," *Bulletin of the American Meteorological Society*, vol. 81, no. 11, pp. 2631–2644, 2000.

[40] G. Flato, J. Marotzke, B. Abiodun et al., "Evaluation of climate models. Contribution of Working Group I to the Fifth Assessment Report of the Intergovernmental Panel on Climate Change," in *Proceedings of the IPCC 2013: Climate Change: The Physical Science Basis*, vol. 5, pp. 741–866, Stockholm, Sweden, September 2013.

[41] K. E. Taylor, R. J. Stouffer, and G. A. Meehl, "A summary of the CMIP5 experiment design," *PCDMI Report*, vol. 33, 2009.

[42] K. E. Taylor, R. J. Stouffer, and G. A. Meehl, "An overview of CMIP5 and the experiment design," *Bulletin of the American Meteorological Society*, vol. 93, no. 4, p. 485, 2012.

[43] P. W. Jones, *A User's Guide for SCRIP: A Spherical Coordinate Remapping and Interpolation Package*, Los Alamos National Laboratory, Los Alamos, NM, USA, 1997.

[44] P. W. Jones, "First- and second-order conservative remapping schemes for grids in spherical coordinates," *Monthly Weather Review*, vol. 127, no. 9, pp. 2204–2210, 1999.

[45] Y. Fujihara, Y. Yamamoto, Y. Tsujimoto, J.-I. Sakagami et al., "Discharge simulation in a data-scarce basin using reanalysis and global precipitation data: a case study of the White Volta Basin," *Journal of Water Resource and Protection*, vol. 6, no. 14, p. 1316, 2014.

[46] H. B. Mann, "Nonparametric tests against trend," *Econometrica: Journal of the Econometric Society*, vol. 13, no. 3, pp. 245–259, 1945.

[47] M. Kendall, "Rank correlation methods," Technical Report, Griffin & Co, London, UK, 1975.

[48] A. Pettitt, "A non-parametric approach to the change-point problem," *Applied Statistics*, vol. 28, no. 2, pp. 126–135, 1979.

[49] R. O. Gilbert, "6.5 Sen's nonparametric estimator of slope," in *Statistical Methods for Environmental Pollution Monitoring*, pp. 217–219, John Wiley and Sons, Hoboken, NJ, USA, 1987.

[50] S. Yue, P. Pilon, and G. Cavadias, "Power of the Mann–Kendall and Spearman's rho tests for detecting monotonic trends in hydrological series," *Journal of Hydrology*, vol. 259, no. 1-4, pp. 254–271, 2002.

[51] R. Modarres and V. d. P. R. da Silva, "Rainfall trends in arid and semi-arid regions of Iran," *Journal of Arid Environments*, vol. 70, no. 2, pp. 344–355, 2007.

[52] S. E. Nicholson, "A revised picture of the structure of the "monsoon" and land ITCZ over West Africa," *Climate Dynamics*, vol. 32, no. 7-8, pp. 1155–1171, 2009.

[53] S. Janicot and B. Sultan, "Intra-seasonal modulation of convection in the West African monsoon," *Geophysical Research Letters*, vol. 28, no. 3, pp. 523–526, 2001.

[54] Z.-Z. Hu, B. Huang, Y.-H. Tseng et al., "Does vertical temperature gradient of the atmosphere matter for El Niño development?," *Climate Dynamics*, vol. 48, no. 5-6, pp. 1413–1429, 2016.

[55] D. J. Seidel, Q. Fu, W. J. Randel, and T. J. Reichler, "Widening of the tropical belt in a changing climate," *Nature Geoscience*, vol. 1, no. 1, pp. 21–24, 2008.

[56] D. A. Randall, R. A. Wood, S. Bony et al., "Climate models and their evaluation. Contribution of Working Group I to the Fourth Assessment Report of the IPCC (FAR)," in *Proceedings of the IPCC 2007: Climate change 2007: The Physical Science Basis*, pp. 589–662, Cambridge University Press, Geneva, Switzerland, 2007.

[57] R. L. Wilby and T. Wigley, "Downscaling general circulation model output: a review of methods and limitations," *Progress in Physical Geography*, vol. 21, no. 4, pp. 530–548, 1997.

[58] P. Laux, H. Kunstmann, and A. Bárdossy, "Predicting the regional onset of the rainy season in West Africa," *International Journal of Climatology*, vol. 28, no. 3, pp. 329–342, 2008.

[59] K. I. Mohr, "Interannual, monthly, and regional variability in the wet season diurnal cycle of precipitation in sub-Saharan Africa," *Journal of Climate*, vol. 17, no. 12, pp. 2441–2453, 2004.

[60] B. Sultan and S. Janicot, "Abrupt shift of the ITCZ over West Africa and intra-seasonal variability," *Geophysical Research Letters*, vol. 27, no. 20, pp. 3353–3356, 2000.

[61] S. E. Nicholson and J. P. Grist, "The seasonal evolution of the atmospheric circulation over West Africa and Equatorial Africa," *Journal of Climate*, vol. 16, no. 7, pp. 1013–1030, 2003.

[62] T. Wigley, P. Jones, K. Briffa, and G. Smith, "Obtaining sub-grid-scale information from coarse-resolution general circulation model output," *Journal of Geophysical Research: Atmospheres*, vol. 95, no. 2, pp. 1943–1953, 1990.

[63] F. Giorgi, "Simulation of regional climate using a limited area model nested in a general circulation model," *Journal of Climate*, vol. 3, no. 9, pp. 941–963, 1990.

[64] M. New, B. Hewitson, D. B. Stephenson et al., "Evidence of trends in daily climate extremes over southern and west Africa," *Journal of Geophysical Research: Atmospheres*, vol. 111, no. 14, 2006.

[65] A. Giannini, R. Saravanan, and P. Chang, "Oceanic forcing of Sahel rainfall on interannual to interdecadal time scales," *Science*, vol. 302, no. 5647, pp. 1027–1030, 2003.

[66] J. G. Charney, "Dynamics of deserts and drought in the Sahel," *Quarterly Journal of the Royal Meteorological Society*, vol. 101, no. 428, pp. 193–202, 1975.

[67] L. M. Druyan, "Studies of 21st-century precipitation trends over West Africa," *International Journal of Climatology*, vol. 31, no. 10, pp. 1415–1424, 2011.

[68] K. E. Taylor, "Summarizing multiple aspects of model performance in a single diagram," *Journal of Geophysical Research: Atmospheres*, vol. 106, no. 7, pp. 7183–7192, 2001.

10

Five Decadal Trends in Averages and Extremes of Rainfall and Temperature in Sri Lanka

G. Naveendrakumar [ID],[1,2] Meththika Vithanage [ID],[3] Hyun-Han Kwon [ID],[4] M. C. M. Iqbal,[5] S. Pathmarajah,[6] and Jayantha Obeysekera[7]

[1]*Postgraduate Institute of Science (PGIS), University of Peradeniya, Peradeniya, Sri Lanka*
[2]*Faculty of Applied Science, Vavuniya Campus of the University of Jaffna, Vavuniya, Sri Lanka*
[3]*Ecosphere Resilience Research Center, Faculty of Applied Sciences, University of Sri Jayewardenepura, Nugegoda, Sri Lanka*
[4]*Department of Civil Engineering, Chonbuk National University, Jeonju, Republic of Korea*
[5]*Plant and Environmental Sciences, National Institute of Fundamental Studies (NIFS), Kandy, Sri Lanka*
[6]*Department of Agricultural Engineering, Faculty of Agriculture, University of Peradeniya, Peradeniya, Sri Lanka*
[7]*Sea Level Solutions Center, Florida International University, Miami, FL, USA*

Correspondence should be addressed to Meththika Vithanage; meththikavithanage@gmail.com

Academic Editor: Roberto Coscarelli

In this study, we used a comprehensive set of statistical metrics to investigate the historical trends in averages and extremes of rainfall and temperature in Sri Lanka. The data consist of 55 years (1961–2015) of daily rainfall, maximum temperature (T_{max}), and minimum temperature (T_{min}) records from 20 stations scattered throughout Sri Lanka. The linear trends were analyzed using the nonparametric Mann–Kendall test and Sen–Theil regression. The prewhitening method was first used to remove autocorrelation from the time series, and the modified seasonal Mann–Kendall test was then applied for the seasonal data. The results show that, during May, 15% of the stations showed a significant decrease in wet days, which may be due to the delayed southwest monsoon (SWM) to Sri Lanka. A remarkable increase in the annual average temperature of T_{min} and T_{max} was observed as 70% and 55% of the stations, respectively. For the entire period, 80% of the stations demonstrated statistically significant increases of T_{min} during June and July. The daily temperature range (DTR) exhibited a widespread increase at the stations located within the southwestern coast region of Sri Lanka. Although changes in global climate, teleconnections, and local deforestation in recent decades at least partially influence the trends observed in Sri Lanka, a formal trend attribution study should be conducted.

1. Introduction

Understanding historical climate trend is the precursor of sound climate change investigations [1]. Assessing historical rainfall and temperature is essential for formulating plausible weather and climate predictions, especially for developing South Asian countries like Sri Lanka. Located at the southern tip of the Indian mainland, Sri Lanka geographically falls within the tropical latitude of 5°55′–9°51′N and longitude of 79°42′–81°53′E [2]. With a population of nearly 21 million and natural forest cover over 29% of the total land area, Sri Lanka has undergone rapid transformations in land use during the past decades due to deforestation for crop and plantation agriculture, increased need for infrastructure, and

significant conversions of wetlands and agriculture land to human settlements. These land use changes are likely to have a feedback effect on the local climate and cannot be ruled out as a factor in historical trends. Increases in extreme weather, both in terms of frequency and magnitude, are a serious challenge for coping with changing climate. The IPCC (Intergovernmental Panel on Climate Change) has determined with high confidence that the intensity of rainfall events will increase with warming temperatures in some parts of the globe [3]. Often, climate change first appears as an increase of intensity and frequency of extreme weather events [4], and therefore, it is essential for historical trend assessments to include extreme events. Sri Lanka is affected by both floods and severe droughts, and the impacts of both have intensified in recent years [5–8].

Assessments of changes in historical trends regarding averages and extremes are essential to introduce adaptation policies to prepare for future extreme events in Sri Lanka. The analyses of extreme rainfall and temperature investigations are also crucial for developing policies associated with long-term planning in agriculture sectors, national heritage sites, tourism sectors, disaster resilience, and many commercial entities which have a direct connection to weather events. Because Sri Lanka is an agricultural country, trend and extreme climate studies will help the agriculture and plantation sectors to introduce appropriate cropping patterns and develop new varieties that can tolerate extreme temperatures.

There have been some prior investigations of rainfall and temperature in Sri Lanka, and such studies have demonstrated significant trends [9–11]. For instance, the dynamic trends in rainfall extremes in Sri Lanka have shown that Colombo experienced the highest value in 2010 among all severe extremes in the historical records [12].

Rather than a parametric analysis, like those used in previous studies to examine rainfall and temperature trends [13–15], the nonparametric investigation used in this study is more appropriate for detecting historical rainfall and temperature trends due to a variety of factors. It is even true in situations with missing/censored data as in Sri Lanka. In particular, such methods are more appropriate for situations in which climatic records include missing/censored data as in Sri Lanka. Prior studies lacked appropriate statistical methods for investigations of extremes. Although nonparametric statistical techniques were previously used by Karunathilaka et al. [11], Herath and Rathnayake [9], and Jayawardene et al. [10], seasonality components were not addressed appropriately for the detection of long-term trends. Recognizing the incomplete nature of prior studies, the objective of this investigation was set to reveal consistent historical linear trends in averages and extremes of rainfall and temperature records of Sri Lanka using both parametric and more robust nonparametric statistical techniques with appropriately handled seasonal components and potential autocorrelation.

In the next section, we review previous studies of factors that influence the climate of Sri Lanka. The sources are described in terms of their necessity and importance. We follow by describing the methods used for the analysis of climatic trends. We finalize by summarizing consistent trends and results and presenting conclusions.

2. An Overview of Climate in Sri Lanka

2.1. Current Climate. Two monsoonal winds primarily influence the climate of Sri Lanka. The first southwest monsoon (SWM) and second northeast monsoon (NEM) reach Sri Lanka during the months of May to September (MJJAS) and December to February (DJF), respectively [16]. During the SWM and NEM seasons, winds come from the northeast and southwest [17], respectively (Figure 1). The periods between these primary monsoons are referred to as intermonsoonal seasons, which usually last for two months. They are called the first intermonsoon (FIM) and second intermonsoon (SIM) and occur during the periods of March-

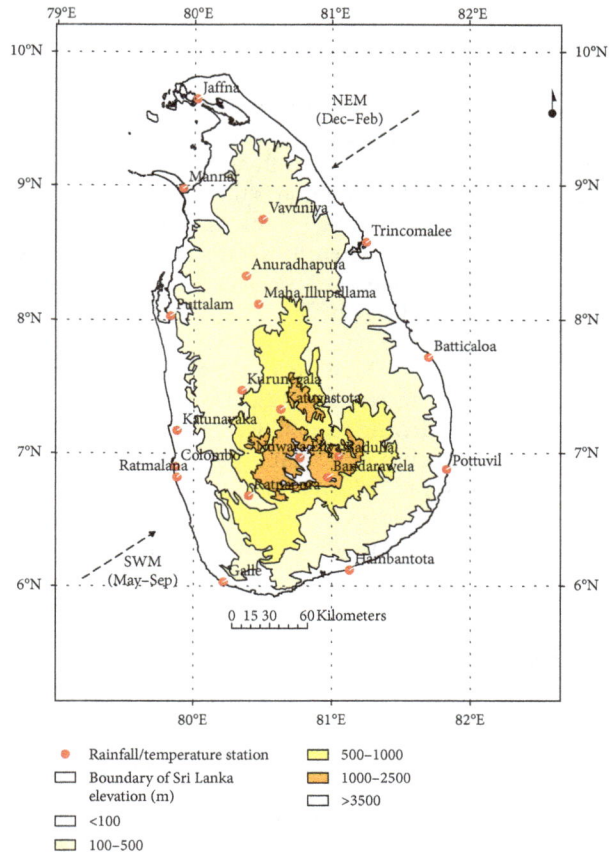

FIGURE 1: Long-term temperature stations used for historical analysis. NEM (northeast monsoon) and SWM (southwest monsoon) indicate the directions of monsoonal winds in Sri Lanka. Topographical variation is shown in colored gradient.

April (MA) and October-November (ON), respectively [16, 18]. Regional climatic patterns in Sri Lanka are primarily influenced by the El Niño-Southern Oscillation (ENSO) [19]. Due to its island geography, seasonal monsoons moderate the climate of Sri Lanka [20, 21]. There is significant spatial variation in temperature due to variations in elevation across the island. Accordingly, two very distinct temperature regimes occur, low-country (or lowlands) and up-country (highlands >300 m MSL), characterized by annual average temperatures of 26.5–28.5°C and 15.9°C, respectively [22]. Due to its tropical climate, evapotranspiration losses are high in the higher temperature regimes of Sri Lanka [23]. In addition to monsoonal seasons, teleconnections influence the climate of Sri Lanka, as rainfall and temperature are strongly influenced by those phenomena. Climate change may cause increases in temperature in Sri Lanka that are more rapid than the rate of global warming [24]. As a result of climate change, drought is inevitable in almost all regions of Sri Lanka [25].

2.2. Rainfall. Previous studies on rainfall in Sri Lanka emphasize that extraordinary rainfall extremes are concentrated in the southwestern part of the country, especially the Colombo and Ratmalana regions [12]. According to Karunathilaka et al. [11], annual rainfall data demonstrate that nearly two-thirds of monitored stations indicate increasing trends, while the remaining one-third showed decreasing trends. High-intensity rainfall events cause flash floods in urban areas, and such flooding has been frequent during recent years. An exception is the Nuwara Eliya station, which showed decreasing trends for all extreme indices, including frequency and intensity of rainfall [12]. Wet-zone cities such as Ratnapura, Ratmalana, and Colombo are profoundly affected by rainfall extremes. The annual maximum consecutive five-day rainfall also shows significant increasing trends in Batticaloa, Colombo, Hambantota, Ratmalana, and Trincomalee [12]. Teleconnections play a significant role in driving rainfall over Sri Lanka [26]. The El Niño-Southern Oscillation (ENSO) phenomenon is the primary climate driver in Sri Lanka and on the Indian mainland. Variation in both means that rainfall intensity and total rainfall across monsoon seasons showed high correlations with ENSO events [16, 27, 28].

2.3. Temperature. Most of the temperature-related investigations in Sri Lanka have demonstrated warming trends [15, 27, 29, 30]. Using monthly mean temperature as baseline data, De Costa [13] detected long-term (1869–2007) annual temperature increases. De Costa [13] also showed that the mean monthly maximum and minimum temperature trends were 2.6°C per 100 years and 1.7°C per 100 years, respectively. According to Zubair et al. [15], the recent period (1961–2000) shows accelerated daytime warming trends. Other studies of Sri Lankan annual mean temperature data for the period 1871–1990 confirmed significant warming trends in most districts [14, 27]. The factors contributing to increasing temperatures in Sri Lanka are anthropogenic activity connected with forest cover depletion, urbanization, natural teleconnections, and global climate change. Although the urban "heat island" effect probably influences temperature increases in Sri Lankan cities, studies are unable to confirm this relationship. It is evident from the above-described trends that temperatures in Sri Lanka are changing, but it is difficult to extract patterns of extreme temperature events from some previous studies because monthly mean and average daily temperatures are used as baseline data [31].

3. Materials and Methods

3.1. Data. We investigated station records of daily total rainfall (mm) and raw (or unadjusted) temperature (°C) for the 55-year period between 1961 and 2015. Daily total rainfall and daily maximum (T_{max}) and minimum (T_{min}) ambient air temperature records at 20 synoptic meteorological stations scattered throughout the island were selected for analysis (Figure 1). These records were obtained from the Meteorological Department of Sri Lanka. The rainfall and temperature data investigated are subject to consistency checks related to station relocations, instrumentation upgrades, and changes in the surrounding ecosystem that are likely to introduce heterogeneity in data, leading to partly

misrepresenting the actual trend. Furthermore, adjustments to such heterogeneities may introduce spurious trends in the data [32]. Therefore, initially, raw (unadjusted) data were chosen to quantify the trends. The statistical trends in rainfall, T_{max}, and T_{min} were investigated using a suite of climate metrics (Table 1). The DTR in Table 1 was computed as the difference between T_{max} and T_{min} [1, 33]. The measures of rainfall and temperature trends were investigated for annual and the four monsoonal seasons in Sri Lanka. Table 1 groups months to represent the distinct monsoonal seasons in Sri Lanka. Such partitioning into seasons has been used widely in previous investigations relevant to the climate of Sri Lanka [16, 18, 19, 34].

3.2. Mann–Kendall Test for Trend Detection. Although it is possible that nonlinear trends describe rainfall and temperature data, in this work, we first aim to reveal the presence or absence of linear trends using the year as the explanatory variable. We used standard methods to assess the statistical significance of the slope parameter in the linear trend. For the majority of the measures in this study, the nonparametric Mann–Kendall trend test and Sen–Theil regression were also utilized. Unlike prior investigations that have focused on monthly and annual sums and averages of rainfall and temperature in Sri Lanka, in this study, we also attempted to investigate trends in extreme events using parametric methods. For this purpose, we used the generalized extreme value (GEV) distribution with and without nonstationary parameters [35]. We performed the analysis in the R-programming environment (hereafter R). R is an integrated, interactive environment for data manipulation and serves as a platform for high-level statistical data analysis [36]. We used multiple statistical libraries available in R for trend detection. A significance level of 0.05 was maintained as default for all statistical tests.

The Mann–Kendall test, which is a nonparametric alternative for ordinary least squares (OLS) regression, was used to detect trends. This test [37–39] is resistant to outlier effects and is considered an improvement to OLS [40]. It is capable of detecting nonlinear trends and is ideal for situations in which data are missing, censored, or nonnormally distributed [1, 40]. This situation is typical in Sri Lanka for the rainfall and temperature data.

The Mann–Kendall test is determined by summing the signs of the differences between pairs of sequential data values in a time series as follows:

$$S = \sum_{k=1}^{n-1} \sum_{j=k+1}^{n} \text{sgn}(x_j - x_k), \quad (1)$$

where x_j and x_k are the sequential data values for times t_j and t_k (here, $j > k$), n is the length of the data set, and sgn is the sign function:

$$\text{sgn}(x_j - x_k) = \begin{cases} +1 & \text{if } (x_j - x_k) > 0, \\ 0 & \text{if } (x_j - x_k) = 0, \\ -1 & \text{if } (x_j - x_k) < 0, \end{cases} \quad (2)$$

TABLE 1: Rainfall, temperature measures, and seasons investigated to determine average and extreme trends in Sri Lanka.

Variable	Measure
Rainfall	Seasonal total rainfall Seasonal wet days[a] Number of extreme rainfall values (>1-in-2) in seasons[b] Seasonal maximum rainfall Monthly total rainfall Mean and maximum rainfall events of duration 2, 3, 5, and 7 days Number of heavy rainfall events (>1-in-2) of duration 2, 3, 5, and 7 days[c] Mean and maximum number of consecutive dry days
Daily temperature (maximum, minimum, and daily temperature range)	Seasonal average temperature Number of extreme temperature values (>1-in-2) in seasons[b] Maximum and minimum seasonal temperatures Monthly average temperature Maximum temperature values for events of duration 2, 3, 5, and 7 days Number of extreme temperature (>1-in-2) events of duration 2, 3, 5, and 7 days[c]
Duration of seasons	Annual (Jan–Dec) Southwest monsoon (May–Sep) First intermonsoon (Mar–Apr) Northeast monsoon (Dec–Feb) Second intermonsoon (Oct–Nov)

[a]Rainfall > trace amount (0.254 mm). [b]Those exceeding the value corresponding to the 1-in-2 recurrence interval for the particular seasonal maximum (Table 1). [c]Numbers of events of heavy rainfall and extreme temperature are those exceeding the value corresponding to a 1-in-2 recurrence interval for the maximum magnitude event of the specific duration 2, 3, 5, and 7 days.

The null (H_0) and alternate (H_a) hypotheses were set as follows:

$$H_0 : S = 0,$$
$$H_a : S \neq 0. \qquad (3)$$

The null hypothesis, H_0, is that the observations, x, represent a sample from n iid (independent and identically distributed random variables) and that there is no trend. The alternate hypothesis, H_a, states that there is a trend in the data (i.e., the distributions of x_k and x_j are not identical for all x, $j \leq n$ and $k \neq j$). Here, H_0 is rejected when the magnitude of S is statistically significant. We used Kendall package in R to perform the Mann–Kendall test.

When $n \geq 10$, S in the Mann–Kendall test approximately follows the normal distribution and is therefore converted to a standard normal variate, Z_S (i.e., $N \sim (0, 1)$) [38]. Here, Z_S is determined using

$$Z_S = \begin{cases} \dfrac{S-1}{\sigma_S} & \text{if } S > 0, \\[2mm] 0 & \text{if } S = 0, \\[2mm] \dfrac{S+1}{\sigma_S} & \text{if } S < 0, \end{cases} \qquad (4)$$

The standard deviation, σ_s, of S is given by Kendall [38] as follows:

$$\sigma_s = \sqrt{\frac{1}{18}\left[n(n-1)(2n+5) - \sum_{p=1}^{q} t_p (t_p - 1)(2t_p + 5) \right]}, \qquad (5)$$

where t_p is the size of the pth tied group and q is the number of tied pairs. H_0 is rejected if $|Z_S|$ is greater than the critical value, $Z_{\alpha/2}$, where α is the level of significance.

The Sen–Theil estimator was used to compute the magnitude of the slope of the trend when it is significant [41, 42]. The fitted regression, or Sen–Theil trend line, is a nonparametric alternative to linear regression that is used in conjunction with the Mann–Kendall test. Unlike linear regression, Sen–Theil regression does not require the data to be normally distributed. However, the method still assumes that the residuals are statistically independent, and that there is a linear relationship between variables. It has the additional advantage of handling censored and missing data, and therefore, it is considered to be robust.

The Sen–Theil slope estimation method consists of computing a simple pairwise estimate as shown in the following equation, for all distinct pairs of observations (x_j, x_k), where $(t_j > t_k)$:

$$S_{kj} = \frac{(x_j - x_k)}{(t_j - t_k)}. \qquad (6)$$

There will be $N = (n(n-1))/2$ slope pairs in the sample size of n. The slope (b) can be determined as the median of all pairwise slopes [$b = \text{median}(S_{kj})$], whereas the intercept (c) is given by $c = \text{median}(x - bt)$. Because it takes median pairwise slopes instead of the average, the extreme pairwise slopes caused by outliers have little impact on the magnitude of the slope.

The Mann–Kendall and Sen–Theil methods rely on the underlying assumption that observations are independent and identically distributed (iid). Therefore, if there is a significant positive autocorrelation, the test tends to overestimate significance, which leads to the rejection of H_0 according to the selected level of significance. The opposite is also true when the test tends to underestimate significance, possibly when negative autocorrelation exists [43].

Since rainfall and temperature data for Sri Lanka show significant autocorrelation, prewhitening is applied to remove positive autocorrelation. In this investigation, we performed an iterative prewhitening method as executed in the *zyp* package in R [44]. We performed the Mann–Kendall test and Sen–Theil slope estimation after prewhitening the data.

A modified version of the Mann–Kendall trend test was also used to analyze the data set when mild serial autocorrelation is present [45]. When serial dependence is absent, the modified Mann–Kendall trend test is less efficient than the original test. Therefore, in this investigation, we applied the modified Mann–Kendall trend test only when lag-1 autocorrelation coefficient was significant, and if it was not significant, then the original seasonal Mann–Kendall test was performed.

The seasonal Sen–Theil slope estimator is computed by calculating pairwise slopes within each season and obtaining slopes from all seasons to compute a median slope for the entire period of record. In this approach, if there are opposing trends, then the power of the test is reduced because such trends may cancel each other out. Therefore, homogeneity between seasons was initially verified using the van Belle and Hughes trend test [46] prior to the application of seasonal Mann–Kendall and Sen–Theil tests. If the data were heterogeneous between seasons, then the original trend test was performed with prewhitening.

In cases of extremes, the data were fitted using the generalized extreme value (GEV) distribution given in [35] as follows:

$$G(z) = \exp\left\{-\left[1 + \zeta\left(\frac{z-\mu}{\sigma}\right)\right]^{-1/\zeta}\right\}, \qquad (7)$$

where μ, σ, and ζ indicate the location, scale, and shape parameters of the GEV, respectively. The GEV is fitted to data by maximum likelihood. The GEV is appropriate for analyzing block-maxima of extreme values [35]. To model block-minima, z is replaced with $(-z)$ in equation (7).

Using a nonstationary GEV model, H_0 was tested for the absence of a trend in location parameter. Under H_a, there exists a trend in the location parameter in the GEV, indicating potential increases in extreme rainfall or temperature. The significance of the trend in the location parameter was evaluated using the likelihood ratio test, which uses the deviance statistic [35] defined as follows:

$$D = 2\{l_1(M_1) - l_0(M_0)\}, \qquad (8)$$

where $l_1(M_1)$ and $l_0(M_0)$ are the log-likelihoods of the models M_1 (with trend) and M_0 (without trend). Here, the distribution of D is approximated by χ^2 with degrees of freedom (df) equal to one. The null hypothesis H_0 is rejected when the calculated D critical value for the $\chi^2_{\text{critical, df}=1}$ with the selected significance level (α) is typically assumed to be 5%.

4. Results and Discussions

The trend results for rainfall, $T_{\max}$, $T_{\min}$, and daily temperature range (DTR) in Sri Lanka over the study period 1961–2015 are presented separately. The number of stations

(out of a total of 20) with significant values for average and extreme measures listed in Table 1 are summarized in Tables 2–6 for rainfall, $T_{\max}$, $T_{\min}$, and DTR, respectively. The statistical significance of all trends was tested against a consistent α level of 0.05.

4.1. Rainfall Trends. Plots of rainfall data show dynamic trends with increases and decreases during the study period (Table 2). Only two stations show significant trends in annual total rainfall, Katunayaka (decreasing) and Bandarawela (increasing) (Figure 2(a)). A recent study by Karunathilaka et al. [11] also reported both increasing and decreasing trends in annual rainfall in Sri Lanka. However, a previous study using a century-long dataset by Jayawardene et al. [10] reported that the annual rainfall calculated using monthly averages did not show consistent increases or decreases in Sri Lanka. Although annual rainfall showed dynamic trends, one station demonstrated an increase for the NEM season (Figure 2(a)). Katugastota was the only station that detected significant rainfall increase for the NEM season. Although the increases were not statistically significant, 80% of the stations showed increasing rainfall during the NEM season. This overall increase in rainfall during the NEM season agrees with the results reported by Karunathilaka et al. [11]. Trends of rainfall patterns are spatially less coherent [16], but most of the stations in the interior of the island documented increases in rainfall, although only one station showed that it was statistically significant. A decrease in monthly total rainfall was observed throughout the island during the month of May, with two stations (Colombo and Ratmalana) showing that the trend was statistically significant (Figure 1). Such decreases in rainfall are likely due to the delay in the SWM seasonal wind in the region [47–49], which may have affected Sri Lanka. In contrast to this decreasing rainfall, during October and November, rainfall appears to have increased at two stations in the peripheral region of Sri Lanka (Jaffna and Pottuvil), showing a statistically significant trend. This finding of surplus rainfall during the SIM season agrees with the results obtained by Hapuarachchi and Jayawardena [50]. Heavy rainfall during the SIM season is mainly due to massive cyclonic thunderstorms that occur over the northern and eastern parts of Sri Lanka [51] during that period. The rainfall received during this season is critical for the off-season irrigation of agricultural fields.

There have been increasing trends in 2-day, 3-day, and 7-day mean rainfall events, with at least two stations showing statistically significant trends. Only one station has experienced significantly decreasing trends in rainfall for 3-day, 5-day, and 7-day durations (Table 2). This pattern indicates increasing tendencies for short duration heavy rainfall to occur during the study period. Extensive studies of rainfall extremes in Sri Lanka are lacking, but some studies have reported increasing trends in the intensity of rainfall [7, 9, 12, 52].

Another measure for detecting increasing rainfall is calculating the number of wet days in a given season. Only one station showed a statistically significant increase in the

TABLE 2: Number of stations exhibiting statistically significant trends in measures listed in Table 1 for daily total rainfall from 1961 to 2015.

	Total		Wet days		Extreme days		Maximum values			Total	
	−	+	−	+	−	+	−	+		−	+
Annual	1	1	0	1	0	0	0	2	Monthlies	1	1
SWM	0	0	3	1	0	0	0	3	Jan	0	0
FIM	0	0	0	1	0	0	0	1	Feb	0	0
NEM	0	1	0	0	0	0	0	1	Mar	0	0
SIM	0	0	0	1	0	0	0	3	Apr	0	0

Mean events for the duration

2 d		3 d		5 d		7 d				
−	+	−	+	−	+	−	+		May	2	0
0	3	1	2	1	0	1	2		Jun	0	0

(continued with monthly column) Jul 1 0; Aug 1 0; Sep 0 0

Max events for the duration

2 d		3 d		5 d		7 d			
−	+	−	+	−	+	−	+	
0	0	0	0	0	1	1	1	

Oct 0 1; Nov 0 1; Dec 0 0

Number of heavy events for the duration

2 d		3 d		5 d		7 d	
−	+	−	+	−	+	−	+
0	0	0	0	0	0	0	0

Dry events

Mean duration		Max duration	
−	+	−	+
0	7	0	4

TABLE 3: Decadal changes in rainfall of Sri Lanka during 1961–2010.

Decade	Trend slopes (mm/decade)
1961–1970	−2.9
1971–1980	−2.7
1981–1990	−0.2
1991–2000	−0.9
2001–2010	+0.6

number of wet days during the annual, SWM, FIM, and SIM periods. For the NEM season, none of the stations showed significant trends. Significant decreasing trends were observed at three stations (15% of the total stations) during the SWM season: Anuradhapura, Batticaloa, and Kurunegala (Table 2 and Figure 3). Further investigations may be necessary to detect whether such reductions in wet days during SWM compared to NEM will persist in the future. Jacobi [48] also reported a similar shift in the rainfall pattern of Sri Lanka. Further analyses may also be required to test these shifting trends.

Seasonal rainfall maxima were investigated using the GEV distribution, and the significance of trends in location parameters was tested using the likelihood ratio test. Such tests demonstrate possible nonstationarity in rainfall extremes [35]. In terms of seasonal maximum rainfall, at least one station showed a statistically significant trend and none showed decreasing trends. During the SWM and SIM seasons, three stations showed significant increases in seasonal maxima rainfall. During SWM, coastal stations (Colombo, Katunayaka, and Ratmalana) and during SIM, stations in the interior parts of Sri Lanka (Badulla, Bandarawela, and Ratnapura) demonstrated significant increases in seasonal maxima rainfall (Figure 4 and Table 2).

The decadal changes in rainfall obtained using the island average values of daily rainfall among meteorological stations in Sri Lanka suggest an overall increasing rainfall trend in recent decades. The decadal changes in trend slopes of the Sen–Theil regression are summarized in Table 3. Switching in the signs of rainfall trend slopes occur in very recent decades. The trend slopes during the 1961–1970 and 2001–2010 periods were found to be −2.9 and +0.6 mm/ decade, respectively. This indicates that island-wide rainfall shows an increasing trend, especially during the most recent decade.

Even though the number of extreme rainfall or precipitation events has increased across the continents, global historical trends in different countries have mixed anomalies in terms of those events [4]. Extreme rainfall events have been observed in a few countries, whereas in many countries, those extremes have been recorded at 70% below average [49]. Trends in Asian rainfall events indicate that South Asian monsoon rainfall extremes are becoming relatively frequent [53]. An increased frequency of intense rainfall has already been reported with an increased likelihood of extreme rainfall in parts of South Asia [54]. Nepal has shown an increasing trend of annual mean rainfall, especially during the months of June and July [55] agrees with rainfall pattern in Sri Lanka. Though generalized trends in averages have been much studied in the countries of South Asia, extreme rainfall trends are scarcely reported in existing literature.

4.2. Temperature Trends. Results of trend analysis showed a general temperature increase across Sri Lanka, which suggests warming of $T_{\max}$ and $T_{\min}$ for most stations,

TABLE 4: Number of stations with statistically significant trends in measures listed in Table 1 for daily maximum temperature (T_{max}) from 1961 to 2015.

	Average −	+	Extreme days −	+	Maximum values −	+	Minimum values −	+		Average −	+
Annual	0	11	0	3	0	9	0	8	Monthlies	0	16
SWM	0	12	1	6	0	12	0	12	Jan	0	11
FIM	1	8	0	3	0	9	0	2	Feb	0	11
NEM	0	9	0	0	0	11	0	5	Mar	0	10
SIM	0	13	0	5	0	9	0	6	Apr	1	7
Maximum events for the duration									May	0	13
2 d		3 d		5 d		7 d			Jun	0	14
−	+	−	+	−	+	−	+		Jul	0	12
1	4	1	4	1	4	0	2		Aug	0	14
Number of extreme events for the duration									Sep	0	12
2 d		3 d		5 d		7 d			Oct	0	15
−	+	−	+	−	+	−	+		Nov	0	10
0	0	0	0	0	0	0	0		Dec	1	10

TABLE 5: Number of stations with statistically significant trends in measures listed in Table 1 for daily minimum temperature (T_{min}) from 1961 to 2015.

	Average −	+	Extreme days −	+	Maximum values −	+	Minimum values −	+		Average −	+
Annual	0	14	0	2	0	10	0	7	Monthlies	0	15
SWM	0	15	0	6	0	11	0	12	Jan	1	10
FIM	0	10	0	0	0	9	0	10	Feb	2	7
NEM	0	13	0	6	0	13	0	7	Mar	1	10
SIM	0	13	0	8	0	13	0	10	Apr	0	9
Maximum events for the duration									May	0	14
2 d		3 d		5 d		7 d			Jun	0	16
−	+	−	+	−	+	−	+		Jul	0	16
0	2	0	2	0	4	0	3		Aug	0	13
Number of extreme events for the duration									Sep	0	14
2 d		3 d		5 d		7 d			Oct	0	14
−	+	−	+	−	+	−	+		Nov	1	13
0	0	0	0	0	0	0	0		Dec	0	12

TABLE 6: Number of stations with statistically significant trends in measures listed in Table 1 for daily temperature range (DTR) from 1961 to 2015.

	Average −	+	Extreme days −	+	Maximum values −	+	Minimum values −	+		Average −	+
Annual	3	3	0	3	0	7	0	4	Monthlies	5	6
SWM	1	3	1	2	0	10	0	3	Jan	2	3
FIM	3	4	1	1	0	6	0	4	Feb	2	3
NEM	2	3	1	1	0	6	0	2	Mar	3	3
SIM	2	3	0	1	0	6	0	2	Apr	3	3
Maximum events for the duration									May	1	3
2 d		3 d		5 d		7 d			Jun	2	3
−	+	−	+	−	+	−	+		Jul	5	4
2	0	2	2	1	2	1	3		Aug	2	5
Number of extreme events for the duration									Sep	2	3
2 d		3 d		5 d		7 d			Oct	1	3
−	+	−	+	−	+	−	+		Nov	3	3
0	0	0	0	0	0	0	0		Dec	2	2

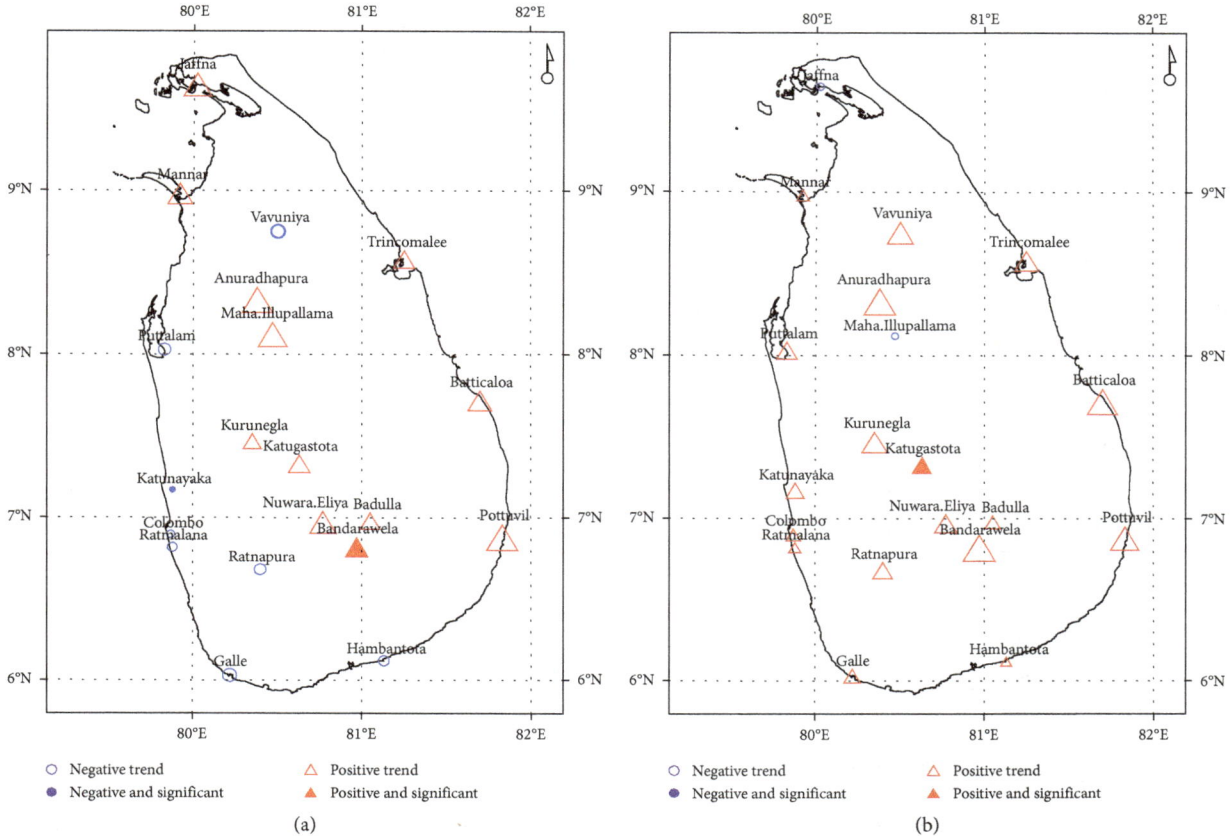

FIGURE 2: Trends in total rainfall: (a) annual and (b) NEM seasons during the study period. Triangles indicate an increasing trend, and circles indicate a decreasing trend. Filled markers indicate significant trends at the α level of 0.05.

except for a few stations in the Southwestern region. Overall, our results show that 55% and 70% of the stations show significant increases in $T_{\max}$ and $T_{\min}$, respectively. The overall trends we observed agree well with the results of previous studies [14, 27, 56] indicating overall warming in Sri Lanka.

Increases of daily $T_{\max}$ during the study period were observed during October, with the highest number of stations (75%) confirming statistically significant increases (Figure 2(a) and Table 4). During the SWM season, we observed that at least 60% of stations showed significant increases in monthly average $T_{\max}$, and this result agrees with seasonal averages of $T_{\max}$. In all months except for April, at least 50% of stations experienced significant increasing trends. In April, only 35% of stations showed statistically significant increases in $T_{\max}$. Moreover, during May, $T_{\max}$ decreased at a small number of stations (10%). This finding may be due to wind effects during the SWM season in Sri Lanka, which might have reduced the ambient temperature. Moreover, during October when SIM is prevalent none of the stations showed significant decreases in monthly averages of either $T_{\max}$ or $T_{\min}$ (Figure 2).

In contrast to increasing $T_{\max}$ trends, Nuwara Eliya and Ratnapura observed decreases during the months of April and December, when the number of stations reporting statistically significant increases in monthly

averages of $T_{\max}$ were low compared with other months (Table 4).

Regarding seasonal maximum temperature, at least 45% of stations showed statistically significant increases, with 60% of the stations doing so during SWM followed by 55% during the NEM season. Fewer stations showed annual and seasonal minimum values with statistically significant trends compared with seasonal maximum and seasonal averages of $T_{\max}$, except during SWM. No station showed a significant decreasing trend in either seasonal maxima or minima.

The main finding drawn from our observations of daily $T_{\min}$ is that more stations reported increases than decreases (Table 5) for almost all the analyzed statistical metrics. At least 50% of stations showed significant increases in averages for all seasons, with 75% of stations doing so during the months of MJJAS coinciding with the SWM season. At least 50% of stations showed significant increases in seasonal maxima, and only 35% of stations did so for seasonal minima. During the NEM and SIM seasons, 65% of stations reported significant increases in seasonal maxima and 60% demonstrated significant increases for the seasonal minima of $T_{\min}$ during the SWM season. Moreover, half of the stations demonstrated increases in annual maxima and one-third of the stations demonstrated increases in annual minima. The monthly averages for $T_{\min}$ show that 80% of

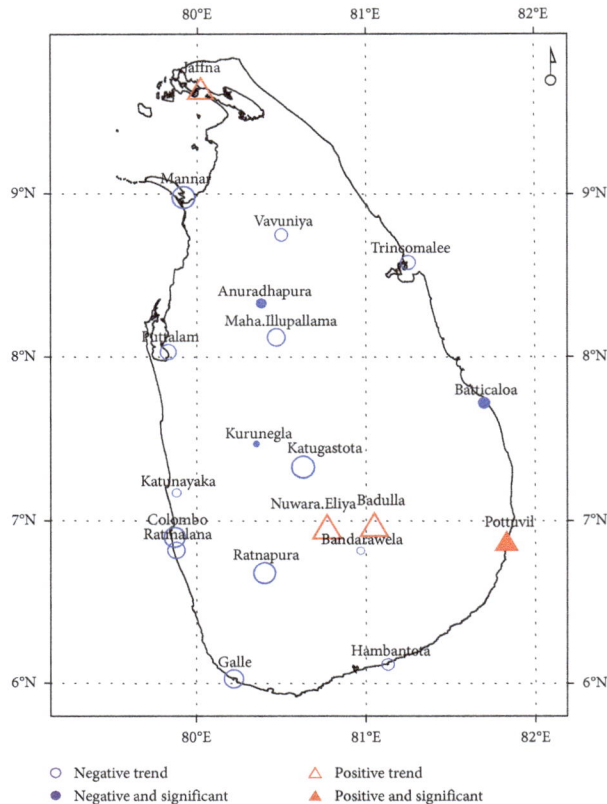

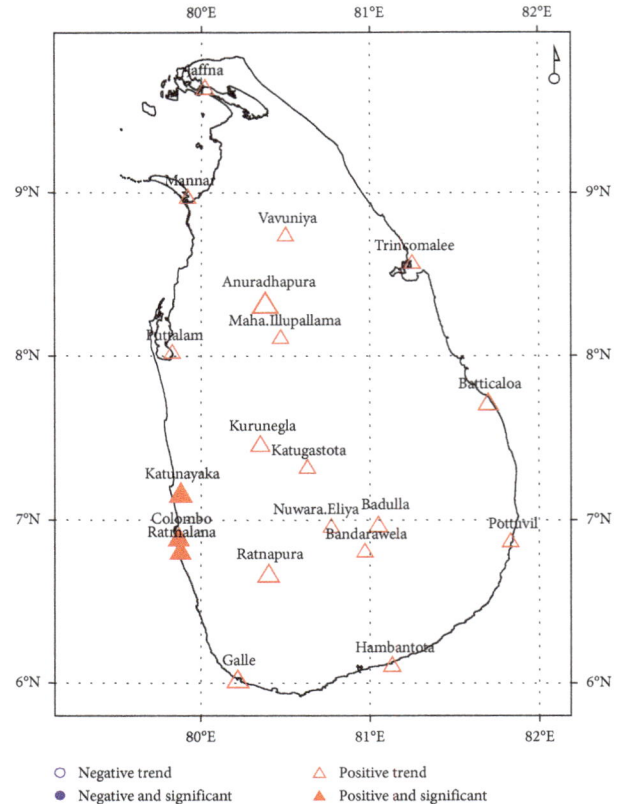

FIGURE 3: Trends in the number of wet days for the SWM season during the study period. Triangles indicate an increasing trend, and circles indicate a decreasing trend. Filled markers indicate significant trends at the α level of 0.05.

FIGURE 4: Trends in seasonal rainfall maxima of SWM during 1961–2015. Triangles indicate an increasing trend, and circles indicate a decreasing trend. Filled markers indicate significant trends at the α level of 0.05.

stations exhibited significant trends during June and July (Figure 5). This general increase in $T_{\min}$ is thought to be caused by warming nighttime temperatures in many cities [57, 58], and is consistent with global trends [33].

Figure 6 presents the annual daily mean temperature over 20 stations, along with a moving average. It is evident that there has been an overall increase in the temperature over time for the island as a whole. Also, there is a considerable degree of interannual variability in the average temperature data. It may be possible due to several factors including the changes in rainfall and large scale climate circulation.

Although averages of DTR show both increasing and decreasing trends, significant increases in seasonal maxima and minima are evident in Table 6. At least 30% and 10% of stations recording maxima and minima, respectively, demonstrate significant increases in DTR. This finding is consistent with trends observed for 2-day maximum events. We observed a general trend demonstrating increasing monthly DTR, although more stations demonstrated decreases during July (Figure 7), including stations at Colombo, Katunayaka, Katugastota, Puttalam, and Ratnapura. The Jaffna, Hambantota, and Trincomalee stations often showed increasing trends in seasonal and monthly averages and in seasonal maxima of DTR. A decreasing trend for DTR was mostly observed at

the Colombo, Katunayaka, and Nuwara Eliya stations. Several factors may have caused decreases in DTR, such as natural variability in the immediate atmosphere. Especially land use and land cover changes in the stations with extended and rapid regional urbanization in Colombo, Katunayaka, Katugastota, Puttalam, and Trincomalee have led to a reduction in forest cover [31] which may have altered the temperature regime.

Research on daily temperature extremes in South Asian countries has revealed the warming of both extreme-cold and extreme-warm distributions [59]. As in Sri Lanka, increases in nighttime temperatures and warmest daytime temperatures have been observed at most weather stations across Nepal, India, and Pakistan [60]. In Pakistan, a greater number of warmer winter months were observed than summer months from 1975 to 2005 [61]. The Punjab province of Pakistan has been shown to have increased numbers of hot days and nights with prolonged summer days [62]. Several countries in South Asia have reported above-normal temperatures in the recent past. In India, temperatures higher than the normal 30-year average have been reported in recent years with a warming of 0.51°C [63, 64]. Also, reports and research studies have indicated rising temperature trends all over the states of India [30, 65]. Available studies suggest that sea surface temperatures and nighttime marine air temperatures over

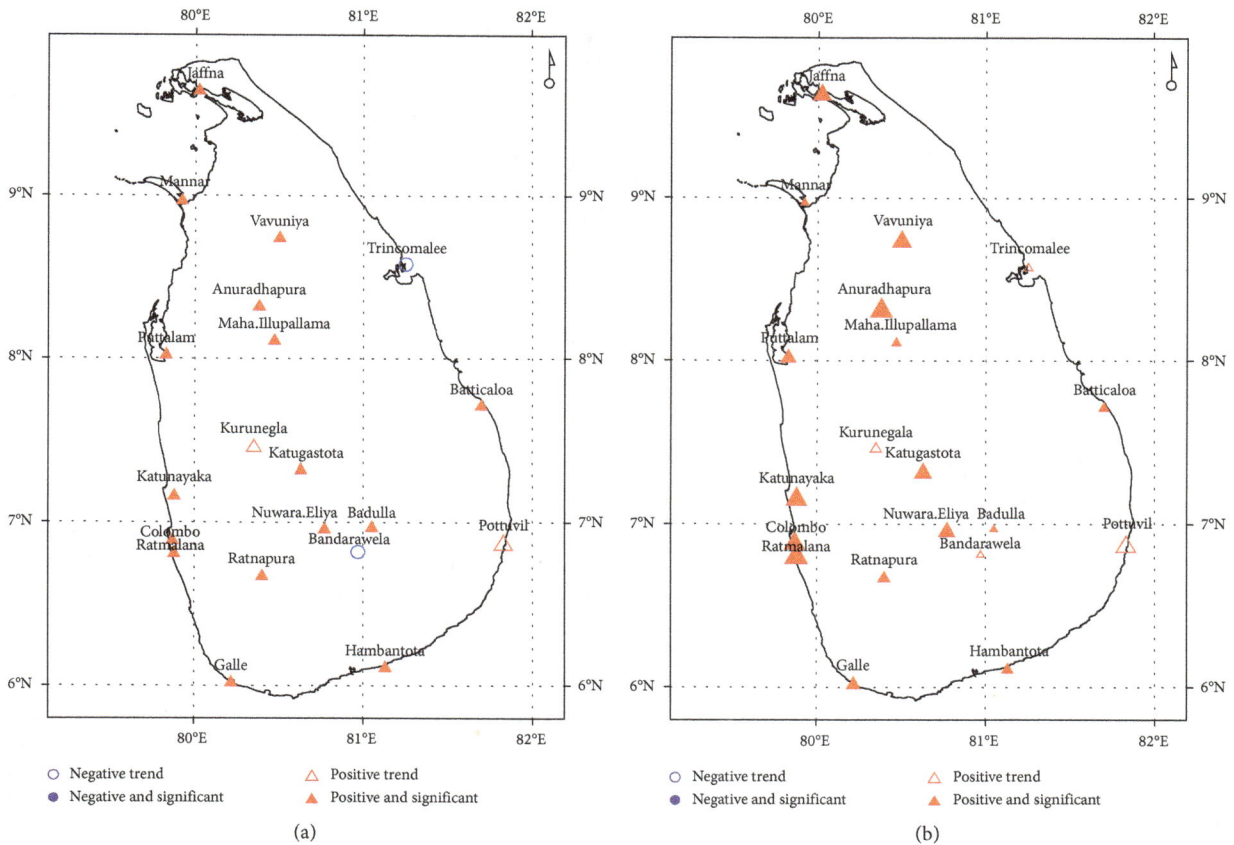

FIGURE 5: Trends in monthly averages in T_{min} for (a) June and (b) July during 1961–2015. Triangles indicate an increasing trend, and circles indicate a decreasing trend. Filled markers indicate significant trends at the α level of 0.05.

oceans worldwide have shown an increase in temperature. Global air temperatures over land surfaces have increased at about double the rate of the oceans [66]. Moreover, in regions of Europe, the high-temperature increase has become more frequent, while low-temperature extremes have become less frequent. The average length of summer days across Western Europe have doubled and the frequency of warm days has almost tripled in recent weather history [67]. Indeed, being an island and not connected to the Indian mainland, Sri Lanka experiences the regional and global increasing temperature trend.

5. Conclusions

Our trend investigations reveal dynamic rainfall trends with both increasing and decreasing patterns in Sri Lanka. Despite these varying trends, in general, we observed significant increases in seasonal rainfall at stations near the coastal regions of Sri Lanka. Decreases in rainfall are most evident for the month of May, likely due to the delay in the SWM winds in Sri Lanka. Furthermore, we observed a decrease in the number of wet days during the SWM season. The reduced number of wet days during the primary monsoonal season will influence agriculture in Sri Lanka. We detected no significant rainfall trends during the NEM season, one of the most important seasons in Sri Lanka. Notable increases

in T_{min} and T_{max} were observed at 70 and 55% of the stations in Sri Lanka, respectively. The increase in T_{min} indicates likely warming of nighttime temperatures in Sri Lanka. Regarding monthly averages in T_{min}, 80% of stations showed significant increases during June and July. The DTR in the southwestern coastal region of Sri Lanka demonstrated a decreasing trend during July. Increases in T_{max} were often detected at stations all across the island.

This investigation, which addresses seasonality and extremes, provides a good understanding of historical temperature and rainfall trends in Sri Lanka. Our methods are ideal for situations in which historical observations consist of missing or censored data, which are common in developing countries such as Sri Lanka. Although the data are subject to general quality control procedures, station displacements, instrumentation upgrades, and changes in station surrounding environment may introduce inhomogeneities. Except for a few, most of the gauging stations in Sri Lanka still operate in the same locations that they historically monitored. Therefore, heterogeneity due to changes in location is not a significant confounder. However, changes in the environments of the station surroundings are challenging to quantify. Therefore, trends in raw (or unadjusted) data were used to quantify temperature trends. In the future, raw and adjusted data at a station may be compared to better understand the homogeneity

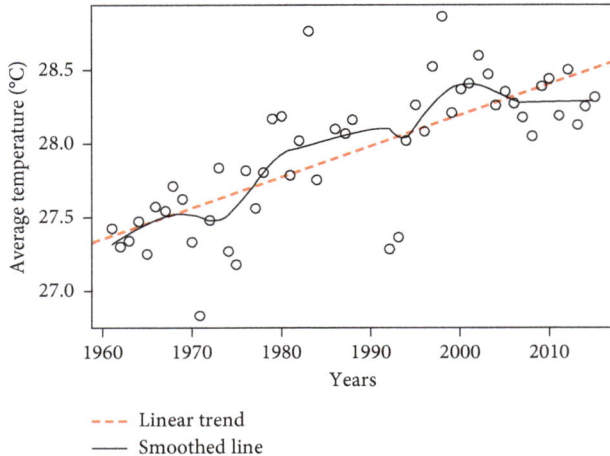

FIGURE 6: Annual daily mean temperature over the 20 stations, along with a moving average, during the period of 1961–2015. The dashed line represents the linear trend from the ordinary least square regression, whereas the solid line is the curve smoothed at the span of 0.25.

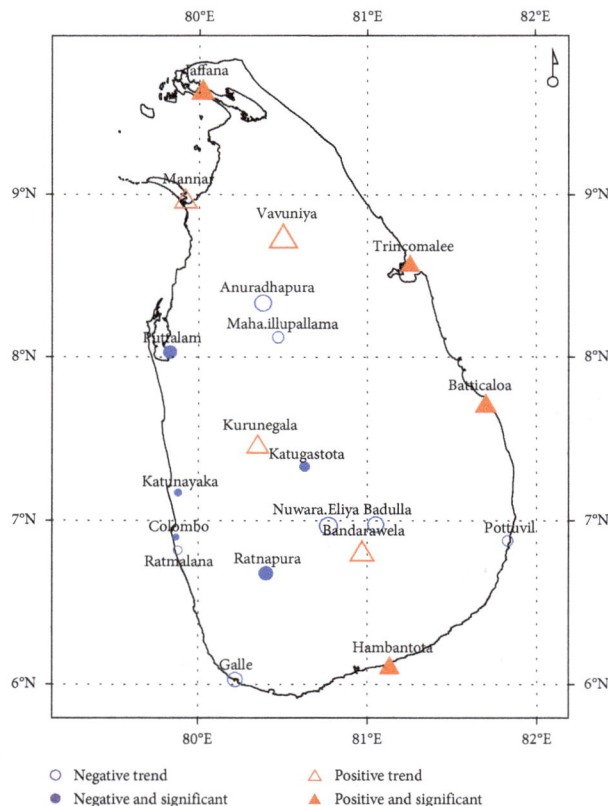

FIGURE 7: Trends in monthly averages of DTR during July from 1961 to 2015. Triangles indicate an increasing trend, and circles indicate a decreasing trend. Filled markers indicate significant trends at the α level of 0.05.

corrections performed and their impact on trends. The trends we report, along with the attribution study, may provide the key to future investigations of heterogeneities in the rainfall and temperature of Sri Lanka.

Conflicts of Interest

The authors declare that there are no conflicts of interest regarding the publication of this paper.

Acknowledgments

The research was supported by the National Research Council (NRC) of Sri Lanka (Grant no. 15-144) and Korea Meteorological Administration Research and Development Program (Grant no. KMI2018-07010).

Supplementary Materials

Supplementary Figure 1. Annual maximum rainfall trend in Sri. Triangles indicate increasing trend, and circles indicate decreasing trend. Filled markers indicate significant trends at the 0.05 level of α. Supplementary Figure 2. Annual maxima of temperature in Sri Lanka (a) T_{max}, (b) T_{min}, and (c) DTR. Triangles indicate increasing trend, and circles indicate decreasing trend. Filled markers indicate significant trends at the 0.05 level of α. Supplementary Table 1. Threshold values of the variables for the statistical measures Supplementary Table 2. Summary of averages in Tave and RH estimated for climatic zones of Sri Lanka for the period of 1996-2015. (*Supplementary Materials*)

References

[1] M. Irizarry Ortiz, J. Obeysekera, J. Park et al., "Historical trends in Florida temperature and precipitation," *Hydrological Processes*, vol. 27, no. 16, pp. 2225–2246, 2013.

[2] National Atlas, *The National Atlas of Sri Lanka*, Department of Survey, Sri Lanka, Colombo, Sri Lanka, 2nd edition, 2007.

[3] IPCC, *Climate Change 2013: the Physical Science Basis. Contribution of Working Group I to the 5th Assessment Report of the Intergovernmental Panel on Climate Change (IPCC)*, Cambridge University Press, Cambridge, UK, 2013.

[4] IPCC, *Climate Change 2007: Impacts, Adaptation and Vulnerability. Contribution of Working Group II to the Fourth Assessment. Report of the Intergovernmental Panel on Climate Change (IPCC)*, Cambridge University Press, Cambridge, UK, 2007a.

[5] CHA, *Impacts of Disasters in Sri Lanka*, Report of the Consortium of Humanitarian Agencies (CHA), Colombo, Sri Lanka, 2016.

[6] S. Chandrasekara, V. Prasanna, and H.-H. Kwon, "Monitoring water resources over the kotmale reservoir in Sri Lanka using ENSO phases," *Advances in Meteorology*, vol. 2017, Article ID 4025964, 9 pages, 2017.

[7] K. A. U. S. Imbulana, N. T. S. Wijesekera, and B. R. Neupane, *Sri Lanka National Water Development Report, MAIandMD, UN-WWAP*, UNESCO and University of Moratuwa, Moratuwa, Sri Lanka, 2006.

[8] L. W. Seneviratne, "Global drought and flood in 2009/10 effected by astronomical conditions," in *Proceedings of the*

International Conference on Sustainable Built Environment (ICSBE-2010), pp. 476–486, Peradeniya, Sri Lanka, December 2010.

[9] S. Herath and U. Rathnayake, "Changing rainfall and its impact on landslides in Sri Lanka," *Journal of Mountain Science*, vol. 2, no. 3, pp. 218–224, 2005.

[10] H. K. W. I. Jayawardene, D. U. J. Sonnadara, and D. R. Jayewardene, "Trends of rainfall in Sri Lanka over the last century," *Sri Lankan Journal of Physics*, vol. 6, pp. 7–17, 2005.

[11] K. L. A. A. Karunathilaka, H. K. V. Dabare, and K. D. W. Nandalal, "Changes in rainfall in Sri Lanka during 1966-2015," *Journal of the Institution of Engineers, Sri Lanka*, vol. 50, no. 2, pp. 39–48, 2017.

[12] R. Sanjeewani and L. Manawadu, "Dynamic trends of rainfall extremes in Sri Lanka," in *Proceedings of International Symposium on ICT for Environmental Sustainability, University of Kelaniya*, Kelaniya, Sri Lanka, August 2014.

[13] W. A. J. M. De Costa, "Adaptation of agricultural crop production to climate change: a policy framework for Sri Lanka," *Journal of National Science Foundation Sri Lanka*, vol. 36, pp. 63–88, 2008.

[14] T. K. Fernando and L. Chandrapala, *Global Warming and Rainfall Variability–Sri Lankan Situation*, Sri Lanka Association for the Advancement of Science (SLASS), Colombo, Sri Lanka, 2002.

[15] L. Zubair, J. Hansen, J. Chandimala et al., *Current Climate and Climate Change Assessments for Coconut and Tea Plantations in Sri Lanka*, START, Washington, DC, USA, 2005.

[16] B. Malmgren, R. Hulugalla, Y. Hayashi, and T. Mikami, "Precipitation trends in Sri Lanka since the 1870s and relationships to El niño–southern oscillation," *International Journal of Climatology*, vol. 23, no. 10, pp. 1235–1252, 2003.

[17] P. Wickramagamage, "Seasonality and spatial pattern of rainfall of Sri Lanka: exploratory factor analysis," *International Journal of Climatology*, vol. 30, pp. 1235–1245, 2010.

[18] E. Nishadi and V. Smakhtin, "How prepared are water and agricultural sectors in Sri Lanka for climate change?: A review," in *Proceedings of Water for Food Conference, International Water Management Institute (IWMI)*, Colombo, Sri Lanka, February 2009.

[19] L. Zubair, M. Siriwardhana, J. Chandimala, and Z. Yahiya, "Predictability of Sri lankan rainfall based on ENSO," *International Journal of Climatology*, vol. 28, no. 1, pp. 91–101, 2008.

[20] K. Abhayasinghe, "Climate," in *National Atlas of Sri Lanka*, Survey Department of Sri Lanka, Colombo, Sri Lanka, 2007.

[21] L. Chandrapala, "Rainfall," in *National Atlas of Sri Lanka*, Survey Department of Sri Lanka, Colombo, Sri Lanka, 2007.

[22] A. Senaratne and C. Rodrigo, *Agriculture Adaptation Practices: Case Study of Sri Lanka*, South Asia Watch on Trade, Economics and Environment (SAWTEE), Kathmandu, Nepal, 2014.

[23] B. Marambe, R. Punyawardena, P. Silva et al., "Climate, climate risk, and food security in Sri Lanka: the need for strengthening adaptation strategies," *Handbook of Climate Change Adaptation*, pp. 1759–1789, 2015.

[24] B. Murphy, "Policy briefing on climate change in Sri Lanka," May 2011, https://www.weadapt.org/knowledge-base/global-initiative-on-community-based-adaptation-gicba/policy-briefing-on-climate-change-in-sri-lanka.

[25] R. D. Chithranayana and B. V. R. Punyawardena, "Identification of drought prone agro-ecological regions in Sri Lanka. Journal of national science foundation identification of drought prone agro-ecological regions in Sri Lanka," *Journal of National Science Foundation Sri Lanka*, vol. 36, no. 2, pp. 117–123, 2008.

[26] T. Burt and K. Weerasinghe, "Rainfall distributions in Sri Lanka in time and space: an analysis based on daily rainfall data," *Climate*, vol. 2, no. 4, pp. 242–263, 2014.

[27] L. Chandrapala, "Long term trends of rainfall and temperature in Sri Lanka," in *Climate Variability and Agriculture*, Narosa Publishing House, New Delhi, India, 1996.

[28] E. Ranatunge, B. A. Malmgren, Y. Hayashi et al., "Changes in the South West Monsoon mean daily rainfall intensity in Sri Lanka: relationship to the el-niño southern oscillation," *Palaeogeography, Palaeoclimatology, Palaeoecology*, vol. 197, no. 1-2, pp. 1–14, 2003.

[29] T. Fernando, "Recent variations of rainfall and air temperature in Sri Lanka," in *Proceedings of 53rd Annual Session of SLASS*, Colombo, Sri Lanka, November 1997.

[30] IPCC, *Climate Change 2007: Synthesis Report. Contribution of Working Groups I, II and III to the 4th Assessment Report of the Intergovernmental Panel on Climate Change Intergovernmental Panel on Climate Change (IPCC)*, Cambridge University Press, Cambridge, UK, 2007b.

[31] L. Zubair, *Climate Change Assessment in Sri Lanka using Quality Evaluated Surface Temperature Data*, FECT, Tech. Rep. 2017-06, Foundation for Environment, Climate and Technology, Digana Village, Sri Lanka, 2017.

[32] R. A. Pielke, C. A. Davey, D. Niyogi et al., "Unresolved issues with the assessment of multidecadal global land surface temperature trends," *Journal of Geophysical Research*, vol. 112, no. 24, 2007.

[33] R. Davy, I. Esau, A. Chernokulsky, S. Outten, and S. Zilitinkevich, "Diurnal asymmetry to the observed global warming," *International Journal of Climatology*, vol. 37, no. 1, pp. 79–93, 2017.

[34] C. S. De Silva, "Impacts of climate change on water resources in Sri Lanka," in *Proceedings of 32nd WEDC International Conference*, pp. 289–295, Colombo, Sri Lanka, 2006.

[35] S. Coles, *An Introduction to Statistical Modeling of Extreme Values*, Springer-Verlag, London, UK, 2001.

[36] R-Core, *R: A Language and Environment for Statistical Computing*, R Foundation for Statistical Computing, Vienna, Austria, 2016.

[37] M. Kendall, "A new measure of rank correlation," *Biometrika*, vol. 30, pp. 81–93, 1938.

[38] M. Kendall, *Rank Correlation Methods*, Charles Griffin, London, UK, 4th edition, 1975.

[39] H. Mann, "Nonparametric tests against trend," *Econometrica*, vol. 13, pp. 245–259, 1945.

[40] J. Obeysekera, J. Park, M. Irizarry-Ortiz et al., "Past and projected trends in climate and sea level for South Florida. Interdepartmental climate change group. South Florida water management district, west palm beach, Florida, hydrologic and environmental systems modeling," Technical report, South Florida Water Management District, West Palm Beach, FL, USA, 2011.

[41] P. Sen, "Estimates of the regression coefficient based on Kendall's Tau," *Journal of American Statistical Association*, vol. 63, no. 324, pp. 1379–1389, 1968.

[42] H. Theil, "A rank-invariant method of linear and polynomial regression analysis III," in *Advanced Studies in Theoretical*

and Applied Econometrics, pp. 1397–1412, Springer, Berlin, Germany, 1950.

[43] S. Yue, P. Pilon, B. Phinney, and G. Cavadias, "The influence of autocorrelation on the ability to detect trend in hydrological series," *Hydrological Processes*, vol. 16, no. 9, pp. 395–429, 2002.

[44] X. Zhang, L. A. Vincent, W. D. Hogg, and A. Niitsoo, "Temperature and precipitation trends in Canada during the 20th century," *Atmosphere-Ocean*, vol. 38, no. 3, pp. 395–429, 2000.

[45] R. Hirsch and J. Slack, "A nonparametric test for seasonal data with serial dependence," *Water Resources Research*, vol. 20, no. 6, pp. 727–732, 1984.

[46] G. van Belle and J. Hughes, "Nonparametric tests for trend in water quality," *Water Resources Research*, vol. 20, no. 1, pp. 127–136, 1984.

[47] P. D. Clift and R. A. Plumb, *The Asian Monsoon: Causes, History and Effects*, Cambridge University Press, Cambridge, UK, 2008.

[48] J. Jacobi, "A changing climate in Sri Lanka: shifts, perceptions, and potential adaptive actions," Ph.D thesis, Vanderbilt University, Nashville, TN, USA, 2014.

[49] Y. Y. Loo, L. Billa, and A. Singh, "Effect of climate change on seasonal monsoon in Asia and its impact on the variability of monsoon rainfall in Southeast Asia," *Geoscience Frontiers*, vol. 6, no. 6, pp. 817–823, 2015.

[50] H. A. S. U. Hapuarachchi and I. M. S. P. Jayawardena, "Modulation of seasonal rainfall in Sri Lanka by ENSO extremes," *Sri Lanka Journal of Meteorology*, vol. 1, pp. 3–11, 2015.

[51] L. Zubair, V. Ralapanawe, U. Tennekoon, Z. Yahiya, and R. Perera, *Chapter 4: Natural Disaster Risks in Sri Lanka: Mapping Hazards and Risk Hotspots*, World Bank, Washington, DC, USA, 2006.

[52] S. Mathanraj and M. I. M. Kaleel, "Rainfall variability in the wet-dry seasons: an analysis in Batticaloa district, Sri Lanka," *World News of Natural Sciences*, vol. 9, pp. 71–78, 2017.

[53] C. Yao, S. Yang, W. Qian, Z. Lin, and M. Wen, "Regional summer precipitation events in Asia and their changes in the past decades," *Journal of Geophysical Research*, vol. 113, no. 17, 2008.

[54] J. H. Christensen, B. Hewitson, A. Busuioc et al., *Regional Climate Projections. Climate Change 2007: The Physical Science Basis. Contribution of Working Group I to the Fourth Assessment Report of the Intergovernmental Panel on Climate Change (IPCC)*, Cambridge University, Cambridge, UK, 2007.

[55] R. M. Shrestha and A. B. Sthapit, "Temporal variation of rainfall in the bagmati river basin, Nepal," *Nepal Journal of Science and Technology*, vol. 16, no. 1, pp. 31–40, 2015.

[56] W. A. J. M. De Costa, "Climate change in Sri Lanka: myth or reality? Evidence from long-term meteorological data," *Journal of National Science Foundation Sri Lanka*, vol. 38, no. 2, pp. 79–89, 2010.

[57] L. Xiaxiang, Z. Xuezhen, and Z. Lijuan, "Observed effects of vegetation growth on temperature in the early summer over the northeast China plain," *Atmosphere*, vol. 8, no. 97, 2017.

[58] L. Zhou, R. E. Dickinson, P. Dirmeyer, A. Dai, and S. K. Min, "Spatiotemporal patterns of changes in maximum and minimum temperatures in multi-model simulations," *Geophysical Research Letters*, vol. 36, no. 2, 2009.

[59] T. A. M. G. Klein, T. C. Peterson, D. A. Quadir et al., "Changes in daily temperature and precipitation extremes in central and south Asia," *Journal of Geophysical Research*, vol. 111, 2006.

[60] M. M. Sheikh, N. Manzoor, J. Ashraf et al., "Trends in extreme daily rainfall and temperature indices over South Asia," *International Journal of Climatology*, vol. 35, no. 7, pp. 1625–1637, 2015.

[61] F. Ikram, M. Afzaal, S. A. A. Bukhari, and B. Ahmed, "Past and future trends in frequency of heavy rainfall events over Pakistan," *Pakistan Journal of Meteorology*, vol. 12, no. 24, pp. 57–78, 2016.

[62] F. Abbas, "Analysis of a historical (1981–2010) temperature record of the Punjab province of Pakistan," *Earth Interactions*, vol. 17, no. 5, pp. 1–23, 2013.

[63] K. S. Anil, J. C. Dagar, A. Arunachalam, R. Gopichandran, and K. N. Shelat, *Climate Change Modelling, Planning and Policy for Agriculture*, Spinger, Berlin, Germany, 2015.

[64] S. Surinder, *The Changing Profile of Indian Agriculture*, TN Ninan for Business Standard Books, New Delhi, India, 2009.

[65] APN, *Water Resources in South Asia: An Assessment of Climate Change-associated Vulnerabilities and Coping Mechanisms*, Asia-Pacific Network for Global Change Research, Kobe, Japan, 2004.

[66] M. V. K. Sivakumar and R. Stefanski, "Chapter 2: climate change in South Asia," in *Climate Change and Food Security in South Asia*, pp. 13–30, Springer, Berlin, Germany, 2011.

[67] EEA, "Air quality in Europe," Report, European Environmental Agency, København, Denmark, 2011.

11

Carbon Exchange between the Atmosphere and a Subtropical Evergreen Mountain Forest in Taiwan

Falk Maneke-Fiegenbaum [ID],[1] Otto Klemm,[1] Yen-Jen Lai [ID],[2] Chih-Yuan Hung [ID],[2] and Jui-Chu Yu [ID][2]

[1]*Climatology Working Group-Institute of Landscape Ecology, University of Münster, 48149 Münster, Germany*
[2]*Experimental Forest, National Taiwan University, 55750 Nantou, Taiwan*

Correspondence should be addressed to Yen-Jen Lai; alanlai@ntu.edu.tw

Academic Editor: Gabriele Buttafuoco

Tropical, temperate, and boreal forests are the subject of various eddy covariance studies, but less is known about the subtropical region. As there are large areas of subtropical forests in the East Asian monsoon region with possibly high carbon uptake, we used three years (2011–2013) of eddy covariance data to estimate the carbon balance of a subtropical mountain forest in Taiwan. Two techniques of flux partitioning are applied to evaluate ecosystem respiration, thoroughly evaluate the validity of the estimated fluxes, and arrive at an estimate of the yearly net ecosystem exchange (NEE). We found that advection is a strong player at our site. Further, when used alone, the nighttime flux correction with the so-called u^* method (u^* = friction velocity) cannot avoid underestimating the nighttime respiration. By using a two-technique method employing both nighttime and daytime parameterizations for flux corrections, we arrive at an estimate of the three-year mean NEE of −561 (±standard deviation 114) $g \cdot C \cdot m^{-2} \cdot yr^1$. The corrected flux estimate represents a rather large uptake of CO_2 for this mountain cloud forest, but the value is in good agreement with the few existing comparable estimates for other subtropical forests.

1. Introduction

There are complex interactions at play between earth's climate and the global forests. For example, properties and processes of the forests such as the albedo, the evapotranspiration, and the carbon cycle are relevant climate drivers and thus have an impact on climate change arising from anthropogenic greenhouse gas emission [1]. For the purpose of studying the carbon cycle of forests, there are hundreds of eddy covariance stations worldwide to quantify the atmosphere-biosphere fluxes of CO_2 [2]. Tropical, temperate, and boreal forests are the subject of numerous eddy covariance studies [3], but less is known about the subtropical region. The quantification of CO_2 fluxes in the subtropical regions is, however, very important for global change studies because these forests—which make up large areas in the East Asian monsoon region (latitude 20–40°N, longitude 100–145°E) [4]—potentially play important roles in the global carbon cycle. Integration of existing eddy covariance studies of subtropical forests in the East Asian monsoon region indicates a high carbon uptake in this region, $-362 \pm 39 \, g \cdot C \cdot m^{-2} \cdot yr^{-1}$ [4]. However, for Taiwan, as part of this region, there are only few studies analysing the CO_2 flux in subtropical forests [5, 6]. These studies analyse processes contributing to the CO_2 fluxes, but they rely only on short experimental periods of a few weeks each, so that annual budgets cannot be derived from these data. With this study, we want to contribute to filling this gap. The aim is to quantify the yearly CO_2 uptake of a subtropical mountain cloud forest in Taiwan and put it into context with other published data.

It is well accepted that the eddy covariance technique is most suitable for quantifying CO_2 fluxes for sites with long fetches, flat terrain, and an overall very homogenous surface [7]. It is, however, a challenge to obtain reliable estimates of yearly carbon uptake by photosynthesis in mountainous forests. Ever-present problems with the eddy covariance approach, such as advection fluxes, temporal storage of air masses within the vegetation, and stably stratified boundary layers, are even more difficult to handle in forests and complex terrain than in flat, homogenously vegetated landscapes. Nevertheless, upon the background of climate change and efforts to mitigate the net emission fluxes of greenhouse gases, it seems critical to investigate the CO_2 fluxes of these forests.

This study is based on a three-year (2011–2013) data set of eddy covariance measurements of CO_2 fluxes above a subtropical mountain forest in Taiwan. For the investigation of the yearly CO_2 net ecosystem exchange (NEE), a detailed analysis of the quality of the measured fluxes is performed. The local meteorology is analysed in detail, and the contribution of the ecosystem respiration to NEE is estimated by using a combination of several flux partitioning techniques. Finally, we develop an improved estimate of the annual carbon uptake of the subtropical forest in complex terrain by including the best estimate of ecosystem respiration into the carbon balance evaluation.

2. Materials and Methods

2.1. Site Description. This study is conducted in the Xitou tract of the Experimental Forest, National Taiwan University, located in the Nantou County, Central Taiwan. The whole forest site has an area of 2,400 ha. Today, about 620 ha are covered by natural hardwood forests [8]. The remaining area has been replaced by plantations of mostly coniferous trees which include Japanese cedar (*Cryptomeria japonica*), Taiwania (*Taiwania cryptomerioides*), Taiwan red cypress (*Chamaecyparis formosensis*), China fir (*Cunninghamia lanceolata*), and Luanta fir (*Cunninghamia konishii*) [8]. The climate in the Xitou area is dominated by a relatively dry season from October to April and a wet season from May to September. The mean annual precipitation is 2,635 mm. The mean annual temperature is 16.6°C, the maximum monthly mean is 20.8°C in July, and the minimum is 12.0°C in January. The mean relative humidity at Xitou exceeds 80%.

The experimental tower XT00 (23°39′52.7″N; 120°47′44.5″E, 1267 m above mean sea level) is located on an incline with a mean slope of 9.4° from south-southwest to north-northeast. The incline lies within complex terrain with higher mountains to the south and east (Figure 1). The tower is surrounded by a plantation composed of Japanese cedars planted in the 1950s; the stand extends up to 500 m north and over 500 m south from the experimental tower site. The plantation had a relatively even canopy height with a mean of 28 m in 2013. The mean diameter at breast height is 35 cm. The tree density is 721 trees ha^{-1}, and the mean annual increment of live-tree biomass in the immediate neighbourhood of the tower, as computed from forest inventory, is 560 g·C·m^{-2}·yr^{-1} (mean by linear interpolation of 2011–2017 data). In the northwest, there are patches of Taiwania, while in the southeast, there are patches of Taiwan red cypress. There are also smaller stands of bamboo forest

(north) and broadleaf forest (northeast). Since the change in forest politics in Taiwan in the 1990s, there has been little forest management, and the former plantation mostly grows without disturbance.

The site has a distinct mountain-valley wind regime with valley breeze from north-northeast in the daytime and mountain breeze from south-southeast during the nighttime. Associated with the wind regime, fog occurs frequently at Xitou. It typically forms with the development of the valley breeze in the mornings. Fog is mostly frequent in the afternoons with a peak in frequency during the late afternoon hours. Over the year, the wet spring and summer seasons have the highest fog frequencies [9].

2.2. Tower, Instrumentation and Data Processing. This study used two eddy covariance systems mounted on XT00 at a 40 m (system 1) and a 32 m (system 2) height above ground level (agl), respectively. Each eddy covariance system consists of a 3D ultrasonic anemometer (Campell Scientific USA) and an open path CO_2 and H_2O gas analyser (LICOR 7500, Licor USA). The signals of the ultrasonic anemometers and the CO_2 and H_2O analysers were logged with 10 Hz temporal resolution on a Campbell CR3000 data logger. The tower is also equipped with a CO_2 and H_2O mixing ratio profile system (Licor LI-840; at heights of: 0.1, 0.5, 1, 2, 4, 8, 16, 25, 30, and 35 m agl) and a 2D wind profile system (Gill WS4 ultrasonic anemometers in the heights 6, 18, 25, 32, and 36 m agl.). Further, net radiometers are mounted at 40 m (Kipp & Zonen CNR4), 12 m, and 35 m (Kipp & Zonen NR-Lite). The photosynthetic active radiation (PAR) is measured at 35 m agl (quantum sensor Licor Li-190). Soil heat flux sensors (Huskeflux HFP01SC) are installed at 0.08 m, and soil temperature sensors (thermocouple T-type) are installed at 0.15, 0.2, and 0.5 m below the ground surface.

The postprocessing of the eddy covariance data was done with EddyPro (version 6.0.0, Licor, USA). The raw data were processed in 30-minute block-averaged intervals. The time lag between the wind sensor and the concentration sensor was compensated for, and the WPL term [10] was added. Standard statistical analyses, including spike removal, examination of amplitude resolution test, detection of dropouts and higher-moment statistics in raw data, were performed [11]. For spectral analysis, a high-pass filtering correction according to Moncrieff et al. [12] and a low-pass filtering correction according to Moncrieff et al. [13] were performed. The coordinate system of the 3D ultrasonic anemometer was rotated with the planar fit method [14]. The method was applied to two sectors separately, one for valley breezes and one for mountain breeze situations. The CO_2 profile data were used to derive the storage-term F_S analogously to suggestions given by Aubinet and Chermanne [15].

Gap filling was performed according to Reichstein et al. [16] and by using an online tool which is provided at http://www.bgc-jena.mpg.de/bgi/index.php/Services/REddyProcWeb. The tool includes a friction velocity (u^*) threshold estimation by the moving point test according to Papale et al. [17] and

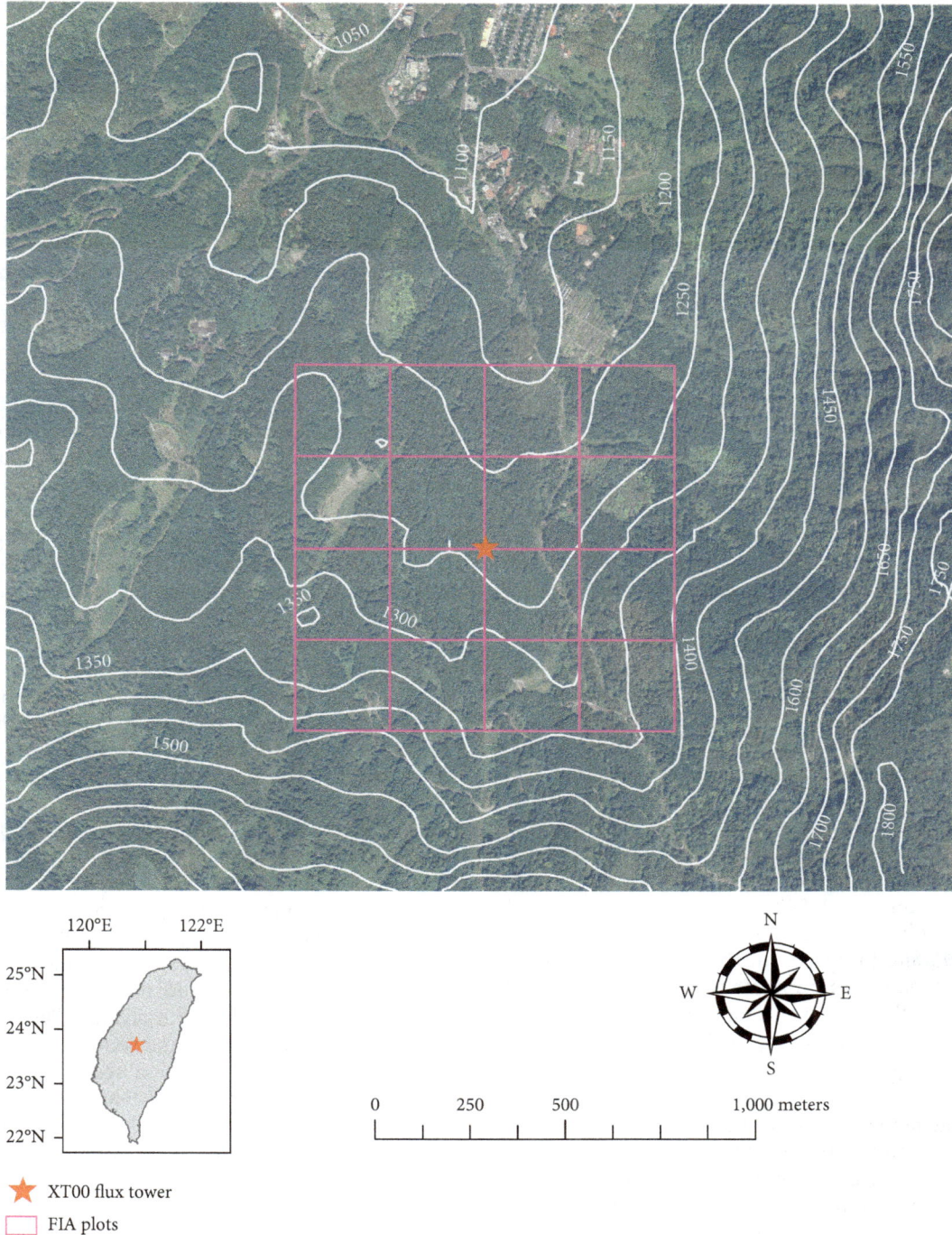

FIGURE 1: Map of the Xitou area with the location of the eddy covariance tower and the Forest Inventory and Analysis (FIA) plot design (purple squares; plot size 250 × 250 meters).

a night-based flux partitioning algorithm according to Reichstein et al. [16].

2.3. NEE Calculation.

The net ecosystem exchange (NEE), as measured with the eddy covariance method, is based on the assumption that the exchange of CO_2 by an ecosystem can be estimated by the sum of the vertical eddy covariance flux (NEE_{EC}) at a specific height and the storage (F_S) between the surface and this height [18]:

$$NEE = NEE_{EC} + F_S. \tag{1}$$

Previous studies found that, during nighttime and under weak mixing conditions, advection cannot be neglected, specifically when the terrain is not flat but has a slope, as in our case [17, 19]. Therefore, the nighttime flux of CO_2 is likely underestimated as long as the nighttime advection is ignored. The advective term (F_{adv}) cannot be quantified directly with a commonly used eddy covariance system, and there is no standard method to evaluate it properly. We address this issue in a separate paragraph below.

2.4. Quality Assurance. Because of the stand's historic background as a plantation, the trees have a similar height, and the area around the tower is covered with the same species almost throughout. The canopy is closed. From this point of view, the forest is very homogenous and thus the terrain is well suited for application of the eddy covariance technique. The terrain exhibits an even slope with only minor variations in the wind direction both during the nights and during the days and no nearby source of flow disturbance.

For quality classification, an overall quality flag system combining a steady state test and integral turbulence characteristics was applied [20]. Data with low quality were excluded from further analysis. Due to the frequent presence of fog and its obstructive effect on the open-path CO_2 concentration measurement, it was additionally necessary to set an analyser-specific threshold for the AGC value (cleanness indicator of the open-path analyser window) and to exclude all data with an AGC value above this threshold. In the end, it was necessary to exclude the highest and lowest flux results (lower bound quantile 0.3%; upper bound quantile of 99.7%) from further analysis, because disproportionally large flux values (negative and positive) remained in the data. After QA/QC and u^* filtering for $NEE_{EC} + F_S$, 56% of available data for system 1 (70% before u^*-filtering) and 65% of available data for system 2 (79% before u^*-filtering) could be used further. For NEE_{EC} analysis only (ignoring F_s), 54% of the data from system 1 (66% before u^*-filtering) and 61% of the data from system 2 (75% before u^*-filtering) were suitable for further analysis.

Three more tools were employed to review the quality of flux data. First, the energy balance of the surface was checked. The energy balance (E_B) is defined as

$$E_B = R_N - B_S - H - LE,\qquad (2)$$

where R_N is the net radiation, B_S the soil heat flux, H is the sensible heat flux, and LE the latent heat flux, which are both as estimated with the eddy covariance method. Generally, the energy balance is not closed (i.e., zero) in forest systems when H and LE are achieved with the eddy covariance method because atmospheric phenomena that cannot be captured with the eddy covariance technique still contribute to a nonperfect closure of the energy balance [18]. A closure of approximately 80% (see method 2 below) has been found for many sites [21, 22]. We conclude that the closure of the energy balance is a helpful indicator of the quality and plausibility of the measured fluxes. If a closure of 80% or more is reached, the measured fluxes are likely of good quality.

Following suggestions given by Wilson et al. [21], we use two methods to evaluate the energy balance closure. The first method includes both linear regression between the

half-hourly data of the dependent flux variables H + LE and the independent estimate of $R_N - B_S$ [21]. The second method is the energy balance ratio over all data for each system: the sum of H + LE divided by the sum of $R_N - B_S$. Table 1 shows that, for each method, the degrees of closure between system 1 and system 2 are similar. However, the linear regression method results in a relatively low degree of closure, as compared to 22 FLUXNET sites [21]. Conversely, the energy balance ratio (method 2) shows high degrees of closure: 92% for system 2 and 95% for system 1.

While the energy balance ratio looks promising, it has been shown to have both strengths and weaknesses. McGloina et al. [23] suggested that the energy balance ratio is probably the better indicator for overall degree of energy balance closure at a particular site, because during low closure conditions (e.g., stable stratification), the terms of the E_B function (equation (2)) are usually very low; while the low values of H, LE and R_N have only a small influence on the regression slope, their influence on the energy balance ratio is noticeable. On the other hand, the weakness of the energy balance ratio lies in the fact that biases are prone to be overlooked [21]. Because of advection, as mentioned above, and because of the likely decoupling of turbulence above and below the canopy, as discussed further below, the nighttime fluxes of sensible and latent heat (H and LE) are likely not fully captured by eddy covariance.

Therefore, we presume that the results of the linear regression (method 1) are more reliable and that although the degree of energy balance closure at Xitou is relatively low, it is still within the range of estimates of other sites.

Because the investigated site is located at an inclined surface, there is a difference between the radiation as measured with horizontally aligned radiometers and the amount of radiation reaching the inclined surface. This could have a nonnegligible impact on E_B [23]. While Olmo et al. [24] proposed a correction algorithm for incoming short wave radiation under such conditions, we presume that the correction is hard to realize properly due to the complexity of the terrain (north-facing valley surrounded by higher mountains to the east and south) and due to the high contribution of diffuse radiation resulting from the high frequency of fog. We therefore doubt that such a correction algorithm, which was developed for general applications, will perform well under these circumstances. Further, a correction would lead to a reduction of R_N because the slope is facing north, away from the sun, and the correction would lead to a better agreement between $R_N - B_S$ and H + LE. For these reasons, we refrain from using this correction, and this approach is a conservative one.

The second approach we used to review the quality of flux data (the first was checking the energy balance of the surface) was assessing the relationship between photosynthetic active radiation (PAR) and $NEE_{EC} + F_S$, which is appropriate for examining site-specific data quality. The response of $NEE_{EC} + F_S$ to PAR is very similar for system 1 and system 2 (Figure 2(a)). At no-light conditions, there is no photosynthesis, resulting in a positive $NEE_{EC} + F_S$ flux. With increasing PAR, the ecosystem takes up more and more CO_2 until a maximum uptake is reached. The daily

TABLE 1: Energy balance closure. Method 1: linear regression between $R_N - B_S$ and $H + LE$. Method 2: energy balance ratio over all data for each system sum of $H + LE$ divided by $R_N - B_S$.

| | Linear regression (method 1) | | | Energy balance ratio (method 2) |
	Slope	Intercept	r^2	%
System 1	0.67	25	0.85	95
System 2	0.68	19	0.89	92

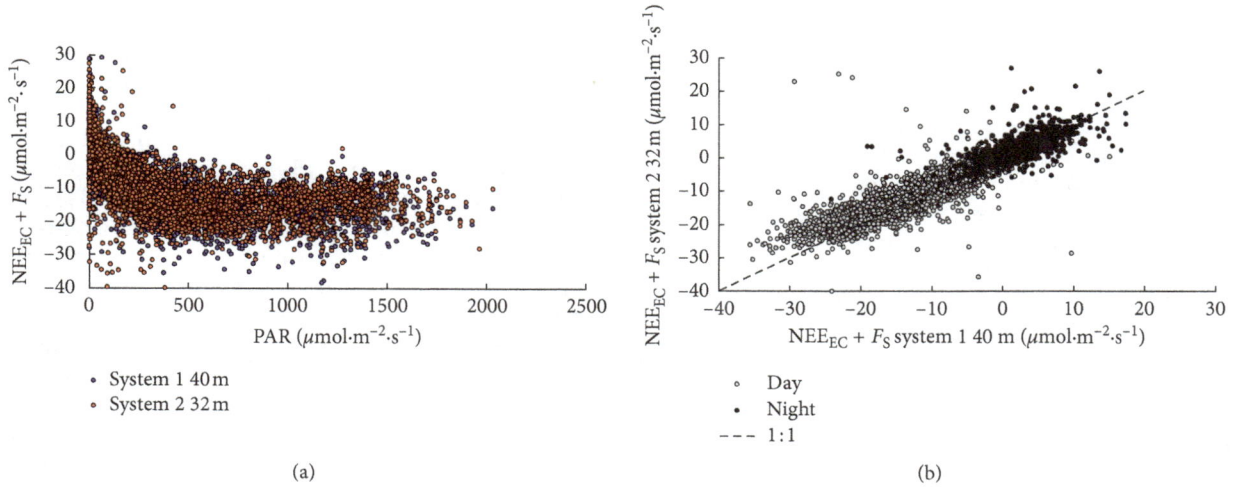

FIGURE 2: (a) Scatter plot of $NEE_{EC} + F_S$ against PAR; blue symbols show data from system 1 at 40 m agl.; red symbols refer to system 2 at 32 m agl. (b) Scatter plot of $NEE_{EC} + F_S$ system 1 against $NEE_{EC} + F_S$ system 2; black is nighttime data; light grey data points represent daytime data; the black dashed line refers to the 1:1 ratio.

mean maximum uptake of all available $NEE_{EC} + F_S$ data is reached at 11:00 a.m. at both systems ($-17\,\mu mol \cdot m^{-2} \cdot s^{-1}$ for system 1 and $-16\,\mu mol \cdot m^{-2} \cdot s^{-1}$ for system 2). Overall, the relationship between $NEE_{EC} + F_S$ shows meaningful flux estimates, except for a few outliers.

In a third approach, we tested the hypothesis that a constant flux layer exists by comparing the results of the two eddy covariance systems at the two heights. The similarity of $NEE_{EC} + F_S$ data for the two systems (which are at different heights) indicates the existence of a constant flux layer. In our data, this is confirmed by the strong correlation between the two data subsets (Figure 2(b)). Note that the relationship between the two systems is stronger during the daytime ($r^2 = 0.80$) than during nighttime ($r^2 = 0.48$).

Overall, all methods employed indicate that the quality of flux data is acceptable even though the site is not located in an ideal environment.

2.5. Advection. It is well accepted that advection is the most common and most important source of errors in eddy covariance applications at night [25]. For mountain and forested sites, the advection issue is particularly important. Especially at nighttime, when turbulence is only weakly developed, phenomena like drainage flow, gravity waves, or even intermittent turbulence may occur and dominate the surface exchange flux [25]. As these phenomena cannot be captured by the eddy covariance measurement, systematic errors may arise when estimating NEE from it. There are different methods for taking into account the influence of weak mixing and the advective flux (F_{adv}) contribution. The most common method is the u^*-filtering approach. Here, a u^* threshold is estimated for the detection of periods when nighttime NEE ($NEE_{EC} + F_S$) becomes sensitive to u^* [17, 19]. Because nighttime flux should be independent of u^*, any dependence should arise from an artefact. In consequence, the periods with u^* values below the threshold, i.e., when NEE is sensitive to u^*, are filtered out. Nonetheless, whether or not the u^*-filtering approach is suitable as a standard quality assurance technique is still controversially discussed [22, 26].

An alternative to the u^* method is given by van Gorsel et al. [27]. They found for different sites that peaks of $NEE_{EC} + F_S$ (defined as R_{max}) occurred during the early evenings and before the full development of the advective processes that lead to an underestimation of the nighttime CO_2 fluxes. They used R_{max} to derive a temperature response function for respiration and eventually an estimate of the nighttime respiration. The results of this method fit well with those of other methods (like independent chamber measurements), and van Gorsel et al. showed that it works for forest sites as well. This method is applicable when storage data is available and if the early-evening peak in $NEE_{EC} + F_S$ occurs.

At the Xitou site, the early-evening peak occurs only during 7 out of 15 months with storage data. During the winter months (except for November 2011), no peak occurred. As the wind direction changes regularly from valley breeze to mountain breeze in the early evenings, these respective transition periods are associated with low wind speeds and low u^*. Further, the quality control requirements of the van Gorsel method reduce the number of available data points at the Xitou site, leading to an unusually high, yet not well documented, peak in September 2013. Overall, our evaluation is in line with the presumption of van Gorsel et al. that complex flow patterns limit the applicability of their approach [27]. For these reasons and for the sheer absence of storage data for large portions of our data set, the van Gorsel method is not useful in this case to estimate yearly carbon balances. In the supplement material, the daily course of each available month of $NEE_{EC} + F_S$ data for system 2 is shown together with the results of the later-described estimate of day-based respiration.

There are also studies which attempt to solve the nighttime advection problem by measuring the advection flux directly. With great effort, Aubinet et al. [28] were able to directly measure advection, but they came to the conclusion that it does not help to solve the nighttime problem. The main problem they identified was that the experimental approach to measure the advection flux was insufficient in terms of spatial representativeness. This led to systematic errors that probably cannot be solved at all [28]. Thomas et al. [26] designed a method to measure the submeso motions below the canopy and to compensate for the advective losses in this way. This method includes eddy covariance measurements below the canopy; however, these measurements are not available at the Xitou site for the experimental period.

To test the data sets for the presence and potential importance of advection fluxes, we use, as a first step, the u^*- filtering method as employed in the gap-filling tool with storage-corrected eddy flux data ($NEE_{EC} + F_S$). The storage term F_S can influence the estimation of the u^* threshold [17]; therefore, it is important to perform the storage correction before u^* threshold estimation. The gap-filling tool identified u^* thresholds between 0.08 and 0.14 m·s^{-1} (0.1 and 0.16 m·s^{-1} without F_S term) for system 1 and system 2, respectively. If a u^* value was below the specific threshold, the corresponding flux value was marked as gap. Overall, 14% of the flux data for each eddy system was marked as a gap because of low u^*. At night, the proportion of data filtered out through the u^* threshold exceeds 25%.

To test whether the u^* filtering approach worked and periods affected by advection were fully excluded, we used two different approaches for partitioning the gap-filled, u^*-filtered $NEE_{EC} + F_S$ data to evaluate ecosystem respiration R_E. First, we used the night-based flux partitioning approach included in the employed gap-filling tool [16]. This sets the nighttime NEE data as R_E and extrapolates this data with the exponential regression model by Lloyd and Taylor [29] to the daytime. In the following, we refer to this approach as night-based respiration R_{E_NB}. Detailed information is given in [16, 30].

Second, we combined one of the day-based approaches (DB all no VPD—all parameters estimated using daytime data, no consideration of vapor pressure deficit (VPD)) of Lasslop et al. [31] with the technique used by Lee et al. [32].

The daytime approach is based on the relationship between the photosynthetically active radiation and the estimated $NEE_{EC} + F_S$. In this model, NEE is described by a combination of the hyperbolic light-response curve [33] and the Lloyd and Taylor model:

$$NEE = \frac{\alpha\beta PAR}{\alpha PAR + \beta} + rb \, \exp\left(E_0\left(\frac{1}{T_{ref} - T_0} - \frac{1}{T - T_0} \right) \right).$$

(3)

The daytime model was fed only with radiation data with PAR > $10\,\mu mol \cdot m^{-2} \cdot s^{-1}$. T_0 and T_{ref} were fixed to the same values as they were in the night-based approach. For T, the air or soil temperatures need to be known. This model was used with a 15-day moving window with the actual day in the middle of the window [32]. The coefficients α, β, rb, and E_0 (Equation (3)) were estimated with a nonlinear curve-fitting method (trust-region-reflective algorithm). The model was quality controlled analogously to suggestions given by Lasslop et al. [31]. The coefficients rb and E_0, as estimated with daytime data, were then used to calculate the day-based respiration R_{E_DB} with the second term on the right-hand side of Equation (3). If the site is not affected or is only a little affected by advection and if the u^* filtering is working properly, the two models should yield similar results.

It is important which temperature (soil or air) is employed for the respective models. Air temperature (T_{air}) is used for both models because the sums of squares between the employed modeling function T_{air} with $NEE_{EC} + F_S$ were better than those between soil temperature and $NEE_{EC} + F_S$. In addition, the physical distance between the anemometer and the soil is large, and the canopy is located between these levels, meaning that a strong correlation between the soil temperature and $NEE_{EC} + F_S$ is not likely to occur.

3. Results and Discussion

Profiles of meteorological parameters and CO_2 concentrations at the tower provide insight into the exchange processes between the air masses above the forest canopy, within the trunk space, and at the soil-surface interface. In Figure 3, median wind speeds and wind directions are shown for different heights within the forest and above the canopy. The median wind speed was generally higher at all heights during nighttime than during the days. There was little diurnal variation of the median wind speed only at the canopy height of 25 m agl. The median wind speed exhibited a minimum in the upper heights during the day, when the wind direction changed in the morning and afternoon. During nighttime, the median wind speed within the denser canopy at 18 m agl and at 25 m agl was lower than in the more open trunk space at 6 m agl. In daytime, when the valley wind was established, there was almost no difference in the median wind speeds at the levels within and below the canopy (25 m, 18 m, and 6 m agl).

The wind direction changed both above and below the canopy due to the valley wind regime. During the mornings, a valley wind developed. The respective change of the

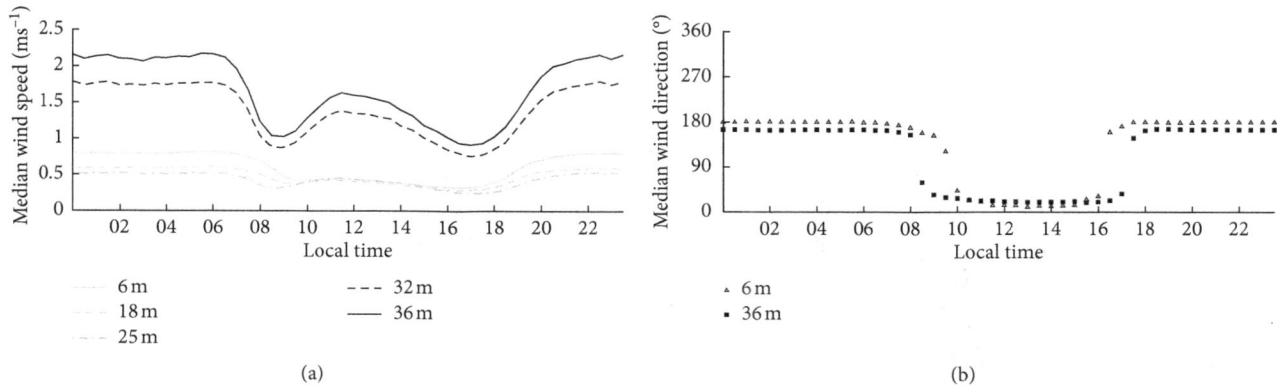

FIGURE 3: Median wind speed and median wind direction at different heights.

median wind direction occurred first above the canopy (36 m agl), and then the change happened about one hour later the canopy (6 m agl). Interestingly, the change from the valley wind (northerly directions) to the mountain wind (southerly directions) in the evenings happened about an hour earlier under the canopy than above it. The temperature profile (not shown) indicates that, in the morning, the air above the canopy heats up earlier than under the canopy, which supports the earlier formation of the valley breeze above the canopy. In the evening, the air under the canopy cools down before the air above, which leads to an earlier downward flow. Note that during nighttime, the wind speed below and above the canopy were higher than during daytime, whereas the wind speed within the canopy differed little between day and night. This strongly indicates that vertical exchange through the canopy is inhibited during the night.

For further insights, we analyzed CO_2 profile data. The CO_2 profile data are available for 2011 (September to December), 2012 (August to December, 84% data), and 2013 (full year). The median CO_2 mixing ratio (Figure 4) shows a well-mixed situation during the days with similar CO_2 mixing ratios throughout almost the entire vertical profile. During nighttime, CO_2 apparently accumulates under the canopy and near the ground.

Combining the wind speed, wind direction, and CO_2 profiles, there are two regimes, which indicate the presence of two separate situations regarding the NEE estimate. In the daytime, the radiation-induced convective processes lead to a mixing of air masses between the layer above the canopy and the trunk space. The air between the soil surface and the eddy systems is well mixed then. During the second situation at night, when the wind direction has changed (or when it is just changing) and a stably stratified boundary layer is being established, there is probably little or no interaction between the flux above the canopy and below. Under these conditions, the canopy builds a barrier that cannot be overcome by mechanically induced turbulence. As a result, the nighttime respiration flux under the canopy is not detected by the eddy covariance system above the canopy. During these times, the respirated CO_2 is not only stored under the canopy (Figure 4) but also transported away by drainage flow, thus bypassing the eddy systems above. We conclude that an unknown amount of respiration flux must be missing in the NEE if it is calculated as u^* corrected $NEE_{EC} + F_S$ during nighttime.

3.1. Comparing R_{E_DB} and R_{E_NB}. Table 2 and Figure 5 show the results of the night-based and the day-based models to estimate ecosystem respiration from $NEE_{EC} + F_S$ data. For $NEE_{EC} + F_S$ data, this comparison could be computed only for a total of 10 months for system 1 and 15 months for system 2 in the period 2011–2013, because profile data are not available for other periods.

Figure 5(b) shows, as an example, the mean daily course of R_E estimates for system 2. The overall mean of the night-based respiration model (R_{E_NB}) for system 2 is $2.2\,\mu mol \cdot m^{-2} \cdot s^{-1}$ ($2.4\,\mu mol \cdot m^{-2} \cdot s^{-1}$ system 1, not shown in detail). For the day-based partitioning (R_{E_DB}), it is $4.6\,\mu mol \cdot m^{-2} \cdot s^{-1}$ ($4.4\,\mu mol \cdot m^{-2} \cdot s^{-1}$ system 1). The Figure 5(a) shows the correlations between air temperature and ecosystem respiration. The minimum of the daily mean air temperature was 6.0°C. The corresponding respiration for system 2 is $1.0\,\mu mol \cdot m^{-2} \cdot s^{-1}$ R_{E_NB} ($0.87\,\mu mol \cdot m^{-2} \cdot s^{-1}$ system 1) and $3.4\,\mu mol \cdot m^{-2} \cdot s^{-1}$ R_{E_DB} ($3.2\,\mu mol \cdot m^{-2} \cdot s^{-1}$ system 1). The maximum of the daily mean air temperature in the analyzed period was 22°C. The corresponding respiration values for system 2 are $3.0\,\mu mol \cdot m^{-2} \cdot s^{-1}$ R_{E_NB} ($3.6\,\mu mol \cdot m^{-2} \cdot s^{-1}$ system 1) and $6.9\,\mu mol \cdot m^{-2} \cdot s^{-1}$ R_{E_DB} ($6.6\,\mu mol \cdot m^{-2} \cdot s^{-1}$ system 1). Table 2 shows the sums of the various R_E model results. The average monthly sums are similar for system 1 and system 2. The estimate of R_{E_DB} is about twice as high as that for R_{E_NB}. This also holds for the hourly median values and the monthly sums.

If there is no or little advection, the results of these models should be similar. Because of the huge differences in the R_E estimations and the described meteorological situation, we presume that, at Xitou, NEE cannot be estimated by $NEE_{EC} + F_S$ only. Our assumption is that during nighttime, the eddy covariance system above the forest and the estimation of the storage cannot detect the full respiration of the forest. Further, the u^* filtering alone does not avoid incorrect nighttime data.

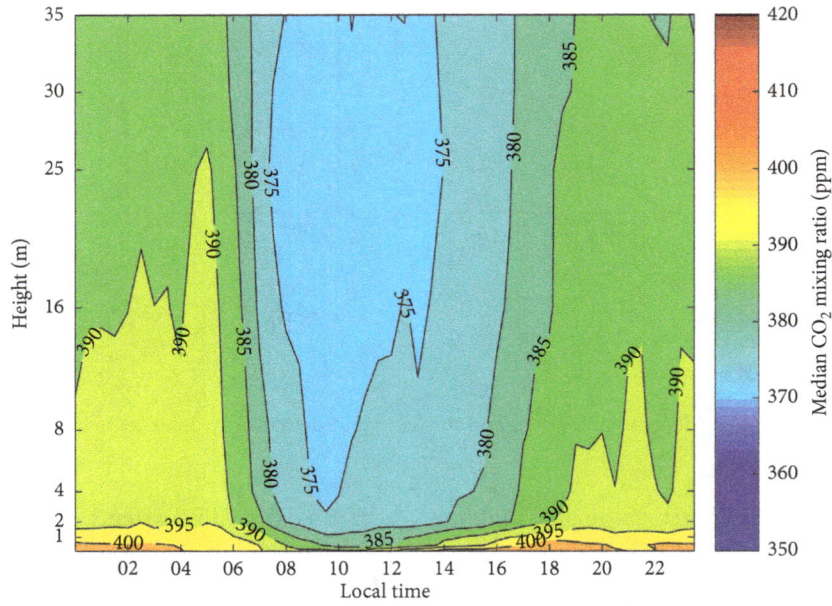

FIGURE 4: Median CO_2 mixing ratios in ppm at different heights.

TABLE 2: Ecosystem respiration (R_E) from the nighttime and the daytime partitioning models R_{E_NB} and R_{E_DB}, respectively. Monthly sums represent times when complete gap-filled data sets are available (available months indicated in brackets).

g·C·m^{-2} per time unit	System 1		System 2	
	R_{E_NB}	R_{E_DB}	R_{E_NB}	R_{E_DB}
2011				
System 1 (10, 11)	144	303	126	289
System 2 (10, 11)				
2012				
System 1 (-)	—	—	233	614
System 2 (9–12)				
2013				
System 1 (2–4; 8–12)	562	1,157	652	1,249
System 2 (1–4; 8–12)				
Average monthly sums	71	146	67	144

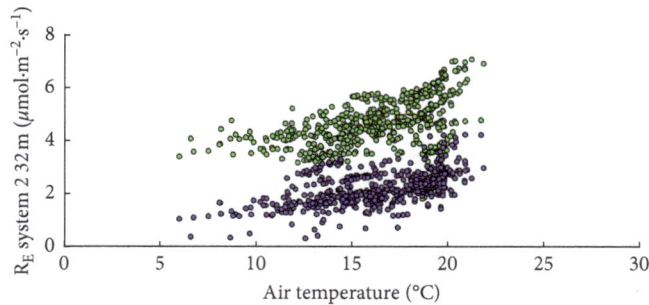

(a)

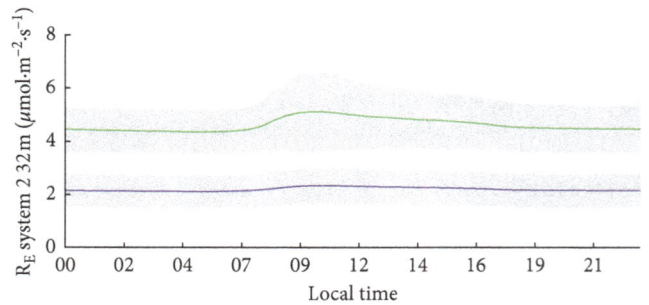

(b)

FIGURE 5: (a) Scatter plot of daily mean R_E system 2 against daily mean air temperature; green symbols show data for the daytime partitioning model R_{E_DB}; blue symbols show data for the nighttime partitioning model R_{E_NB}. (b) Median daily course of R_{E_DB} and R_{E_NB} for system 2; grey areas indicate one standard deviation.

3.2. Advanced Methodology to Estimate NEE. In the quality assurance section above, we showed that this site is, in principle, suitable for the application of the eddy covariance technique even though it is located in not ideal heterogenic topography. It was also shown that advection cannot be neglected. In order to develop an advanced estimate of the NEE, we combine the measured NEE_{EC} and the results of the day-based partitioning approach.

However, it is necessary to evaluate the influence of the storage term. We compare the storage-corrected and not-storage-corrected gap-filled fluxes with the u^*-filtered NEE. A linear regression (not shown in detail) shows a strong relationship between NEE_{EC} and $NEE_{EC} + F_S$ with $r^2 = 0.98$ for system 1 and $r^2 = 0.97$ for system 2. The summed-up flux of system 1 shows almost identical results for storage-corrected and not-storage-corrected fluxes. For system 2, the summed-up storage-corrected fluxes are 1% lower than the fluxes without storage correction. Note that, for system 1, there are no respective data available for 2012. In addition, the storage data of 2011 and 2012 are only available for the last months of these years. It cannot be excluded that for a full data set, the storage term might have a bigger influence on the monthly and yearly sum of NEE. Nevertheless, because of missing profile measurements and because the contribution of storage is limited, we ignore this term during further analysis in order to achieve an estimate for the whole NEE_{EC} data set.

Based on the presumption that the nighttime NEE_{EC} data need to be corrected, we arrive at NEE_X, which is an advanced estimate of NEE at the Xitou site. NEE_X for every half hour is defined as

$$NEE_X = \begin{cases} NEE_{EC}, & \text{not mountain breeze} \\ NEE_{DB}, & \text{mountain breeze} \end{cases},$$ (4)

$$NEE_{DB} = NEE_{EC} - R_{E\_NB} + R_{E\_DB}.$$

For a mountain breeze (wind directions 135°–225°), NEE is defined as NEE_{DB}. During the changes of the wind regime and during valley winds (wind directions 315°–45°), NEE is defined as NEE_{EC}. For gap-filling purposes, the missing wind direction data are defined as the weekly or monthly means for each half hour. Missing data were primarily filled with the weekly means of the missing half hours; only in cases when no weekly mean data points were available for gap filling were the gaps filled with the monthly means.

3.3. Results of NEE Estimate. Table 3 lists the results of the nighttime and daytime partitioning approaches and the NEE estimates including the daytime respiration. Negative NEE values indicate carbon uptake.

The results of R_{E_DB} shown in Table 3 are (like in Table 2) nearly twice as high as the results for R_{E_NB}. The results of the R_{E_NB} estimate, with a range from 938 to 1,038 $g \cdot C \cdot m^{-2} \cdot y^{-1}$, seem very low for this kind of ecosystem, whereas the results of R_{E_DB} (1,754 to 1,961 $g \cdot C \cdot m^{-2} \cdot y^{-1}$) are more reasonable. As a consequence of low R_{E_NB}, the estimate of NEE_{EC} is highly negative ranging from −1,027 to −1,101 $g \cdot C \cdot m^{-2} \cdot y^{-1}$. Based on the estimate given in Equation (4), the magnitude

of NEE is reduced to a range from −434 to −652 $g \cdot C \cdot m^{-2} \cdot y^{-1}$ (NEE_X) with an average of −561 $g \cdot C \cdot m^{-2} \cdot y^{-1}$.

Table 4 presents results of NEE estimates at forest sites comparable to ours. For a subtropical mountain forest with frequent occurrence of fog like at Xitou, central Taiwan, there are only a few studies representing a similar ecosystem of forest sites. Therefore, a number of studies from Japan with a more temperate climate are included for comparison (Table 4).

Table 4 lists a number of comparable studies with NEE and R_E estimates and results of the increase of live-tree biomass evaluated with independent forest inventory methods. The largest NEE estimate found is by Tan et al. [34] with about −900 $g \cdot C \cdot m^{-2} \cdot yr^{-1}$. Tan et al. [34] identified two potential drivers for the large carbon uptake found at their study site. First, the low temperature at the high-altitude site may have reduced the respiration, and secondly, the high proportion of indirect, diffuse solar radiation due to the extended presence of clouds may have enhanced photosynthesis. In the other studies, the NEE ranges between −330 and −630 $g \cdot C \cdot m^{-2} \cdot yr^{-1}$. The respective respiration rates R_E range between 991 and 3061 $g \cdot C \cdot m^{-2} \cdot y^{-1}$.

If we compare our results to those of these other studies, the estimates of NEE_X and R_{E_DB} (−561 and 1,828 $g \cdot C \cdot m^{-2} \cdot yr^{-1}$, respectively) seem reasonable. It is evident that the more direct estimates NEE_{EC} and R_{E_NB} (−1,060 and 983 $g \cdot C \cdot m^{-2} \cdot yr^{-1}$) provide no valid estimates of the forest-atmosphere exchange at the Xitou site. Further, the estimate for the mean increment in live-tree biomass in the direct vicinity of the flux tower (560 $g \cdot C \cdot m^{-2} \cdot yr^{-1}$; mean 2011–2017; interpolated data) and the estimate of Cheng et al. [38] (265 $g \cdot C \cdot m^{-2} \cdot yr^{-1}$) for the whole Xitou area are in rather good agreement to the NEE_X estimates. Yu et al. [39] estimated an annual soil respiration of 1,003 $g \cdot C \cdot m^{-2} \cdot yr^{-1}$ derived from monthly data at Xitou in 2012. They used automated chamber systems, which operate independently of the eddy covariance method. A comparison of this estimate of soil respiration only with R_{E_NB} of the entire ecosystem (938 $g \cdot C \cdot m^{-2} \cdot yr^{-1}$ in 2012) again leads to the presumption that R_{E_NB} does not fully capture the ecosystem respiration. R_{E_NB} should be larger than the soil respiration because it also includes the respiration of the above-ground vegetation. It is reasonable to assume that this discrepancy applies to years other than 2012 as well. The listed data support our understanding that the NEE cannot be estimated with u^*-corrected NEE_{EC} data at the Xitou site. Due to the fact that the contribution of the storage term F_S is limited, NEE cannot be estimated by u^*-corrected $NEE_{EC} + F_S$ either. In essence, NEE_X is the best estimate for the net ecosystem exchange flux (NEE) at the Xitou site.

4. Conclusions

This study analyzed CO_2 flux data of a subtropical mountain forest site in central Taiwan from 2011 through 2013. The analysis of the local wind regime and a detailed study of NEE and R_E estimates, based on various partitioning models fed with u^* and storage-corrected flux data, provided evidence

TABLE 3: Yearly ecosystem respiration (R_E) from the nighttime (gap-filling tool) and daytime (Equation (3)) partitioning models; net ecosystem exchange defined as NEE_{EC} after QA/QC, gap filling and u^* filtering; net ecosystem exchange including NEE_{DB} for mountain breeze (NEE_X based on equation (4)).

$g \cdot C \cdot m^{-2} \cdot yr^{-1}$	System 1				System 2			
	R_{E_NB}	R_{E_DB}	NEE_{EC}	NEE_X	R_{E_NB}	R_{E_DB}	NEE_{EC}	NEE_X
2011	972	1,754	−1,102	−652	—	—	—	—
2012	—	—	—	—	938	1,961	−1,027	−434
2013	—	—	—	—	1,038	1,769	−1,051	−598

TABLE 4: Yearly sums of NEE and R_E from this study and other studies. Estimates of yearly live-tree biomass with forest inventory methods (positive values represent carbon sequestration) around the flux tower and in the whole Xitou area. Numbers in parentheses are single standard deviations.

Reference	NEE $g \cdot C \cdot m^{-2} \cdot yr^{-1}$	R_E
This study: mean 2011–2013	NEE_{EC} = −1,060 (±38) NEE_X = −561 (±114)	R_{E_NB} = 983 (±51) R_{E_DB} = 1,828 (±115)
Tan et al. [34] quantify the carbon uptake of a 300-year-old subtropical evergreen broadleaved forest	~−900	—
Yu et al. [4], East Asian monsoon region	−362 (±39)	—
Takanashi et al. [35], Japanese cedar forest in Japan 2001 and 2002 [35]	−477, −480	991, 1,129
Saitoh et al. [36], mostly Japanese cedar and Japanese cypress forest in Japan; 2006 and 2007	−330, −350	1,740, 1980
Kosugi et al. [37] analyse 7 years of data; 50-year-old forest consisting Japanese cypress (method 1-> night-based gap-filling; method 2-> day-based gap filling)	m1 = −490 m2 = −630	m1 = 1,555 m2 = 1,554
Luyssaert et al. [3], combining 29 tropical humid evergreen forest sites	−403 (±102)	3061 (±56)
	Yearly increase in live-tree biomass ($g \cdot C \cdot m^{-2} \cdot yr^{-1}$)	
Forest biomass survey in Xitou		
Forest inventroy and analysis (FIA) plot design; mean of 4 plots from 2011–2017 linear interpolated	560	—
Cheng et al. [38], forest inventory methods, complete analysis of Xitou area	265	—

of a large underestimate of the positive nighttime fluxes. Advection is a strong player at this site. Below-canopy drainage flow leads to a net downhill transport of CO_2 from nighttime respiration. This process cannot be detected by the eddy covariance systems above the canopy because there is an effective nighttime decoupling of air masses below the canopy from those above. When using a day-based partitioning model for the whole data set of NEE_{EC} data, we arrived at an improved estimate of the carbon uptake of the forest (Equation (4)). According to our estimate, NEE_X is −561 (±114) $g \cdot C \cdot m^{-2} \cdot yr^{-1}$ (±one standard deviation). The mean increment of live-tree biomass in the direct neighbourhood of the tower was estimated to be 560 $g \cdot C \cdot m^{-2} \cdot yr^{-1}$ (mean 2011–2017; interpolated data). Consequently, the growth of above-ground live-tree biomass is the main factor for carbon taken up by the forest. We trust that this estimate of the yearly carbon uptake is the best that can be achieved with the available data sets. The use of several flux partitioning models for the purpose of quality assurance proved to be a helpful approach to quantify the yearly flux estimates at the mountain forest site under study. Nonetheless, the remaining uncertainty is hard to quantify. It is planned to compare this approach with other methods like those proposed by Thomas et al [26], which require additional eddy covariance measurements within the trunk space.

Conflicts of Interest

The authors declare that there are no conflicts of interest regarding the publication of this paper.

Acknowledgments

We thank Celeste Brennecka for language editing. The provision of travel funds from the German Academic Exchange Service (DAAD) through funds of the German Federal Ministry of Education and Research (BMBF), the Taiwanese funds from the Ministry of Science and Technology (MOST) under Grant 105-2911-I-002-529-MY2, and

Experimental Forest, National Taiwan University, are all gratefully acknowledged.

References

[1] G. Bonan, "Forests and climate change: forcings, feedbacks, and the climate benefits of forests," *Science*, vol. 320, no. 5882, pp. 1444–1449, 2008.

[2] FLUXNET, *FLUXNET—Global Network of Micrometeorological Tower Sites that Use Eddy Covariance*, 2018, https://fluxnet.ornl.gov/.

[3] S. Luyssaert, I. Inglima, M. Jung et al., "CO$_2$ balance of boreal, temperate, and tropical forests derived from a global database," *Global Change Biology*, vol. 13, no. 12, pp. 2509–2537, 2007.

[4] G. Yu, Z. Chen, S. Piao et al., "High carbon dioxide uptake by subtropical forest ecosystems in the East Asian monsoon region," *Proceedings of the National Academy of Sciences*, vol. 111, no. 13, pp. 4910–4915, 2014.

[5] K. Mildenberger, E. Beiderwieden, Y.-J. Hsia, and O. Klemm, "CO$_2$ and water vapor fluxes above a subtropical mountain cloud forest-the effect of light conditions and fog," *Agricultural and Forest Meteorology*, vol. 149, no. 10, pp. 1730–1736, 2009.

[6] T. S. El-Madany, H. F. Duarte, D. J. Durden et al., "Low-level jets and above-canopy drainage as causes of turbulent exchange in the nocturnal boundary layer," *Biogeoscience*, vol. 11, no. 16, pp. 4507–4519, 2014.

[7] D. Baldocchi, "Assessing the eddy covariance technique for evaluating carbon dioxide exchange rates of ecosystems: past, present and future," *Global Change Biology*, vol. 9, no. 4, pp. 479–492, 2003.

[8] Y.-L. Liang, T.-C. Lin, J.-L. Hwong, N.-H. Lin, and C.-P. Wang, "Fog and precipitation chemistry at a mid-land forest in central Taiwan (in Chinese with english abstract)," *Journal of Environment Quality*, vol. 38, no. 2, pp. 627–636, 2009.

[9] T. H. Wey, Y. J. Lai, C. S. Chang et al., "Preliminary studies on fog characteristics at Xitou region of central Taiwan," *Journal of the Experimental Forest of National Taiwan University*, vol. 25, pp. 149–160, 2011.

[10] E. K. Webb, G. I. Pearman, and R. Leuning, "Correction of flux measurements for density effects due to heat and water vapor transfer," *Quarterly Journal of the Royal Meteorological Society*, vol. 106, no. 447, pp. 85–100, 1980.

[11] D. Vickers and L. Mahrt, "Quality control and flux sampling problems for tower and aircraft data," *Journal of Atmospheric and Oceanic Technology*, vol. 14, no. 3, pp. 512–526, 1997.

[12] J. B. Moncrieff, R. Clement, J. Finnigan, and T. Meyers, "Averaging, detrending and filtering of eddy covariance time series," in *Handbook of Micrometeorology: A Guide for Surface Flux Measurements*, pp. 7–13, Springer Science & Business Media, Berlin, Germany, 2004.

[13] J. Moncrieff, J. Massheder, H. deBruin et al., "A system to measure surface fluxes of momentum, sensible heat, water vapour and carbon dioxide," *Journal of Hydrology*, vol. 188-189, pp. 589–611, 1997.

[14] J. Wilczak, S. Oncley, and S. Stage, "Sonic anemometer tilt correction algorithms," *Boundary-Layer Meteorology*, vol. 99, no. 1, pp. 127–150, 2001.

[15] M. Aubinet and B. Chermanne, "Long term carbon dioxide exchange above a mixed forest in Belgium," *Agricultural and Forest Meteorology*, vol. 108, no. 4, pp. 293–315, 2001.

[16] M. Reichstein, E. Falge, and D. Baldocchi, "On the separation of net ecosystem exchange into assimilation and ecosystem respiration: review and improved algorithm," *Global Change Biology*, vol. 11, no. 9, pp. 1424–1439, 2005.

[17] D. Papale, M. Reichstein, M. Aubinet et al., "Towards a standardized processing of net ecosystem exchange measured with eddy covariance technique: algorithms and uncertainty estimation," *Biogeoscience*, vol. 3, no. 4, pp. 571–583, 2006.

[18] T. Foken, R. Leuning, S. R. Oncley, M. Mauder, and M. Aubinet, "Corrections and data quality control," in *Eddy Covariance—A Practical Guide to Measurement and Data Analysis*, pp. 85–131, Springer Science & Business Media, Berlin, Germany, 2012.

[19] M. Goulden, J. Munger, S. Fan, B. Daube, and S. Wofsy, "Measurements of carbon sequestration by long-term eddy covariance: methods and a critical evaluation of accuracy," *Global Change Biology*, vol. 2, no. 3, pp. 169–182, 1996.

[20] M. Mauder and T. Foken, *Documentation and Instruction Manual of the Eddy-Covariance Software Package TK3*, 2004, https://epub.uni-bayreuth.de/342/1/ARBERG046.pdf.

[21] K. Wilson, A. Goldstein, E. Falge et al., "Energy balance closure at FLUXNET sites," *Agricultural and Forest Meteorology*, vol. 113, no. 1–4, pp. 223–243, 2002.

[22] T. Foken, F. Meixner, E. Falge et al., "Coupling processes and exchange of energy and reactive and non-reactive trace gases at a forest site results of the EGER experiment," *Atmospheric Chemistry and Physics*, vol. 12, no. 4, pp. 1923–1950, 2012.

[23] R. McGloina, L. Šiguta, K. Havránková, J. Dušeka, M. Pavelkaa, and P. Sedláka, "Energy balance closure at a variety of ecosystems in Central Europe with contrasting topographies," *Agricultural and Forest Meteorology*, vol. 248, pp. 418–431, 2018.

[24] F. Olmo, J. Vida, I. Foyo, Y. Castro-Diez, and L. Alados-Arboledas, "Prediction of global irradiance on inclined surfaces from horizontal global irradiance," *Energy*, vol. 24, no. 8, pp. 689–704, 1999.

[25] M. Aubinet, "Eddy covariance CO$_2$ flux measurements in nocturnal conditions: an analysis of the problem," *Ecological Applications*, vol. 18, no. 6, pp. 1368–1378, 2008.

[26] C. Thomas, J. G. Martin, B. E. Law, and K. Davis, "Toward biologically meaningful net carbon exchange estimates for tall: multi-level eddy covariance canopy coupling regimes in a mature Douglas fir forest Oregon," *Agricultural and Forest Meteorology*, vol. 173, pp. 14–27, 2013.

[27] E. van Gorsel, N. Delpierre, R. Leuning et al., "Estimating nocturnal ecosystem respiration from the vertical turbulent flux and change in storage of CO$_2$," *Agricultural and Forest Meteorology*, vol. 149, no. 11, pp. 1919–1930, 2009.

[28] M. Aubinet, C. Feigenwinter, B. Heinesch et al., "Direct advection measurements do not help to solve the night-time CO$_2$ closure problem: evidence from three different forests," *Agricultural and Forest Meteorology*, vol. 150, no. 5, pp. 655–664, 2010.

[29] J. Lloyd and J. Taylor, "On the temperature dependence of soil respiration," *Functional Ecology*, vol. 8, no. 3, pp. 315–323, 1994.

[30] M. P. I. f. Biogeochemistry, *Data Products, Online Services & Open Software*, 2018, https://www.bgc-jena.mpg.de/bgi/index.php/Services/Overview.

[31] G. Lasslop, M. Reichstein, D. Papale et al., "Separation of net ecosystem exchange into assimilation and respiration using a light response curve approach: critical issues and global evaluation," *Global Change Biology*, vol. 16, no. 1, pp. 187–208, 2010.

[32] X. Lee, J.-D. Fuentes, R. Staebeler, and H. Neumann, "Long-term observation of the atmospheric exchange of CO_2 with a temperate deciduous forest in southern Ontario, Canada," *Journal of Geophysical Research: Atmospheres*, vol. 104, no. 13, pp. 15975–15984, 1999.

[33] E. Falge, D. Baldocchi, and R. Olson, "Gap filling strategies for defensible annual sums of net ecosystem exchange," *Agricultural and Forest Meteorology*, vol. 107, no. 1, pp. 43–69, 2001.

[34] Z.-H. Tan, Y.-P. Zhang, D. Schaefer, G.-R. Yu, N. Liang, and Q.-H. Song, "An old-growth subtropical Asian evergreen forest as a large carbon sink," *Atmospheric Environment*, vol. 45, no. 8, pp. 1548–1554, 2011.

[35] S. Takanashi, Y. Kosugi, Y. Tanaka et al., "CO_2 exchange in a temperate Japanese cypress forest compared with that in a cool-temperate deciduous broad-leaved forest," *Ecological Research*, vol. 20, no. 3, pp. 313–324, 2005.

[36] T. M. Saitoh, I. Tamagawa, H. Muraoka, N.-Y. M. Lee, Y. Yashiro, and H. Koizumi, "Carbon dioxide exchange in a cool-temperate evergreen coniferous forest over complex topography in Japan during two years with contrasting climates," *Journal of Plant Research*, vol. 123, no. 4, pp. 473–483, 2010.

[37] Y. Kosugi, S. Takanashi, M. Ueyama et al., "Determination of the gas exchange phenology in an evergreen coniferous forest from 7 years of eddy covariance flux data using an extended big-leaf analysis," *Ecological Research*, vol. 28, no. 3, pp. 373–385, 2013.

[38] C.-H. Cheng, C.-Y. Hung, C.-P. Chen, and C.-W. Pei, "Biomass carbon accumulation in aging Japanese cedar plantations in Xitou, central Taiwan," *Botanical Studies*, vol. 54, no. 1, p. 60, 2013.

[39] J. Yu, Y. Lai, P. Chiang, N. Liang, Y. Wang, and C. Horng, "Soil respiration dynamic in subtropical old-plantation forest," in *Proceedings of International Conference on Global Changes, Forest Adaption and CO_2 Flux Monitoring*, Islamabad, Pakistan, May 2012.

12

Severe Weather Events over Southeastern Brazil during the 2016 Dry Season

Amanda Rehbein,[1] **Lívia Márcia Mosso Dutra,**[1] **Tercio Ambrizzi ⓘ,**[1]
Rosmeri Porfírio da Rocha,[1] **Michelle Simões Reboita,**[2]
Gyrlene Aparecida Mendes da Silva ⓘ,[3] **Luiz Felipe Gozzo,**[4]
Ana Carolina Nóbile Tomaziello,[1] **José Leandro Pereira Silveira Campos,**[1]
Victor Raul Chavez Mayta,[1] **Natália Machado Crespo,**[1] **Paola Gimenes Bueno ⓘ,**[1]
Vannia Jaqueline Aliaga Nestares,[1] **Laís Tabosa Machado,**[1] **Eduardo Marcos De Jesus,**[1]
Luana Albertani Pampuch,[5] **Maria de Souza Custódio,**[4] **and Camila Bertoletti Carpenedo**[6]

[1]*Departamento de Ciências Atmosféricas, Instituto de Astronomia,*
 Geofísica e Ciências Atmosféricas da Universidade de São Paulo, São Paulo, SP, Brazil
[2]*Instituto de Recursos Naturais da Universidade Federal de Itajubá, Itajubá, MG, Brazil*
[3]*Departamento de Ciências do Mar da Universidade Federal de São Paulo, São Paulo, SP, Brazil*
[4]*Departamento de Física da Universidade Estadual Paulista Júlio de Mesquita Filho, Campus de Bauru, SP, Brazil*
[5]*Instituto de Ciência e Tecnologia da Universidade Estadual Paulista Júlio de Mesquita Filho, Campus de São José dos Campos,*
 São Paulo, SP, Brazil
[6]*Instituto de Geografia da Universidade Federal de Uberlândia, Uberlândia, MG, Brazil*

Correspondence should be addressed to Tercio Ambrizzi; ambrizzi@model.iag.usp.br

Academic Editor: Anthony R. Lupo

Southeastern Brazil is the most populated and economically developed region of this country. Its climate consists of two distinct seasons: the dry season, extending from April to September, the precipitation is significantly reduced in comparison to that of the wet season, which extends from October to March. However, during nine days of the 2016 dry season, successive convective systems were associated with atypical precipitation events, tornadoes and at least one microburst over the southern part of this region. These events led to flooding, damages to buildings, shortages of electricity and water in several places, many injuries, and two documented deaths. The present study investigates the synoptic and dynamical features related to these anomalous events. The convective systems were embedded in an unstable environment with intense low-level jet flow and strong wind shear and were supported by a sequence of extratropical cyclones occurring over the Southwest Atlantic Ocean. These features were intensified by the Madden–Julian oscillation (MJO) in its phase 8 and by intense negative values of the Pacific South America (PSA) 2 mode.

1. Introduction

Climate and weather components that affect directly the population and economy of Southeastern Brazil [1, 2] have been widely studied in recent years, as well as the large-scale forcings from tropical and extratropical origins. It is well known that the climate of Southeastern Brazil is influenced by the South America monsoon system, where during the summer (Dec-Jan-Feb), there is a predominance of intense convective precipitation due to the availability of plentiful heat and moisture over the tropical region [3]. This intense convective precipitation delineates a cloud corridor known as the South Atlantic Convergence Zone (SACZ), which extends from the southwest Amazon Basin and through Southeast Brazil, reaching the Atlantic Ocean [4]. From the beginning of autumn until midspring, the frequency of SACZ episodes decreases, initiating the dry period over Southeastern Brazil [5]. Moreover, it is during this period that the South Atlantic Subtropical Anticyclone (SASA) reaches its most westerly position, extending over Southeastern Brazil, which impedes the passage of frontal systems [6]. Therefore, during this period, the precipitation events are normally quick, isolated, and not intense.

During the dry season of 2016, there were 9 consecutive atypical days (May 30 to June 07, 2016) with thunderstorms, tornadoes, and at least one microburst over Southeastern Brazil. These phenomena caused floods, smashed houses, personal injuries, and two documented deaths (http:// g1.globo.com/sao-paulo/sorocaba-jundiai/noticia/2016/06/ meteorologistas-analisam-se-tornado-causou-destruicao-em-jarinu.html and http://www.saoroquenoticias.com.br/ noticia.asp?idnoticia=16053). These atypical weather events affected mainly the southern part of Southeastern Brazil, with the most severe conditions occurring over the cities of Campinas, Jarinu, and São Roque, which are close to city of São Paulo in São Paulo State. In Campinas, a probable microburst occurred on June 05 between 00:00 and 00:30 Local Time (LT; Rachel Ifanger Albrecht, personal communication, 2016). At Jarinu on June 05 at about 21 LT and at São Roque on June 06 in the late afternoon, the occurrences of tornadoes were confirmed by analysis of damage by local meteorological institutes and civil defense, besides being observed in the meteorological radar data (Rachel Ifanger Albrecht, personal communication, 2018). Precipitation anomalies from May 30 to June 07 reached values of around 200 mm in Southeastern Brazil, in a region comprising São Paulo State and parts of other surrounding states (Figure 1). For instance, at the meteorological station of the Institute of Astronomy, Geophysics and Atmospheric Science of the University of Sao Paulo (IAG/USP), located in the southern part of the city of São Paulo, the climatological precipitation for the period of 1981 to 2010 is 55.5 mm for the entire month of June. In the first 7 days of June 2016, at this station, the total precipitation was 175.4 mm (316% of the climatological value for the entire month).

The aims of the present study are (a) to investigate the dynamic forcings associated with those severe weather and extreme rainfall events over Southeastern Brazil in the dry season of 2016 and (b) to verify whether or not the most-used forecast models in Brazil predicted this period of intense precipitation. The datasets and methodology used are described in Section 2; Section 3 presents the synoptic discussion, low-frequency analysis, and the model forecasts results; and the concluding remarks are given in Section 4.

2. Data and Methods

The period of analysis is May 30 to June 07, 2016, and the region of interest covers the area between latitudes 25°S to 19°S and longitudes 53°W to 42°W (red box in Figure 1), where the most intense precipitation and severe weather events were registered.

2.1. Synoptic and Thermodynamic Analysis. The synoptic fields were constructed using data from the ERA-Interim

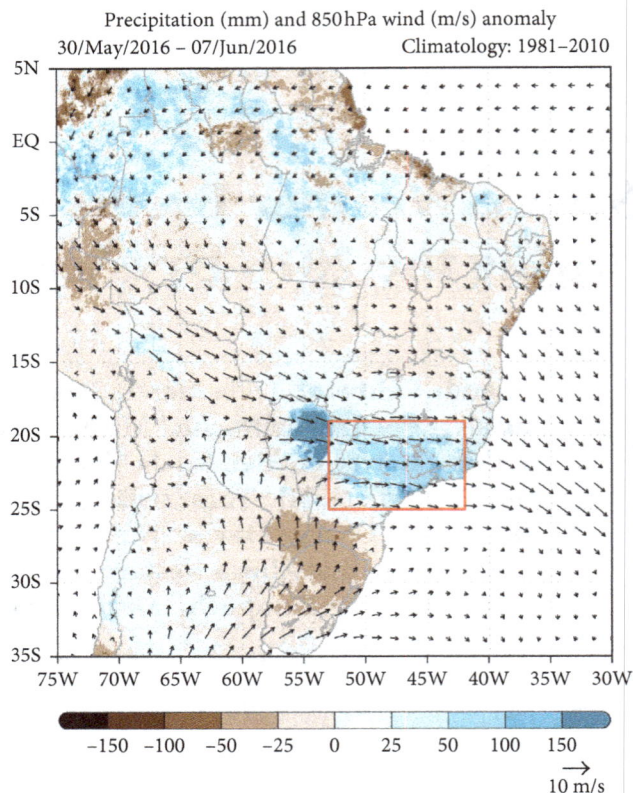

Figure 1: Precipitation anomalies (mm) and 850 hPa wind composite (m·s^{-1}) for May 30 to June 07, 2016. The blue (brown) colors indicate the positive (negative) rain anomalies, and the red box indicates the area of study. The anomalies were calculated using the 30-year period, 1981 to 2010.

reanalysis [7] from the European Centre for Medium-Range Weather Forecasts (ECMWF). These data are available every six hours (0000, 0006, 1200, and 1800 UTC) with spatial resolution of 0.75°, for various pressure levels [7]. We analyzed the synoptic fields at low, middle, and upper levels at each available time; however, for brevity, only the 1200 UTC fields are presented here.

Infrared satellite images (about 10.7 μm) with 4 km and 30 minutes of spatial and temporal resolution, respectively, are from the Geostationary Operational Environmental Satellite (GOES-13; Janowiak et al. [8]) and were made available by the CPC/NCEP/NWS (Climate Prediction Center/National Centers for Environmental Prediction/National Weather Service) via ftp://ftp.cpc.ncep.noaa.gov/precip/global_full_res_IR/.

Five thermodynamic indices (Convective Available Potential Energy—CAPE, Convective Inhibition—CIN, K index, Total Totals—TT, and Showalter) are used to characterize the environmental instability. Moreover, two kinematic indices (sweat and vertical shear of horizontal wind—here termed "wind shear") are also presented because when their values are strong, the environment is favorable to severe weather events [9, 10] and to the formation of stronger convective supercells [11]. The instability and kinematic indices were obtained for a point (23°S/47°W) representative of the severe storm sites—Campinas, Jarinu, and São Roque. This point is located less than 60 km from these sites. In the literature, a distance of up to 180 km is used for the representativeness of such surveys [12, 13]. The Convective Available Potential Energy (CAPE) and Convective Inhibition (CIN) were obtained from the Global Forecast System (GFS) model analysis with spatial resolution of 0.5° and available for 0000, 0600, 1200, and 1800 UTC. The K index [14], Total Totals (TT; Miller [15]), Showalter [16], wind shear, and Sweat index were calculated using the GFS analysis data. The Sweat index is adapted from Miller [15] to Southern Hemisphere wind conditions following Nascimento [10].

2.2. Climate Analysis. The weather and climate in South America are influenced by relatively well-known teleconnection patterns of tropical and extratropical origins that we can observe and measure through indices and statistical analysis. In this study, we investigated the influence of the most important atmospheric and oceanic phenomena that can affect the weather over the Southeastern Brazil: Madden Julian Oscillation [17], Pacific South America pattern, first and second modes [18, 19]; Indian Ocean Dipole (IOD; [20–22]); Southern Annular Mode [23]; and blocking events [24, 25].

The MJO is triggered in the Indian and Pacific Oceans and propagates eastward over the tropical region with a cycle of about 30 to 60 days [26]. During its propagation, it comprises regions with enhanced and suppressed convection. In São Paulo, the most favorable conditions for convection occur with suppression of convection over Indonesia, when the MJO is in its phases 8 and 1, as shown by Jones and Carvalho [17]. Here, to better understand the influence on the extreme rainfall variability over Southeastern Brazil by the eastward-propagating MJO-related large-scale convective and circulation envelope, we have constructed lagged/lead composites for the 0.21 sigma-level (approximately 200 hPa) velocity potential and outgoing longwave radiation (OLR) anomalies.

The velocity potential was obtained from National Centers for Environmental Prediction/National Center for Atmospheric Research (NCEP/NCAR; Kalnay et al. [27]) and the OLR field from the High Resolution Infrared Radiation Sounder (HIRS; Lee et al. [28]). Daily anomalies of OLR and velocity potential were calculated at every grid point by subtracting the long-term average (1979–2015) in order to remove the seasonal cycle. The intraseasonal signals are isolated from the OLR daily anomalies by applying Lanczos bandpass filter [29] using cutoff frequencies at 20 and 96 days. To assemble the composites, we considered the Wheeler and Hendon [30] real-time multivariate MJO (RMM) index for our period of analysis. This index is available at the Centre for Australian Weather and Climate Research website (see: http://www.bom.gov.au/climate/mjo/) and is based on a pair of empirical orthogonal functions (EOFs) of the combined fields of near-equator averaged 850 hPa zonal wind, 200 hPa zonal wind, and satellite-observed outgoing longwave radiation (OLR) data [30]. The evolution of these anomalies from "day −12" to "day +9", where "day 0" represents the active phase (enhanced convection) over tropical South America, is shown in Figure 8. RMM amplitude in phase 8 reaches its maximum value at "day 0," which means that the association between rainfall anomalies and MJO passage over Southeastern Brazil was strong.

The PSA modes are teleconnection patterns extending poleward and eastward over the Pacific Ocean [31], modulating the circulation and precipitation anomalies over South America [32]. PSA teleconnection patterns consist of two distinct modes: PSA 1, related to the El Niño Southern Oscillation (ENSO; Karoly [31]) and PSA 2, associated with the MJO during the winter [18]. Both of them have impacts on the climate of South America, and consequently on the rainfall intensity and distribution over São Paulo state. The PSA modes are defined as the first and second leading rotated principal component modes of the 200 hPa stream function anomaly, respectively [18, 32]; these patterns are also presented in other time scales such as pentads and annual [32–34]. In this study, both PSA modes were computed using ERA-Interim reanalysis for the 200 hPa pentad stream function anomaly data. The covariance matrix was obtained through the extraction of the annual cycle computed with the climatology of 1981–2010 as a basis period.

Saji et al. [35] showed that the anomalous warming of the tropical Indian Ocean due to low level evaporation can lead to divergence in the upper troposphere, sourcing Rossby wave trains propagating from the Indo-Pacific region towards the South Atlantic Ocean in an arch-like trajectory. Taschetto and Ambrizzi [22] showed that anomalous warming throughout the Indian Ocean Basin can excite Rossby wave trains moving towards the South Atlantic, and also amplifying El Niño patterns in the precipitation over the South American continent, for the austral autumn season (March–May). In order to explore the effects of the Indian Ocean on South American precipitation, the Indian Ocean Dipole (IOD; Saji et al. [36]; Webster et al. [37]), that is, the difference between the Eastern and Western Basin sea surface temperature anomaly (SSTa), is computed through the extraction of the annual cycle based on the 1981–2010 climatology, for 36 years (1980–2016) of ERA-Interim data.

The SAM, also known as Antarctic Oscillation (AAO), is the main mode of extratropical circulation variability in the Southern Hemisphere. It consists of zonally symmetric structures, with geopotential height perturbations of opposing signs in Antarctica and in the surrounding zonal ring centered near 45° latitude [38]. Reboita et al. [23] observed that during negative SAM phases, the cyclone trajectories are northward of their positions during the positive phase, and in the South America and South Atlantic sectors, there is intense frontogenetic activity and a positive precipitation anomaly over southeastern South America, which influences the weather in São Paulo. To monitor SAM, we used the daily AAO index available on the Climate Prediction Center/National Oceanic and Atmospheric Administration (CPC/NOAA) website (http://www.cpc.ncep.noaa.gov/products/precip/CWlink/daily_ao_index/aao/aao.shtml). This index is constructed using 700 hPa geopotential height anomalies projected onto the leading EOF mode [39]. To define the phase of the SAM, we use a methodology similar

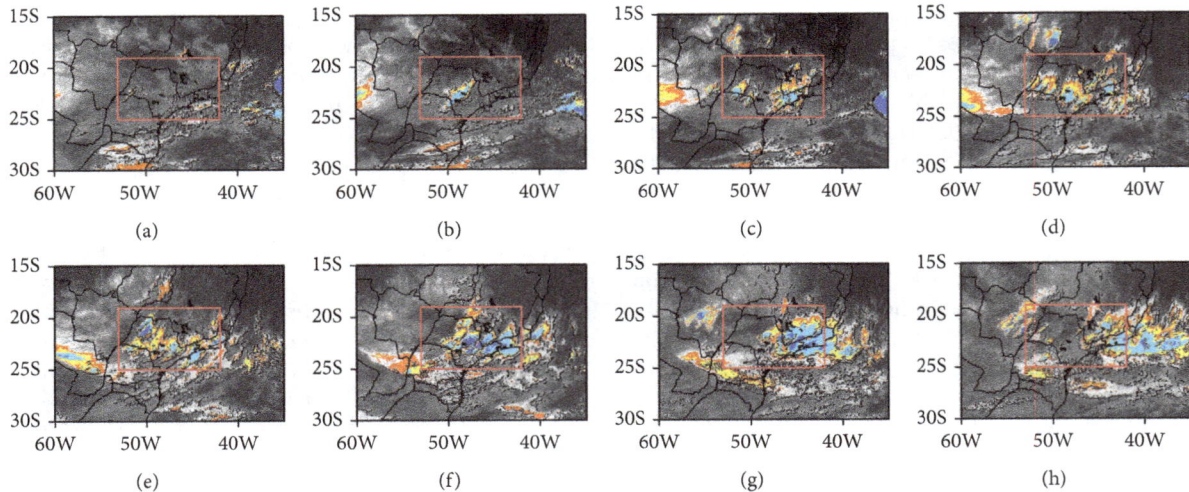

FIGURE 2: GOES-13 enhanced infrared images on June 04, 2016, at (a) 1200 UTC, (b) 1500 UTC, (c) 1800 UTC, and (d) 2100 UTC; and on June 05, 2016, at (e) 0000 UTC, (f) 0300 UTC, (g) 0600 UTC, and (h) 0900 UTC.

to Reboita et al. [23], in which values above (below) one standard deviation indicate the positive (negative) phase. The standard deviation value of the daily SAM time series from 1979 to 2015 is equal to 1.4, and thus values between ±1.4 indicate the neutral phase.

Atmospheric blocking episodes are due to quasi-stationary planetary waves of large amplitude [40], persisting from days to a few weeks, leading to episodes of prolonged extreme weather conditions over some areas. Over the Southeastern Pacific, Southern Atlantic and Oceania, the low-pressure anomalies occurring on the equatorial flank of the blocking pattern favor the development of transient systems that may cause precipitation as they move eastward (Mendes et al. [41]). The resulting impacts on temperature and precipitation are most frequently observed over Southern Brazil, but they can also influence our region of interest (Southeastern Brazil; Mendes et al. [41]). In the latter case, Mendes et al. [41] observed that southeastern Pacific blocking has higher impact on precipitation in austral summer and spring (wet season), while the Atlantic blocking affects precipitation in austral autumn and winter (dry season).

For the identification of blocking events over the Southern Hemisphere, we used the objective method of Tibaldi et al. [42], modified from Lejeñas [43]. This method was adapted to a smaller horizontal spacing of the ERA-Interim reanalysis ($1.5° × 1.5°$ of horizontal resolution) instead of $3.75° × 3.75°$ used by Tibaldi et al. [42] and stratified into five bands of latitudes, according to Oliveira et al. [25]. For an episode to be characterized as a blocking event, it must persist for at least 3 days [25, 44, 45].

The ENSO signal was not evaluated in this work because the São Paulo (SP) region is located in between the two sectors of South America in El Niño (EN), and La Niña (LN) episodes usually affect the observed precipitation with opposing contributions. For instance, during EN conditions, there is increased precipitation over the southeastern sector of South America (including Southern Brazil) and reduced precipitation over the northern/northeastern sector of South

America (including northern/northeastern Brazil; Grimm and Ambrizzi [46]; da Rocha et al. [47]). Given its location, the SP region is considered to be a transition region where the effect of ENSO could be either to increase or reduce precipitation [48].

3. Results and Discussions

3.1. Synoptic Analysis. A rainfall anomaly averaging 47 mm occurred over the SP region (red box in Figure 1) during the period May 30 to June 07, 2016. Figure 1 shows that in specific regions, rainfall anomalies reached more than 100 mm over these 9 days. The satellite images show convective systems forming in the western SP region and moving eastward throughout their life cycle (see, e.g., Figure 2). In addition, some convective systems were generated northwest of the SP region propagating along the low-level mean flow and growing as they moved into the region. Each system had its own lifetime, starting, and developing preferentially during early afternoon (1200 to 1500 LT). Figure 1 also shows that the predominant wind at 850 hPa flowed from the southern Amazon Basin into the SP region. This is the same as the configuration observed by Morales et al. [49] on days with thunderstorms in the city of São Paulo.

Figures 3(a)–3(d) show vertical profiles of horizontal temperature advection, divergence of the horizontal wind, pressure vertical velocity (omega), and moisture convergence averaged over the SP region, from May 30 to June 07, 2016, every 6 hours. Overall, warm-air advection occurred at all levels of the troposphere, peaking on June 05 (Figure 3(a)) when two of the most severe events were reported (microburst at Campinas, SP, and tornado at Jarinu, SP). This warm advection contributes to increased instability over the SP region and is associated with a northerly mean flow, as further discussed in this section. On June 07, the last day of the observed anomalous precipitation over the SP region, intense cold advection occurred in the lower troposphere (Figure 3(a)), associated with a change in the direction of the

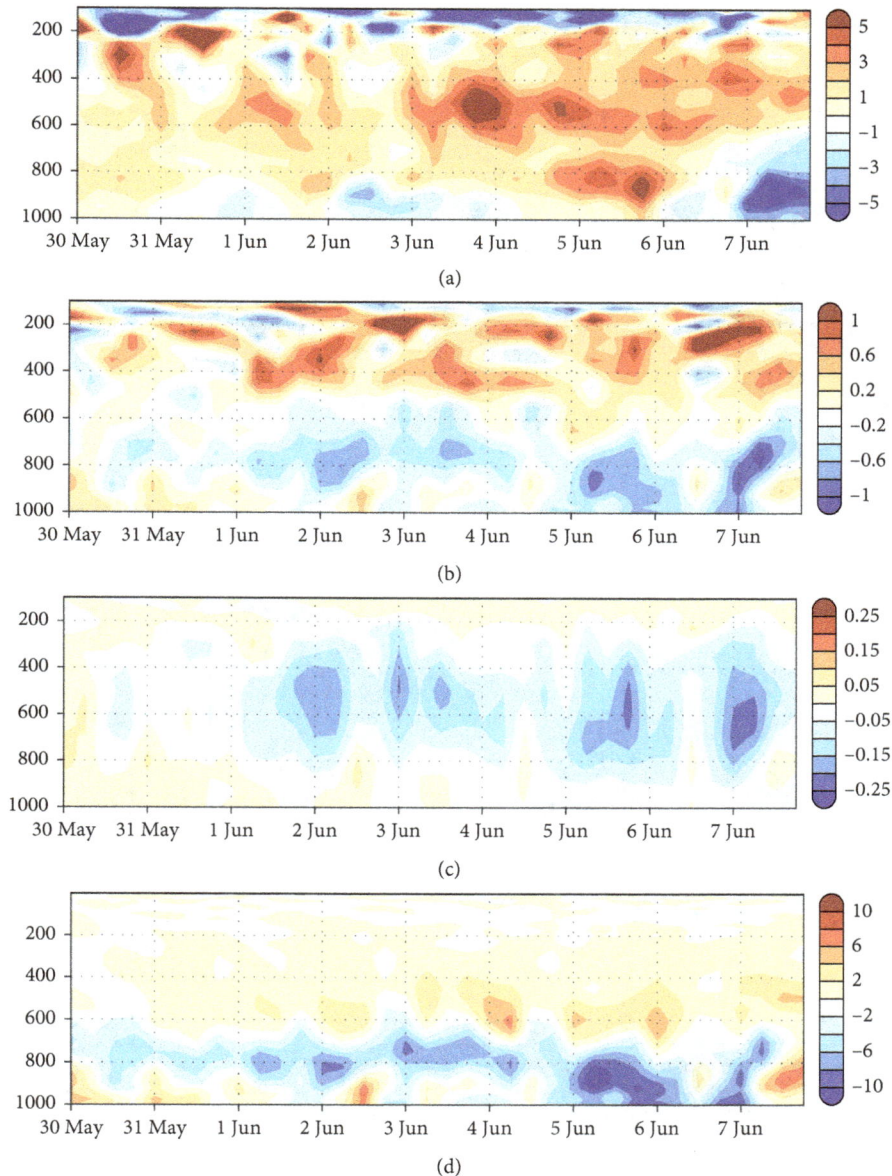

FIGURE 3: Vertical profiles of (a) horizontal temperature advection (K·day^{-1}), (b) divergence of the horizontal wind (10^{-5}·s^{-1}), (c) pressure vertical velocity (or omega; Pa·s^{-1}), and (d) moisture divergence (10^{-5}·g·kg^{-1}·s^{-1}) averaged over the SP region, from May 30 to June 07, 2016, every 6 hours.

mean flow over the region at these levels. Convergence in the lower troposphere and divergence in the upper troposphere occurred during all days analyzed (Figure 3(b)), favoring upward motion over the SP region (Figure 3(c)). At lower and middle levels (up to 600 hPa), moisture convergence occurred on all days (except June 07; Figure 3(d)), indicating favorable conditions for the formation of convective systems in the SP region. The general pattern described above was the reverse of that of the week prior to May 30 and after June 07 (not shown), when there was no anomalous precipitation over the SP region.

The lower-level moisture convergence and warm-air advection over the SP region occurred in association with the South American low-level jet (SALLJ) (east of the Andes),

which comes from tropical latitudes over the south Amazon Basin towards the subtropics, exiting over the SP region. This SALLJ developed around 0600 UTC on May 30 (not shown) and was sustained until 0000 UTC on 07 June. It can be seen as a northwesterly band of maximum wind intensity at 850 hPa in association with a poleward transport of warm and moist air (Figure 4). Commonly, days with thunderstorms over the city of São Paulo (about 100 km away from where the severe events occurred) are accompanied by strong northerly winds [49].

The SALLJs are observed throughout the year but are more frequent and intense during the warm season (NDJF) when the northeast trade winds in the equatorial western Atlantic flow towards the Amazon Basin [50]. The trade

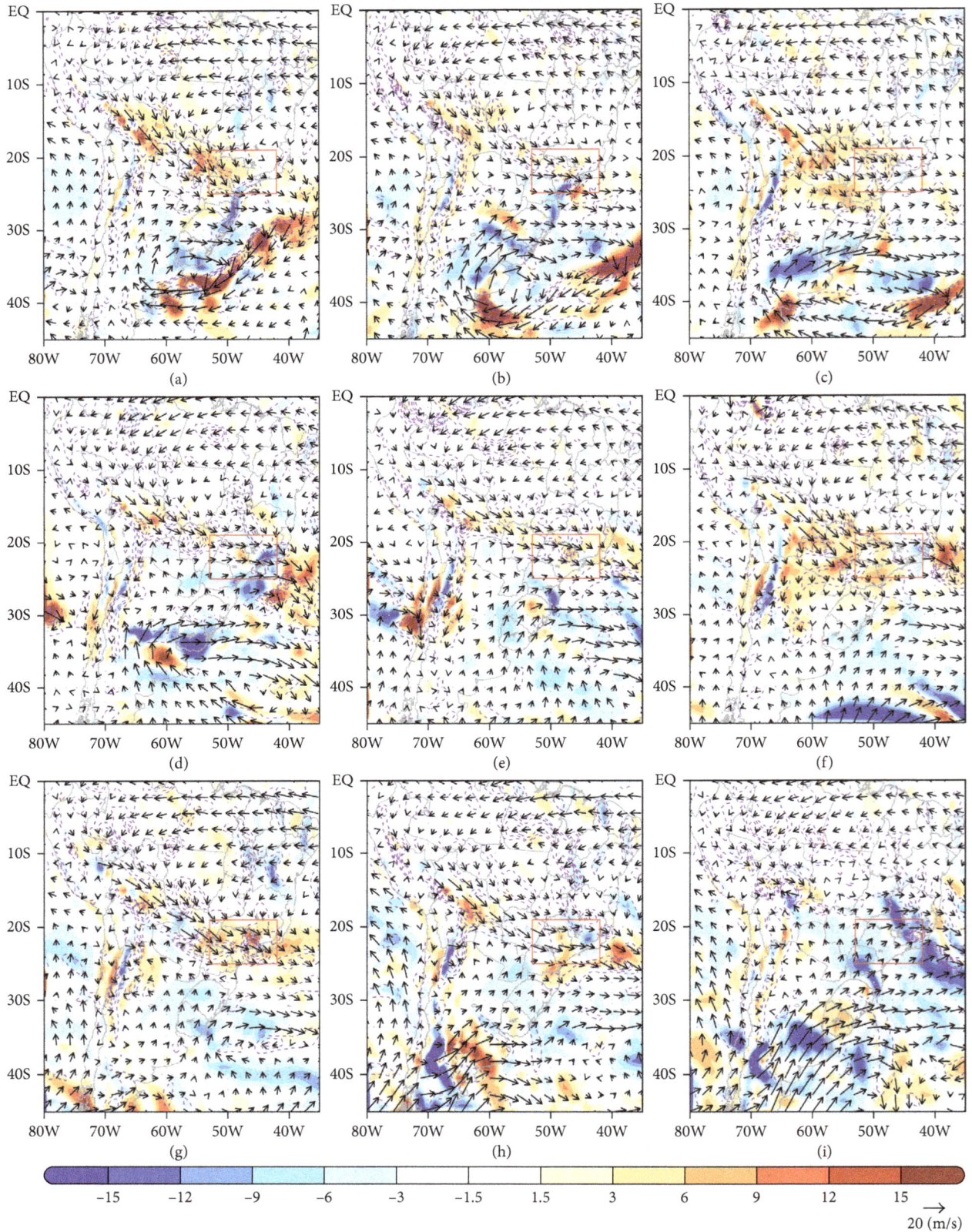

FIGURE 4: Horizontal temperature advection (shaded; K·day^{-1}), moisture flux convergence (purple contours every 10×10^{-5} g·kg^{-1}·s^{-1} from 5×10^{-5} to 55×10^{-5} g·kg^{-1}·s^{-1}), and horizontal wind (vectors; m·s^{-1}) at 850 hPa at 1200 UTC on (a) 30 May, (b) 31 May, (c) 01 June, (d) 02 June, (e) 03 June, (f) 04 June, (g) 05 June, (h) 06 June, and (i) 07 June, 2016.

winds are deflected toward the southeast as they approach the mountain barriers and then converge with the flow from the western branch of the South Atlantic Subtropical High (SASH), producing strong wind speeds at low levels and convective development over the exit region of the jet (see, conceptual model presented in Marengo et al. [51] and their Figure 1). This exit region is located typically over Southern Brazil-Northern Argentina as described in Marengo et al. [51]. During May 30 to June 06, 2016, however, the SALLJ was active northeastward of its typical position, such that its exit was located over the SP region instead of Southern Brazil (Figure 4). This displacement in the jet direction was likely driven by the intense extratropical cyclonic activity over the Atlantic Ocean adjacent to the South American coast during the days analyzed (Figure 5). This activity consisted of two main extratropical cyclones plus three secondary cyclones forming and acting along the southeastern coast of South America. The cyclonic (clockwise) circulation predominant over the southwestern Atlantic Oceanfavored the blocking—by the northwesterly SALLJ—of the advection of cold and dry air from higher latitudes into eastern Argentina and Southern Brazil (Figure 4). At 0000 UTC on June 07, a cold front finally penetrates the continent (figure not shown), reaching the southern Amazon region and weakening the SALLJ, by means of advection from a colder and dryer air mass into the rear of the frontal zone (Figure 4(i)). These persistent low-level patterns (Figures 4 and 5) all seem to be unrelated to a persistent subtropical jet at high-levels (Figure 6).

During the period of interest, atmospheric instability was calculated in terms of various thermodynamic (Figures 7(a)–7(e)) and kinematic (Figure 7(f) and 7(g)) indices. The K index indicates high probability (above 80%) of storm occurrence from 1200 UTC on May 31 to 0000 UTC on June 07 (except at 1200 UTC on June 02 and 03; Figure 7(a)). After this period, the K index decreased substantially. The Total Totals index was high (above 46°C) during most of our period of interest (Figure 7(b)), indicating some scattered storms. From 0000 UTC on June 04 to 1200 UTC on June 06, the warm air advection at middle levels was greater than on other days (Figure 3(a)), decreasing the values of the Total Totals during this period. The Showalter index indicated the possibility of storms, remaining most of the time below 1°C (Figure 7(c)). However, it did not match the thresholds for tornados (below −6°C) in the days in which this phenomenon happened (June 05 and 06). This is possibly due to the strong midlevel warm advection over the SP region during these days (Figure 3(a)), which contributed to raising the temperature at 500 hPa. It is interesting to note that the CAPE was relatively low (Figure 7(d)), CIN was relatively intense (Figure 7(e)), and the wind shear was above the traditional threshold to favor rotating supercells (15 m·s^{-1}; Figure 7(f) [10]). The severe weather observed during these days is the characteristic of the cold season, where there is usually low thermodynamic convective potential but strong wind shear (an example of such a type of event in the United States is presented in Markowski and Straka [52]). The wind shear was high even after the period of interest because of the influence of the jet

streams on those days (Figure (6)). The Sweat index increased from May 30 to June 6; however, it did not reach the thresholds for the development of severe storms in the United States (above 300; Figure 7(g)). Other studies in Brazil have also found that some thermodynamic and kinematic indices are indicative of the instability of a given region; however, they may not reach established thresholds of severity even when severe weather occurs [53, 54].

3.2. Low-Frequency Analysis. The following analysis and discussions cover the contribution of climate indices (Table 1) and planetary-scale influence on the severe rain event studied in the present paper.

3.2.1. The Influence of the Madden–Julian Oscillation. Anomalous upper-level velocity divergence (represented by negative velocity potential anomalies—dashed lines in Figure 8—had been established over tropical South America by "day −9" (May 28) and persisted to "day +3" (June 09), favoring upward motion mainly between "day −3" (June 03) and "day 0" (June 06), the period of the most intense rainfall events over the state of Sao Paulo. Thus, the OLR pattern showed enhanced convection over southeastern South America and the adjacent Atlantic Ocean toward the southern Amazon basin, resembling the austral winter situation with the active MJO phase over South America (e.g., [55]). In addition, "day 0" was when the RMM reached phase 8 and its amplitude began to increase. As pointed out in previous studies (e.g., Jones and Carvalho 2012 [17, 56, 57]), phase 8 and phase 1 favor convection over tropical South America. Even though phase 1 favors convection over tropical South America and the following days were in that phase, it was not sufficient to favor convection over the SP region because, as shown by the synoptic analysis, and on June 07, the atmospheric environment over this region began to stabilize, after the cold front passage, with cold and dry air being advected from the south.

Upper-level tropical convergence (represented by positive velocity potential anomalies (Figure 8)) progresses eastward from the western tropical Pacific from "day −12" (May 25) and arrives in tropical South America by "day +9" (June 15), indicating the end of the active MJO period over tropical South America. It is noteworthy that the evolution of the MJO influence (onset-peak-demise) has a period of approximately 10 days (from "day −6" to "day +3"), average duration of the MJO passage over South America [58].

3.2.2. Pacific South America Mode. The dominant mode of low-frequency climate variability for the last pentad of May and first pentad of June 2016 was the PSA 2 mode, presenting negative values below 2 standard deviations (Table 1); this possibly results from a combination of ENSO influence in the interannual band [31] and tropical convection associated with the MJO in the intraseasonal band [18]. This strongly negative value of PSA 2 during late May/early June 2016 likely contributed to the persistent cyclogenetic activity over the Atlantic near the South

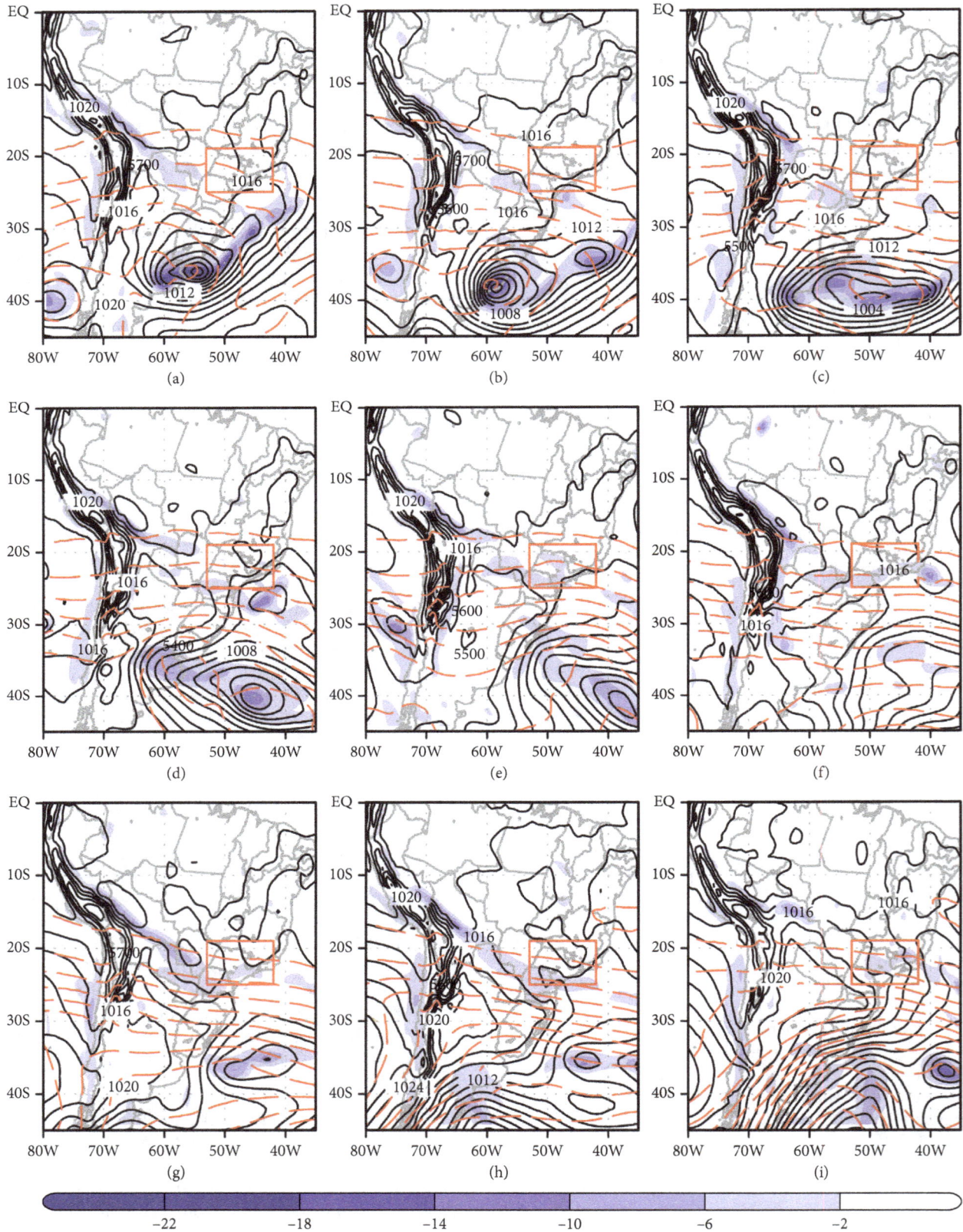

FIGURE 5: Mean sea level pressure (black contours; hPa), geopotential height (red-dashed contours; gpm), and relative vorticity (shaded; $10^{-5} \cdot s^{-1}$) at 850 hPa at 12 UTC on (a) 30 May, (b) 31 May, (c) 01 June, (d) 02 June, (e) 03 June, (f) 04 June, (g) 05 June, (h) 06 June, and (i) 07 June, 2016.

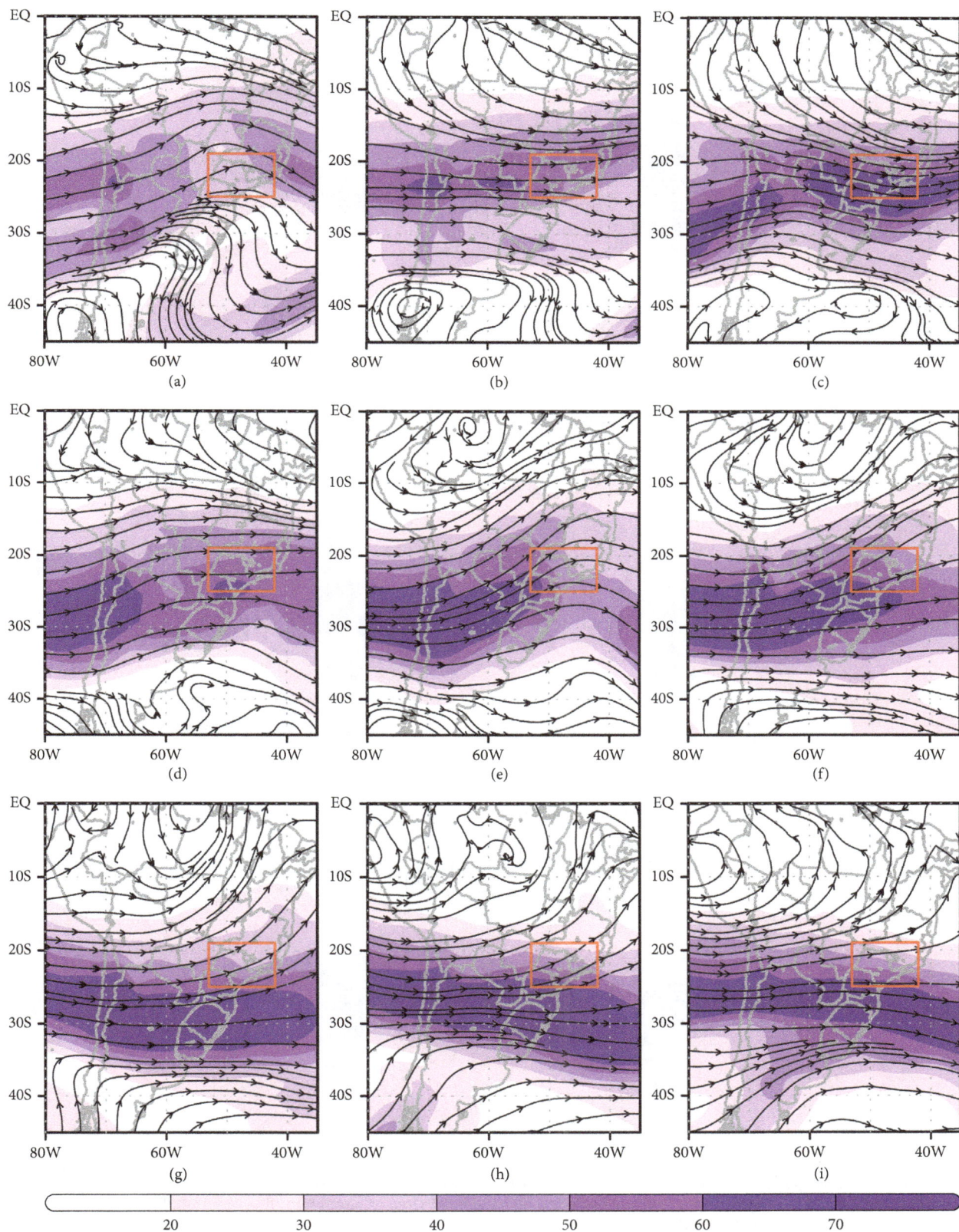

Figure 6: 200 hPa streamlines and isotachs (shaded; m·s^{-1}) at 12 UTC on (a) 30 May, (b) 31 May, (c) 01 June, (d) 02 June, (e) 03 June, (f) 04 June, (g) 05 June, (h) 06 June, and (i) 07 June, 2016.

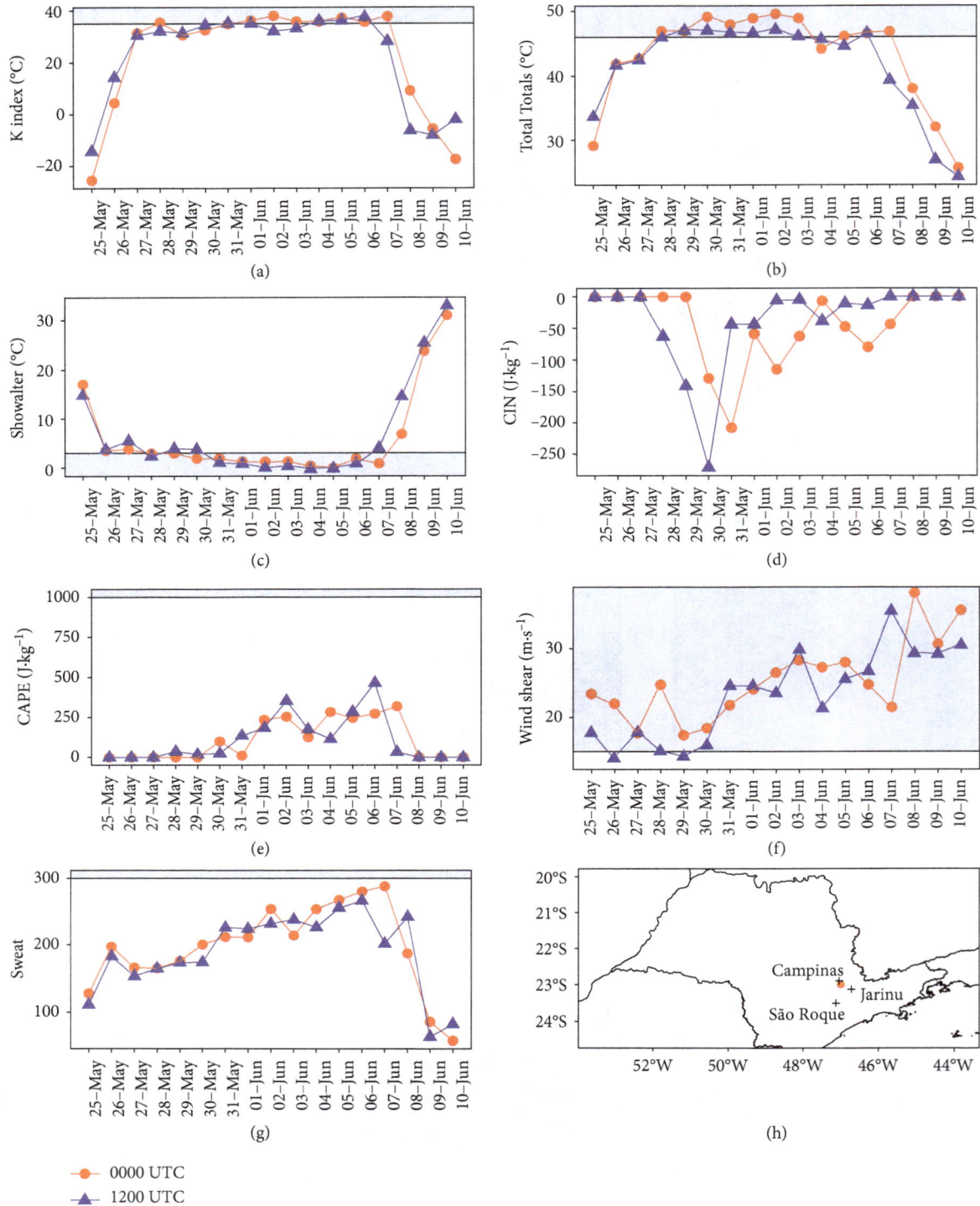

FIGURE 7: Temporal evolution of the instability indices at 0000 UTC (red) and 1200 UTC (blue): (a) *K* index (°C), (b) Total Totals (°C), (c) Showalter index (°C), (d) CAPE (J·kg⁻¹), (e) CIN (J·kg⁻¹), (f) Wind shear (m·s⁻¹), (g) Sweat index, and (h) Map of São Paulo state, with the location of where these indices were calculated (red dot) and the three cities where the most severe thunderstorms occurred (black crosses). The shaded areas from (a) to (g) indicate the values for instability.

TABLE 1: Values of the climate indices analyzed. σ is the standard deviation.

Index	Values	Period
PSA 1	+4.0	June 01–05, 2016
PSA 2	−59.0 (−2.3σ)	June 01–05, 2016
IOD	−0.61	June 2016
AAO	+1.759	May 30 to June 07, 2016

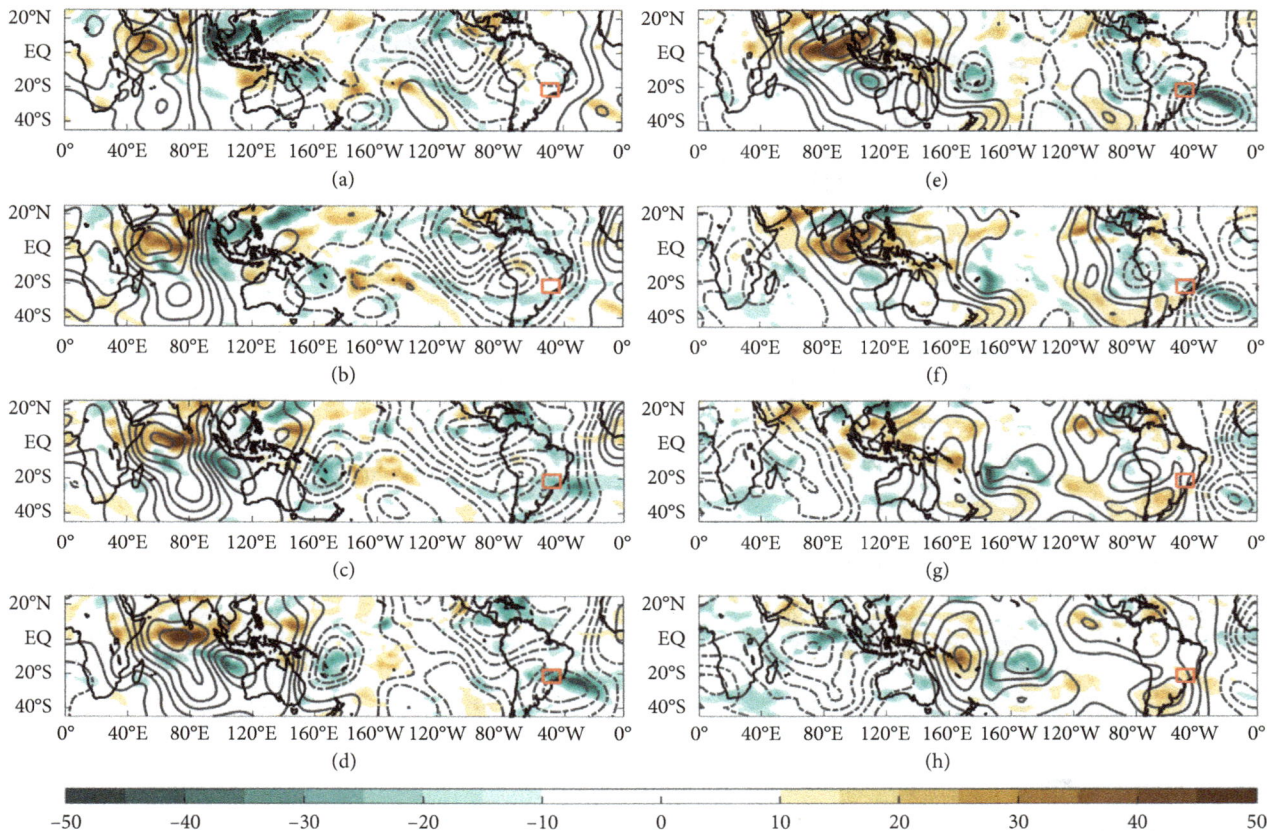

FIGURE 8: Lagged composite maps of filtered OLR anomaly (W·m^{-2}; shaded) and sigma-level 0.21 velocity potential (black contours every 0.75×10^{6}·m^{2}·s^{-1}; negative values are dashed) for a frequency band of 20 to 96 days. Composites are centered on day −12 to day +9 in which day 0 represents the most intense precipitation events over SP region (June 06). (a) Day −12 (May 25), (b) Day −9 (May 28), (c) Day −6 (May 31), (d) Day −3 (Jun 03), (e) Day 0 (Jun 06), (f) Day +3 (Jun 09), (g) Day +6 (Jun 12), and (h) Day +9 (Jun 15).

American coast that influenced the weather in the SP region during the period analyzed.

The PSA 1 mode in June 2016 was not significant, staying between ±1 standard deviation (Table 1).

3.2.3. Indian Ocean Dipole and Wave Source Analysis.
For May/June 2016, the IOD presents negative values, configuring a negative dipole event (western basin colder and eastern basin warmer). Through the analysis of the wave activity flux divergence (Figure 9) for the May-June 2016 basic state [59], a wave like pattern coming from the Indian Ocean eastern basin (Indo-Pacific region) is not found; thus, this region is probably not a Rossby wave train source influencing the South America—even though it is found to be an upper-level divergence source (see the red shaded area over the eastern basin).

3.2.4. Southern Annular Mode.
The AAO index, which measures the phase of the SAM, was positive during the period of study, with an average of +1.759 (Table 1), indicating the predominance of negative anomalies of geopotential height at southern high latitudes and positive anomalies in the middle latitudes [38, 60, 61]. Reboita et al. [23] showed that the low-pressure circumpolar belt is shifted south during the positive phase of the AAO in relation to that in the negative phase, which is unfavorable for the propagation of cyclonic systems to the north that could propagate to Southeastern Brazil. Therefore, the AAO did not interfere in the analyzed extreme event.

3.2.5. Blocking Events.
No blocking events affecting the weather over South America were found during the period evaluated (figure not shown).

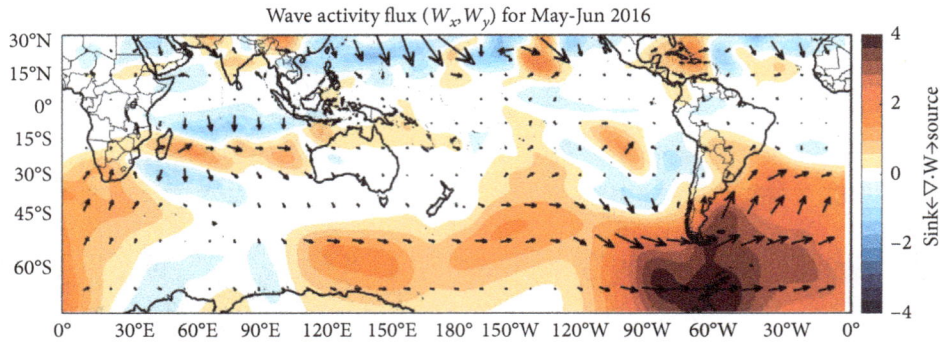

FIGURE 9: Wave activity flux ($\times 10^6$) for May-Jun 2016. The arrows represent the wave flux W_x and W_y, and the shading represents the wave flux divergence, wave flux divergence indicates Rossby wave sources and convergence indicates Rossby wave sinks. Following Takaya and Nakamura [59], the wave flux is parallel to the phase velocity of the Rossby wave packages.

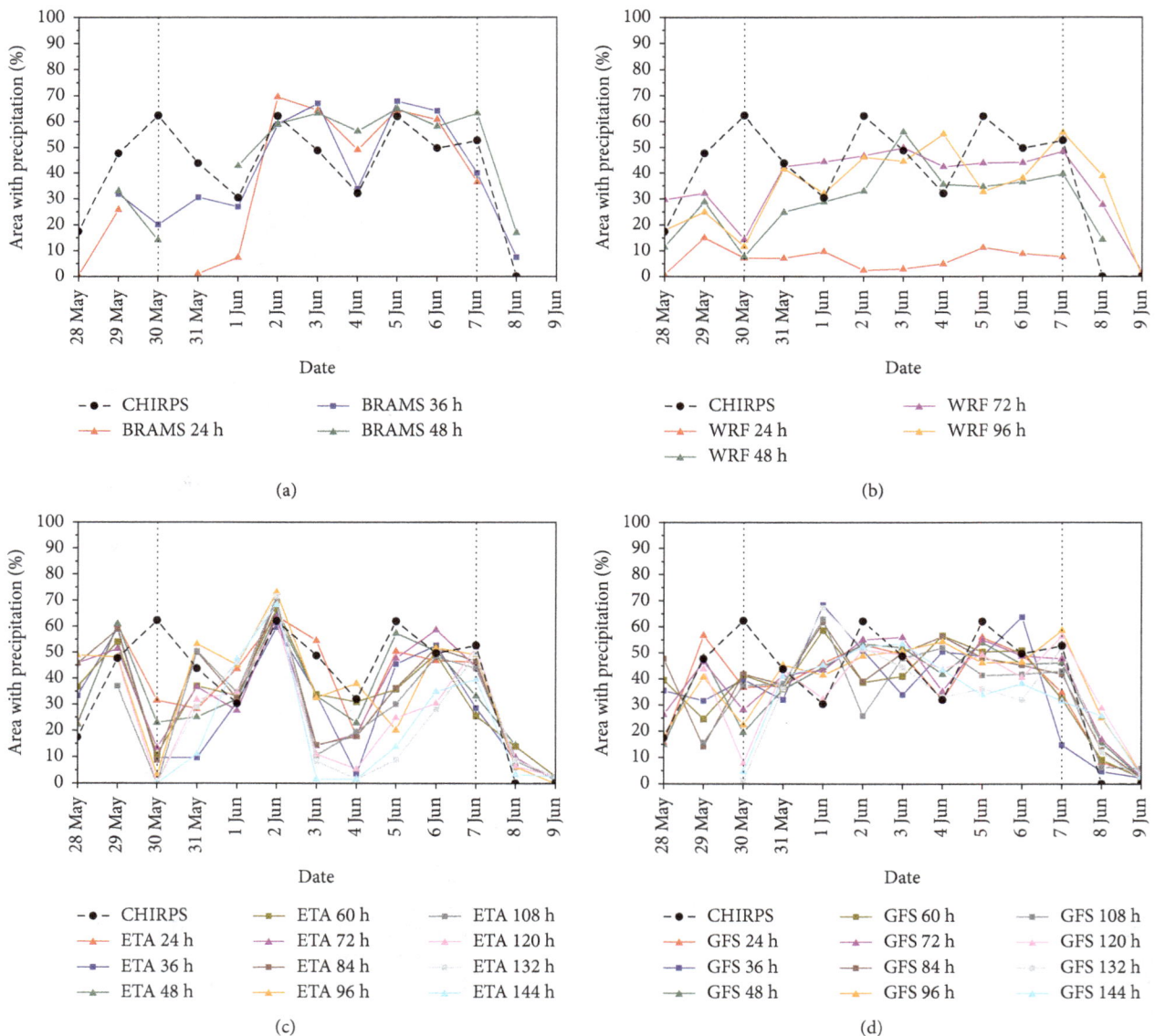

FIGURE 10: Temporal evolution of the percentage of the SP region area with precipitation above the threshold of 5 mm·day^{-1} (%) for CHIRPS and for different forecast times of the models (a) BRAMS, (b) WRF, (c) Eta, and (d) GFS, from 29 May to 9 June 2016. Dotted lines indicate the start (30 May 2016) and end (07 June 2016) of the period of interest.

4. Conclusions

A sequence of successive convective systems favoring atypical precipitation events, with high volumes of rain, occurrences of tornadoes, and a microburst over Southeastern Brazil took place during the dry season of 2016, more specifically, from May 30 to June 07, 2016 (9 days). These anomalous events caused flooding, damages to houses and buildings, and shortages of electricity and water in several places, with many injuries and two deaths documented.

These severe weather events were associated with a daily sequence of convective systems that formed preferentially during early afternoon in the western part of SP state (red box in Figure 1) and also northwest of the SP region, propagating along the low-level mean flow and growing as they moved into the region. The convective systems were embedded in an instable environment (high K index, high Total Totals, low Showalter, and high wind shear) with an intense and persistent South American low-level jet (SALLJ) (east of the Andes) advecting heat and moisture from the Amazon Basin into Southeastern Brazil. The exit region of the SALLJ was located over the SP region instead of Southern Brazil-northern Argentina (which is its typical exit location according to climatological studies). This displacement along the direction of the jet was likely driven by a sequence of extratropical cyclones that formed to the south of the region of interest over the Southwest Atlantic Ocean during the 9-day rainy period. The SALLJ was identified as the main mechanism contributing to the observed precipitation; when this jet weakened (in association with a colder and drier air mass advected over its exit region), the convective activity decreased over the SP region, characterizing the end of the atypical precipitation events.

The effects of the local dynamic and thermodynamic processes mentioned in the previous paragraph were intensified by the MJO in its phase 8 and by the intense negative PSA 2 mode. There was no evidence of a constructive or destructive contribution from the PSA 1, from the IOD or from atmospheric blocking. Moreover, the AAO index was positive (+1.759) during the 9-day rainy period, which would disfavor cyclone propagation to the north; thus, the AAO was not a constructive factor in the analyzed extreme event.

Given the above analysis, one might ask if numerical models were able to predict this anomalous precipitation event. A preliminary analysis considering four different numerical models indicate that they all failed to forecast the beginning of the precipitation period. However, they successfully forecasted its end (Figure 10). Overall, the Global Forecasting System (GFS) was the best model over the 9-day rainy period, followed by the Brazilian Regional Atmospheric Modeling System (BRAMS), the Eta Model, and the Weather Research and Forecasting Model (WRF). The 24-hour forecasts of the WRF were the worst among all forecasts, underestimating both the area and intensity (not shown) of precipitation. Among the possible reasons why these models did not present good performance is that they might have not represented the main dynamic forcings (e.g., SALLJ), leading to this anomalous rainy event. Results

of a detailed verification will be presented and discussed in a future paper.

Atypical precipitation events like the one analyzed in the present paper can happen again in the future, causing further significant impacts for society and the economy. Studies to detect and evaluate the mechanisms contributing to these anomalous events are important for the improvement of the forecasts and mitigation of the associated consequences. Suggestions for future studies include investigating why the models did not predict the beginning of the precipitation period and whether this type of severe weather will be frequent in the coming warmer climate.

Conflicts of Interest

The authors declare that there are no conflicts of interest regarding the publication of this paper.

Acknowledgments

The authors are grateful to Dr. José Roberto Rozante, Dr. Jorge Luís Gomes, and Vinicius Matoso Silva from CPTEC/INPE for providing BRAMS and Eta datasets and to the Universidade Federal de Itajubá (UNIFEI) for providing WRF datasets. Tercio Ambrizzi was supported by CNPq and FAPESP.

References

[1] C. A. S. Coelho, C. P. de Oliveira, T. Ambrizzi et al., "The 2014 southeast Brazil austral summer drought: regional scale mechanisms and teleconnections," *Climate Dynamics*, vol. 46, no. 11-12, pp. 1–16, 2015.
[2] A. Seth, K. Fernandes, and S. J. Camargo, "Two summers of São Paulo drought: origins in the western tropical Pacific," *Geophysical Research Letter*, vol. 42, no. 24, pp. 10816–10823, 2015.
[3] C. S. Vera, W. Higgins, J. Amador et al., "Towards a unified view of the American Monsoon Systems," *Journal of Climate*, vol. 19, no. 20, pp. 4977–5000, 2006.
[4] Y. M. Kodama, "Large-scale common features of sub- tropical precipitation zones (the Baiu Frontal Zone, the SPCZ, and the SACZ). Part I: characteristics of subtropical frontal zones," *Journal of the Meteorological Society of Japan*, vol. 70, no. 4, pp. 813–835, 1992.
[5] T. Ambrizzi and S. E. T. Ferraz, "An objective criterion for determining the South Atlantic Convergence Zone," *Frontiers in Environmental Science*, vol. 3, pp. 3–23, 2015.
[6] M. S. Reboita, M. A. Gan, R. P. da Rocha, and T. Ambrizzi, "Regimes de precipitação na América do Sul: uma revisão bibliográfica," *Revista Brasileira de Meteorologia*, vol. 25, no. 2, pp. 185–204, 2010.
[7] D. P. Dee, S. M. Uppala, A. J. Simmons et al., "The ERA-Interim reanalysis: configuration and performance of the data assimilation system," *Quarterly Journal of the Royal Meteorological Society*, vol. 137, no. 656, pp. 553–597, 2011.
[8] J. E. Janowiak, R. J. Joyce, and Y. Yarosh, "A real-time global half-hourly pixel-resolution infrared dataset and its applications," *Bulletin of the American Meteorological Society*, vol. 82, no. 2, pp. 205–217, 2001.
[9] J. P. Craven and H. E. Brooks, "Baseline climatology of sounding derived parameters associated with deep moist convection," *National Weather Digest*, vol. 28, pp. 13–24, 2004.

[10] E. L. Nascimento, "Prediction of severe storms using convective parameters and mesoscale models: an operational strategy adopted in Brazil," *Brazilian Journal of Meteorology*, vol. 20, pp. 121–140, 2005.

[11] H. Bluestein, "Advances in applications of the physics of fluids to severe weather systems," *Reports on Progress in Physics*, vol. 70, no. 8, pp. 1259–1323, 2007.

[12] J. P. Monteverdi, C. A. Doswell, and G. S. Lipari, "Shear parameter thresholds for forecasting tornadic thunderstorms in northern and central California," *Weather and Forecasting*, vol. 18, no. 2, pp. 357–370, 2003.

[13] S. Grünwald and H. E. Brooks, "Relationship between sounding derived parameters and the strength of tornados in Europe and the USA from reanalysis data," *Atmospheric Research*, vol. 100, no. 4, pp. 479–488, 2011.

[14] J. J. George, *Weather Forecasting for Aeronautics*, Academic Press, Cambridge, MA, USA, 1960.

[15] R. C. Miller, "Notes on analysis and severe storm forecasting procedures of the Air Force Global Weather Central," Technical Report 200, p. 190, Air Weather Service, United States Air Force Washington, DC, USA, 1972.

[16] A. K. Showalter, "A stability index for forecasting thunderstorms," *Bulletin of the American Meteorological Society*, vol. 34, pp. 250–252, 1947.

[17] C. Jones and L. M. V. Carvalho, "Stochastic simulations of the Madden–Julian oscillation activity," *Climate Dynamics*, vol. 36, no. 1-2, pp. 229–246, 2011.

[18] K. C. Mo and R. W. Higgins, "The Pacific-South American modes and tropical convection during the Southern Hemisphere winter," *Monthly Weather Review*, vol. 126, no. 6, pp. 1581–1596, 1998.

[19] D. L. Herdies, A. Da Silva, M. A. F. Silva Dias, and R. Nieto-Ferreira, "Moisture budget of the bimodal pattern of the summer circulation over South America," *Journal Geophysical Research*, vol. 107, no. 20, pp. 42/1–42/10, 2002.

[20] A. R. M. Drumond and T. Ambrizzi, "The role of the South Indian and Pacific oceans in South American monsoon variability," *Theoretical and Applied Climatology*, vol. 94, no. 3-4, pp. 125–137, 2008.

[21] S. C. Chan, S. K. Behera, and T. Yamagata, "Indian ocean dipole influence on South American rainfall," *Geophysical Research Letters*, vol. 35, no. 14, p. L14S12, 2008.

[22] A. S. Taschetto and T. Ambrizzi, "Can Indian Ocean SST anomalies influence South American rainfall?," *Climate Dynamics*, vol. 38, no. 7-8, pp. 1615–1628, 2012.

[23] M. S. Reboita, T. Ambrizzi, and R. P. Rocha, "Relationship between the Southern Annular Mode and Southern Hemisphere Atmospheric Systems," *Revista Brasileira de Meteorologia*, vol. 24, no. 1, pp. 48–55, 2009.

[24] M. C. D. Mendes, R. Trigo, I. Cavalcanti, and C. Camara, "Climatologia de bloqueios sobre o Oceano Pacífico Sul: período de 1960 a 2000," *Revista Brasileira de Meteorologia*, vol. 20, pp. 175–190, 2005.

[25] F. N. M. Oliveira, L. M. V. Carvalho, and T. Ambrizzi, "A new climatology for Southern Hemisphere blockings in the winter and the combined effect of ENSO and SAM phases," *International Journal of Climatology*, vol. 34, pp. 1676–1692, 2013.

[26] R. Madden and P. Julian, "Detection of a 40–50 day oscillation in the zonal wind in the tropical Pacific," *Journal of the Atmospheric Sciences*, vol. 28, no. 5, pp. 702–708, 1971.

[27] E. Kalnay, M. Kanamitsu, R. Kistler et al., "The NCEP/NCAR 40-year reanalysis project," *Bulletin of the American Meteorological Society*, vol. 77, no. 3, pp. 437–471, 1996.

[28] H.-T. Lee, A. Gruber, R. G. Ellingson, and I. Laszlo, "Development of the HIRS outgoing longwave radiation climate dataset," *Journal of Atmospheric and Oceanic Technology*, vol. 24, no. 12, pp. 2029–2047, 2007.

[29] C. E. Duchon, "Lanczos filtering in one and two dimensions," *Journal of Applied Meteorology*, vol. 18, no. 8, pp. 1016–1022, 1979.

[30] M. C. Wheeler and H. H. Hendon, "An all-season real-time multivariate MJO index: development of an index for monitoring and prediction," *Monthly Weather Review*, vol. 132, no. 8, pp. 1917–1932, 2004.

[31] D. J. Karoly, "Southern Hemisphere circulation features associated with El Nino–Southern Oscillation events," *Journal of Climate*, vol. 2, no. 11, pp. 1239–1251, 1989.

[32] K. C. Mo and J. N. Paegle, "The Pacific-South American modes and their downstream effects," *International Journal of Climatology*, vol. 21, no. 10, pp. 1211–1229, 2001.

[33] M. Ghil and K. C. Mo, "Intraseasonal oscillations in the global atmosphere. Part II: Southern Hemisphere," *Journal of the Atmospheric Sciences*, vol. 48, no. 5, pp. 780–790, 1991.

[34] M. K. Lau, P. J. Sheu, and I. S. Kang, "Multiscale low-frequency circulation modes in the global atmosphere," *Journal of the Atmospheric Sciences*, vol. 51, no. 9, pp. 1169–1193, 1994.

[35] N. H. Saji, T. Ambrizzi, and S. E. T. Ferraz, "Indian Ocean Dipole mode events and austral surface air temperature anomalies," *Dynamics of Atmospheres and Oceans*, vol. 39, no. 1-2, pp. 87–101, 2005.

[36] N. H. Saji, B. N. Goswami, P. N. Vinayachandran, and T. Yamagata, "A dipole mode in the tropical Indian Ocean," *Nature*, vol. 401, no. 6751, pp. 360–363, 1999.

[37] P. J. Webster, A. M. Moore, J. P. Loschnigg, and R. R. Leben, "Coupled ocean-atmosphere dynamics in the Indian Ocean during 1997–98," *Nature*, vol. 401, no. 6751, pp. 356–360, 1999.

[38] D. W. J. Thompson, and J. M. Wallace, "Annular modes in the extratropical circulation. Part I: month-to-month variability," *Journal of Climate*, vol. 13, no. 5, pp. 1000–1016, 2000.

[39] K. C. Mo, "Relationships between low-frequency variability in the Southern Hemisphere and sea surface temperature anomalies," *Journal of Climate*, vol. 13, no. 20, pp. 3599–3610, 2000.

[40] D. J. Karoly, "Rossby wave propagation in a barotropic atmosphere," *Dynamics of Atmospheres and Oceans*, vol. 7, no. 2, pp. 111–125, 1983.

[41] M. D. Mendes, R. Trigo, I. Cavalcanti, and C. Camara, "Blocking episodes in the southern hemisphere: impact on the climate of adjacent continental areas," *Pure and Applied Geophysics*, vol. 165, pp. 1941–1962, 2008.

[42] S. Tibaldi, E. Tosi, A. Navarra, and L. Pedulli, "Northern and Southern Hemisphere variability of blocking frequency and predictabilty," *Monthly Weather Review*, vol. 122, no. 9, pp. 1971–2003, 1994.

[43] H. Lejenäs, "Characteristics of Southern Hemisphere blocking as determined from a Time series of observational data," *Quarterly Journal of the Royal Meteorological*, vol. 110, no. 9, pp. 967–979, 1984.

[44] M. R. Sinclair, "Reply," *Monthly Weather Review*, vol. 124, no. 11, pp. 2615–2618, 1996.

[45] R. F. C. Marques and V. B. Rao, "A diagnosis of a long-lasting blocking event over the Southeast Pacific Ocean," *Monthly Weather Review*, vol. 127, no. 8, pp. 1761–1776, 1999.

[46] A. M. Grimm and T. Ambrizzi, "Teleconnections into South America from the tropics and extratropics on interannual and intraseasonal timescales," in *Past Climate Variability in South America and Surrounding Regions: Developments in Paleoenvironmental Research*, F. Vimeux, F. Sylvestre, and

M. Khodri, Eds., vol. 14, Springer, Dordrecht, Netherlands, 2009.

[47] R. P. da Rocha, M. S. Reboita, L. M. M. Dutra et al., "Interannual variability associated with ENSO: present and future climate projections of RegCM4 for South America-CORDEX domain," *Climatic Change*, vol. 125, no. 1, pp. 95–109, 2014.

[48] C. A. S. Coelho, C. B. Uvo, and T. Ambrizzi, "Exploring the impacts of the Tropical Pacific SST on the precipitation patterns over South America during ENSO periods," *Theoretical and Applied Climatology, Austria*, vol. 71, no. 3-4, pp. 185–197, 2002.

[49] C. A. R. Morales, R. P. da Rocha, and R. Bombardi, "On the development of summer thunderstorms in the city of São Paulo: mean meteorological characteristics and pollution effect," *Atmospheric Research*, vol. 96, no. 2-3, pp. 477–488, 2010.

[50] J. A. Marengo, M. Douglas, and P. Silva Dias, "The South American low-level jet east of the Andes during the LBATRMM and WETAMC campaign of January–April 1999," *Journal of Geophysical Research*, vol. 107, no. 20, 2002.

[51] J. A. Marengo, W. R. Soares, C. Saulo, and M. Nicolini, "Climatology of the low-level jet east of the Andes as derived from the NCEP–NCAR reanalyses: characteristics and temporal variability," *Journal of Climate*, vol. 17, no. 12, pp. 2261–2280, 2004.

[52] P. M. Markowski and J. M. Straka, "Some observations of rotating updrafts in a low-buoyancy, highly sheared environment," *Monthly Weather Review*, vol. 128, no. 2, pp. 449–461, 2000.

[53] A. C. N. Tomaziello and A. W. Gandu, *Análise Estatística de Índices de Instabilidade Termodinâmica em São Paulo, in: XV Congresso Brasileiro de Meteorologia, São Paulo, SP. Anais. Soc. Bras. Meteorologia. DVD*, 2008.

[54] R. Hallak and A. J. Pereira Filho, "Análise de desempenho de índices de instabilidade atmosférica na previsão de fenômenos convectivos de mesoescala na região metropolitana de São Paulo entre 28 de janeiro e 04 de fevereiro de 2004," *Revista Brasileira de Meteorologia*, vol. 27, no. 2, pp. 173–206, 2012.

[55] C. S. Vera, M. S. Alvarez, P. L. M. Gonzalez, B. Liebmann, and G. N. Kiladis, "Seasonal cycle of precipitation variability in South America on intraseasonal timescales," *Climate Dynamics*, pp. 1–11, 2017.

[56] C. Jones and L. M. V. Carvalho, "Spatial–intensity variations in extreme precipitation in the contiguous United States and the Madden–Julian oscillation," *Journal of Climate*, vol. 25, no. 14, pp. 4898–4913, 2012.

[57] M. S. Alvarez, C. S. Vera, G. N. Kiladis, and B. Liebmann, "Influence of the Madden Julian Oscillation on precipitation and surface air temperature in South America," *Climate Dynamics*, vol. 46, no. 1-2, pp. 245–262, 2016.

[58] E. B. Souza and T. Ambrizzi, "Modulation of the intraseasonal rainfall over tropical Brazil by the Madden–Julian oscillation," *International Journal of Climatology*, vol. 26, pp. 1759–1776, 2006.

[59] K. Takaya and H. Nakamura, "A formulation of a phase-independent wave-activity flux for stationary and migratory quasigeostrophic eddies on a zonally varying basic flow," *Journal of the Atmospheric Sciences*, vol. 58, no. 6, pp. 608–627, 2001.

[60] D. W. J. Thompson and S. Solomon, "Interpretation of recent Southern Hemisphere climate change," *Science*, vol. 296, no. 5569, pp. 895–899, 2002.

[61] N. P. Gillett, T. D. Kell, and P. D. Jones, "Regional climate impacts of the Southern Annular Mode," *Geophysical Research Letters*, vol. 33, no. 23, pp. 1–4, 2006.

13

Homogeneity Test and Correction of Daily Temperature and Precipitation Data (1978–2015) in North China

Lingling Shen ⓘ,[1] Li Lu,[1] Tianjie Hu,[1] Runsheng Lin,[1] Ji Wang,[2] and Chong Xu ⓘ[3]

[1]*Beijing Meteorological Information Center, Beijing Meteorological Service, Beijing 100089, China*
[2]*Beijing Climate Center, Beijing Meteorological Service, Beijing 100089, China*
[3]*Institute of Geology, China Earthquake Administration, Beijing 100029, China*

Correspondence should be addressed to Lingling Shen; wudahe2010@gmail.com

Academic Editor: Jorge E. Gonzalez

Homogeneity of climate data is the basis for quantitative assessment of climate change. By using the MASH method, this work examined and corrected the homogeneity of the daily data including average, minimum, and maximum temperature and precipitation during 1978–2015 from 404/397 national meteorological stations in North China. Based on the meteorological station metadata, the results are analyzed and the differences before and after homogenization are compared. The results show that breakpoints are present pervasively in these temperature data. Most of them appeared after 2000. The stations with a host of breakpoints are mainly located in Beijing, Tianjin, and Hebei Province, where meteorological stations are densely distributed. The numbers of breakpoints in the daily precipitation series in North China during 1978–2015 also culminated in 2000. The reason for these breakpoints, called inhomogeneity, may be the large-scale replacement of meteorological instruments after 2000. After correction by the MASH method, the annual average temperature and minimum temperature decrease by 0.04°C and 0.06°C, respectively, while the maximum temperature increases by 0.01°C. The annual precipitation declines by 0.96 mm. The overall trends of temperature change before and after the correction are largely consistent, while the homogeneity of individual stations is significantly improved. Besides, due to the correction, the majority series of the precipitation are reduced and the correction amplitude is relatively large. During 1978–2015, the temperature in North China shows a rise trend, while the precipitation tends to decrease.

1. Introduction

North China is an important base of grain, cotton, and edible oil production, as well as a major wheat-growing region. The study of meteorological elements such as precipitation and temperature in North China is an important part of meteorological research, which is of great significance for agriculture. Such research requires high-quality meteorological data, so it is necessary to test and correct its homogeneity so as to be able to objectively reflect the true processes of climate change.

Methods commonly used ones for the work aforementioned include the SNHT (standard normal homogeneity test), Rhtests, and MASH method (multiple analysis of series for homogenization).

The SNHT was first proposed by Alexandersson [1] for homogeneity tests of precipitation data, which was improved later by [2]. Further development to this method was made by Khaliq and Ouarda [3], in which the sample interval corresponding to the key *T* value was expanded from the original [10, 250] to [10, 50000]. The SNHT is widely used in research of meteorological series homogeneity because of its simplicity and intuition [4–12]. While it is widely applied to the meteorological data with normal distribution such as annual average temperature and annual precipitation, it is not suitable for the data with obvious differences such as seasonal or daily temperature data.

The need of homogeneity tests and correction to the meteorological data on month or day scales sparked the generation of the Rhtests and MASH methods.

The Rhtests, established by the Climate Research Centre of the Environment Ministry of Canada, can test and correct the time series of meteorological data on year, month, and day scales. It is based on the penalized maximal T test (PMT) [13] and the penalized maximal F test (PMFT) [14]. Generally, in order to improve the accuracy and rationality of Rhtests, it is necessary to further confirm the detected breakpoints in combination with the metadata [15]. Thus, for day-scale time series, it takes a lot of labor and time to compare the Rhtests results with metadata station by station.

The MASH procedure was developed by Szentimrey [16, 17] of the Hungarian Meteorological Service, which is one of the methods recommended by "COST (European Cooperation in Science and Technology) Action ES0601: Advances in homogenization methods of climate series: an integrated approach (HOME)" for testing the homogeneity of meteorological data series [18]. This method does not assume that the reference series is homogeneous and fixed. Comparing the series between stations in the same climatic region permits to determine the possible breakpoints and correct the meteorological data of year, month, and day scales. For different meteorological elements, additive (e.g., temperature) or multiplicative (e.g., precipitation) models can be applied, and both can also be converted mutually by taking logarithms. As the MASH method is concerned with relative homogeneity, it does not depend on metadata. Previous study [19] shows that the metadata does not have a large effect on the results of MASH method tests, while the choice of reference series does. The relative homogeneity methods also have some drawbacks [20]. The assumption that climatic patterns are identical within a geographical region does not well stand because different terrains can lead to a local microclimate, especially when there are mountains and plains. The representative studies in this aspect include the following examples. Li and Yan [19] applied the MASH method to test the homogeneity of daily temperature data from 1960 to 2006 in Beijing. Li et al. [21] used the MASH method to test the homogeneity of daily average, maximum, and minimum temperature of 545 stations in China from 1960 to 2011. Lakatos et al. [22] applied this method to the meteorological data in the Carpathian region.

In this study, the MASH method is used to test the homogeneity of the temperature and precipitation data in North China. Compared with previous studies, first, we used a different station selection. In previous studies [19, 23], in order to ensure the consistent length of the meteorological series participating in the homogenization test, the stations with shorter series were eliminated, resulting in sparse stations participating in the test. However, the MASH method is concerned with relative homogeneity, which relies on mutual comparison between stations. In order to balance the number and distribution density of the participating stations, and considering the length of the series, the starting time point of the series was set in 1978. Secondly, this study also tried the homogenization test on the precipitation data. In order to allow more stations to participate in the homogenization test process, the temperature data of stations with insufficient sequence length were added in turn. Last,

the study area of previous work was either Beijing or China [19, 23]. The area selected in this study was North China.

Homogenization of spatially and temporally discontinuous daily precipitation data has always been a difficult issue [24, 25]. Previous studies on the homogenization of precipitation data were based on continuous monthly precipitation or annual precipitation data [1, 5, 6, 7, 12]. The MASH method uses a multiplicative model for precipitation data and then applies logarithmization to convert the raw data into nonzero data, converting discretely distributed raw data into continuously distributed data. Compared with the SNHT method, the MASH method is more reasonable in the selection of reference stations, and the practice of selecting reference stations in turn is more rigorous and scientific. The MASH procedure has a rigorous mathematic principle and perfect system. It also has some innovations in the selection of reference series, thus being able to check and correct the homogeneity of a large number of time series in a relatively short time.

Therefore, this work employs this method to test the homogeneity of the temperature and precipitation data in North China, which were recorded at 404/397 meteorological stations during 1978–2015 and correct the inhomogeneity when detected.

2. Data and Methodology

2.1. Data. Figure 1 shows the area of North China and locations of weather stations. The data used in this paper include (1) latitudes, longitudes, and elevations of 404 meteorological stations in North China; (2) daily average temperature, maximum temperature, and minimum temperature and daily precipitation of these stations from 1978 to 2015, which are from China National Meteorological Information Center; and (3) metadata of these stations.

2.2. Methodology. This work uses the MASH method to homogenize the aforementioned data, i.e., examining the inhomogeneities in the time series of data first and then correcting them. As a relative method, the MASH procedure establishes the optimal difference series of the reference series and the candidate series. By point estimate and confidence interval, the breakpoints and the shifts of the difference series can be detected and estimated. Then, the breakpoints and the shifts of the candidate series are obtained. Finally, based on the above results, the candidate series are corrected. Since the composition of the difference series is not fixed, which is calculated by each station in turn, the entire calculation process will iterate many times to find the most probable breakpoints and shifts of the candidate series. This process can also be achieved by using the maximum likelihood method. The procedures are presented briefly below.

A single candidate series can be expressed as

$$X_j(t) = \mu(t) + E_j + \text{IH}_j(t) + \varepsilon_j(t)$$
$$(j = 1, 2, \ldots, N; \ t = 1, 2, \ldots, n), \tag{1}$$

where $X_j(t)$ is the candidate series, $\mu(t)$ is the unknown climate change signal (temporal trend), E_j is the spatial

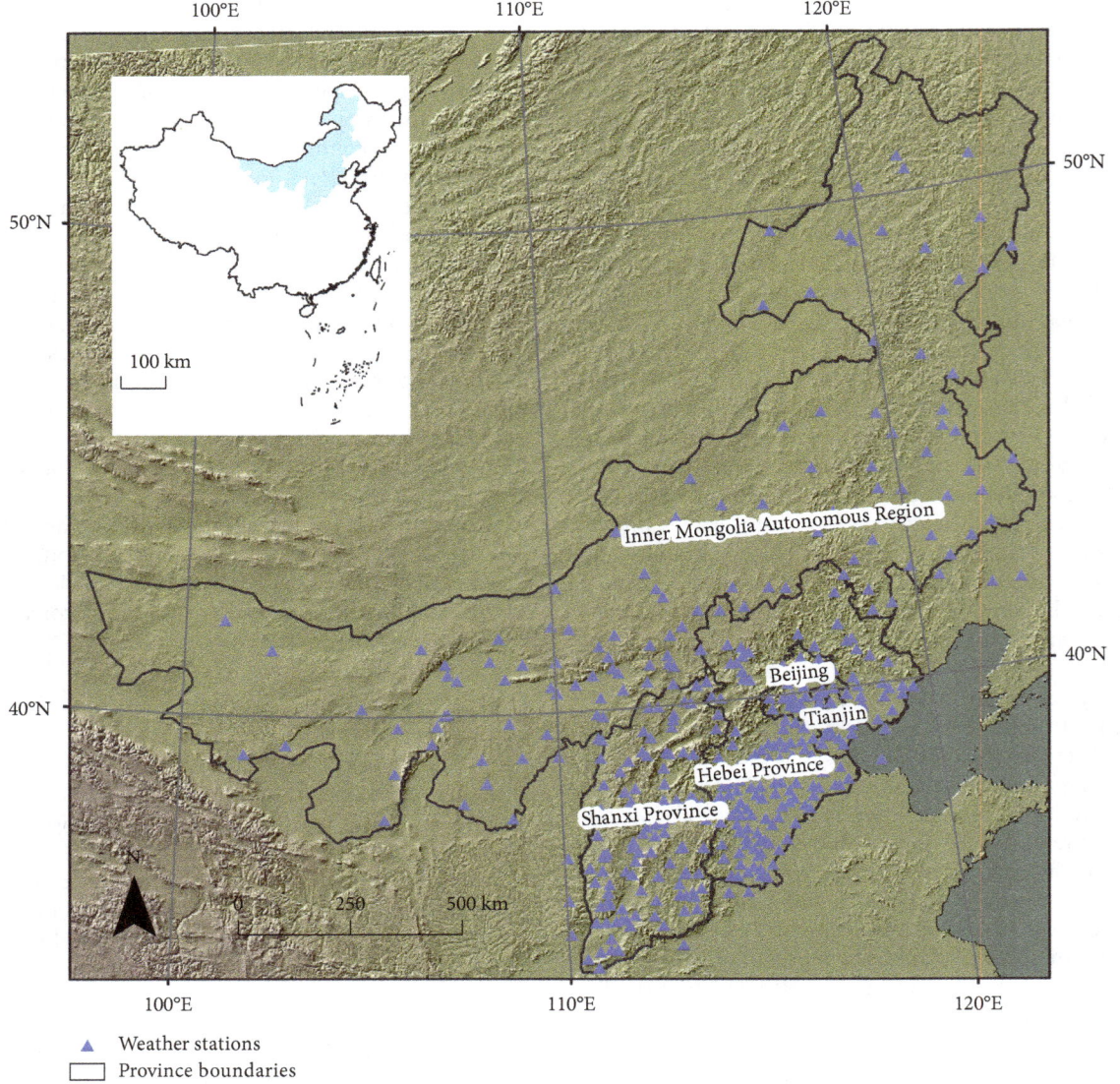

FIGURE 1: Locations of the weather stations and North China.

expected values (spatial trend), $IH_j(t)$ is the inhomogeneity signals, and $\varepsilon_j(t)$ is the normal white noise series.

Then, the reference series can be expressed by $X_i(t)$, $i \in N_{-j}$. Thus, the difference series of the candidate series and the reference series is

$$Z_j(t) = X_j(t) - X_i(t) \quad \left(i \in N_{-j}\right). \tag{2}$$

Substituting Equation (1) into (2) yields

$$Z_j(t) = IH_j(t) - \sum_{i \in J} \lambda_{ji} IH_i(t) + \varepsilon_{Z_j}(t) \quad \left(J \subseteq N_{-j}\right), \tag{3}$$

where λ_{ji} is the weighting factor, $\sum_{i \in J} \lambda_{ji} = 1$. By establishing the difference series, the common climate change signal $\mu(t)$ is filtered out. In order to ensure the maximum correction effect, the inhomogeneity signals are removed by letting the following item in Equation (3):

$$\mathrm{var}\left(Z_j\right) = \mathrm{var}\left(\varepsilon_{Z_j}\right). \tag{4}$$

To realize Equation (4), the weighting factor $\lambda_{ji}(i \in J \in N_{-j})$ is written in the vector form as

$$\lambda_{j,J} = C_{J,J}^{-1}\left(c_{j,J} + \frac{\left(1 - 1^T C_{J,J}^{-1} c_{j,J}\right)}{1^T C_{J,J}^{-1} 1}1\right), \tag{5}$$

where $c_{j,J}$ is the candidate-reference covariance vector, $C_{J,J}$ is the reference-reference covariance matrix [16, 26], and the covariance matrix C determines the optimal weighting factors that minimize the variance. The optimal difference series thus obtained can be used for detection and correction of inhomogeneities efficiently. For the candidate series $X_j(t)$, there is a total of $2^{N-1} - 1$ difference series with optimal weight by changing the composition of the reference series $X_i(t)$ $(i \in J \in N_{-j})$. Through iteration in the MASH procedure, applying point estimates and confidence intervals, breakpoints and shifts in these difference series can be detected, which further can be attained in candidate series. By these processing, the candidate series are corrected [27].

2.3. Data Preprocessing. The sorting and statistics of the original meteorological data in North China show that stations with complete meteorological data after 1978 account for 99.02% of the total. In order to ensure the accuracy of the correction and maximize the number of series involved in the correction, the meteorological data of North China from 1978 to 2015 are selected. Of it, there are still more than one month data missing at 18 stations for daily precipitation, which can be interpolated in the MASH procedure [16].

First, the data are transformed into a software-readable format. MASH method does not assume that the reference series is homogeneous, since some researchers argued that the assumption that the created reference series is homogeneous is false and without theoretical basis [27]. The reference series is not fixed either. Through the comparison of stations in the same climatic region, possible breakpoints can be found and corrected. Applying the additive model to correct the temperature data, through multiple iterative calculations, finally we get the homogeneity temperature data. For precipitation data, we use the multiplicative model. Since the daily precipitation data are discontinuous with values of 0, in order to avoid its influence on the calculation process, the multiplicative model is used to process the data into nonzero data.

The length of each series is kept consistent during the calculation. A total of 397 meteorological stations have reached the required length, and the remaining 7 series with insufficient length are added in turn. Each time a series is added and a new calculation is carried out. The closer to the year 2016, the more the stations involved. However, due to too much missing data, the precipitation series cannot be cyclically added, so only 397 series are finally involved in the calculation.

3. Results

3.1. Homogenization of Temperature Series. Because of the loop join, each new series is added at a different length of time, resulting in a new breakpoints file. Taking the initial 397 stations and the average temperature series from 1978 to 2015 as an example, under the confidence level of 0.01, about 98% of the series are detected a varying number of breakpoints, of which over 75% of the series contains 4 to 11 breakpoints. As the number of series involved in the calculation increases, the number and frequency of breakpoints also change (Figures 2(a)–2(c)). The years with breakpoints detected in each series during each calculation are accumulated, and the repeated years are removed to obtain the total years with breakpoints. Consequently, a total of 5,485 breakpoints are present in the daily average temperature

series, 5833 breakpoints in the daily minimum temperature series, and 4,509 breakpoints in the daily maximum temperature series, respectively. The frequency of breakpoints in the daily average temperature series is in the range of [12, 16], while in the daily minimum temperature series, it is in the range of [10, 20] and in the maximum temperature series, it is in the range of [7, 15]. It can be seen from Figure 3(a) that most of the breakpoints are after 2000, especially culminating from 2006 to 2007, which are consistent with the minimum temperature series, the maximum temperature series, and the average temperature series. The statistics of the total breakpoints also shows that the frequency of breakpoints is relatively high after 2000. From the spatial distribution of breakpoints at 404 meteorological stations, the Beijing-Tianjin-Hebei region with dense meteorological stations has more breakpoints and more stations with breakpoints (Figure 4).

Breakpoints are the highest in the minimum temperature series and the lowest in the maximum temperature series, respectively. The maximum temperature generally occurs around noon, when the solar radiation makes the ground warming fast, resulting in a high vertical decline rate of temperature and poor atmospheric stability; thus, the various scales of the atmosphere are fully mixed, which at last leads to a small temperature difference among stations. By contrast, the minimum temperature generally appears before or after the sunrise, when the atmospheric stability is higher, the temperature mixing between different stations is not sufficient, resulting in large temperature differences among stations. Inferring from the calculation principle of relative homogeneity methods, these differences are difficult to be considered signals of climate change and are more likely to be identified as inhomogeneous signals through comparisons of stations. This may be the reason that the breakpoints are the most in the minimum temperature series and the minimum in the maximum temperature series. However, this inference still needs further research to prove.

The metadata show that after 2000, more than 40% of the meteorological stations in North China have been moved. And around 2000 is also the time when China Meteorological Administration implemented the construction of automatic meteorological stations to replace the original manual measurement acquisition [28]. At that time, temperature monitoring throughout the country was gradually changed from the original manual observation of glass liquid thermometers to platinum resistance temperature sensors. Previous study generally agrees that the discrepancies caused by this large-scale instrument change are within the allowable range of automatic station error [29]. However, it inevitably resulted in some degree of inhomogeneity in observational data [23]. Since 2003, the vast majority of meteorological stations gradually began to move from two-track observation of artificial stations and automatic meteorological stations to single-track observation of automatic stations. Most of the meteorological stations in 2006 replaced the manual observation data with automatic meteorological station data to prepare reports. Large-scale replacement of instruments and the change of the observation mode could lead to the significant increase of

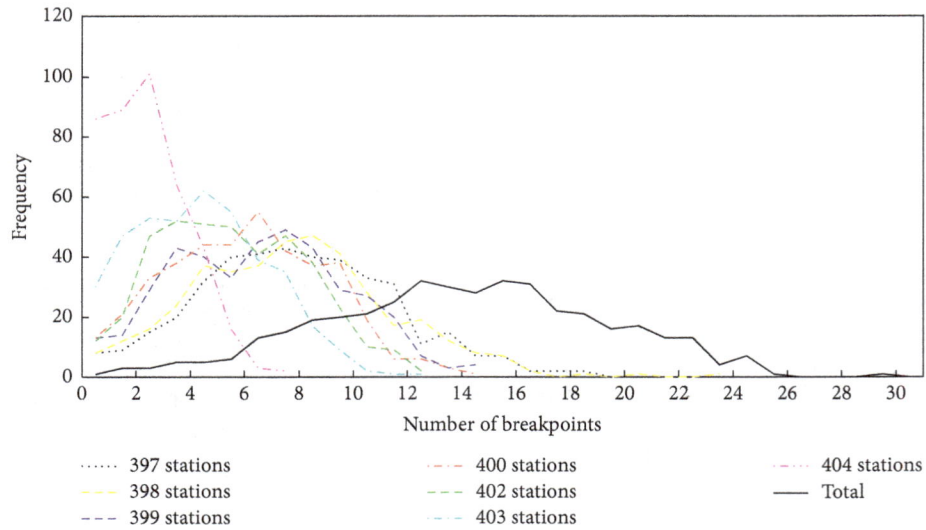

(a)

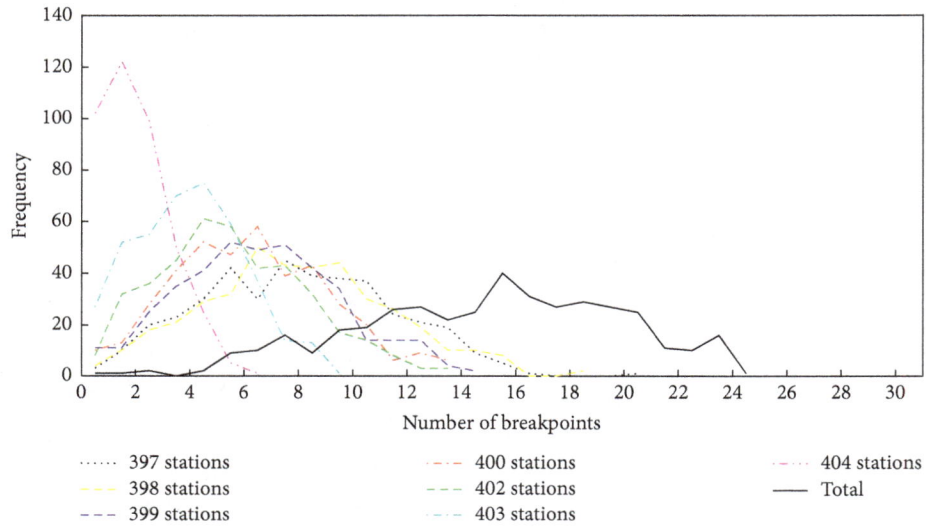

(b)

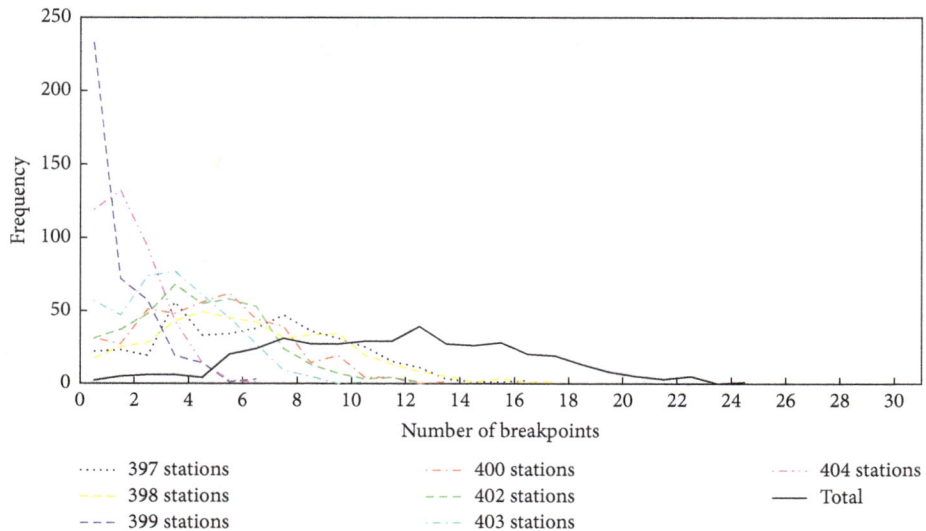

(c)

FIGURE 2: Frequency of breakpoint numbers in the average (a), minimum (b), and maximum (c) temperature series at 397~404 mete-orological stations in North China.

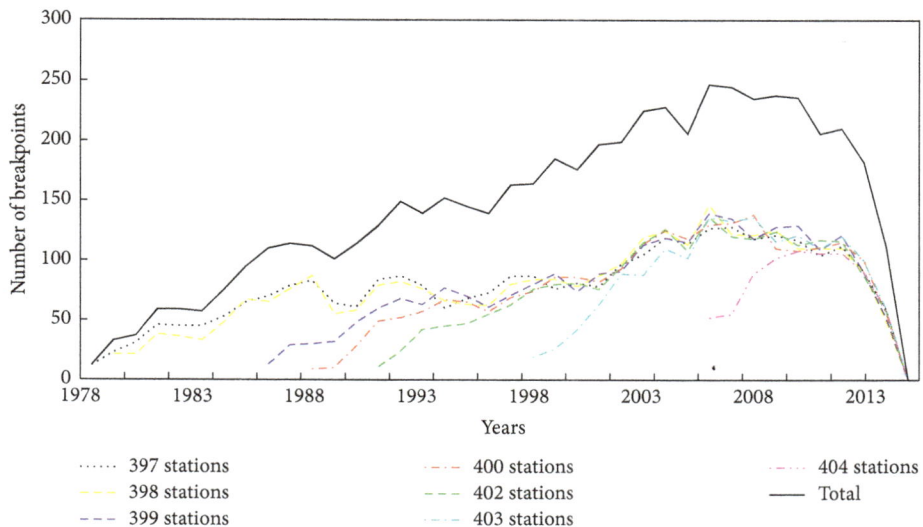

(a)

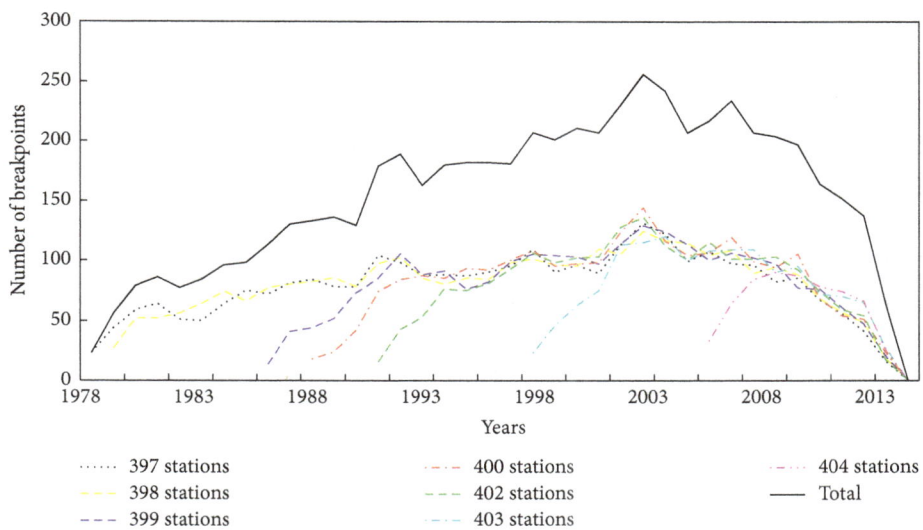

(b)

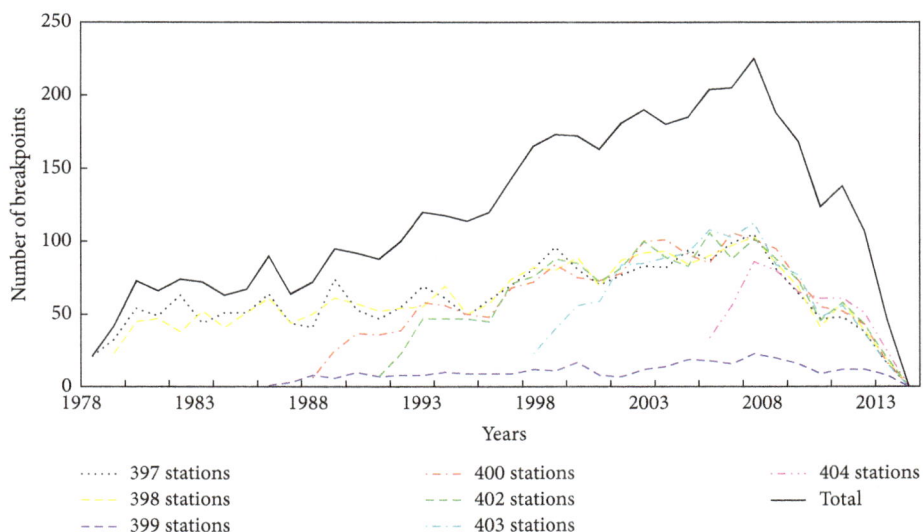

(c)

FIGURE 3: Annual breakpoint numbers in the average (a), minimum (b), and maximum (c) temperature series of 397~404 meteorological stations in North China.

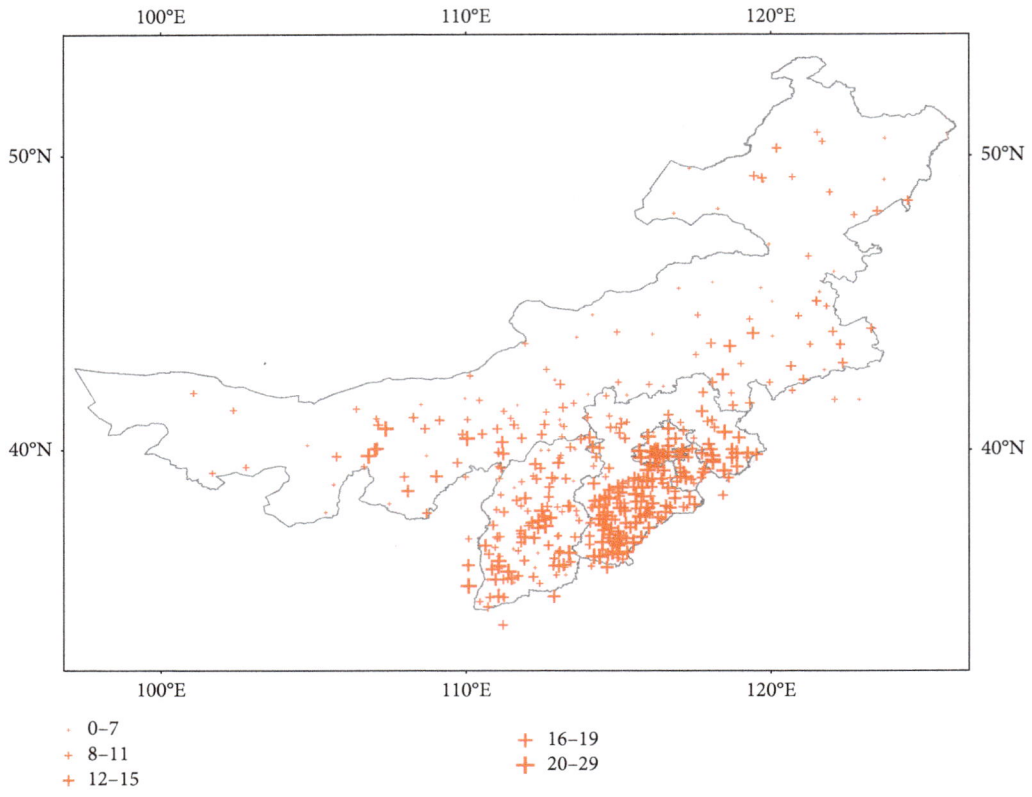

(a)

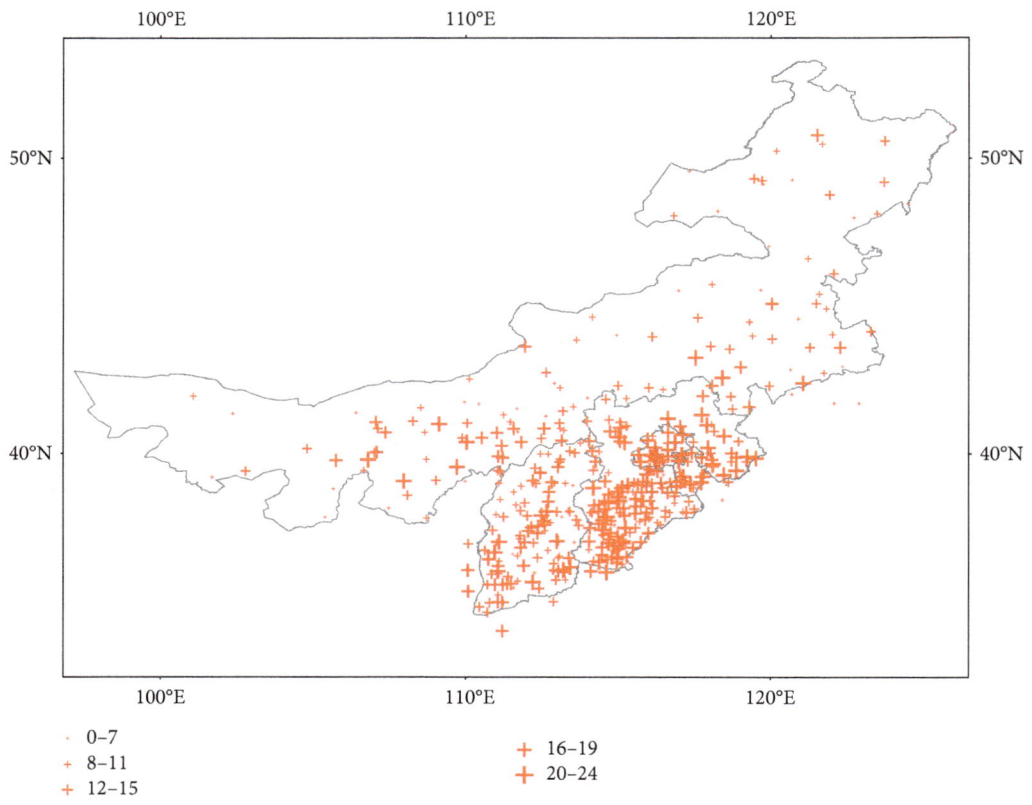

(b)

FIGURE 4: Continued.

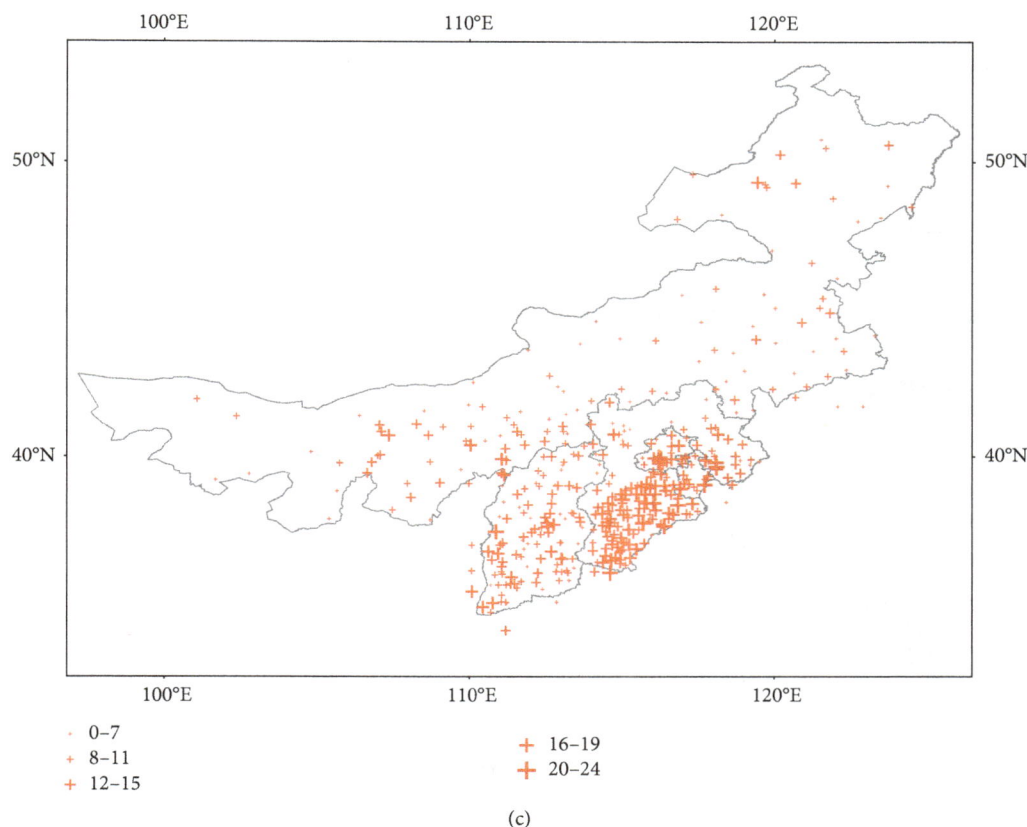

FIGURE 4: Spatial distribution of breakpoints in (a) the average temperature series (T_{ave}), (b) the minimum temperature series (T_{min}), and (c) the maximum temperature series (T_{max}) of 404 meteorological stations in North China.

breakpoints after 2000 in the temperature series of North China.

With the strict algorithm in the MASH procedure, no breakpoint has been detected in the daily average temperature series of Xinbarag Left Banner (50618) station in Inner Mongolia. Probably because few stations are around this station, and thus, the reference series has been assigned a low weight. Correcting the inhomogeneous series of all other stations in North China, the overall annual average temperature drops by 0.04°C, which shows an upward trend from 2012 (Figure 5(a)). All the daily minimum temperature series have been detected breakpoints. The average annual minimum temperature decreases by 0.06°C after the correction, in which the correction amplitude in 1992 is relatively larger, showing an upward trend from 2012 (Figure 5(b)). No breakpoint has been detected in the daily maximum temperature series of Xinbarag Left Banner (50618) station either. After correcting the inhomogeneous series of this parameter, its average increases by 0.01°C, also with a large correction amplitude in 1992 and an upward trend since 2012 (Figure 5), which is consistent with the work of Zhang [30]. As shown in Figures 2 and 3, the number of breakpoints detected before and after 1992 is not many. From the meteorological station metadata, few stations were moved and no large-scale instrument replacement was done in 1992. The

significant increase of the correction amplitude for minimum and maximum temperature series in 1992 is an issue that needs further study.

From the spatial distribution of differences before and after the correction, the average temperature and the minimum temperature series have much larger correction amplitude than the maximum temperature (Figure 6). Meteorological stations located in the Taihang Mountains, Yinshan Mountains, and the junction area between Hebei Province and Shandong Province have relatively larger correction amplitude. After the adjustment, many series of annual average and minimum temperature drop. Among them, 222 and 182 average temperature series have negative and positive differences, respectively. For the minimum temperature, 228 series have negative differences and 176 have positive differences, respectively. For the maximum temperature series, 209 have negative differences and 195 have positive differences, respectively. Although many temperature series have lower values after correction, the series with positive differences are assigned large correction amplitude. As a result, the overall trends of temperature changes before and after correction are largely the same, implying that the inhomogeneity of data from individual stations has been significantly improved.

After adjustment, the linear trends of 253 series rise and 151 series decline in the average temperature series,

(a)

(b)

(c)

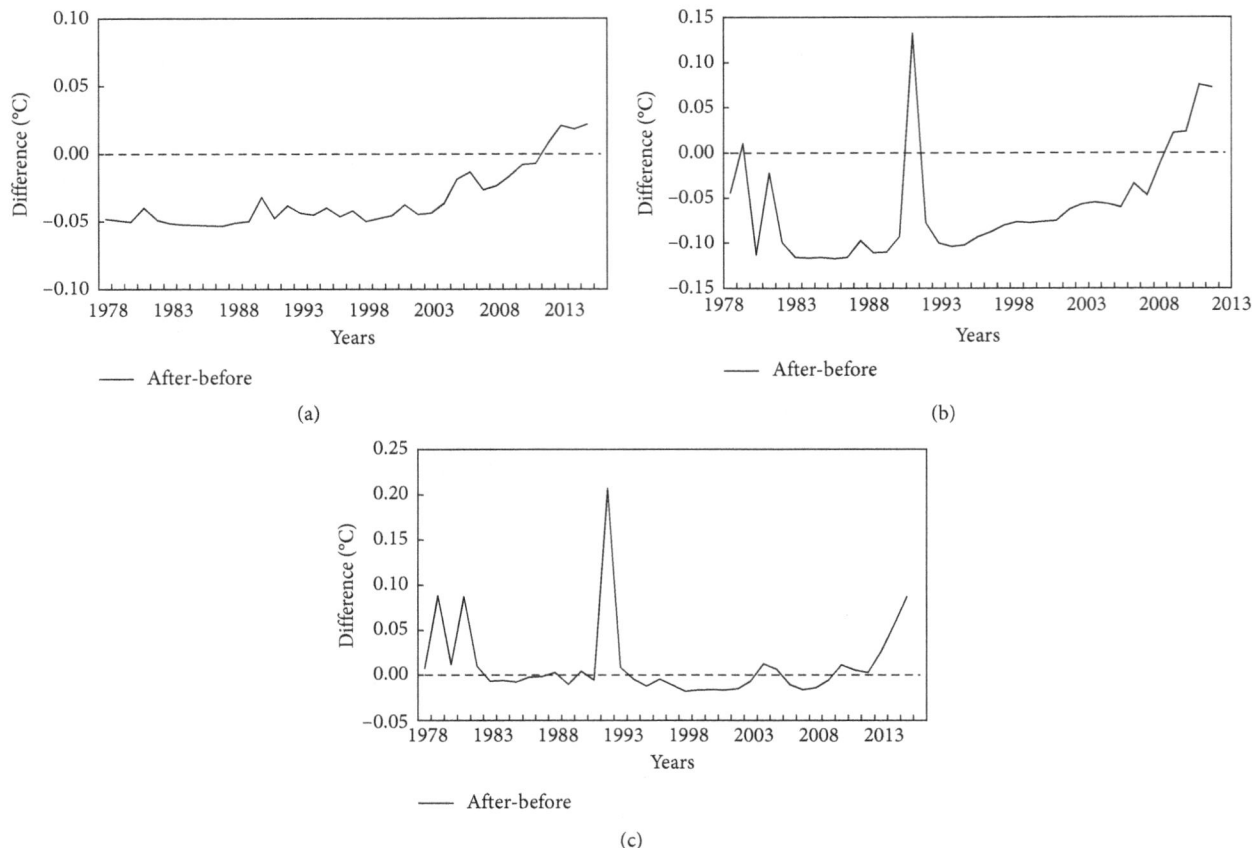

FIGURE 5: Differences of average annual temperature (a), average annual minimum temperature (b), and average annual maximum temperature (c) before and after correction.

respectively, while in the minimum temperature series, the linear trends of 240 series rise and 164 series drop, respectively, and in the maximum temperature series, the linear trends of 218 series rise and 186 series decline, respectively.

The linear trends of the average and maximum temperature series after adjustment are positive except for the shorter series of 54519 (with data beginning on January 1, 1998) and 50924 (with data beginning on January 1, 2006). The 16 negative series with downward trends in the minimum temperature series have been adjusted to positive values with upward trend. The overall trends of the adjusted temperature series are upward. The adjusted series are apparently more in line with the trend of climate warming. At the same time, it proves that the MASH method performs well at middle-high latitudes.

The spatial variations of the trends before and after adjustment are shown by using the inverse distance weighting method to perform spatial interpolation (Figure 7). Here, we take the Wutai Mountains station and Huairou station with the average temperature series as examples. The Wutai Mountains station (53588) was built in October 1955 at the top of Zhongtai of the Wutai Mountains (39°02′N and 113°32′7″E) with an elevation of 2895.8 m. On January 1, 1998, this station was moved to the Muyu mountain of the

Wutai Mountains (38°57′7″N, 113°31′7″E), with an elevation 2208.3 m. The relocation of this station drops nearly 687.5 m in altitude and is 20 km away from the old site, causing obvious rise of temperature since 1998. After adjustment with the MASH method, the upward trend of the temperature of the Wutai Mountains station dropped significantly. The Huairou station (54419) was relocated in 1996 from 40°18′N and 116°37′E to 40°22′N and 116°38′E with horizontal distance about 7.5 km, while its elevation changed from 63.1 m to 75.7 m. The relocation led to the linear trend of the annual average temperature to drop from 0.046°C/a before 1996 to −0.000097°C/a. After adjustment with the MASH method, the trend is 0.034°C/a. These data variations show that inhomogeneity of the two series is well improved.

3.2. Homogenization of Precipitation Series. Homogeneity tests show that 135 of the all 397 precipitation series have breakpoints, majority of which have 1 to 2 breakpoints each (Figure 8). The years with breakpoints are centered around 2000 (Figure 9). Since 2000, the automatic meteorological observation system has been gradually put into operation at the nationwide meteorological stations. The precipitation observation has accordingly been gradually shifted from the artificial way to the automatic approach. Such an instrument

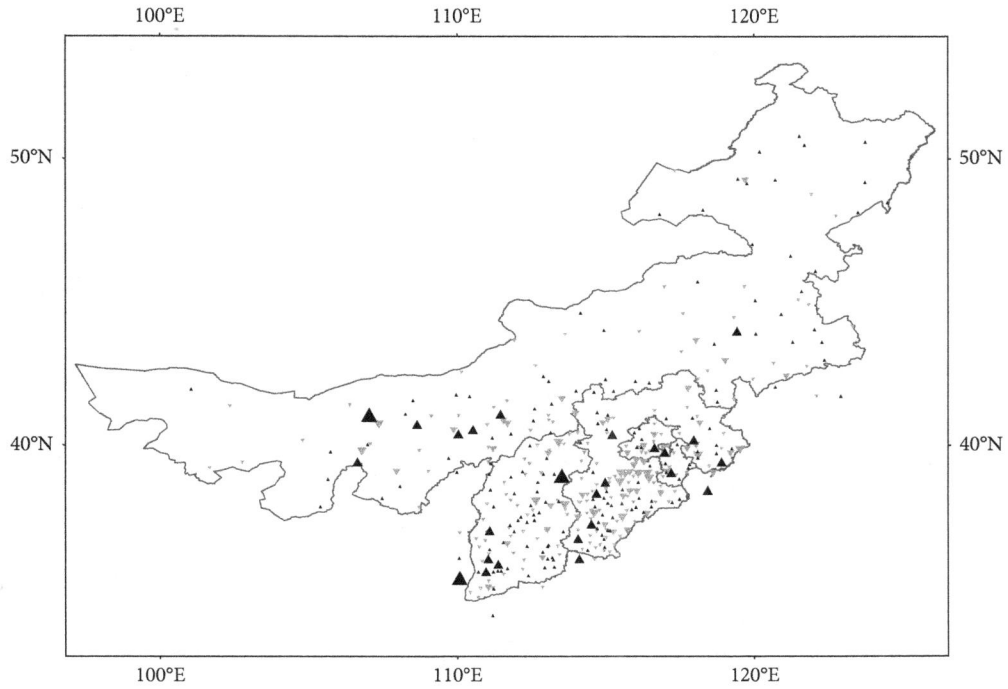

After-before
- · 0 to 0.20
- ▲ 0.20 to 0.50
- ▲ 0.50 to 2.08

- ▽ −0.96 to −0.50
- ▽ −0.50 to −0.20
- · −0.20 to 0

(a)

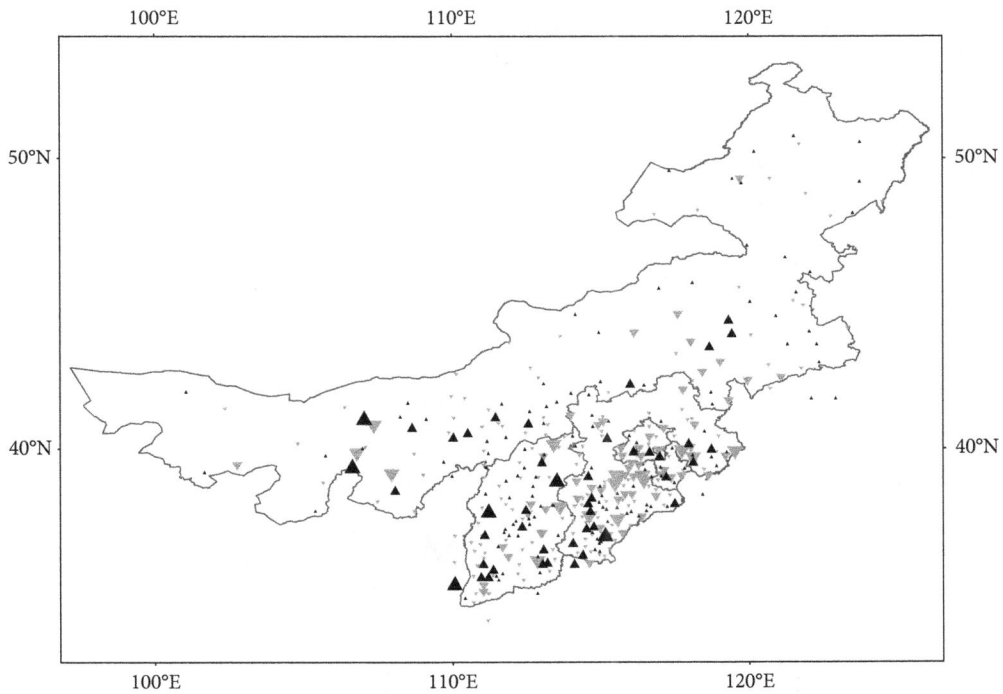

After-before
- · 0 to 0.30
- ▲ 0.30 to 0.90
- ▲ 0.90 to 2.40

- ▽ −2.00 to −0.90
- ▽ −0.90 to −0.30
- · −0.30 to 0

(b)

FIGURE 6: Continued.

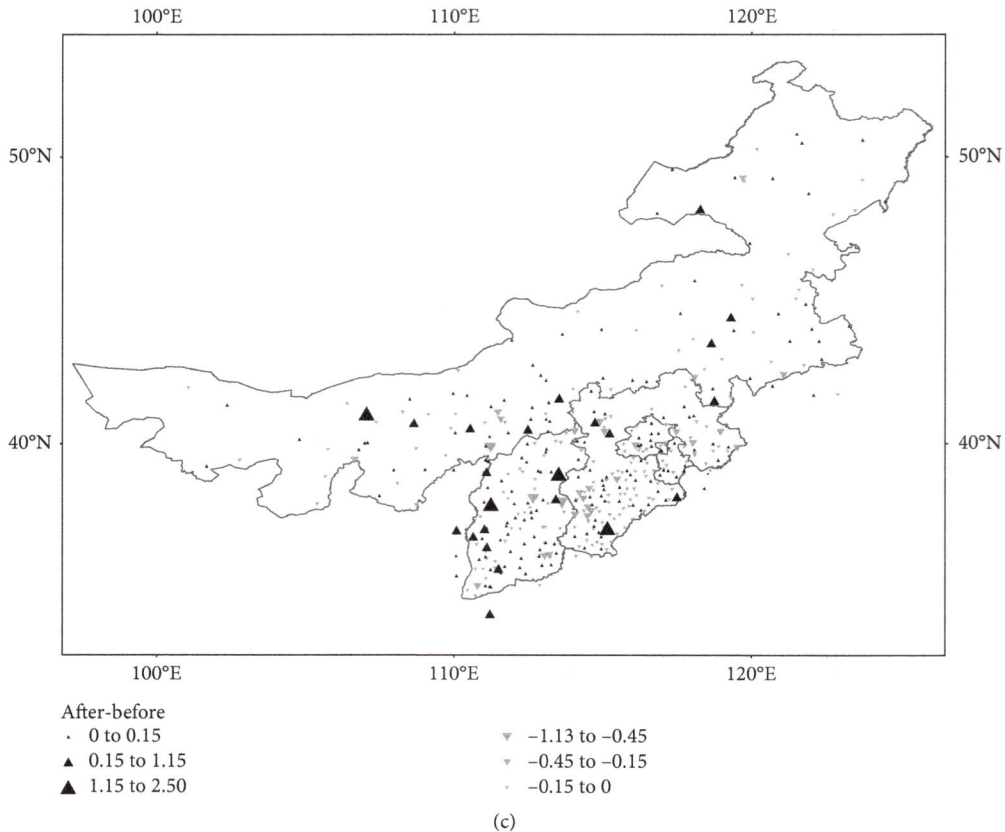

(c)

FIGURE 6: Spatial distribution of differences before and after correction of (a) annual average temperature series (T_{ave}), (b) annual minimum temperature series (T_{min}), and (c) annual maximum temperature series (T_{max}) from 404 meteorological stations in North China.

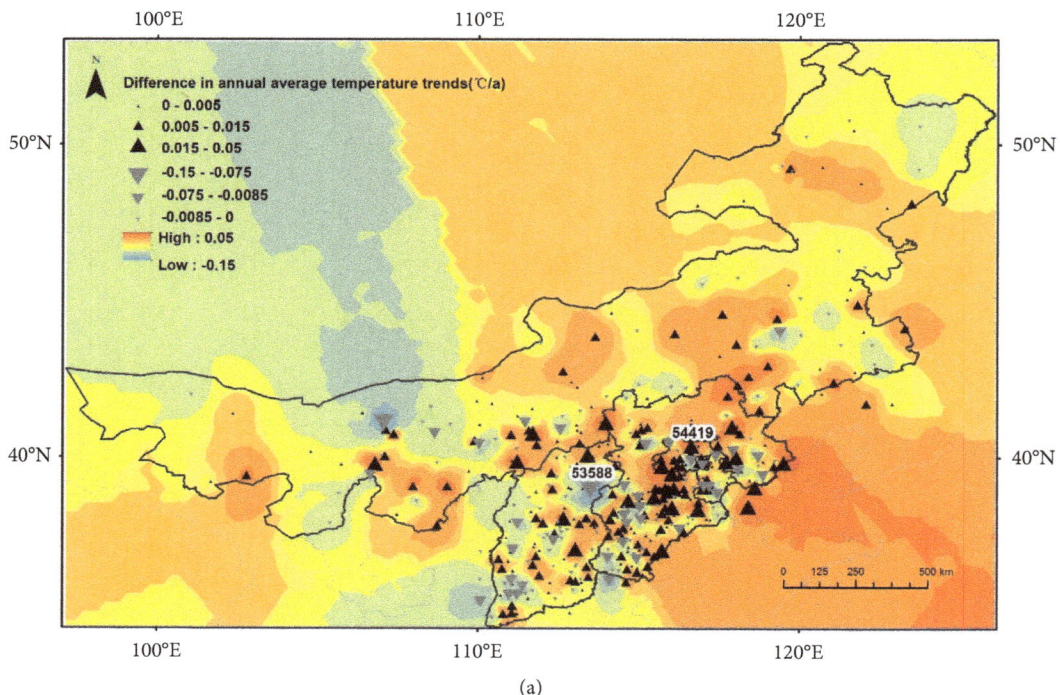

(a)

FIGURE 7: Continued.

(b)

(c)

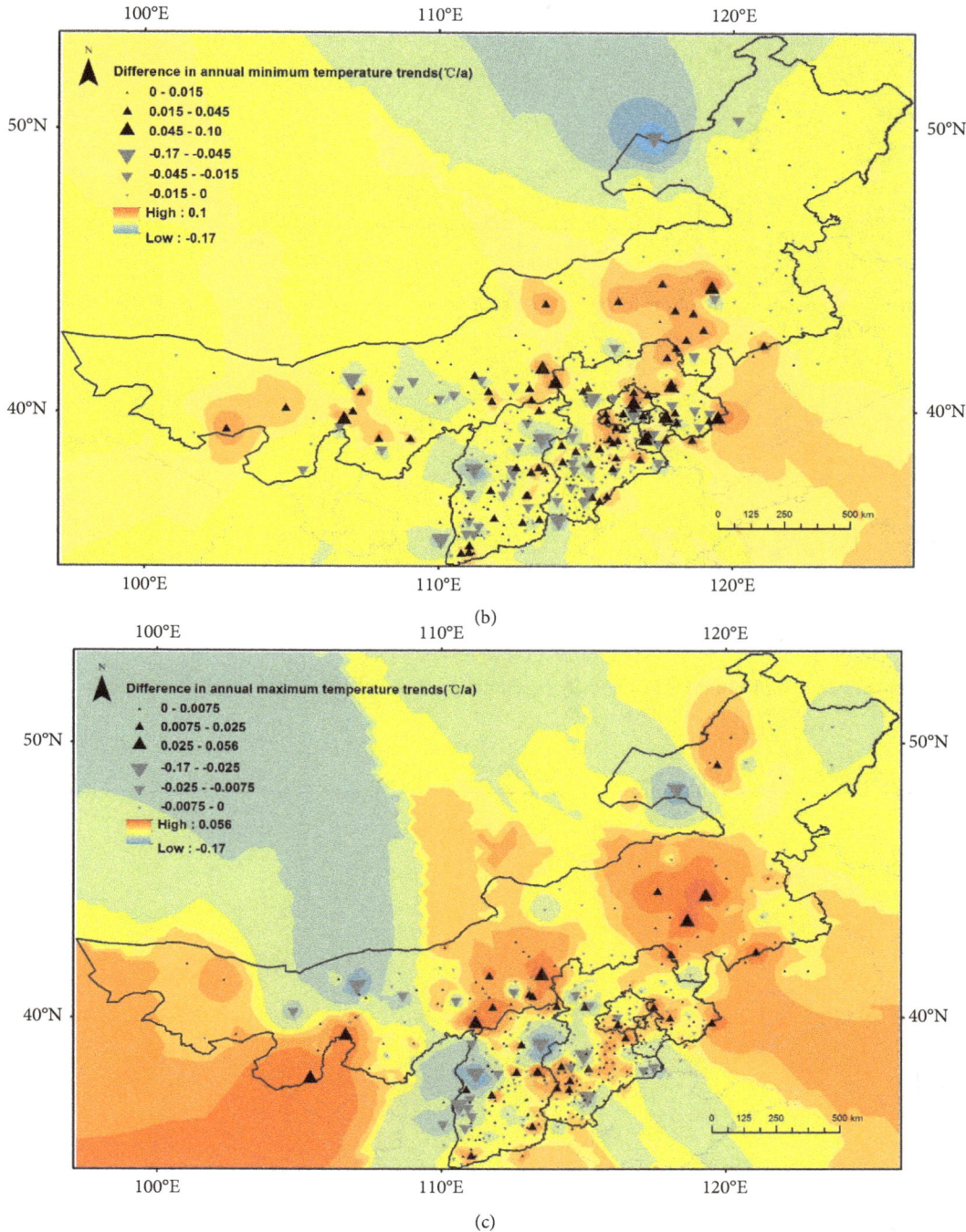

FIGURE 7: Linear trend changes (°C/a) before and after the adjustment of the average, minimum, and maximum temperature of 404 stations in North China.

change led to the inhomogeneous feature in daily precipitation series. Compared with temperature data, precipitation data have lager spatial and temporal variations, especially remarkable spatial discontinuity; thus, the detected breakpoints of temporal precipitation series are relatively less, which makes the homogeneity test difficult to obtain satisfied results. With the improvement of test methods, perhaps more breakpoints could be detected in the precipitation series. Previous study shows that it is difficult

to reveal the implicit impact on precipitation data after a typical horizontal short-distance relocation [31]. From the spatial distribution of the breakpoints (Figure 10), the detected breakpoints are more present in the areas with densely distributed meteorological stations, which is similar to the temperature data.

The corrected annual precipitation is 0.96 mm less than before. The correction amplitude is the largest for 1998, reaching −3 mm (Figure 11). By referring to the

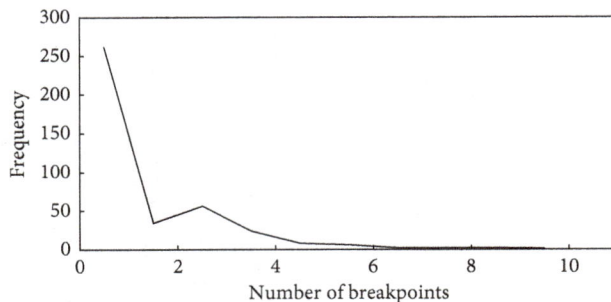

FIGURE 8: Frequency of breakpoint numbers for the daily precipitation series of 397 meteorological stations in North China.

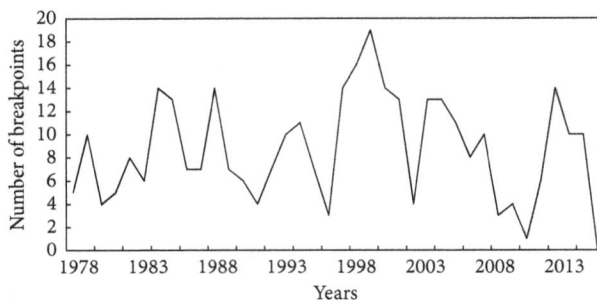

FIGURE 9: Annual breakpoint numbers in the daily precipitation series of 397 meteorological stations in North China.

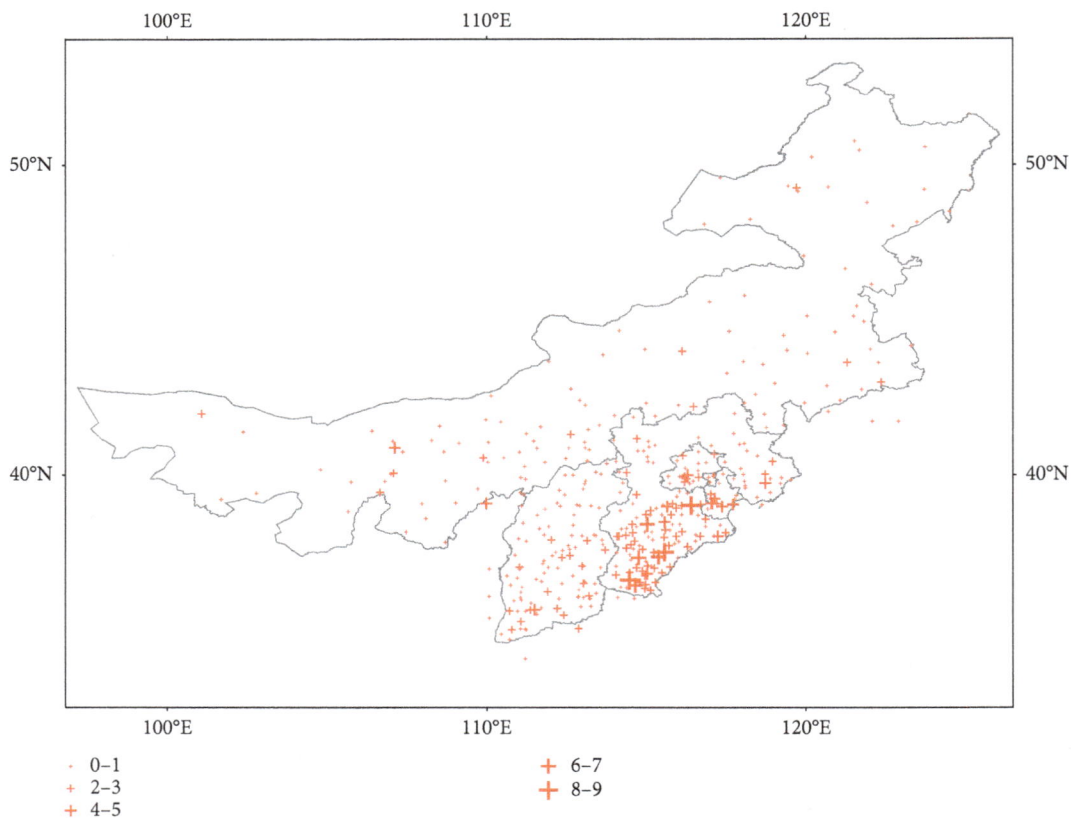

FIGURE 10: Spatial distribution of breakpoints in the daily precipitation data of 397 meteorological stations in North China.

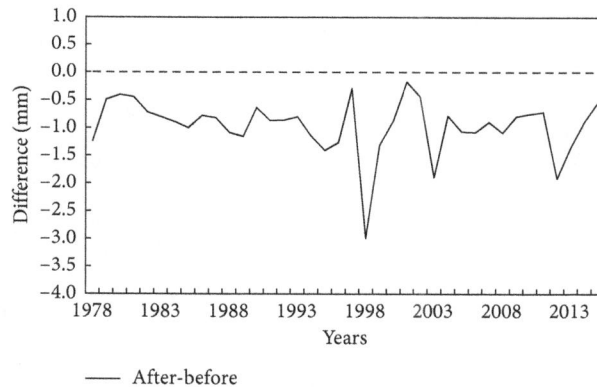

FIGURE 11: Differences of annual precipitation before and after data correction.

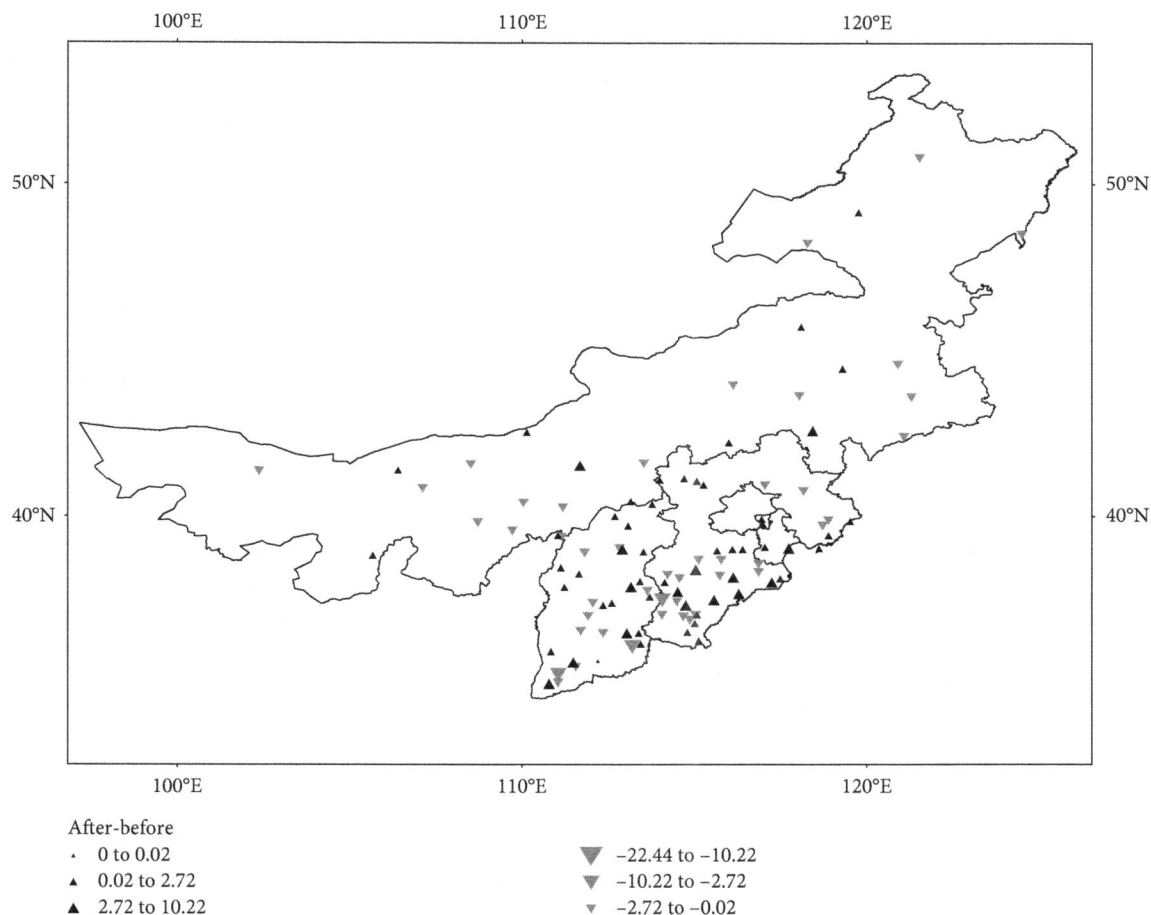

After-before
· 0 to 0.02   ▼ −22.44 to −10.22
▲ 0.02 to 2.72   ▼ −10.22 to −2.72
▲ 2.72 to 10.22   ▼ −2.72 to −0.02

FIGURE 12: Spatial distribution of differences before and after correction of annual precipitation series of 397 meteorological stations in North China.

meteorological station metadata, it is found that the stations that were moved around 1998 are few, and large-scale instrument replacement is also less then. Such fluctuations in the correction amplitude still need further research.

From the spatial distribution of the differences of annual precipitation before and after adjustment, the series with large correction amplitude are mostly distributed in the southern Hebei Province and Taihang Mountains (Figure 12). The values of 339 precipitation series decrease after the correction and 58 series increase after the correction. The corrected series with lower precipitation have relatively larger correction amplitude, which is consistent with the trend of decreasing precipitation in North China in the past 50 years [32].

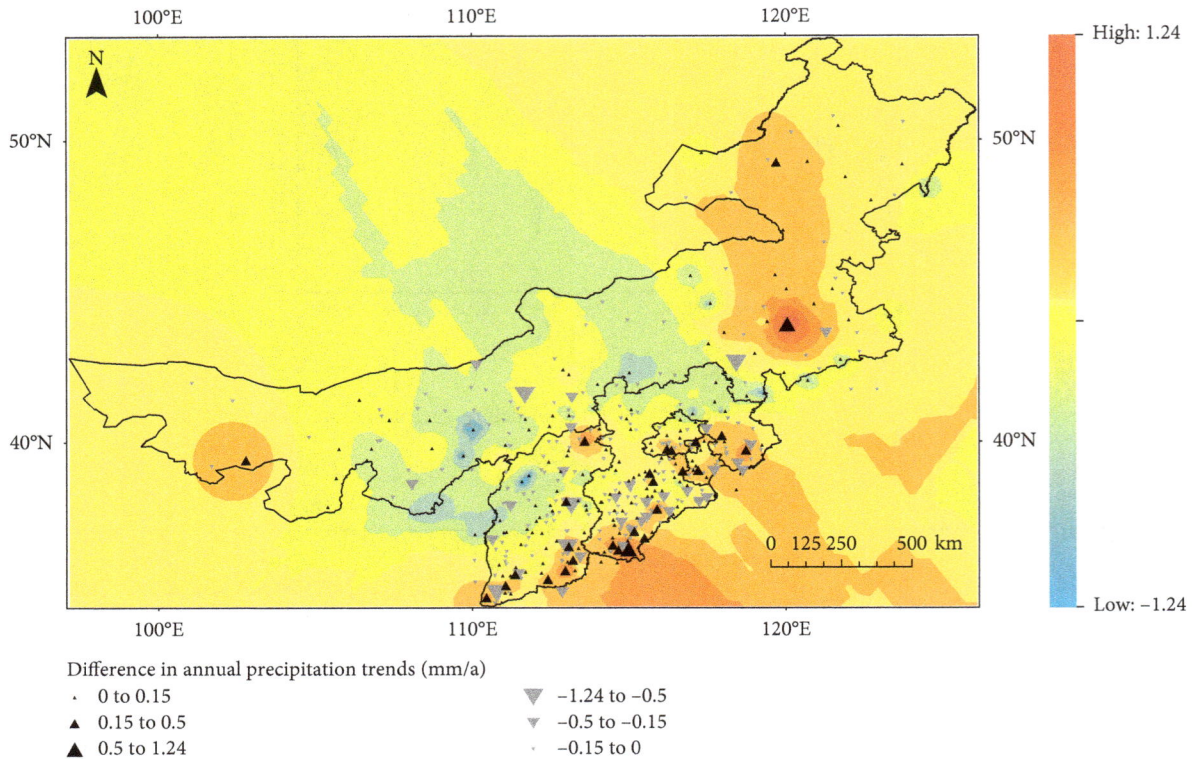

Difference in annual precipitation trends (mm/a)

- · 0 to 0.15
- ▲ 0.15 to 0.5
- ▲ 0.5 to 1.24

- ▼ −1.24 to −0.5
- ▼ −0.5 to −0.15
- · −0.15 to 0

FIGURE 13: Linear trend changes (mm/a) before and after adjustment of annual precipitation series of 397 stations in North China.

Of the adjusted annual precipitation series, the linear trends of 187 series rise and 217 ones decline. After the adjustment, the linear trends of 204 series are negative, compared with 201 before adjustment. The overall trends before and after the adjustment are negative with −0.15 and −0.16, respectively, which shows that the precipitation in North China was gradually decreasing. Compared with the temperature series, although the precipitation series are detected a smaller number of breakpoints, the correction amplitude is much larger. It can be seen from Figure 13 that the areas with large differences before and after the correction are mainly distributed in the vicinity of the Taihang Mountains, Yanshan Mountains, and Yinshan Mountains.

4. Discussion

In Section 3.1, we find that the breakpoints in the minimum temperature series are highest, next is in the average temperature series, and last is in the maximum temperature series. However, in actual observations, the observation method and instruments of the three are unified. In fact, in China, the daily maximum temperature and daily minimum temperature data are derived from minute observation data, while the daily average temperature data are the average of four hourly (02, 08, 14, and 20) observations. Before the emerging of the minute data, the daily maximum temperature and daily minimum temperature data are derived from the artificially observed maximum temperature thermometer and minimum temperature thermometer.

Therefore, in theory, if there are inhomogeneities caused by nonclimate factors, the number of breakpoints in the maximum temperature, average temperature, and minimum temperature data should be consistent. However, majority of existing studies, whether using SNHT or Rhtest methods, have drawn similar conclusions that breakpoints present in the maximum temperature data are less than in the minimum temperature data on different time scales [30, 33]. Li and Deng [34] hold that the reason for this phenomenon may be station migration. Because the minimum temperature is more sensitive to changes in distance. Based on the data extracted from the US Climate Reference Network (USCRN), they found that the minimum temperature variations between stations at short distances are much larger than the maximum temperature variations.

In this study, we calculate the breakpoints in the minimum temperature series and in the maximum temperature series. Combined with the station metadata, in the 244 stations with more breakpoints in the minimum temperature series than in the maximum temperature series, only 54 stations have the years of station migrations coinciding with the years before and after breakpoints, accounting for 22.1%. That is, to say, the change in distance caused by the station migrations may not be the reason why breakpoints in the minimum temperature series are more than in the maximum temperature series.

Combined with the principle of the homogenization algorithm, we infer that under the same calculation conditions, the minimum temperature series with variation and regional differences, whether using the absolute homogenization method or the relative homogenization method, are likely to distinguish some normal values as breakpoints mistakenly. The minimum temperature is more sensitive to distance, and another consequence may be that among the reference series and the candidate series, the differences between the two kinds of series caused by distance are more likely to be detected as breakpoints by mistakes.

Based on the inference above, it is necessary to improve the homogenization algorithm according to the characteristics of different meteorological elements.

5. Conclusions

In this paper, the MASH method is used to test and correct the inhomogeneity of the daily average temperature, maximum temperature, and minimum temperature and daily precipitation series of 404/397 national meteorological stations in North China. Combined with the meteorological station metadata, the results are analyzed and the overall differences before and after the correction are compared:

(1) After 2000, the change of the artificial methods into automatic measurement, and the replacement of the instruments caused the inhomogeneity of the temperature and precipitation series. After 2006, report preparation by hand at most of the meteorological stations was replaced by the automatic manner, making the inhomogeneity of the temperature series more prominent.

(2) The overall annual average temperature after the correction decreases by 0.04°C; the annual minimum temperature decreases by 0.06°C and the annual maximum temperature increases by 0.01°C. The corrected temperature series with lower temperature are more than before, while the corrected temperature series with higher temperature have lager variation amplitude. The temperature in North China during 1978–2015 shows an upward trend. The trend of overall corrected temperature change is largely the same as before, though the homogeneity of the series of individual stations has been significantly improved.

(3) The MASH method-corrected annual precipitation is 0.96 mm less than before. After the correction, most of the precipitation series have less precipitation and relatively larger correction amplitude. Precipitation in North China tends to decrease gradually from 1978 to 2015.

(4) Compared with the original series, the MASH method-corrected annual minimum temperature and annual maximum temperature have abnormal increases in 1992, while the MASH method-corrected annual precipitation has unusually large change in 1998. The causes for these anomalies need further study.

Conflicts of Interest

The authors declare that there are no conflicts of interest regarding the publication of this paper.

Acknowledgments

This research was supported by an International Cooperation Project (41661144037) of National Natural Science Foundation of China (NSFC) and International Centre for Integrated Mountain Development (ICIMOD); Beijing Science and Technology Plan Project "Research on Winter Meteorological Service Technology of Complex Terrain (D171100000717002)"; and the Natural Science Foundation of Anhui Province "Generalised variational assimilation of AIRS water vapor channel brightness temperature and the application study in severe" (1708085QD89). We wish to thank the timely help given by Liu Yu in data processing.

References

[1] H. Alexandersson, "A homogeneity test applied to precipitation data," *International Journal of Climatology*, vol. 6, no. 6, pp. 661–675, 1986.

[2] H. Alexandersson and A. Moberg, "Homogenization of Swedish temperature data. Part I: homogeneity test for linear trends," *International Journal of Climatology*, vol. 17, no. 1, pp. 25–34, 1997.

[3] M. N. Khaliq and T. B. M. J. Ouarda, "A note on the critical values of the Standard Normal Homogeneity Test (SNHT)," *International Journal of Climatology*, vol. 27, no. 5, pp. 681–687, 2007.

[4] M. Andrighetti, D. Zardi, and M. D. Franceschi, "History and analysis of the temperature series of Verona (1769-2006)," *Meteorology and Atmospheric Physics*, vol. 103, no. 1–4, pp. 267–277, 2009.

[5] M. Firat, F. Dikbas, A. C. Koç, and M. Gungor, "Missing data analysis and homogeneity test for Turkish precipitation series," *Sadhana*, vol. 35, no. 6, pp. 707–720, 2010.

[6] I. Hanssenbauer and E. J. Førland, "Homogenizing long norwegian precipitation series," *Journal of Climate*, vol. 7, no. 6, pp. 1001–1013, 1994.

[7] S. Kundu, D. Khare, A. Mondal, and P. K. Mishra, "Analysis of spatial and temporal variation in rainfall trend of Madhya Pradesh, India (1901-2011)," *Environmental Earth Sciences*, vol. 73, no. 12, pp. 8197–8216, 2015.

[8] T. Likso, "Inhomogeneities in temperature time series in Croatia," *Croatian Meteorological Journal*, vol. 38, pp. 3–9, 2003.

[9] K. S. Pandzic and T. Likso, "Homogeneity of average annual air temperature time series for Croatia," *International Journal of Climatology*, vol. 30, no. 8, pp. 1215–1225, 2010.

[10] B. Rudolf, C. Kateřina, D. Petr, and T. Radim, "Climate fluctuations in the Czech Republic during the period 1961-2005," *International Journal of Climatology*, vol. 29, pp. 223–242, 2008.

[11] M. Staudt, M. J. Esteban-Parra, and Y. Castro-Díez, "Homogenization of long-term monthly Spanish temperature

data," *International Journal of Climatology*, vol. 27, no. 13, pp. 1809–1823, 2007.

[12] S. Sugahara, R. P. Da Rocha, R. Y. Ynoue, and R. B. Da Silveira, "Homogeneity assessment of a station climate series (1933-2005) in the Metropolitan Area of São Paulo: instruments change and urbanization effects," *Theoretical and Applied Climatology*, vol. 107, no. 3-4, pp. 361–374, 2012.

[13] Y. Wang, X. L. Liu, and X. H. Jv, "Differences between automatic and manual observation," *Journal of Applied Meteorological Science*, vol. 18, pp. 849–855, 2007.

[14] X. L. Wang, "Accounting for autocorrelation in detecting mean shifts in climate data series using the penalized maximal t or F test," *Journal of Applied Meteorology and Climatology*, vol. 47, no. 9, pp. 2423–2444, 2008.

[15] G. J. Zhang, J. H. He, Z. J. Zhou, and L. J. Cao, "Homogeneity study of precipitation data over China using RHtest method," *Meteorological Science and Technology*, vol. 40, pp. 914–921, 2012.

[16] T. Szentimrey, "Multiple analysis of series for homogenization (MASH)," in *Proceedings of Second Seminar for Homogenization of Surface Climatological Data*, Budapest, Hungary, 1999.

[17] T. Szentimrey, "Multiple analysis of series for homogenization (MASH); verification procedure for homogenized time series," in *Proceedings of 4th Seminar for Homogenization and Quality Control in Climatological Databases*, Budapest, Hungary, 2003.

[18] V. K. C. Venema, O. Mestre, E. Aguilar et al., "Benchmarking homogenization algorithms for monthly data," *Climate of the Past*, vol. 8, no. 1, pp. 89–115, 2012.

[19] Z. Li and Z. Yan, "Application of multiple analysis of series for homogenization to Beijing daily temperature series (1960-2006)," *Advances in Atmospheric Sciences*, vol. 27, no. 4, pp. 777–787, 2010.

[20] A C Costa and A Soares, "Homogenization of climate data: review and new perspectives using geostatistics," *Mathematical Geosciences*, vol. 41, no. 3, pp. 291–305, 2009.

[21] Z. Li, Z. Yan, and H. Wu, "Updated homogenized chinese temperature series with physical consistency," *Atmospheric and Oceanic Science Letters*, vol. 8, pp. 17–22, 2015.

[22] M. Lakatos, T. Szentimrey, Z. Bihari, and S. Szalai, "Creation of a homogenized climate database for the Carpathian region by applying the MASH procedure and the preliminary analysis of the data," *Idojaras*, vol. 117, no. 1, pp. 143–158, 2013.

[23] Y. L. Li, Z. H. Ren, G. F. Chen, Q. L. Xia, Y. He, and P. Yu, "Causes and impact analysis of errors between temperatures obtained by automatic and manual observations at 143 national automatic benchmark stations," *Meteorological Monthly*, vol. 41, pp. 1007–1016, 2015.

[24] Q. Li, Z. Jiang, Q. Huang, and Y. You, "The experimental detecting and adjusting of the precipitation data homogeneity in the Yangtze Delta," *Journal of Applied Meteorological Science*, vol. 19, pp. 219–226, 2008.

[25] Y. Su and Q. Li, "Improvement in homogeneity analysis method and update of China precipitation data," *Progressus Inquisitiones de Mutatione Climatis*, vol. 10, pp. 276–281, 2014.

[26] N. A. C. Cressie, *Statistics for Spatial Data*, Wiley, Hoboken, NJ, USA, 2015.

[27] T. Szentimrey, "Methodological questions of series comparison," in *Proceedings of the Sixth Seminar for Homogenization and Quality Control in Climatological Databases*, pp. 1–7, Budapest, Hungary, 2008.

[28] M. Chen, X. B. Gai, and X. Y. Fan, "Analysis on the influence of automatic station data on the sequence continuity of meteorological historical data," *Journal of Anhui Agricultural Sciences*, vol. 39, pp. 12356-12357, 2011.

[29] X. L. Wang, Q. H. Wen, and Y. Wu, "Penalized maximal t test for detecting undocumented mean change in climate data series," *Journal of Applied Meteorology and Climatology*, vol. 46, no. 6, pp. 916–931, 2007.

[30] G. J. Zhang, *Homogeneity Study of the Temperature and Precipitation Data over China using Multi-Methods*, Nanjing University of Information Science and Technology, Nanjing, China, 2012.

[31] R. Qin et al., "Data comparison and analysis of kashgar national reference climatological station before and after moving," *Desert and Oasis Meteorology*, vol. 9, pp. 55–61, 2015.

[32] L. S. Hao, *Spatial-Temporal Variation of the Precipitation in North China and the Impact Factors of Precipitation Reduction*, Nanjing University of Information Science and Technology, Nanjing, China, 2011.

[33] Q. Li, M. J. Menne, C. N. Williams, and B. Sun, "Detection of discontinuities in chinese temperature series using a multiple test approach," *Climatic and Environmental Research*, vol. 10, pp. 736–742, 2005.

[34] Q. Li and W. Deng, "Detection and adjustment of undocumented discontinuities in Chinese temperature series using a composite approach," *Advances in Atmospheric Sciences*, vol. 26, no. 1, pp. 143–153, 2009.

14

Spatial Distribution and Temporal Trend of Tropospheric NO$_2$ over the Wanjiang City Belt of China

Yu Xie [ID],[1] Wei Wang [ID],[2] and Qinglong Wang[1]

[1]*Department of Electronic Information and Electrical Engineering, Hefei University, Hefei 230601, Anhui, China*
[2]*Key Laboratory of Environmental Optics and Technology, Anhui Institute of Optics and Fine Mechanics, Chinese Academy of Sciences, Hefei 230031, Anhui, China*

Correspondence should be addressed to Wei Wang; wwang@aiofm.ac.cn

Academic Editor: Enrico Ferrero

We utilize the tropospheric NO$_2$ columns derived from the observations of Ozone Monitoring Instrument (OMI) onboard AURA to analyze the spatial distributions and temporal trends of NO$_2$ in Wanjiang City Belt (WCB) of China from 2005 to 2016. The aim of this study is to assess the effect of industrial transfer policy on the air quality in WCB. Firstly, we used the surface in situ NO$_2$ concentrations to compare with the OMI-retrieved tropospheric NO$_2$ columns in order to verify the accuracy of the satellite data over the WCB area. Although it is difficult to compare the two datasets directly, the comparison results prove the accuracy of the OMI-retrieved tropospheric NO$_2$ columns in cities of WCB. Then, the spatial distributions of the annual averaged tropospheric NO$_2$ total columns over Anhui Province show that NO$_2$ columns were considerably higher in WCB than those in other areas of Anhui. Also, we compared the spatial distributions of the total NO$_2$ columns in 2005 through 2010 and in 2011 through 2016 and found that the total NO$_2$ columns in WCB increased by 19.9%, while the corresponding value increased only 13.9% in other Anhui areas except the WCB area. Furthermore, the temporal variations of NO$_2$ columns show that although the NO$_2$ columns over WCB and Anhui increased significantly from 2005 to 2011, they decreased sharply from 2011 to 2016 due to the strict emission reduction measures in China. Finally, the HYSPLIT model was used to analyze the origins of NO$_2$ and transport pathways of air masses in a typical city, Ma'anshan city.

1. Introduction

Nitrogen dioxide (NO$_2$) is a reactive, short-lived atmospheric trace gas with both natural and anthropogenic sources. Major sources of NO$_2$ are fossil fuel combustion, biomass burning, soil emissions, and lightning [1]. NO$_2$ is a toxic air pollutant on the condition of high concentration and plays an important role in tropospheric chemistry as a precursor of tropospheric ozone and secondary aerosols [2]. Observations of the spatiotemporal variations of NO$_2$ form the basis of understanding the spatial distributions and temporal trends of NO$_2$.

Many techniques and methods have been successfully used in monitoring atmospheric NO$_2$ based on surface in situ measurements, remote sensing from satellite sensors, and ground-based instruments [3–8]. Although the in situ measurements and remote sensing from ground-based instruments show high accuracy and precision, their usefulness in determining the spatiotemporal distributions of trace gases is limited due to their sparse spatial and temporal coverage. Space-based measurements provide information on NO$_2$ distributions at a large scale and over areas where in situ and ground-based systems cannot be easily deployed [9].

A series of sun-synchronous satellites were launched with spectrometers, which allowed scientists to observe the global distribution of several important tropospheric trace gases including NO$_2$, SO$_2$, and O$_3$. Satellite observations make it easy to understand the spatiotemporal variations of atmospheric NO$_2$ [10–13]. Lamsal et al. examined the seasonal variation in lower tropospheric NO$_2$ by the observation of the OMI, in situ surface measurements, and a global GEOS-Chem model [9]. Ul-Haq et al. applied the linear regression model for tropospheric NO_2 and the anthropogenic NO_x emissions using OMI data [14]. Gu et al. used the NO_2 columns observed from OMI and the Community Multiscale Air Quality (CMAQ) model to derive the ground-level NO_2 concentrations in China [15]. Sharma et al. presented the temporal variations of surface NO_x during 2012 to 2014 at an urban site of Delhi, India [16]. Varotsos et al. found a progressive increase of mean values of NO_2/NO_x versus NO_x when the level of NO_x increases in Athens, Greece [17].

Han et al. used the tropospheric NO_2 columns observed from OMI to compare with the bottom-up emissions of NO_2 derived from the CMAQ model and three emission inventories over East Asia [18]. Liu et al. [19] analyzed the NO_x emission trends and the major reasons for changes over China from satellite observations and the Multi-resolution Emission Inventory for China (MEIC). The emissions derived from the bottom-up method based on the inventory and the top-down method based on OMI observations showed good agreement. Lamsal et al. used aircraft and surface in situ measurements as well as ground-based remote sensing data to validate the OMI retrieval of tropospheric NO_2 [20]. Ialongo et al. compared the OMI NO_2 total columns with the ground-based remote sensing data collected by the Pandora spectrometer to evaluate the satellite data product at high latitudes [21]. Tong et al. utilized OMI observations and Air Quality System (AQS) data to study the long-term NO_x trends over eight large US cities [22]. McLinden et al. combined OMI observations with a regional-scale air quality model to monitor the air quality of the Canadian Oil Sands [23]. Kim et al. used three regression models in conjunction with OMI tropospheric NO_2 columns to estimate the surface NO_2 volume mixing ratio in five cities of South Korea [24].

Tropospheric NO_2 vertical columns obtained from satellite instruments have been widely used to study NO_x pollutions over China [25–27]. Lin found that the anthropogenic emissions are the dominant source of NO_x over East China [28]. Understanding global and regional distributions and temporal trends of the pollution gases provides a basis for development of mitigation strategies. Most studies focus on the air quality of North China Plain, Pearl River Delta, and Yangtze River Delta [28, 29], which are the economic development centers of China and have regional heavy pollutions. But we pay little attention to mideastern China. Mideastern China is experiencing significant socioeconomic changes following the national industrial transfer strategies. Excessive development in a limited number of regions tends to be unsustainable because of limited resources. So industrial transfer is performed from the coastal areas to the inland areas [30]. Wanjiang City Belt (WCB) was established in January 2010 by National Development and Reform Commission (NDRC) of China to make the industrial transfer from the Yangtze River Delta and other mega-regions to Anhui Province [31]. Anhui is located in the mideastern region of China.

The aim of this study is to describe the spatial distributions and temporal trends of tropospheric NO_2 based on satellite observations in twelve years in Anhui, in order to

assess the effect of industrial transfer policy on the air quality in WCB. This paper is organized as follows. Firstly, the materials and methods used are described in Section 2. The area of Wanjiang City Belt, satellite data, surface in situ data, and HYSPLIT model used in the analysis are introduced. Secondly, results and discussion are presented in Section 3. Comparisons of satellite data with surface in situ data for NO_2 in WCB are made in Section 3.1. The spatial distributions of the annual averaged tropospheric NO_2 total columns in Anhui Province are shown in Section 3.2. The variation of the total tropospheric NO_2 columns before and after establishment of the WCB is discussed in Section 3.3. Also, the seasonal variations of tropospheric NO_2 are analyzed in Section 3.4. Finally, conclusions are presented in Section 4.

2. Materials and Methods

2.1. The Introduction of Wanjiang City Belt. Industrial transfer is one of the important national strategies in China. The construction of WCB is the first approved demonstration area for industrial transfer on the national level [32]. WCB comprises 59 counties in Anhui Province along the Yangtze River, including Anqing, Chaohu, Chizhou, Chuzhou, Hefei, Ma'anshan, Tongling, Wuhu, Xuancheng, Jin'an District, and Shucheng County of Lu'an [33]. Figure 1 shows the location of Anhui Province, and the green area represents WCB.

Anhui Province has diverse topography, as shown in Figure 2. The north of Anhui belongs to the North China Plain, while the north-central areas are part of the Huai River Plain. The two regions are flat with dense population. The south of the province is characterized by uneven topography. The Yangtze River runs through the south of Anhui between the Dabie Mountains and a series of hills.

2.2. Satellite Data. OMI is an ultraviolet/visible spectrometer aboard the NASA's EOS Aura satellite. The instrument provides information on trace gases, such as ozone (O_3), sulfur dioxide (SO_2), and nitrogen dioxide (NO_2), and other pollutants retrieved from the spectral region between 270 and 500 nm [34]. EOS Aura circles in a polar sun-synchronous orbit with a 98.2° inclination to the equator, at an altitude of around 705 km. The overpass times are about 13:45 mean local solar time [11, 34].

In the present study, we collect the OMI-retrieved tropospheric NO_2 columns from Royal Netherlands Meteorological Institute (KNMI) DOMINO v2.0 products from 2005 to 2016, which are available at http://www.temis.nl/airpollution/no2col/no2regioomimonth_v2.php. The spatial resolution is 0.125 × 0.125° latitude-longitude, which has been widely used for scientific applications [21, 35, 36]. We used the monthly mean data to analyze the spatial distribution and temporal trends.

2.3. Surface In Situ Data. The Chinese Ministry of Environmental Protection issued "construction scheme of National Environmental Monitoring Network (in cities at the

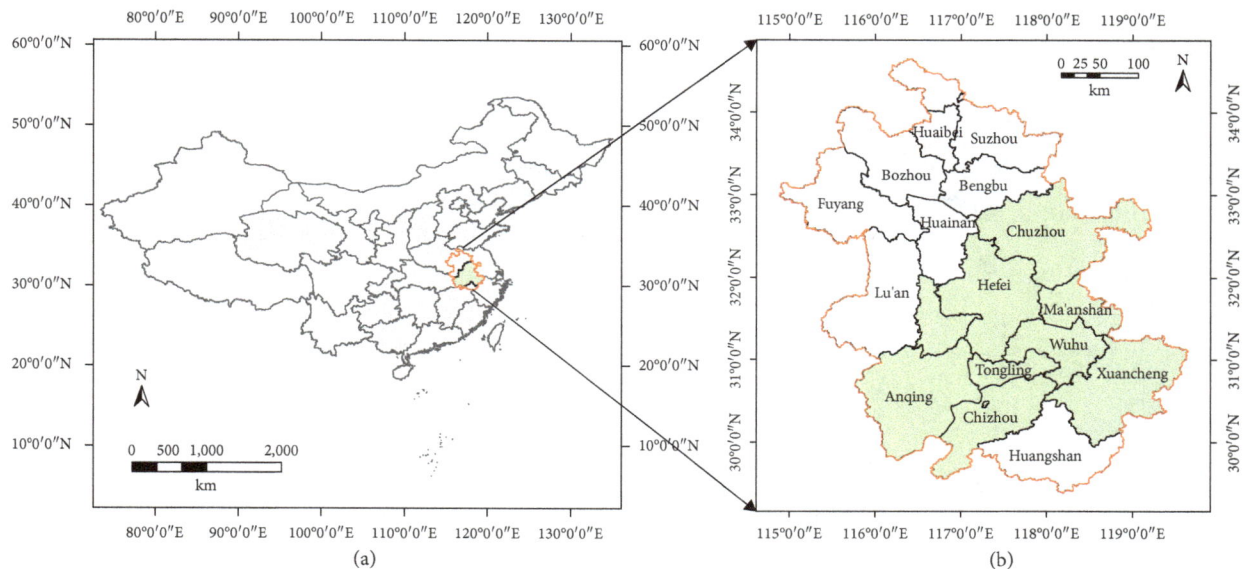

FIGURE 1: The location of (a) Anhui Province and (b) Wanjiang City Belt (WCB).

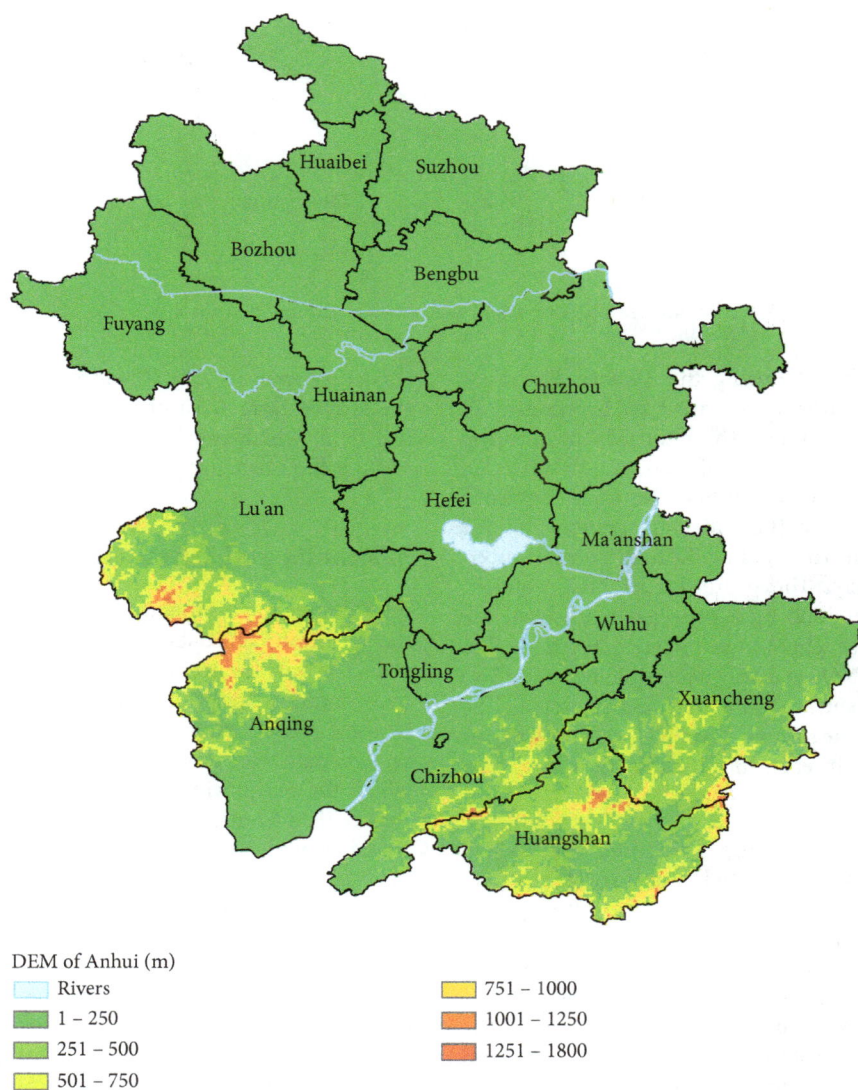

FIGURE 2: Digital elevation model (DEM) of Anhui (m).

Figure 3: Distribution of CNEMC stations in Anhui.

prefecture level and above) during the Twelfth Five-Year Plan" in 2012. 1436 monitoring stations have been set up in 338 cities in China since then. These surface monitoring stations provide the concentrations of NO_2, SO_2, PM10, CO, O_3, and PM2.5 and visibility. Chinese National Environmental Monitoring Center (CNEMC) is responsible for publishing the near-real-time data collected from all monitoring stations publicly. The ground-level NO_2 concentrations are mainly obtained by a nitrogen oxide analyzer based on the gas-phase chemiluminescence method. The surface in situ data of NO_2 are only accessible from 2015 to 2016, so we use the surface data in the two years. The temporal resolution of the ground-level NO_2 concentrations is one datum per hour in each monitoring station.

2.4. HYSPLIT Model. In this study, we used the Hybrid Single-Particle Lagrangian Integrated Trajectory (HYSPLIT) model developed by National Oceanic and Atmospheric Administration (NOAA) to simulate the back trajectories of air mass [37]. The HYSPLIT model is a complete system, which has been extensively used in calculation of air mass trajectories, atmospheric transport, and dispersion. The model is often used to locate the origin of air masses and build the relationships between source and receptor by back

trajectory analysis [38]. The input for the HYSPLIT model is the Global Data Assimilation System (GDAS) meteorological data, which are available at the GDAS website (ftp://arlftp.arlhq.noaa.gov/pub/archives/gdas1).

3. Results and Discussion

3.1. Comparison of Satellite Data with Surface Data. We used the surface in situ data to compare with the satellite data in order to verify the accuracy of the OMI-retrieved tropospheric NO_2 columns. The ground-level NO_2 concentrations observed by the CNEMC stations in 2015 and 2016 were utilized. The satellite data were extracted corresponding to the data grid in which the monitoring stations are located. We collected the surface data from 13:00 to 14:00 everyday, as this time period coincides with the OMI overpass local time. Figure 3 shows the selected CNEMC stations in Anhui Province.

The monthly averaged data from each CNEMC station in Anhui are compared with satellite data. Figure 4 shows the comparison results of the two datasets. From Figure 4, the two data show almost the same variation trend of NO_2 in each city. The Pearson linear correlation coefficients of the data for each area are high, as listed in Table 1. We used the 2-tailed test to test the statistical significance of the

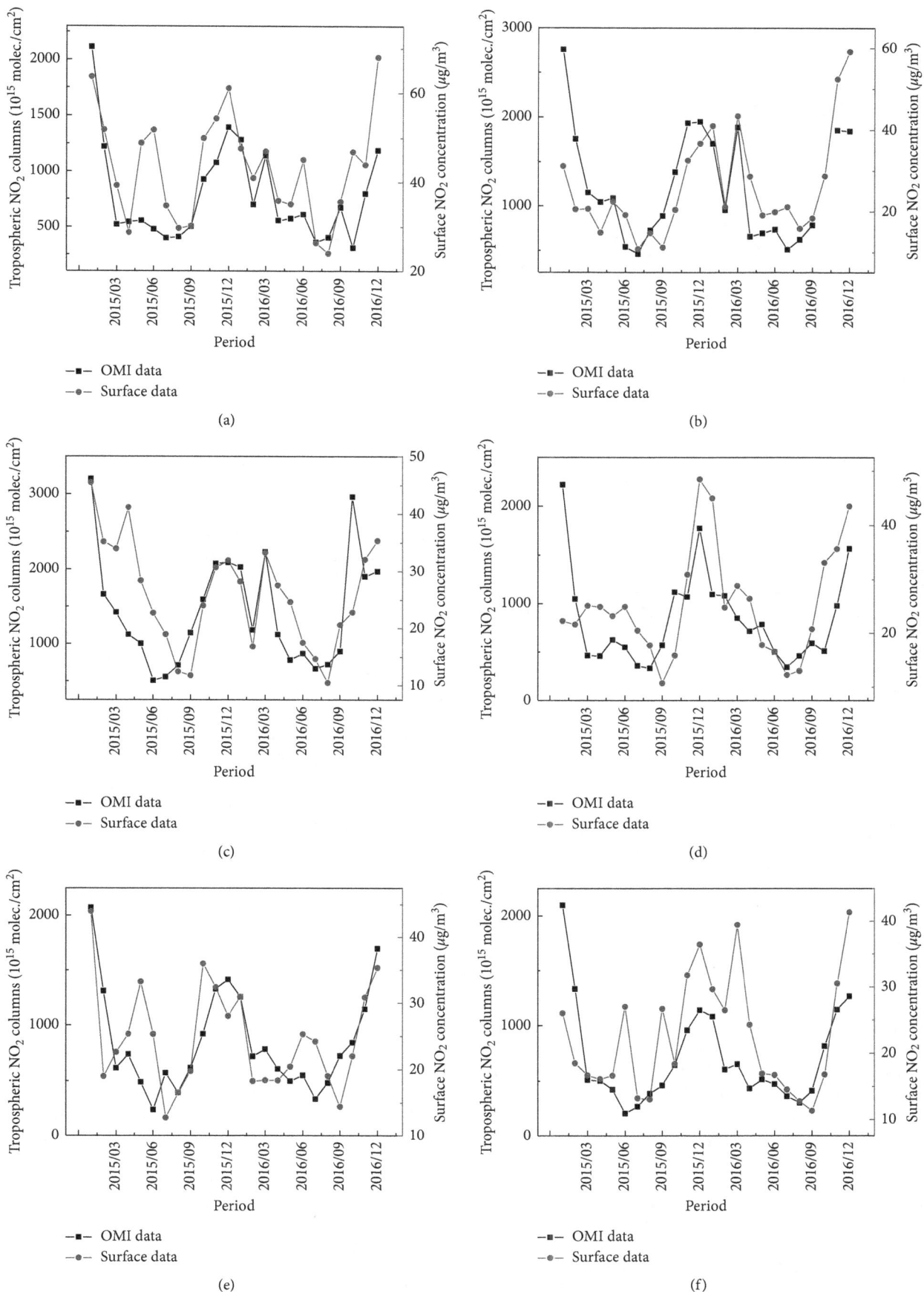

FIGURE 4: Continued.

(g)

(h)

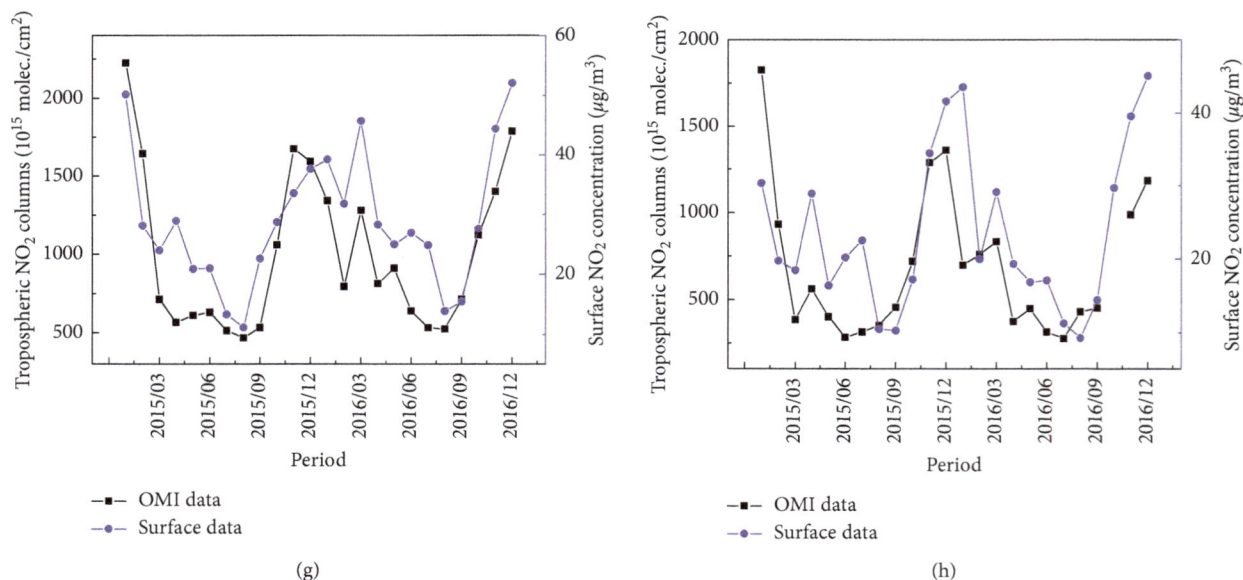

FIGURE 4: Comparison of OMI-retrieved tropospheric NO_2 columns and surface in situ concentrations for eight WCB cities: (a) Hefei; (b) Wuhu; (c) Ma'anshan; (d) Chuzhou; (e) Chizhou; (f) Anqing; (g) Tongling; (h) Xuancheng.

TABLE 1: The Pearson correlation coefficients of OMI-retrieved tropospheric NO_2 columns and surface in situ concentrations for eight WCB cities.

City	Correlation coefficient
Hefei	0.75
Wuhu	0.69
Ma'anshan	0.63
Chuzhou	0.53
Chizhou	0.65
Anqing	0.50
Tongling	0.83
Xuancheng	0.69

correlation coefficient, and the correlation is significant at the 0.05 level. The two datasets have different spatial scale representativeness and surface sensitivity, so it is unreasonable to compare the two datasets directly. However, our comparison results prove the accuracy of the OMI-retrieved tropospheric NO_2 columns in the WCB area.

3.2. Spatial Distributions of Tropospheric NO_2 in WCB. Figure 5 plots the spatial distributions of the annual averaged tropospheric NO_2 total columns over Anhui throughout the years from 2005 to 2016. It is found that the NO_2 columns were considerably higher in WCB than those in other areas of Anhui. The annual averaged NO_2 column reached 811×10^{13} molec./cm^2 in the WCB region from 2005 to 2016, while the annual averaged NO_2 column was 733×10^{13} molec./cm^2 in other areas of Anhui during the same period. The p value is 0.09221 when the two-sample t test is used. Figure 6 is the plot of the histogram of the annual averaged tropospheric NO_2 total columns in each city of Anhui Province. As can be seen from Figure 6, the highest NO_2 columns appeared in the Ma'anshan city, where iron and steel industry is the

major industry with high emission of pollutions. The spatial distributions of annual averaged tropospheric NO_2 columns in this province agree with the results of satellite-retrieved NO_2 emissions in eastern China in other studies [1, 10].

3.3. Temporal Trends of Tropospheric NO_2 in WCB. The WCB region is established in 2010, so we compared the tropospheric NO_2 columns in this region before and after the year of 2010. Figure 7 shows the spatial distributions of the total NO_2 columns in 2005 through 2010 and in 2011 through 2016 as well as the difference between the two periods. The difference between the two periods represents the change in the total NO_2 columns. The total NO_2 columns in WCB increased from 531×10^{15} molec./cm^2 to 637×10^{15} molec./cm^2, and the relative increase rate is about 19.9% between the two periods. The total NO_2 columns in other Anhui areas except the WCB area increased from 494×10^{15} molec./cm^2 to 563×10^{15} molec./cm^2, and the relative increase rate is about 13.9%. It is clear that the total columns in WCB increased more than those in other areas, which may result from the construction of the WCB and the policy of industrial transfer from eastern coastal areas to inland areas. Furthermore, the fraction of tropospheric NO_2 columns in WCB to the total tropospheric NO_2 columns in Anhui is up to 59.3% in 2016, while this value is 56.6% in 2011 (Figure 8). The increased fraction of tropospheric NO_2 columns after the year of 2011 in WCB also reflects the effect of the construction of WCB on the air quality.

Furthermore, the temporal variations of NO_2 columns are studied. The NO_2 columns of Anhui and WCB from 2005 to 2016 are plotted in Figure 9. Fortunately, it is found that the NO_2 columns over WCB and Anhui increased significantly from 2005 to 2011 and then decreased sharply from 2011 to 2016. The statistical significance of all the linear fits in Figure 9

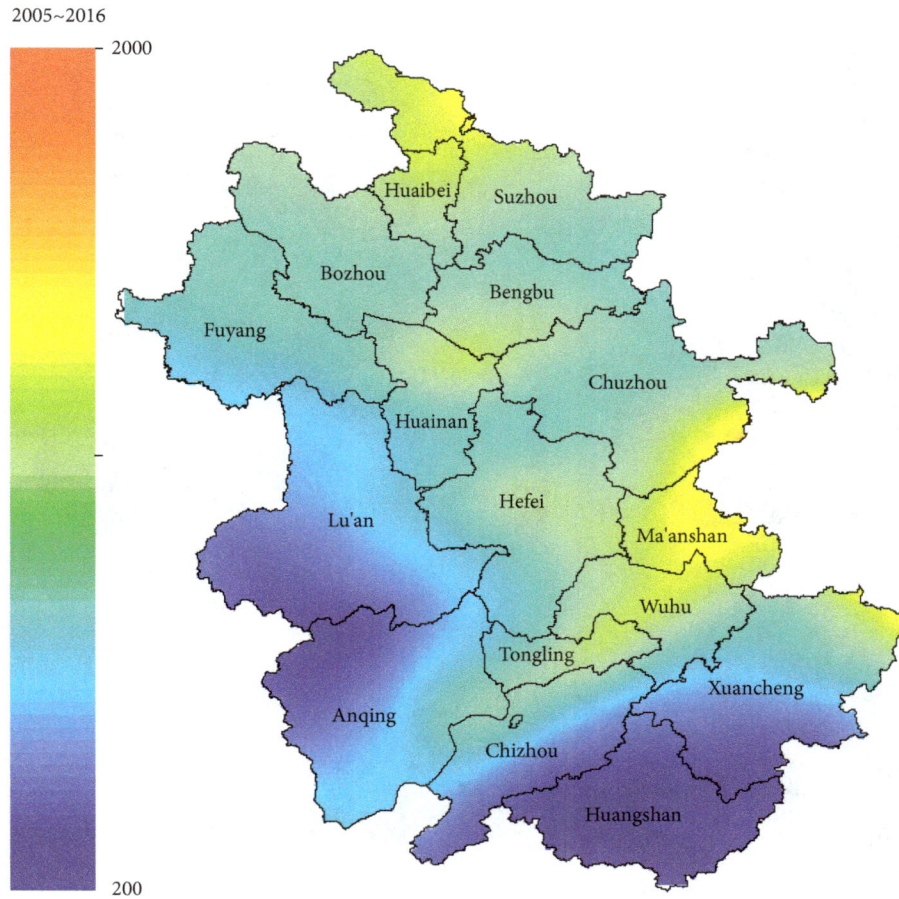

Figure 5: Annual averaged OMI-retrieved NO_2 columns ($\times 10^{13}$ molec./cm^2) from 2005 to 2016 in Anhui Province.

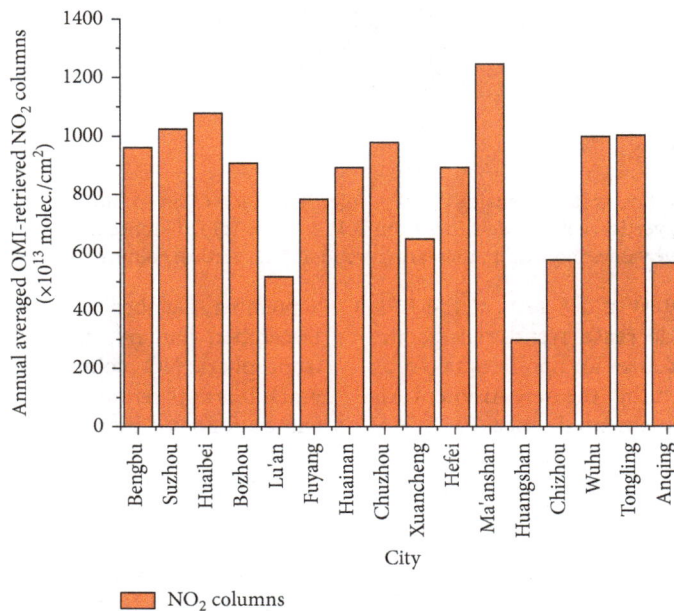

Figure 6: Annual averaged OMI-retrieved NO_2 columns ($\times 10^{13}$ molec./cm^2) from 2005 to 2016 of cities in Anhui Province.

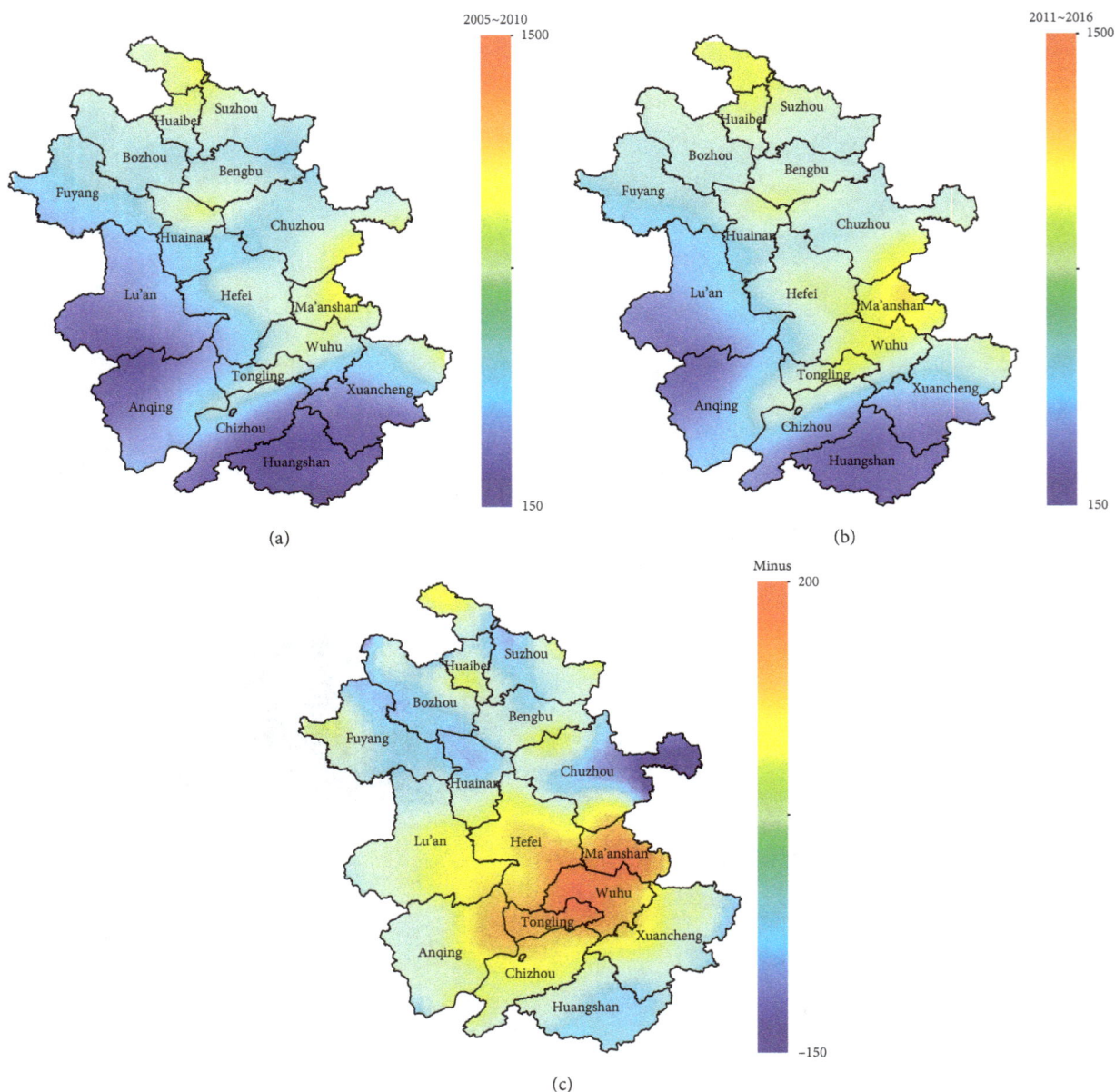

FIGURE 7: Total of OMI-retrieved vertical column densities ($\times 10^{15}$ molec./cm^2) of NO$_2$ in Anhui. (a) Total of NO$_2$ vertical column densities from 2005 to 2010. (b) Total of NO$_2$ vertical column densities from 2011 to 2016. (c) Difference in the total of NO$_2$ vertical column densities between the two periods. The positive values indicate an increasing trend of NO$_2$ vertical column densities from 2005 to 2016, and vice versa.

was confirmed by using the t test and F test, at the 95% confidence level. The recent decrease trend reflects the impact of emission control measures and policies taken by the government. It is well known that the new Ambient Air Quality Standard has been implemented since 2012. It is a stricter air quality standard than the previous standard, especially for NO$_2$ and fine particles in the atmosphere [39, 40].

3.4. Seasonal Variations of Tropospheric NO$_2$ in WCB. Seasonal variations of tropospheric NO$_2$ were analyzed. In Anhui, spring includes March, April, and May, summer comprises June, July, and August, autumn includes

September, October, and November, while winter comprises December, January, and February. Figure 10 displays the tropospheric NO$_2$ columns in different seasons during the 12 years. It is apparent that the highest NO$_2$ column occurred in winter, followed by autumn and spring, while summer had the lowest NO$_2$. Also, the seasonal variation shows the same trend during all twelve years. This seasonal trend may be due to the combined effect of the emission source, sink, and weather conditions. Emissions from power plants increase due to domestic heating in winter. In addition, the weather of winter is characterized by lower temperature and more overcast days than that of other seasons, which results in the reduction of the photochemical reaction of NO$_2$ with volatile organic compounds (VOCs) [41].

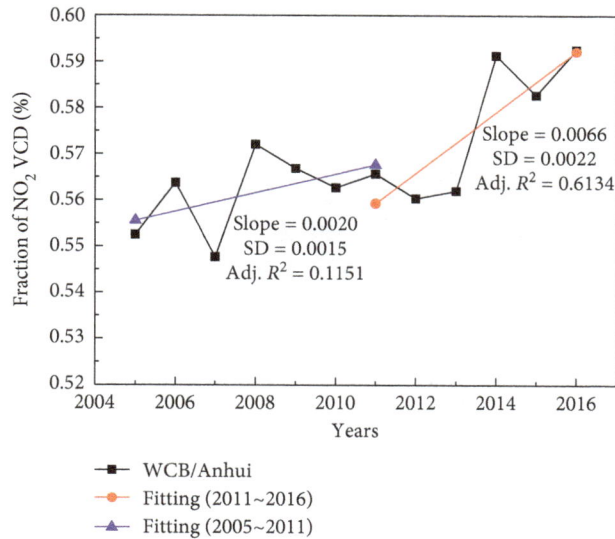

Figure 8: Fractions of OMI-derived tropospheric NO$_2$ columns over WCB in the total tropospheric NO$_2$ columns of Anhui from 2005 to 2016. Fraction of tropospheric NO$_2$ columns is calculated as the ratio of the sum of annual averaged tropospheric NO$_2$ columns in WCB to the sum of annual averaged tropospheric NO$_2$ columns in Anhui from 2005 to 2016 (%).

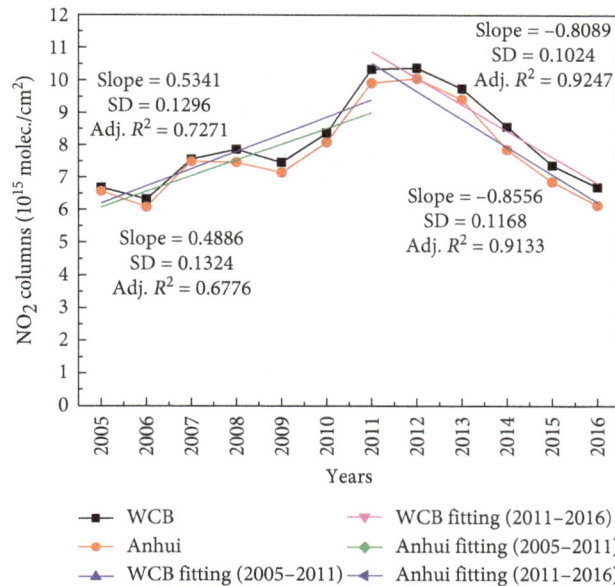

Figure 10: Average tropospheric NO$_2$ columns for four seasons from 2005 to 2016 in WCB.

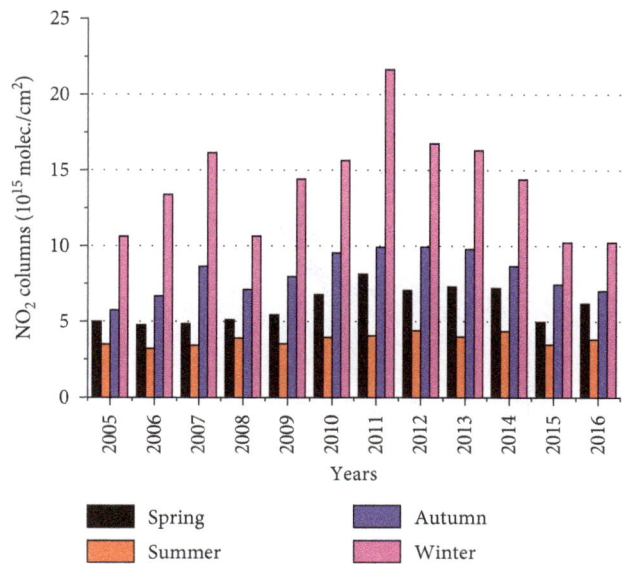

Figure 9: Average tropospheric NO$_2$ columns from 2005 to 2016 in Anhui and WCB.

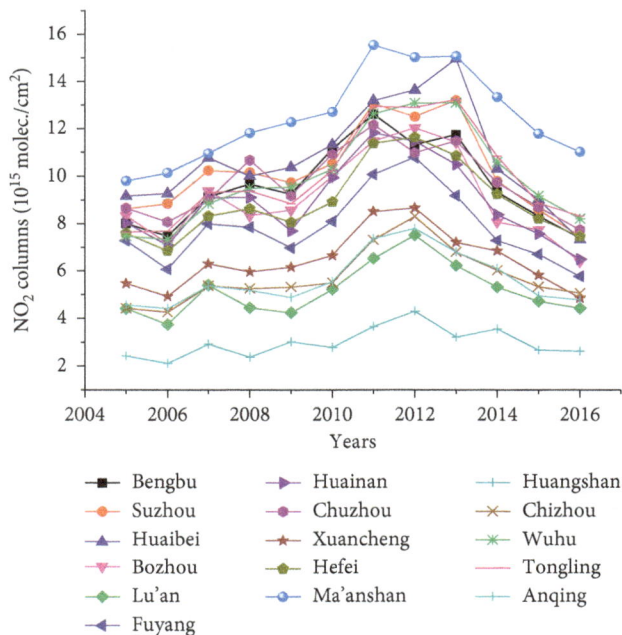

Figure 11: The average tropospheric NO$_2$ columns of cities in Anhui from 2005 to 2016.

Figure 11 illustrates the time series of the annual averaged tropospheric NO$_2$ columns for each city in Anhui Province from 2005 to 2016. It can be seen from Figure 11 that Ma'anshan, Bengbu, Huainan, and Chuzhou showed the maximum NO$_2$ level in 2011. Bozhou, Lu'an, Fuyang, Xuancheng, Hefei, Huangshan, Chizhou, and Anqing displayed the maximum NO$_2$ level in 2012. Suzhou, Huaibei, Wuhu, and Tongling showed the maximum NO$_2$ level in 2013. As mentioned earlier, the new Ambient Air Quality

Standard of China has been implemented since 2012. The effect of Ambient Air Quality Standard often lags behind the policy itself, so some cities in Anhui Province reached their maximum of NO$_2$ columns in 2012 or 2013.

We used the HYSPLIT model to analyze the origins of NO$_2$ and transport pathways of air masses in the typical city of WCB, Ma'anshan city. Ma'anshan city is in the west of Nanjing area and about 40 km from the center of Nanjing city, whereas Nanjing is one of the industrial centers of Yangtze Delta.

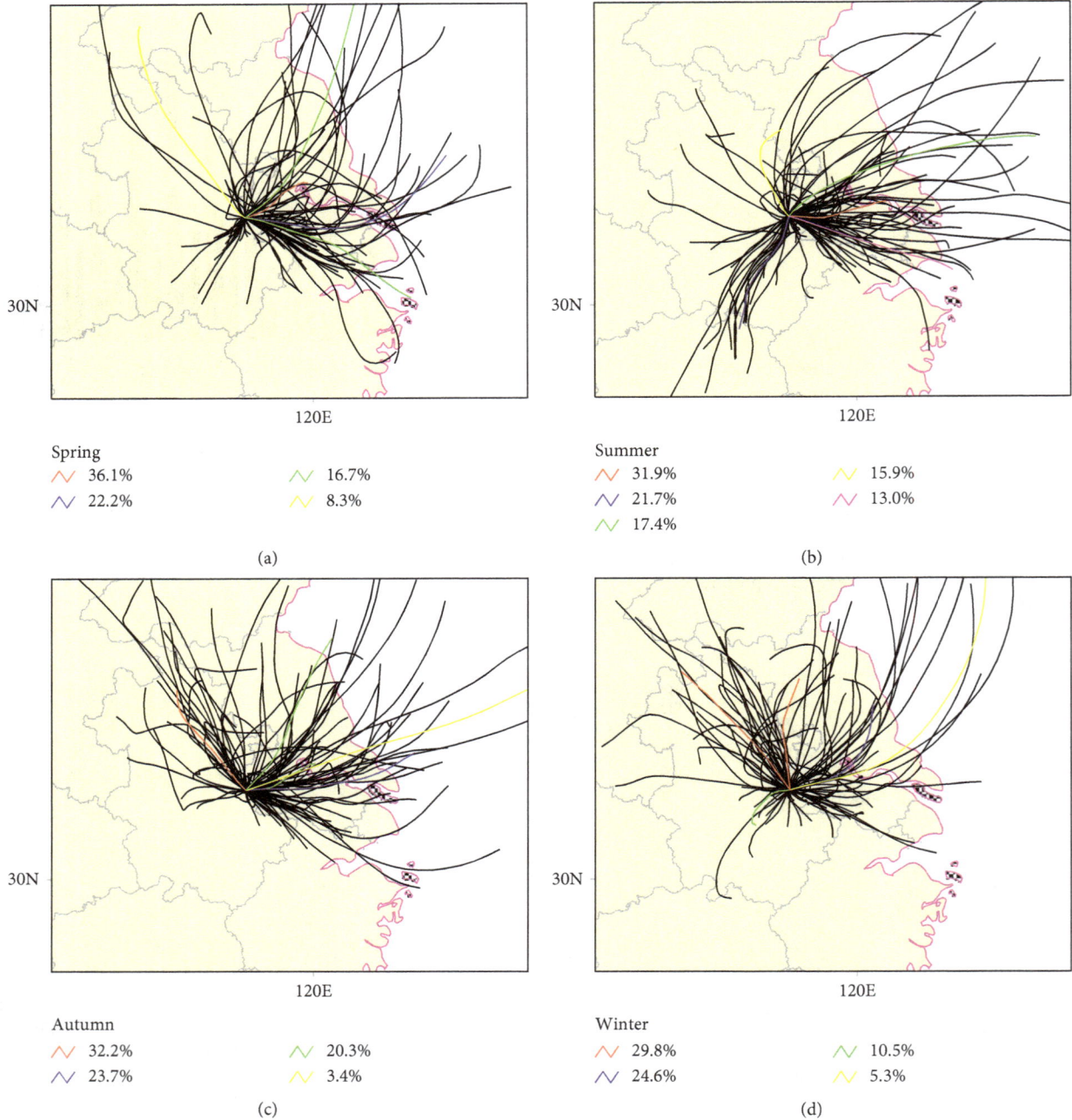

FIGURE 12: The cluster of air mass backward trajectories in different seasons in Ma'anshan based on the HYSPLIT model: (a) spring; (b) summer; (c) autumn; (d) winter. The black triangle represents the location of Ma'anshan.

We performed the cluster analysis of the 24 h air mass back trajectories starting at 500 m for the full year of 2015. Figure 12 shows five major types of backward trajectory clusters for different seasons in Ma'anshan in 2015. During the spring, summer, and autumn, air masses are mainly from the eastern regions, so the high emissions of NO_2 in the Nanjing area may influence the concentration of atmospheric NO_2 in the Ma'anshan area. In the winter, the prevailing wind is from north (>50%), where the tropospheric NO_2 columns are relatively low. This means that the high level of NO_2 in Ma'anshan in winter is not from the

transport but caused by the local emissions. The high level of tropospheric NO_2 in the Ma'anshan area results from the rapid industrial development and the increase of vehicles on the road.

4. Conclusions

Atmospheric nitrogen dioxide plays an important role in tropospheric chemistry and air quality. Satellite observations have great potential for understanding the spatial distributions and temporal variations in atmospheric NO_2 on a regional scale, with high spatial and temporal resolutions. We utilize the tropospheric NO_2 columns observed from OMI to analyze the spatial distributions and temporal trends of NO_2 in Wanjiang City Belt (WCB) of China from 2005 to 2016. The objective of this study is to describe the spatial distributions and temporal trends of tropospheric NO_2 based on satellite observations in twelve years, in order to assess the effect of industrial transfer policy on the air quality in WCB.

Firstly, we used the surface NO_2 concentrations to compare with the OMI-retrieved tropospheric NO_2 columns in order to verify the accuracy of the satellite data over the WCB area. Although the two datasets have different spatial scale representativeness and surface sensitivity, the comparison results prove the accuracy of the OMI-retrieved tropospheric NO_2 columns in cities of WCB.

Then, we examined the spatial distributions of the annual averaged tropospheric NO_2 total columns over Anhui Province. The results show that NO_2 columns were considerably higher in WCB than those in other areas of Anhui. The annual averaged NO_2 column reached 811×10^{13} molec./cm^2 in the WCB region from 2005 to 2016, while the annual averaged NO_2 column was 733×10^{13} molec./cm^2 in other areas of Anhui during the same period. Also, the spatial distributions of annual averaged tropospheric NO_2 columns in this area agree with the results of satellite-retrieved NO_2 emissions in eastern China in other studies.

In order to evaluate the effect of industrial transfer policy on the air quality in WCB, we compared the spatial distributions of the total NO_2 columns in 2005 through 2010 and in 2011 through 2016. It is obvious that the total NO_2 columns in WCB increased more significantly than those in other areas between the two periods. The total NO_2 columns in WCB increased by 19.9%, while the corresponding value increased only 13.9% in other Anhui areas except the WCB area. Furthermore, the increased fraction of tropospheric NO_2 columns in WCB to the total value in Anhui after the year of 2011 also reflects the effect of the construction of WCB on the air quality. Fortunately, the temporal variations of NO_2 columns show that although the NO_2 columns over WCB and Anhui increased significantly from 2005 to 2011, they decreased sharply from 2011 to 2016 due to the strict emission reduction measures in China.

Furthermore, the seasonal variations of tropospheric NO_2 were analyzed in detail. As is seen from the results, the highest NO_2 column occurred in winter, followed by autumn and spring, while summer had the lowest NO_2 during all twelve years. The seasonal trend may be due to the combined effect of the emission source, sink, and weather conditions as well as air mass transport. We used the HYSPLIT model to analyze the origins of NO_2 and transport pathways of air masses in the typical city of WCB, Ma'anshan city. The outcome shows that the high level of NO_2 in Ma'anshan in winter is not from the air mass transport but from the local emissions. Although the study involves only one important trace gas, the results offer a useful tool for policy-makers to plan and implement pollution control regulations.

Conflicts of Interest

The authors declare that they have no conflicts of interest.

Authors' Contributions

Yu Xie and Wei Wang conceived, designed, and performed the experiments. Qinglong Wang provided valuable comments in revising the manuscript.

Acknowledgments

We acknowledge the free use of tropospheric NO_2 column data from the OMI sensor from http://www.temis.nl. This research was funded by the National Natural Science Foundation of China (grant numbers 41775025 and 41405134), the National Key Technology R&D Program of China (2018YFC0213201), and the Hefei University Foundation (grant number 16-17RC21).

References

[1] N. A. Krotkov, C. A. McLinden, C. Li et al., "Aura OMI observations of regional SO_2 and NO_2 pollution changes from 2005 to 2015," *Atmospheric Chemistry and Physics*, vol. 16, no. 7, pp. 4605–4629, 2016.

[2] D. J. Jacob, E. G. Heikes, S. M. Fan et al., "Origin of ozone and NO_x in the tropical troposphere: a photochemical analysis of aircraft observations over the south Atlantic basin," *Journal of Geophysical Research: Atmospheres*, vol. 101, no. 19, pp. 24235–24250, 1996.

[3] N. Blond, K. F. Boersma, H. J. Eskes et al., "Intercomparison of sciamachy nitrogen dioxide observations, in situ measurements and air quality modeling results over western Europe," *Journal of Geophysical Research: Atmospheres*, vol. 112, no. 10, 2007.

[4] C. Ordóñez, A. Richter, M. Steinbacher et al., "Comparison of 7 years of satellite-borne and ground-based tropospheric NO_2 measurements around Milan, Italy," *Journal of Geophysical Research*, vol. 111, no. 5, 2006.

[5] T. N. Knepp, R. Querel, P. Johnston, L. Thomason, D. Flittner, and J. M. Zawodny, "Intercomparison of pandora stratospheric NO_2 slant column product with the NDACC-certified M07 spectrometer in Lauder, New Zealand," *Atmospheric Measurement Techniques*, vol. 10, pp. 4363–4372, 2017.

[6] H. Liu, C. Liu, Z. Xie et al., "A paradox for air pollution controlling in China revealed by "APEC blue" and "PARADE blue"," *Scientific Reports*, vol. 6, no. 1, p. 34408, 2016.

[7] K. Xiao, Y. Wang, G. Wu, B. Fu, and Y. Zhu, "Spatiotemporal characteristics of air pollutants (PM10, PM2.5, SO_2, NO_2, O_3, and CO) in the inland basin city of Chengdu, southwest China," *Atmosphere*, vol. 9, no. 2, p. 74, 2018.

[8] T. Drosoglou, M. E. Koukouli, N. Kouremeti et al., "MAX-DOAS NO_2 observations over Guangzhou, China; ground-based and satellite comparisons," *Atmospheric Measurement Techniques*, vol. 11, no. 4, pp. 2239–2255, 2018.

[9] L. N. Lamsal, R. V. Martin, A. V. Donkelaar et al., "Indirect validation of tropospheric nitrogen dioxide retrieved from the OMI satellite instrument: insight into the seasonal variation of nitrogen oxides at northern Midlatitudes," *Journal of Geophysical Research*, vol. 115, no. 5, pp. 458–473, 2010.

[10] L. Liu, X. Zhang, W. Xu et al., "Temporal characteristics of atmospheric ammonia and nitrogen dioxide over China based on emission data, satellite observations and atmospheric transport modeling since 1980," *Atmospheric Chemistry and Physics*, vol. 17, no. 15, pp. 9365–9378, 2017.

[11] K. F. Boersma, H. J. Eskes, J. P. Veefkind et al., "Near-real time retrieval of tropospheric NO_2 from OMI," *Atmospheric Chemistry and Physics*, vol. 7, no. 8, pp. 2103–2118, 2007.

[12] Z. Xiao, H. Jiang, X. Song, and X. Zhang, "Monitoring of atmospheric nitrogen dioxide using ozone monitoring instrument remote sensing data," *Journal of Applied Remote Sensing*, vol. 7, no. 1, article 073534, 2013.

[13] K. Y. Kondratyev and C. A. Varotsos, "Global tropospheric ozone dynamics," *Environmental Science and Pollution Research International*, vol. 8, no. 2, pp. 113–119, 2001.

[14] Z. Ul-Haq, A. D. Rana, S. Tariq, K. Mahmood, M. Ali, and I. Bashir, "Modeling of tropospheric NO_2 column over different climatic zones and land use/land cover types in South Asia," *Journal of Atmospheric and Solar-Terrestrial Physics*, vol. 168, pp. 80–99, 2018.

[15] J. Gu, L. Chen, C. Yu et al., "Ground-level NO_2 concentrations over China inferred from the satellite OMI and CMAQ model simulations," *Remote Sensing*, vol. 9, no. 6, p. 519, 2017.

[16] A. Sharma, T. K. Mandal, S. K. Sharma, D. K. Shukla, and S. Singh, "Relationships of surface ozone with its precursors, particulate matter and meteorology over Delhi," *Journal of Atmospheric Chemistry*, vol. 74, no. 4, pp. 451–474, 2016.

[17] C. A. Varotsos, J. M. Ondov, M. N. Efstathiou, and A. P. Cracknell, "The local and regional atmospheric oxidants at Athens (Greece)," *Environmental Science and Pollution Research*, vol. 21, no. 6, pp. 4430–4440, 2014.

[18] K. M. Han, S. Lee, L. S. Chang, and C. H. Song, "A comparison study between CMAQ-simulated and OMI-retrieved NO_2 columns over east asia for evaluation of NO_x emission fluxes of INTEX-B, CAPSS, and REAS inventories," *Atmospheric Chemistry and Physics*, vol. 15, no. 4, pp. 1913–1938, 2015.

[19] F. Liu, Q. Zhang, D. A. Van et al., "Recent reduction in NO_x emissions over China: synthesis of satellite observations and emission inventories," *Environmental Research Letters*, vol. 11, no. 11, article 114002, 2016.

[20] L. N. Lamsal, R. V. Martin, A. Van Donkelaar et al., "Ground-level nitrogen dioxide concentrations inferred from the satellite-borne ozone monitoring instrument," *Journal of Geophysical Research*, vol. 113, no. 16, 2008.

[21] I. Ialongo, J. Herman, N. Krotkov et al., "Comparison of OMI NO_2 observations and their seasonal and weekly cycles with ground-based measurements in Helsinki," *Atmospheric Measurement Techniques*, vol. 9, no. 10, pp. 5203–5212, 2016.

[22] D. Q. Tong, L. Lamsal, L. Pan et al., "Long-term NO_x trends over large cities in the United States during the great recession: comparison of satellite retrievals, ground observations, and emission inventories," *Atmospheric Environment*, vol. 107, pp. 70–84, 2015.

[23] C. A. McLinden, V. Fioletov, K. F. Boersma et al., "Improved satellite retrievals of NO_2 and SO_2 over the Canadian oil sands and comparisons with surface measurements," *Atmospheric Chemistry and Physics*, vol. 14, no. 7, pp. 3637–3656, 2014.

[24] D. Kim, H. Lee, H. Hong, W. Choi, Y. Lee, and J. Park, "Estimation of surface NO_2 volume mixing ratio in four metropolitan cities in Korea using multiple regression models with OMI and airs data," *Remote Sensing*, vol. 9, no. 6, p. 627, 2017.

[25] Y. Cui, J. Lin, C. Song et al., "Rapid growth in nitrogen dioxide pollution over western China, 2005–2013," *Atmospheric Chemistry and Physics*, vol. 16, no. 10, pp. 6207–6221, 2016.

[26] F. Liu, S. Beirle, Q. Zhang, S. Dörner, K. He, and T. Wagner, "NO_x lifetimes and emissions of cities and power plants in polluted background estimated by satellite observations," *Atmospheric Chemistry and Physics*, vol. 16, no. 8, pp. 5283–5298, 2016.

[27] F. Liu, S. Beirle, Q. Zhang et al., "NO_x emission trends over Chinese cities estimated from OMI observations during 2005 to 2015," *Atmospheric Chemistry and Physics*, vol. 17, no. 15, pp. 9261–9275, 2017.

[28] J. T. Lin, "Satellite constraint for emissions of nitrogen oxides from anthropogenic, lightning and soil sources over east China on a high-resolution grid," *Atmospheric Chemistry and Physics*, vol. 12, no. 6, pp. 2881–2898, 2012.

[29] B. N. Duncan, L. N. Lamsal, A. M. Thompson et al., "A space-based, high-resolution view of notable changes in urban NO_x pollution around the world (2005–2014)," *Journal of Geophysical Research: Atmospheres*, vol. 121, no. 2, pp. 976–996, 2016.

[30] H. Zou, X. Duan, L. Ye, and L. Wang, "Locating sustainability issues: identification of ecological vulnerability in Mainland China's mega-regions," *Sustainability*, vol. 9, no. 7, p. 1179, 2017.

[31] Approval of Wanjiang City Belt (WCB) as demonstration area for industrial transfer from the Yangtze River Delta and other megaregions to Anhui Province by the National Development and Reform Commission (NDRC), January 2010, http://www.ndrc.gov.cn/zcfb/zcfbghwb/201003/t20100324_585471.html.

[32] B. Liu, S.-J. Lee, Z. Jiao, and L. Wang, *Contemporary Logistics in China: An Introduction*, World Scientific, Singapore, 2012.

[33] S. Tang, "Wanjiang city belt in the making," *China Today*, vol. 2011, pp. 40–43, 2011.

[34] P. F. Levelt, G. H. J. V. D. Oord, M. R. Dobber et al., "The ozone monitoring instrument," *IEEE Transactions on Geoscience and Remote Sensing*, vol. 44, no. 5, pp. 1093–1101, 2006.

[35] J. Z. Ma, S. Beirle, J. L. Jin, R. Shaiganfar, P. Yan, and T. Wagner, "Tropospheric NO_2 vertical column densities over Beijing: results of the first three years of ground-based MAX-DOAS measurements (2008–2011) and satellite validation," *Atmospheric Chemistry and Physics*, vol. 13, no. 3, pp. 1547–1567, 2013.

[36] P. Castellanos, K. F. Boersma, O. Torres, and J. F. De Haan, "OMI tropospheric NO_2 air mass factors over south America: effects of biomass burning aerosols," *Atmospheric Measurement Techniques*, vol. 8, no. 9, pp. 3831–3849, 2015.

[37] R. R. Draxler and G. Hess, "An overview of the hysplit_4 modelling system for trajectories," *Australian Meteorological Magazine*, vol. 47, pp. 295–308, 1998.

[38] A. F. Stein, R. R. Draxler, G. D. Rolph, B. J. B. Stunder, M. D. Cohen, and F. Ngan, "NOAA's HYSPLIT atmospheric transport and dispersion modeling system," *Bulletin of the American Meteorological Society*, vol. 96, no. 12, pp. 2059–2077, 2015.

[39] Z. Ling, T. Huang, Y. Zhao et al., "OMI-measured increasing SO_2 emissions due to energy industry expansion and relocation in northwestern China," *Atmospheric Chemistry and Physics*, vol. 17, no. 14, pp. 9115–9131, 2017.

[40] *Air Pollution Prevention and Control Action Plan*, 2018, http://www.gov.cn/zhengce/content/2013-09/13/content_4561.htm.

Permissions

List of Contributors

Yaolin Lin
School of Civil Engineering and Architecture, Wuhan University of Technology, Wuhan 430070, China

Wei Yang
College of Engineering and Science, Victoria University, Melbourne, VIC 8001, Australia
School of Civil Engineering and Architecture, Wuhan University of Technology, Wuhan 430070, China

Chun-Qing Li
School of Engineering, RMIT University, Melbourne, VIC 3000, Australia

Masoud Irannezhad and Bjørn Kløve
Water Resources and Environmental Engineering Research Unit, Faculty of Technology, University of Oulu, 90014 Oulu, Finland

Masoud Irannezhad
School of Environmental Science and Engineering, Southern University of Science and Technology, Shenzhen 518055, China

Hamid Moradkhani
Center for Complex Hydrosystems Research, Department of Civil, Construction and Environmental Engineering, University of Alabama, Tuscaloosa, AL 35487, USA

Jang Hyun Sung
Ministry of Environment, Han River Flood Control Office, Seoul, Republic of Korea

Hyung-Il Eum
Alberta Environment and Parks, Calgary, Canada

Junehyeong Park
Civil, Construction, and Environmental Engineering, University of Alabama, Tuscaloosa, AL, USA

Jaepil Cho
Climate Services and Research Department, APEC Climate Center, Busan, Republic of Korea

Wei Wei, Baitian Wang and Kebin Zhang
College of Water and Soil Conservation, Beijing Forestry University, Beijing 100083, China

Zhongjie Shi, Genbatu Ge and Xiaohui Yang
Institute of Desertification Studies, Chinese Academy of Forestry, Beijing 100091, China

Jing Zou and Tong Hu
Institute of Oceanographic Instrumentation, Qilu University of Technology (Shandong Academy of Sciences), Qingdao 266001, China

Chesheng Zhan
Institute of Geographic Sciences and Natural Resources Research, Chinese Academy of Sciences, Beijing 100101, China

Ruxin Zhao
Beijing Normal University, Beijing 100875, China

Peihua Qin
Institute of Atmospheric Physics, Chinese Academy of Sciences, Beijing 100029, China

Feiyu Wang
School of Geography and Tourism, Shaanxi Normal University, Xi'an 710119, China

Lenin Campozano, Rolando Célleri, Esteban Samaniego and Cristóbal Albuja
Departamento de Recursos Hídricos y Ciencias Ambientales, Universidad de Cuenca, Cuenca, Ecuador

Lenin Campozano, Rolando Célleri and Esteban Samaniego
Facultad de Ingeniería, Universidad de Cuenca, Cuenca, Ecuador

Lenin Campozano
Depto. de Ingeniería Civil y Ambiental, Escuela Politecnica Nacional, Quito, Ecuador

Katja Trachte
Laboratory for Climatology and Remote Sensing (LCRS), Faculty of Geography, Philipps-University Marburg, Deutschhausstraße 10, 35032 Marburg, Germany

John F. Mejia
Department of Atmospheric Sciences, Desert Research Institute, Reno, NV, USA

Yongwei Liu and Yuanbo Liu
Key Laboratory of Watershed Geographic Sciences, Nanjing Institute of Geography & Limnology, Chinese Academy of Sciences, Nanjing 210008, China

Wen Wang
State Key Laboratory of Hydrology-Water Resources and Hydraulic Engineering, Hohai University, Nanjing 210098, China

Edvinas Stonevicius, Gintautas Stankunavicius and Egidijus Rimkus
Institute of Geosciences, Vilnius University, Vilnius, Lithuania

Befikadu Esayas, Belay Simane and Ermias Teferi
Center for Environment and Development Studies, Addis Ababa University, Addis Ababa, Ethiopia

Victor Ongoma
Department of Meteorology, South Eastern Kenya University, Kitui, Kenya

Nigussie Tefera
The United Nations, World Food Programme (WFP), Addis Ababa, Ethiopia

Jacob Agyekum, Thompson Annor and Emmannuel Quansah
Department of Physics, Kwame Nkrumah University of Science and Technology (KNUST), Kumasi, Ghana

Benjamin Lamptey
African Centre of Meteorological Applications for Development (ACMAD), Niamey, Niger

Richard Yao Kuma Agyeman
Numerical Weather Prediction Unit, Ghana Meteorological Agency (GMet), Accra, Ghana

G. Naveendrakumar
Postgraduate Institute of Science (PGIS), University of Peradeniya, Peradeniya, Sri Lanka
Faculty of Applied Science, Vavuniya Campus of the University of Jaffna, Vavuniya, Sri Lanka

Meththika Vithanage
Ecosphere Resilience Research Center, Faculty of Applied Sciences, University of Sri Jayewardenepura, Nugegoda, Sri Lanka

Hyun-Han Kwon
Department of Civil Engineering, Chonbuk National University, Jeonju, Republic of Korea

M. C. M. Iqbal
Plant and Environmental Sciences, National Institute of Fundamental Studies (NIFS), Kandy, Sri Lanka

S. Pathmarajah
Department of Agricultural Engineering, Faculty of Agriculture, University of Peradeniya, Peradeniya, Sri Lanka

Jayantha Obeysekera
Sea Level Solutions Center, Florida International University, Miami, FL, USA

Falk Maneke-Fiegenbaum and Otto Klemm
Climatology Working Group-Institute of Landscape Ecology, University of Münster, 48149 Münster, Germany

Yen-Jen Lai, Chih-Yuan Hung and Jui-Chu Yu
Experimental Forest, National Taiwan University, 55750 Nantou, Taiwan

Amanda Rehbein, Lívia Márcia Mosso Dutra, Tercio Ambrizzi, Rosmeri Porfírio da Rocha, Ana Carolina Nóbile Tomaziello, José Leandro Pereira Silveira Campos, Victor Raul Chavez Mayta, Natália Machado Crespo, Paola Gimenes Bueno, Vannia Jaqueline Aliaga Nestares, Laís Tabosa Machado and Eduardo Marcos De Jesus
Departamento de Ciências Atmosféricas, Instituto de Astronomia, Geofísica e Ciências Atmosf éricas da Universidade de São Paulo, São Paulo, SP, Brazil

Michelle Simões Reboita
Instituto de Recursos Naturais da Universidade Federal de Itajubá, Itajubá, MG, Brazil

Gyrlene Aparecida Mendes da Silva
Departamento de Ciências do Mar da Universidade Federal de São Paulo, São Paulo, SP, Brazil

Luiz Felippe Gozzo and Maria de Souza Custódio
Departamento de Física da Universidade Estadual Paulista Júlio de Mesquita Filho, Campus de Bauru, SP, Brazil

Luana Albertani Pampuch
Instituto de Ciência e Tecnologia da Universidade Estadual Paulista Júlio de Mesquita Filho, Campus de São José dos Campos, São Paulo, SP, Brazil

Camila Bertoletti Carpenedo
Instituto de Geografia da Universidade Federal de Uberlândia, Uberlândia, MG, Brazil

Lingling Shen, Li Lu, Tianjie Hu and Runsheng Lin
Beijing Meteorological Information Center, Beijing Meteorological Service, Beijing 100089, China

Ji Wang
Beijing Climate Center, Beijing Meteorological Service, Beijing 100089, China

Chong Xu
Institute of Geology, China Earthquake Administration, Beijing 100029, China

Yu Xie and Qinglong Wang
Department of Electronic Information and Electrical Engineering, Hefei University, Hefei 230601, Anhui, China

Wei Wang
Key Laboratory of Environmental Optics and Technology, Anhui Institute of Optics and Fine Mechanics, Chinese Academy of Sciences, Hefei 230031, Anhui, China

Index